T0344875

Environmental Pest Management

Environmental Pest Management

Challenges for Agronomists, Ecologists, Economists and Policymakers

Edited by

Moshe Coll

Department of Entomology
The Robert H. Smith Faculty of Agriculture, Food and Environment
The Hebrew University of Jerusalem
Rehovot, Israel

Eric Wajnberg

INRA, Sophia Antipolis
France

This edition first published 2017

Registered Office(s)
John Wiley and Sons, Inc., 111 River Street, Hoboken, NJ 07030, USA
John Wiley and Sons Ltd, The Atrium, Southern Gate, Chichester, West Sussex, PO19 8SQ, UK

Editorial Office
9600 Garsington Road, Oxford, OX4 2DQ, UK

For details of our global editorial offices, customer services, and more information about Wiley products visit us at www.wiley.com.

Wiley also publishes its books in a variety of electronic formats and by print-on-demand. Some content that appears in standard print versions of this book may not be available in other formats.

Library of Congress Cataloging-in-Publication data applied for

ISBN: 9781119255550

Cover Design: Wiley
Cover Image: Courtesy of Dr Yael Mandelik

Set in 10/12pt Warnock by SPi Global, Pondicherry, India
Printed and bound in Malaysia by Vivar Printing Sdn Bhd

10 9 8 7 6 5 4 3 2 1

Contents

List of Contributors

David Adamson
The Centre for Global Food and Resources
University of Adelaide
SA
Australia

Nigel R. Andrew
Centre of Excellence for Behavioural and Physiological Ecology, Natural History Museum
University of New England
Armidale
NSW
Australia

Bruce Auld
School of Agricultural and Wine Sciences
Charles Sturt University
Orange
NSW
Australia

Barbara I.P. Barratt
AgResearch
Invermay Agricultural Centre
Private Bag
Mosgiel
New Zealand

Nir Becker
Department of Economics and Management
Tel-Hai College
Israel

Ana Maria Calderón de la Barca
Centro de Investigación en Alimentación y Desarrollo
Hermosillo
México

Stefano Colazza
Department of Agricultural and Forest Sciences
University of Palermo
Palermo
Italy

Moshe Coll
Department of Entomology
The Robert H. Smith Faculty of Agriculture, Food and Environment
The Hebrew University of Jerusalem
Rehovot
Israel

Antonino Cusumano
Laboratory of Entomology
Wageningen University
Wageningen
The Netherlands

Antonio DiTommaso
Soil and Crop Sciences Section
School of Integrative Plant Science
Cornell University
Ithaca
NY
USA

Jian J. Duan
USDA-ARS
Beneficial Insects Introduction
Research Unit
Newark
DE
USA

Clark A.C. Ehlers
Environmental Protection Authority
Private Bag
Wellington
New Zealand

Margaret I. FitzSimmons
Department of Environmental Studies
University of California, Santa Cruz
Santa Cruz
CA
USA

Peter A. Follett
USDA-ARS
US Pacific Basin Agricultural Research
Center
Nowelo St.
Hilo
USA

Mark A.K. Gillespie
Department of Engineering and Natural
Sciences
Western Norway University of Applied
Science
Sogndal
Norway

Felix Herzog
Agroscope
Zürich
Switzerland

Sarah J. Hill
Centre of Excellence for Behavioural and
Physiological Ecology
Natural History Museum
University of New England
Armidale
NSW
Australia

Jane A. Hoppin
North Carolina State University
Department of Biological Sciences
Center for Human Health and the
Environment
Raleigh
NC
USA

Katja Jacot
Agroscope
Zürich
Switzerland

David E. Jennings
Department of Entomology
University of Maryland
College Park
MD
USA

Catherine E. LePrevost
North Carolina State University
Department of Applied Ecology
Center for Human Health and the
Environment
Raleigh
NC
USA

Deborah K. Letourneau
Department of Environmental Studies
University of California, Santa Cruz
Santa Cruz
CA
USA

John Losey
Department of Entomology
Cornell University
Ithaca
NY
USA

Javier Magaña-Gómez
Universidad Autónoma de Sinaloa
Culiacán
México

Maria Navajas
Institut National de la Recherche Agronomique, INRA
UMR CBGP
Montferrier-sur-Lez
France

Helle Ørsted Nielsen
Aarhus University
Department of Environmental Science
Roskilde
Denmark

Diego J. Nieto
Department of Environmental Studies
University of California, Santa Cruz
Santa Cruz
CA
USA

Anders Branth Pedersen
Aarhus University
Department of Environmental Science
Roskilde
Denmark

Ezio Peri
Department of Agricultural and Forest Sciences
University of Palermo
Palermo
Italy

George K. Roderick
Department of Environmental Science, Policy and Management
University of California
Berkeley
CA
USA

Matthew Ryan
Soil and Crop Sciences Section
School of Integrative Plant Science
Cornell University
Ithaca
NY
USA

Francisco Sánchez-Bayo
School of Life & Environmental Sciences
The University of Sydney
Eveleigh
NSW
Australia

Morgan W. Shields
Bio-Protection Research Centre
Lincoln University
Lincoln
New Zealand

Pieter Spanoghe
Ghent University
Department of Crop Protection
Laboratory of Crop Protection Chemistry
Ghent
Belgium

Janice Thies
Soil and Crop Sciences Section
School of Integrative Plant Science
Cornell University
Ithaca
NY
USA

Clement A. Tisdell
School of Economics
University of Queensland
Brisbane St Lucia
QLD
Australia

Matthias Tschumi
Lund University
Lund
Sweden

Eric Wajnberg
INRA
Sophia Antipolis
France

Thomas Walter
Agroscope
Zürich
Switzerland

Peter B. Woodbury
Soil and Crop Sciences Section
School of Integrative Plant Science
Cornell University
Ithaca
NY
USA

Steve D. Wratten
Bio-Protection Research Centre
Lincoln University
Canterbury
New Zealand

Preface

With the rapid growth of awareness and concern regarding adverse effects of pest management activities on human and environmental health, researchers and, to a lesser extent, policymakers have recently begun to appreciate these impacts as well as the influence of environmental factors on our ability to manage pest populations. In this respect, we were surprised to find that no single volume has as yet been devoted to these complex interactions. In addition, economic and societal considerations have been largely neglected while other topics, such as pesticide toxicity, have been the focus of much attention.

This volume is aimed at filling these gaps by addressing these pressing issues. It is designed to help develop and improve environmental pest management policies and agro-environmental schemes so that they encompass all major elements operating between pest management practices and the environment. It provides up-to-date fundamental information as well as recent research findings and current thinking on each topic so that complex issues are made available to readers across disciplines. It overviews major agronomic, ecological and human health aspects of pest management–environment interactions, discusses economic tools and caveats, and assesses shortcomings of various agro-environmental policies. Finally, taken together, it proposes a new framework for the development of effective, sustainable and environmentally compatible pest management programmes.

We believe that this timely treatment of the topic in a single, interdisciplinary volume will be of interest to an unusually wide readership. The book should be valuable for everyone interested in agriculture, ecology, entomology, pest control, public health, environmental economics and ecotoxicology, as well as policymakers worldwide. It will also be useful as a versatile teaching resource. Teachers of undergraduate and graduate courses in related fields will find the book useful as both a reference and background reading ahead of group discussions on controversial issues. Finally, we hope the book will promote interdisciplinary discussion and co-ordination between pest management stakeholders, conservation ecologists and environmentalist groups.

After a short introductory chapter (Chapter 1), the first part of the book provides general background to Integrated Pest Management (Chapter 2) and to pest management economics (Chapter 3). The second part addresses environmental concerns surrounding various pest management tactics, such as pesticide use (Chapter 4), biological control (Chapter 5) and the use of transgenic crops (Chapter 6). The third section discusses positive and negative ecosystem services provided by natural areas to influence pest management (Chapters 7 and 8, respectively). Then, the fourth section addresses effects of global processes such as climate change (Chapter 9) and biological invasions (Chapter 10) on pest suppression. The fifth section covers the influence of pesticide use and the consumption of genetically modified foods on public health (Chapters 11 and 12, respectively). The sixth section then discusses policies related to pesticide use (Chapter 13), importation of biological control agents (Chapter 14), food safety (Chapter 15), externalizing economic drivers (Chapter 16) and agro-environmental schemes (Chapter 17). In the concluding chapter (Chapter 18), we summarize take-home messages and propose a new framework for future research, extension and legislative work.

We thank the following referees for their critical comments on the book's chapters: Nir Becker, Dale G. Bottrell, Ephraim Cohen, Antonio Cusumano, Georges de Sousa, Roy van Driesche, Peter Follett, Fred Gould, Isaac Ishaaya, Hagai Levine, Philippe Nicot, Yvan Rahbé, Helen Roy, Clement Tisdell, Linda Thomson, and Steve Wratten. However, all information, results, views and discussions are the sole responsibility of the respective authors. Finally, we express our sincere thanks to the people at Wiley for their efficient help and support in the production of this book.

November 2016

Moshe Coll
Eric Wajnberg

1

Environmental Pest Management: A Call to Shift from a Pest-Centric to a System-Centric Approach

Moshe Coll and Eric Wajnberg

1.1 Introduction

According to a United Nations Food and Agriculture Organization estimate, about 795 million people suffered from chronic undernourishment in 2015 (FAO, IFAD and WFP 2015), indicating that one in nine people is deficient in calories, protein, iron, iodine or vitamin A, B, C or D, or any combination thereof (Sommer and West 1996). Such high levels of global food insecurity make many human societies vulnerable to health problems, reduced productivity and geopolitical unrest. A crop loss due to pest activity is a major contributor to food insecurity: 30–40% of potential world crop production is destroyed by pests (Natural Resources Institute 1992; Oerke *et al.* 1994). Of all pests, insects cause an estimated 14% of crop losses, plant pathogens 13% and weeds 13% (Pimentel 2007). An additional 30% of the crop is destroyed by postharvest insect pests and diseases, particularly in the developing world (Kumar 1984).

Humans have probably struggled with pestiferous insects, mites, nematodes, plant pathogens, weeds and vertebrates since the dawn of agriculture some 10 000 years ago (Figure 1.1). The earliest approaches employed were probably hand removal of pests and weeds, scaring away seed-consuming birds and trapping of granivorous rodents. Crop rotation, intercropping and selection of pest-resistant cultivars soon followed. The earliest recorded use of chemical pesticides dates back to 2500 BC, when the Sumerians used sulphur compounds as insecticides (see Figure 1.1). The use of botanical compounds, such as nicotine and pyrethrum, was later reported. However, pesticide application became common practice only in the 19th century, with increased agricultural mechanization.

1.2 Modern Developments in Pest Control

In the 20th century, the discovery of synthetic compounds with insecticidal and herbicidal properties, such as DDT and 2,4-D in 1939 and 1940, respectively, quickly made chemical control the predominant method of pest control. In most cropping systems, this has remained the case to this day, in spite of growing awareness of the negative impacts of pesticides on human health and the environment. In fact, many of our current serious pest problems have been brought about by intensification

Environmental Pest Management: Challenges for Agronomists, Ecologists, Economists and Policymakers, First Edition. Edited by Moshe Coll and Eric Wajnberg.

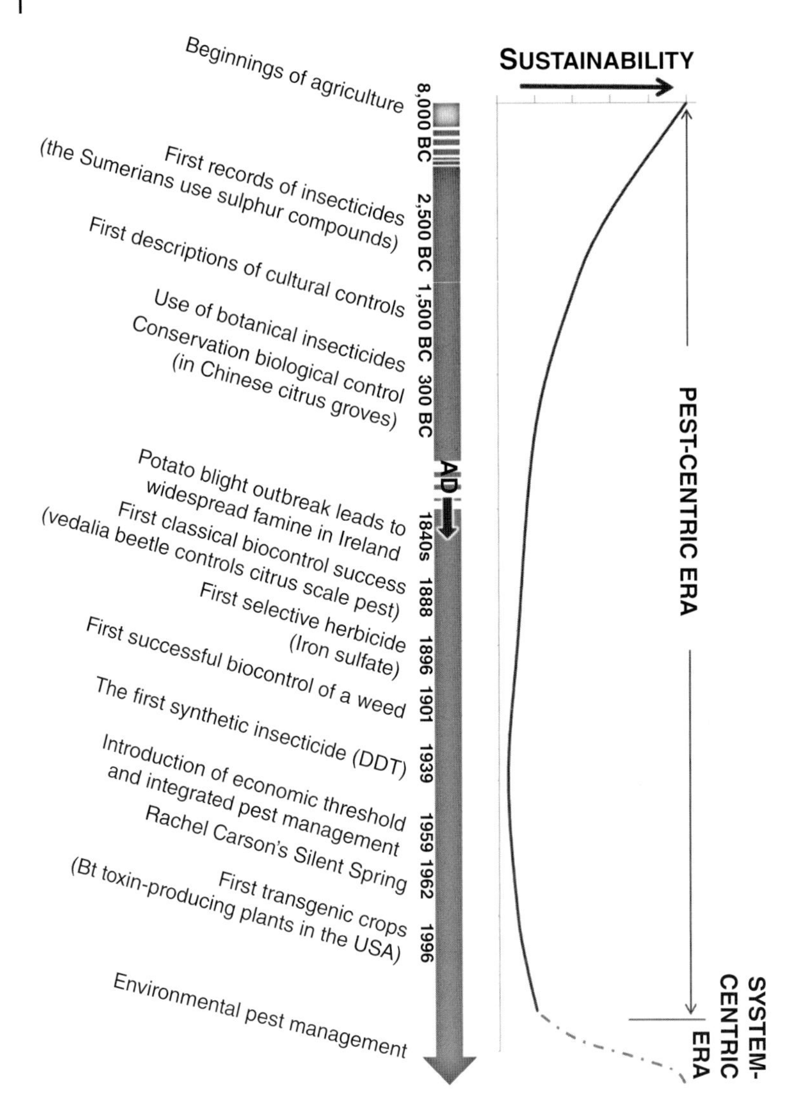

Figure 1.1 The history of pest management and changes in agro-ecosystem sustainability. Historic data are based on Abrol and Shankar (2012) and https://courses.cit.cornell.edu/ipm444/lec-notes/extra/ipm-history.html.

of cropping systems, mechanization, selection for high yielding but pest-susceptible crop genotypes, fertilization and irrigation inputs, and frequent application of pesticides (Thomas 1999; Waage 1993). Therefore, since the middle of the 20th century, most pest control measures have targeted specific pests on particular crops within single fields. Although reliance on a single tactic, usually the application of chemical pesticides, provides only a short-term solution (Thomas 1999), such a bottom-up approach has remained dominant is spite of widespread promotion of Integrated Pest Management (IPM) (Ehler 2006).

Integrated Pest Management has been accepted worldwide as the strategy of choice for pest population management. Since the United Nations Conference on the Environment in 1992 in Rio de Janeiro, Brazil, it has been the global policy in agriculture, natural resource management and trade. As a result, most of the world's population now lives in countries with IPM-guided policies for the production of most of the world's staple foods (Vreysen *et al.* 2007). Nonetheless, the definition of IPM has remained vague and highly inconsistent for more than 55 years (Table 1.1) (Bajwa and Kogan 2002). Van den Bosch and Stern (1962) stated that 'it is the entire ecosystem and its components that are of primary concern and not a particular pest.' Yet only 24% (16 of 67) of IPM definitions surveyed by Bajwa and Kogan (2002) included the term 'system' as the implementable programme or ecological unit. Furthermore, none of the surveyed definitions presented the term 'integrated' (in IPM) to indicate the integration of different measures employed simultaneously against several taxa across pest types (plant pathogens, insects, mites, nematodes, weeds, etc.). Since IPM is not legislatively defined, its definitions seem to reflect the respective interests and points of view of different individuals and organizations. Therefore, IPM is not a distinct, well-defined crop production strategy.

In spite of the original intent, IPM, as practised today, cannot be considered a holistic, system-wide approach. As pointed out by Ehler and Bottrell (2000) in the online periodical of the US National Academy of Sciences, 'despite three decades of research, there is very little "I" in IPM.' Instead, the vast majority of 'IPM' programmes are dominated by single technologies, a few of them by biological control, host plant resistance or biopesticides that are used as replacements for synthetic chemicals. All other programmes rely primarily on pesticides to suppress pest populations. Furthermore, these so-called IPM programmes rarely integrate different technologies. Their compatibility and the potential for interactive effects among control measures are not being explored. Therefore, the vast majority of IPM systems are not currently based upon the truly integrated, ecosystem-based strategy envisioned by, for example, researches and extension officers at the University of California (UC-IPM 2008). Furthermore, surveys completed between 2003 and 2006 (USDA NRCS Conservation Effects Assessment Project 2016) found that multiple IPM tactics are employed in only about 6% of cropland in the Mid-Western United States.

1.3 The Disillusionment with Integrated Pest Management

Much like the situation throughout the history of pest control, IPM programmes have generally focused on single pest species rather than on whole agro-ecosystems (Ehler 2006). Moreover, reduction in pesticide use is not indicated as a goal even in the 'true' ecosystem-based IPM approach (UC-IPM 2008), and pesticide reduction is not mentioned as a defining component of successful IPM (Kogan 1998). Therefore, it is not surprising that 'IPM' has had only a limited impact in reducing overall use of pesticides. Actually, pesticide use increased between 1970 and 2015 (see Chapter 2). It is disturbing that after decades of research, extension and legislation promoting true IPM programmes, the vast majority of current so-called 'IPM programmes' are 'nothing more than a reinvention of the supervised control of 50 [now 55] years ago' (Ehler and Bottrell 2000). The 'supervised control' approach, developed shortly after World War II, merely promoted the idea that decisions concerning insecticide application should be based on

Table 1.1 Selected definitions of Integrated Pest Management proposed or used by prominent authorities, arranged in chronological order (based in part on Bajwa and Kogan 2002).

Year	Definition	Source
1959	Applied pest control which combines and integrates biological and chemical control. Chemical control is used as necessary and in a manner which is least disruptive to biological control. Integrated control may make use of naturally occurring biological control as well as biological control affected by manipulated or induced biotic agents.	Stern *et al.* (1959)
1966	A pest population management system that utilizes all suitable techniques in a compatible manner to reduce pest populations and maintain them at levels below those causing economic injury.	Smith and Reynolds (1966)
1967	A pest management system that, in the context of the associated environment and the population dynamics of the pest species, utilizes all suitable techniques and methods in as compatible a manner as possible and maintains the pest populations at levels below those causing economic injury.	FAO (1967)
1969	Utilization of all suitable techniques to reduce and maintain pest populations at levels below those causing injury of economic importance to agriculture and forestry, or bringing two or more methods of control into a harmonized system designed to maintain pest levels below those at which they cause harm – a system that must rest on firm ecological principles and approaches.	National Academy of Science (1969)
1972	An approach that employs a combination of techniques to control the wide variety of potential pests that may threaten crops. It involves maximum reliance on natural pest population controls, along with a combination of techniques that may contribute to suppression – cultural methods, pest-specific diseases, resistant crop varieties, sterile insects, attractants, augmentation of parasites or predators, or chemical pesticides as needed.	Council on Environmental Quality (1972)
1978	A multidisciplinary, ecological approach to the management of pest populations, which utilizes a variety of control tactics compatibly in a single co-ordinated pest management system.	Smith (1978)
1979	The selection, integration and implementation of pest control based on predicted economic, ecological and sociological consequences.	Bottrell (1979)
1979	The optimization of pest control in an economically and ecologically sound manner, accomplished by the co-ordinated use of multiple tactics to assure stable crop production and to maintain pest damage below the economic injury level while minimizing hazards to humans, animals, plants and the environment.	Office of Technology Assessment (1979)
1980	An interdisciplinary approach incorporating the judicious application of the most efficient methods of maintaining pest populations at tolerable levels. Recognition of the problems associated with widespread pesticide application has encouraged the development and utilization of alternative pest control techniques. Rather than employing a single control tactic, attention is being directed to the co-ordinated use of multiple tactics, an approach known as integrated pest management.	FAO (1980)

Table 1.1 (Continued)

Year	Definition	Source
1981	An ecologically based pest control strategy that relies heavily on natural mortality factors, such as natural enemies and weather, and seeks out control tactics that disrupt these factors as little as possible. IPM uses pesticides, but only after systematic monitoring of pest populations and natural control factors indicate a need. Ideally, an integrated pest management programme considers all available pest control actions, including no action, and evaluates the potential interaction among various control tactics, cultural practices, weather, other pests, and the crop to be protected.	Flint and van den Bosch (1981)
1982	The use of two or more tactics in a compatible manner to maintain the population of one or more pests at acceptable levels in the production of food and fiber while providing protection against hazards to humans, domestic animals, plants and the environment.	Council for Agricultural Science and Technology (1982)
1984	A strategy for keeping plant damage within bounds by carefully monitoring crops, predicting trouble before it happens, and then selecting the appropriate controls – biological, cultural or chemical control as necessary.	Yepsen (1984)
1987	A pest population management system that anticipates and prevents pests from reaching damaging levels by using all suitable techniques, such as natural enemies, pest-resistant plants, cultural management and judicious use of pesticides.	National Coalition on Integrated Pest Management (1987)
1989	An ecologically based pest control strategy that relies on natural mortality factors such as natural enemies, weather and crop management and seeks control tactics that disrupt these factors as little as possible.	National Academy of Science, Board of Agriculture (1989)
1989	A pest control strategy based on the determination of an economic threshold that indicates when pest population is approaching the level at which control measures are necessary to prevent a decline in net returns. In principle, IPM is an ecologically based strategy that relies on natural mortality factors and seeks control tactics that disrupt these factors as little as possible.	National Research Council, Board of Agriculture (1989)
1989	A comprehensive approach to pest control that uses combined means to reduce the status of pests to tolerable levels while maintaining a quality environment.	Pedigo (1989)
1990	A systematic approach to crop protection that uses increased information and improved decision-making paradigms to reduce purchased inputs and improve economic, social and environmental conditions on the farm and in society. Moreover, the concept emphasizes the integration of pest suppression technologies that include biological, chemical, legal and cultural controls.	Allen and Rajotte (1990)
1991	An approach to pest control that utilizes regular monitoring to determine if and when treatments are needed and employs physical, mechanical, cultural, biological and educational tactics to keep pest numbers low enough to prevent intolerable damage or annoyance. Least-toxic chemical controls are used as a last resort.	Olkowski and Daar (1991)

(Continued)

Table 1.1 (Continued)

Year	Definition	Source
1992	The co-ordinated use of pest and environmental information along with available pest control methods, including cultural, biological, genetic and chemical methods, to prevent unacceptable levels of pest damage by the most economical means, and with the least possible hazard to people, property and the environment.	Sorensen (1992)
1992	An ecologically based pest control strategy which is part of the overall crop production system. 'Integrated' because all appropriate methods from multiple scientific disciplines are combined into a systematic approach for optimizing pest control. 'Management' implies acceptance of pests as inevitable components, at some population level of agricultural system.	Zalom *et al.* (1992)
1993	A management approach that encourages natural control of pest populations by anticipating pest problems and preventing pests from reaching economically damaging levels. All appropriate techniques are used such as enhancing natural enemies, planting pest-resistant crops, adapting cultural management and using pesticides judiciously.	United States Department of Agriculture, Agricultural Research Service (1993)
1993	Management activities that are carried out by farmers that result in potential pest populations being maintained below densities at which they become pests, without endangering the productivity and profitability of the farming system as a whole, the health of the family and its livestock, and the quality of the adjacent and downstream environments.	Wightman (1993)
1994	The use of all economically, ecologically and toxicologically justifiable means to keep pests below the economic threshold, with the emphasis on the deliberate use of natural forms of control and preventive measures.	Dehne and Schonbeck (1994)
1994	Integrated Pest Management is the use of a variety of pest control methods designed to protect public health and the environment, and to produce high-quality crops and other commodities with the most judicious use of pesticides.	Co-operative Extension System, University of Connecticut (1994)
1994	An effective and environmentally sensitive approach to pest management that relies on a combination of common-sense practices. IPM programmes use current, comprehensive information on the life cycles of pests and their interactions with the environment. This information, in combination with available pest control methods, is used to manage pest damage by the most economical means, and with the least possible hazard to people, property and the environment. IPM takes advantage of all pest management options possible, including, but not limited to, the judicious use of pesticides.	Leslie (1994)
1994	A control strategy in which a variety of biological, chemical and cultural control practices are combined to give stable long-term pest control.	Ramalho (1994)
1995	A pest management system that, in the socioeconomic context of farming systems, the associated environment and the population dynamics of the pest species, utilizes all suitable techniques in as compatible a manner as possible and maintains the pest population levels below those causing economic injury.	Dent (1995)

Table 1.1 (Continued)

Year	Definition	Source
1996	A sustainable approach to managing pests by combining biological, cultural, physical and chemical tools in a way that minimizes economic, health and environmental risks.	Food Quality Protection Act (1996)
1996	A crop protection system which is based on rational and unbiased information leading to a balance of non-chemical and chemical components moving pesticide use levels away from their present political optimum to a social optimum defined in the context of welfare economics.	Waibel and Zadoks (1996)
1997	An ecosystem-based strategy that focuses on long-term prevention of pests or their damage through a combination of techniques such as biological control, habitat manipulation, modification of cultural practices and use of resistance varieties. Pesticides are used only after monitoring indicates they are needed according to established guidelines, and treatments are made with the goal of removing only target organisms. Pest control materials are selected and applied in a manner that minimizes risks to human health, beneficial and non-target organisms and the environment.	University of California (1997)
1998	A decision support system for the selection and use of pest control tactics, singly or harmoniously co-ordinated into a management strategy, based on cost/benefit analyses that take into account the interests of and impacts on producers, society and the environment.	Kogan (1998)
2000	An approach to the management of pests in public facilities that combines biological, cultural, physical and chemical tools in a way that minimizes economic, health and environmental risks.	Children's Health Act (2000)
2002	A broad ecological approach to pest management utilizing a variety of pest control techniques targeting the entire complex of a crop ecosystem. This approach promises to ensure high-quality agricultural production in a sustainable, environmentally safe and economically sound manner.	Bajwa and Kogan (2002)
2009	The rational application of a combination of biological, biotechnical, chemical, cultural or plant-breeding measures, whereby the use of plant protection products is limited to the strict minimum necessary to maintain the pest population at levels below those causing economically unacceptable damage or loss.	European Union, Directive 91/414/EEC (2009)
2013	A science-based, decision-making process that identifies and reduces risks from pests and pest management-related strategies. IPM co-ordinates the use of pest biology, environmental information and available technology to prevent unacceptable levels of pest damage by the most economical means, while minimizing risk to people, property, resources and the environment. IPM provides an effective strategy for managing pests in all arenas from developed agricultural, residential and public lands to natural and wilderness areas. IPM provides an effective, all-encompassing, low-risk approach to protect resources and people from pests.	USDA national road map for integrated pest management (2013)
2015	A system based on three main principles: (1) the use and integration of measures that discourage the development of populations of harmful organisms (prevention), (2) the careful consideration of all available plant protection methods, and (3) their use to levels that are economically and ecologically justified.	Lefebvre *et al.* (2015)

(*Continued*)

Table 1.1 (Continued)

Year	Definition	Source
2016	A sustainable approach to managing pests by combining biological, cultural, physical and chemical tools in a way that minimizes economic, health and environmental risks. IPM emphasizes the growth of a healthy crop with the least possible disruption to agricultural ecosystems and encourages natural pest control mechanisms.	Department of Agriculture, Environment and Rural Affairs, UK (2016)
2016	Socially acceptable, environmentally responsible and economically practical crop protection.	IPM Centers (2016)
2016	Management of agricultural and horticultural pests that minimizes the use of chemicals and emphasizes natural and low-toxicity methods (as the use of crop rotation and beneficial predatory insects).	Merriam-Webster Dictionary (2016)
2016	An ecosystem approach to crop production and protection that combines different management strategies and practices to grow healthy crops and minimize the use of pesticides.	UN-FAO (2016)
2016	The implementation of diverse methods of pest controls, paired with monitoring to reduce unnecessary pesticide applications.	US Department of Agriculture (2016)
2016	An environmentally friendly, common-sense approach to controlling pests that is focused on pest prevention, the use of pesticides only as needed, the integration of multiple control methods based on site information obtained through inspection, monitoring, and reports.	US Environmental Protection Agency (2016)

routine pest monitoring rather than on calendar-based treatments (Smith and Smith 1949). For the most part, this is the current situation: efforts are largely limited to pesticide management (Ehler 2006), in line with a World Bank (2005) report that concluded that IPM adoption level is low with no indication of change in pesticide use.

1.3.1 Causes for IPM Failure

Why, then, did the IPM approach largely fail to provide growers, and society at large, with effective, safe and sustainable pest management systems? It was clear from the outset that successful IPM is 'knowledge intensive': it requires in-depth ecological understanding of the structure and function of agro-ecosystems, particularly the food webs and species associations and interactions through which energy flows in the system (Barfield and Swisher 1994; Wood 2002). IPM also requires a good grasp of economic, public health and consumer concerns, as well as an appreciation of environmental conservation. These complexities, and the multidisciplinary nature of IPM in the field, are evidently unsuited to the bottom-up manner in which IPM has evolved. Furthermore, the idiosyncratic behaviour of many agro-ecosystems, as well as the site-specific nature of most pest problems, often makes predetermined thresholds operationally intractable (Ehler and Bottrell 2000). Moreover, a field-by-field IPM approach is often insufficient, particularly when pests are mobile. Finally, the cost of generating ecological information needed for development and implementation of functional IPM systems for local situations is prohibitive (Morse and Buhler 1997).

The use of multiple pest control tactics, a fundamental paradigm underlying IPM, presents additional levels of complications, especially when multiple pest types, such as plant pathogens, insects, mites and nematodes, are targeted. This is particularly important because simply combining different management tactics is not sufficient for the implementation of true IPM programmes (Ehler and Bottrell 2000). Control measures often interact in their effects on various organisms in the field. Furthermore, reliance on a single control tactic rarely yields satisfactory results and often causes environmental degradation, food contamination and resistance development in both target and non-target species, seriously impairing agro-ecosystem sustainability (Abrol and Shankar 2012). In general, the use of multiple pest control tactics provides more reliable, efficient and cost-effective solutions. However, mixing control measures employed against one pest without determining their compatibility or effects on other organisms in the system may actually aggravate pest problems or bring about unintended results. Clearly, integrating tactics across different groups of pests – insects, plant pathogens, weeds, etc. – presents even greater challenges than integrating several tactics against a single pest. Combining harmonious – and not antagonistic – tactics to achieve the best long-term control of individual pests or groups of different pests, while ensuring compatibility with the local ecological community, requires considerable research. This integrated study on different pest classes may be discouraged by the organizational structure of research institutions, as departments are often arranged by pest disciplines (Ehler 2006). As a result, perhaps, only a few field-tested examples exist to show how two tactics can be optimally integrated to suppress a single pest in large-scale cropping systems, and studies of the combination of a wider array of tactics are even rarer (Thomas 1999).

The spatial scale to be considered imposes additional constraints on the development of holistic IPM programmes. First, it is unclear what defines the IPM boundary in the farming landscape. Properties of the focal and neighbouring crop fields and their distribution pattern in the landscape, dispersal capacity of the pests, climatic and topographic considerations and many other factors will together determine the distance at which a particular operational IPM system is effective. Second, successful management of some pests may require collective action by neighbouring farmers, especially when the farm holdings are small and close together and pests are mobile. An IPM programme involving migrant pests that function as metapopulations may have to extend over a huge expanse of land. Such area-wide control of agricultural pests would require a centrally managed top-down approach with a regulatory component to ensure full participation and compliance of stakeholders within the region (Vreysen *et al.* 2007). This stands in sharp contrast to the bottom-up approach that has been the operational mode for IPM at the farm and community levels for years.

The dramatic impact of ecological complexity on the efficacy of IPM programmes is evident even when broad pest occurrence patterns, such as the effects of vegetation diversification on pest populations in the IPM landscape, have been demonstrated. The scientific literature generally suggests that plant diversification is a viable strategy for suppressing pests, in part by increasing the level of biological control (see meta-analysis by Letourneau *et al.* 2011). This positive impact of plant diversification was observed, for example, when blast-susceptible rice varieties were planted in combination with resistant varieties: the fungus *Pyricularia oryzae* was 94% less severe in mixtures than in pure rice stands (Zhu *et al.* 2000). However, many diversification schemes slightly but significantly reduce crop yields, in part because intercropping, or the inclusion of non-crop plants, removes some land area from production. Therefore, the potential ecosystem services (benefits) as well as disservices (costs) of vegetation diversification must be quantified for the management of harmful organisms, even though the positive effects usually outweigh the negative.

Another hindrance to the development and implementation of successful IPM programmes is limited and short-term governmental commitment. For the most part, IPM programmes rely on know-how that cannot be commercialized. As such, these programmes are developed by researchers in governmental organizations and public research institutes, such as universities, that are funded mostly by governments, grower associations and other public sources. Many programmes are then implemented through governmental extension services, farmer participatory research, and demonstration and educational programmes (Matteson 2000). Such programmes are the most effective way to disseminate good farming practices, especially, but not only, in developing countries. However, funding constraints, privatization of extension services and shifting attention to other sectors such as urban populations have reduced overall resources devoted to IPM research and implementation in many countries. This global trend is exemplified in the FAO-IPM programme in South and South-East Asian rice crops. This programme was extremely successful for some 20 years. It encompassed training farmers in 13 different countries and educational programmes supported by the respective governments to promote IPM and discourage unnecessary use of pesticides. But when public funding for these programmes dried up, farmers, in response to advocating chemical companies, were quick to revert to pesticide-dependent plant protection practices (Bottrell and Schoenly 2012; Heong and Hardy 2009). Although some IPM efforts have stood the test of time, many others have not, thus allowing the agrochemical industry to sway plant protection away from true IPM and back to the 'supervised control' of the 1950s.

An additional weakness aspect of plant protection research is the need to respond to constant changes in technology, production practices, markets and ecosystem conditions. New, higher yielding crops and cultivars that are more susceptible to pest attacks; novel cultivation practices such as irrigation technologies, no-till cultivation and fertilizer formulation; genetically modified crops; new pesticides and other pest control tools and other innovations force applied scientists to devise solutions to continuously emerging pest problems. Likewise, markets for agricultural produce are constantly in flux, with seasonal price changes, increased demands for produce free of pesticide residues and environmentally friendly food production practices, shifts in global trade in fruit, vegetable and flower crops, and other elements contributing to instability. All these factors influence both economic threshold levels and the arsenal of available pest control measures. In addition, major changes take place due to global warming and desertification, pest invasions, new regulatory actions and many additional factors.

Under these conditions, plant pathologists, weed scientists and entomologists have often only responded to the changes in their attempts to minimize pest-induced yield losses, instead of driving the field toward predetermined goals. In addition, applied scientists, perhaps because of their need to specialize and their appreciation of the uniqueness of their research objects (Rosenheim and Coll 2008), have found it difficult to view the agricultural production system as a whole. As a result, applied researchers rarely integrate multiple scales in their studies, be they multiple pests, several control tactics, several crops, larger spatial scales or long-term dynamics. They instead seek solutions to specific problems, responding to needs only at the local level. Unfortunately, such an approach may not be an optimal way to utilize limited resources and may even conflict with existing research incentives and institutional structures (Waage 1998).

1.3.2 The Impact of the Agro-Chemical Industry

The characteristics of pest management research described above leave the field highly susceptible to the influence of various powerful interest groups, particularly the agro-chemical industry. Until now, IPM has evolved in a bottom-up manner so that even public funding is highly sensitive to crises and is therefore not stable. When funding for research and extension is reduced, chemical companies increase pesticide use again. Similarly, plant protection scientists and professionals may influence national policy, sometimes even working against true IPM. As a case in point, in November 2012, the three professional societies most involved in pest management in the USA (Weed Science Society of America, American Phytopathological Society and Entomological Society of America) released a joint policy statement which clearly rejects the notion that pesticide use in IPM should be restricted to least toxic compounds, and that even those should be used only when no other options exists. They argue that 'suggesting that only "least toxic pesticides" be used, as a "last resort" ignores the extensive research, regulatory, educational and stewardship efforts that make important pesticide tools available and define their proper and safe use in Integrated Pest Management programmes' (www.entsoc.org/press-releases/issues-associated-least-toxic-pesticides-applied-last-resort). This statement appears to be heavily weighted in favour of the agro-chemical industry, and this approach may serve to hamper any effort to implement IPM on the ground.

Given all the obstacles described above, it is not surprising that sustainable IPM systems are extremely rare globally and pesticides use is once again on the increase. Commonly employed IPM practices offer no viable alternatives that would reduce pesticides use and farmers are easily swayed by the pesticide industry. The rate at which farmers revert to 'supervised control' has accelerated in recent years, particularly as inexpensive generic compounds have become available. Therefore, farmers are driven to apply these pesticides rather than scouting their fields. Scouting, after all, is more costly than applying pesticides manufactured in less developed countries where, generally speaking, few environmental, human health and labour regulations are enforced. As a result, global average pesticide use has increased by 8.1% over the last 15 years (Abrol and Shankar 2012). Interestingly, proportionate use of insecticides of all used pesticides is much higher in developing countries than in developed ones, whereas in the latter countries, proportionally more herbicides are used, likely because of the higher prevalence of herbicide-tolerant transgenic crops (Abrol and Shankar 2012).

1.4 A Call for Environmental Pest Management

The pesticide industry clearly has its own incentives and huge endowments to ensure that farmers buy its products. These should be countered by externalizing pesticide-inflicted costs: external costs to human health, the environment and society at large should be levied onto manufacturers, dealers and users of pesticides. The sustainable support of public sector-driven IPM must be guaranteed so that researchers and extension officers stay intimately involved on a long-term basis. The ultimate challenge is to harmonize IPM systems with the farming and consumer communities to ensure that it is compatible with the social, economic, marketing and political considerations that affect IPM adoption (Prokopy and Croft 1994). Toward this goal, constantly evolving scientific, social and economic constraints must be overcome to enable plant protection to become a sustainable component of agriculture with maximum value to farmers, society and the environment. It is apparent that these challenges cannot be met through the traditional, bottom-up approach to the development and implementation of IPM.

We argue that the way in which we approach agricultural pest management must change if we are to develop truly sustainable, environmentally compatible, safe and effective plant protection systems. We need to make the transition from a conventional pest- and crop-centric, bottom-up approach to a more holistic, system-centric, top-down scheme. The time has come to employ top-down tools through regulatory action, positive and negative incentive systems, and by imposing accountability for external costs. The external costs of pesticides have been estimated at US\$ 4–19 kg^{-1} of applied active ingredient (Pretty and Bharucha 2015). Adding these costs to the price of pesticides could help to reduce excessive applications. Such an approach would set desirable overall, ecosystem-wide goals and then devise ways to achieve them on the ground. Theoretical and empirical research will of course still be needed to generate predictive and practical tools, respectively.

While system-wide approaches of this sort are beginning to emerge and even mature in some countries, many of these agro-environmental schemes fail to consider the full range of mutual impacts between pest management and the environment, including effects on human health. A top-down approach would also address the most frequently cited obstacles to the adoption of IPM in developing countries, namely the 'lack of favourable government policies and support' and the need for 'collective action within a farming community' (Parsa *et al.* 2014).

This volume is intended to aid in the development and improvement of agro-environmental systems encompassing all major interactions between pest management practices and the environment. We argue that grassroots research, extension and farmer training efforts must be backed by legislative, regulatory and enforcement actions taken by governments. Governmental inputs acting to promote sustainable pest management practices and nature conservation should have four main objectives that are currently missing in most legislation: (1) the establishment of goal-based agro-environmental schemes that include pest management objectives, (2) externalizing true costs of pesticide use, (3) strengthening of the public extension service, and (4) soliciting goal-specific plant protection research.

Properties and methods used for the implementation of these objectives would certainly vary greatly among countries. Governmental and social structures, economic forces, traditions and other factors will shape needs, impose constraints and determine feasibility of means, and thus influence goals and approaches. However, in some cases, the required infrastructure already exists and needs only to be adjusted to the new objectives. For example, the State of California, USA, charges a "Mill Assessment" fee on pesticide sales that could be adjusted upward in order to discourage pesticide use and cover health and environmental costs related to pesticide application.

For practical, marketing or ideological reasons, growers should be allowed to meet regulatory requirements in different ways: through organic farming, permaculture, IPM, or by adopting just a few practices which promote desirable outcomes. Governmental involvement would also facilitate co-ordination and communication between landowners within a landscape and a thorough understanding of local and regional patterns of multi-scale ecosystem services and disservices. These are essential for sustainable pest management. Finally, centralized schemes and policies could be amended and fine-tuned as more information becomes available and with changes in agricultural production and market conditions. These continuous adjustments are crucial for the sustainability of safe and environmentally compatible pest management practices.

Acknowledgements

We thank Ruth-Ann Yonah for her help in manuscript preparation.

References

Abrol, D.P. and Shankar, U. (2012) History, overview and principles of ecologically-based pest management. In: Abrol, D.P. and Shankar, U. (eds) *Integrated Pest Management: Principles and Practice.* CABI, Croydon, UK, pp. 1–26.

Allen, W.A and Rajotte, E.G. (1990) The changing role of extension entomology in the IPM era. *Annual Review of Entomology* **35**: 379–397.

Bajwa, W.I. and Kogan, M. (2002) Compendium of IPM Definitions (CID). What is IPM and how is it defined in the worldwide literature? Integrated Plant Protection Center (IPPC), Oregon State University, Corvallis, OR, USA. Available at: www.ipmnet.org/ipmdefinitions/index.pdf (accessed 28 February 2017).

Barfield, C.S. and Swisher, M.E. (1994) Integrated pest management: ready for wood export? Historical context and internationalization of IPM. *Food Reviews International* **10**: 215–267.

Bottrell, D.G. (1979) *Integrated Pest Management.* Council on Environmental Quality, US Government Printing Office, Washington, DC, USA.

Bottrell, D.G. and Schoenly, K.G. (2012) Resurrecting the ghost of green revolutions past: the brown planthopper as a recurring threat to high-yielding rice production in tropical Asia. *Journal of Asia-Pacific Entomology* **15**: 122–140.

Children's Health Act (2000) Public Law 106-310, Title V, Section 511. Available at: www.congress.gov/106/bills/hr4365/BILLS-106hr4365enr.pdf (accessed 28 February 2017).

Co-operative Extension System, University of Connecticut (1994) *Integrated Pest Management Programs.* University of Connecticut, Mansfield, CT, USA.

Council for Agricultural Science and Technology (1982) *Integrated Pest Management.* Report No. 93. Council for Agricultural Science and Technology, Ames, IA, USA.

Council on Environmental Quality (1972) *Integrated Pest Management*. US Government Printing Office, Washington, DC, USA.
Dehne, H.W. and Schonbeck, F. (1994) Crop protection – past and present. In: Oerke, E.C., Dehne, H.W., Schonbeck, F. and Weber, A. (eds) *Crop Production and Crop Protection*. Elsevier, Amsterdam, The Netherlands, pp. 45–71.
Dent, D.R. (1995) *Integrated Pest Management*. Chapman and Hall, London, UK.
Department of Agriculture, Environment and Rural Affairs, UK (2016) Integrated Pest Management Guide. Available at: www.daera-ni.gov.uk/publications/integrated-pest-management-guidance (accessed 28 February 2017).
Ehler, L.E. (2006) Integrated pest management (IPM): definition, historical development and implementation, and the other IPM. *Pest Management Science* **62**: 787–789.
Ehler, L.E. and Bottrell, D.G. (2000) The illusion of integrated pest management. *Issues in Science and Technology* **16** (3). Available at: http://issues.org/16-3/ehler/ (accessed 28 February 2017).
European Union, Directive 91/414/EEC (2009) Available at: http://ec.europa.eu/environment/archives/ppps/pdf/final_report_ipm.pdf (accessed 28 February 2017).
FAO (1967) *Report of the First Session of the FAO Panel of Experts on Integrated Pest Control*. FAO, Rome.
FAO (1980) Research Summary. Integrated pest management. EPA-600/8-80-044. Available at: https://nepis.epa.gov/Exe/ZyPDF.cgi/91013QAV.PDF?Dockey=91013QAV.PDF (accessed 28 February 2017).
FAO, IFAD and WFP (2015) The state of food insecurity in the world 2015. Meeting the 2015 international hunger targets – taking stock of uneven progress. FAO, Rome. Available at: www.fao.org/3/a-i4646e.pdf (accessed 28 February 2017).
Flint, M.L. and van den Bosch, R. (1981) *Introduction to Integrated Pest Management*. Plenum Press, New York, USA.
Food Quality Protection Act (1996) Public Law 104- 170, Title II, Section 303, Enacted August 3, 1996. Codified in: Title 7, U.S. Code, Section 136r-1. Integrated Pest Management. Pg 1512. US Government Publishing Office. Available at: www.gpo.gov/fdsys/pkg/PLAW-104publ170/pdf/PLAW-104publ170.pdf (accessed 28 February 2017).
Heong, K.L. and Hardy, B. (2009) *Planthoppers: New Threats to the Sustainability of Intensive Rice Production Systems in Asia*. International Rice Research Institute, Los Baños, Philippines.
IMP Centers (2016) www.ipmcenters.org/index.cfm/about-centers/what-is-ipm/ (accessed 28 February 2017).
Kogan, M. (1998) Integrated pest management: historical perspectives and contemporary developments. *Annual Review of Entomology* **43**: 243–270.
Kumar, R. (1984) *Insect Pest Control with Special Reference to African Agriculture*. Edward Arnold, London.
Lefebvre, M., Langrell, S.R.H. and Gomez-y-Paloma, S. (2015) Incentives and policies for integrated pest management in Europe: a review. *Agronomy for Sustainable Development* **35**: 27–45.
Leslie, A.R. (1994) Preface. In: Leslie, A.R. (ed.) *Pest Management for Turf and Ornamentals*. Lewis Publishers, London, UK.
Letourneau, D.K., Armbrecht, I., Rivera, B.S., *et al.* (2011) Does plant diversity benefit agroecosystems? A synthetic review. *Ecological Applications* **21**: 9–21.
Matteson, P.C. (2000) Insect pest management in tropical Asian irrigated rice. *Annual Review of Entomology* **45**: 549–574.

Merriam-Webster Dictionary (2016) www.merriam-webster.com/dictionary/integrated%20pest%20management (accessed 18 April 2017).
Morse, S. and Buhler, W. (1997) *Integrated Pest Management: Ideals and Realities in Developing Countries*. Lynne Riener Publishers, Boulder, CO, USA.
National Academy of Science (1969) Insect-pest management and control. In: Palm C.E. and Subcommittee on Insect Pests (eds) *Principles of Plant and Animal Pest Control, Vol. 3*. National Academy of Sciences, Washington, DC, USA, pp. 448–449.
National Academy of Science, Board on Agriculture (1989) *Alternative Agriculture*. National Academy Press, Washington, DC, USA.
National Coalition on Integrated Pest Management (1987) *Integrated Pest Management*. National Coalition on Integrated Pest Management, Austin, TX, USA.
National Research Council, Board on Agriculture (1989) *Alternative Agriculture*. National Academy Press, Washington, DC, USA.
Natural Resources Institute (1992) *A Synopsis of Integrated Pest Management in Developing Countries in the Tropics*. Natural Resources Institute, Chatham, UK.
Oerke, E.C., Dehne, H.W., Schonbeck, F. and Weber, A. (1994) *Crop Production and Crop Protection: Estimated Losses in Major Food and Cash Crops*. Elsevier, Amsterdam, The Netherlands.
Office of Technology Assessment (1979). *Pest Management Strategies Crop Protection*. Vol. 1. Congress of the United States, Washington, DC, USA.
Olkowski, W. and Daar, S. (1991) *Common Sense Pest Control*. Taunton Press, Newtown, CT, USA.
Parsa, S., Morse, S., Bonifacio, A., *et al.* (2014) Obstacles to integrated pest management adoption in developing countries. *Proceedings of the National Academy of Sciences* **111**: 3889–3894.
Pedigo, L.P. (1989) *Entomology and Pest Management*. Macmillan Publishing, New York, USA.
Pimentel, D. (2007) Area-wide pest management: environmental, economic and food issues, In: Vreysen, M.J.B., Robinson, A.S. and Hendrichs, J. (eds) *Area-Wide Control of Insect Pests From Research to Field Implementation*. Springer, Dordrecht, The Netherlands, pp. 35–47.
Pretty, J. and Bharucha, Z.P. (2015) Integrated pest management for sustainable intensification of agriculture in Asia and Africa. *Insects* **6**: 152–182.
Prokopy, R.J. and Croft, B.A. (1994) Apple insect management, In: Metcalf, R.L. and Luckmann, W.H. (eds) *Introduction to Insect Pest Management*, 3rd edn. Wiley, NewYork, pp. 543–589.
Ramalho, F.S. (1994) Cotton pest management. *Annual Review of Entomology* **39**: 563–578.
Rosenheim, J. A., and Coll, M. (2008). Pest-centric versus process-centric research approaches in agricultural entomology. *American Entomologist* **54**: 70–72.
Smith, R.F. (1978) History and complexity of integrated pest management. In: Smith, E.H. and Pimentel, D. (eds) *Pest Control Strategies*. Academic Press, New York USA, pp. 41–53.
Smith, R F. and Reynolds, H.T. (1966) Principles, definitions and scope of integrated pest control. Proceedings of the FAO Symposium on Integrated Pest Control. FAO, Rome, pp. 11–17.
Smith, R.F. and Smith, G.L. (1949) Supervised control of insects: utilizes parasites and predators and makes chemical control more efficient. *California Agriculture* **3**: 3–12.

Sommer, A. and West, K.P. (1996) *Vitamin Deficiency: Health Survival and Vision*. Oxford University Press, New York, USA.
Sorensen, A.A. (1992) Proceedings of the National Integrated Pest Management Forum, Arlington, Virginia, USA, June 17–19, 1992. American Farmland Trust, Center for Agriculture in the Environment, DeKalb, GA, USA.
Stern, V.M., Smith, R.F., van den Bosch, R. and Hagen, K.S. (1959) The integrated control concept. *Hilgardia* **29**: 81–101.
Thomas, M.B. (1999) Ecological approaches and the development of 'truly integrated' pest management. *Proceedings of the National Academy of Sciences* **96**: 5944–5951.
UC-IPM (2008) What is IPM? Statewide IPM Program, University of California, Agriculture and Natural Resources. Available at: www.ipm.ucdavis.edu/WATER/U/ipm.html (accessed 28 February 2017).
UN-FAO (2016) www.fao.org/agriculture/crops/thematic-sitemap/theme/pests/ipm/en/ (accessed 28 February 2017). University of California (1997) Wide integrated pest management project annual reports. University of California, Berkeley, CA, USA.
US Department of Agriculture, Agricultural Research Service (1993) *USDA Programs Related to Integrated Pest Management*. USDA Program Aid 1506. Available at: https://archive.org/details/CAT31312787 (accessed 28 February 2017).
US Department of Agriculture (2016) https://nifa.usda.gov/program/integrated-pest-management-program-ipm (accessed 28 February 2017).
USDA national road map for integrated pest management (2013) Available at: http://ipmcenters.org/Docs/IPMRoadMap.pdf (accessed 28 February 2017).
USDA NRCS Conservation Effects Assessment Project (2016) Cropland National Assessment. Available at: www.nrcs.usda.gov/wps/portal/nrcs/detail/national/technical/nra/ceap/na/?cid=nrcs143_014144 (accessed 28 February 2017).
US Environmental Protection Agency (2016) www.epa.gov/managing-pests-schools/introduction-integrated-pest-management (accessed 28 February 2017).
Van den Bosch, R. and Stern, V.M. (1962) The integration of chemical and biological control of arthropod pests. *Annual Review of Entomology* **7**: 367–386.
Vreysen, M.J.B., Robinson, A.S., Hendrichs, J. and Kenmore, P. (2007) Area-wide Integrated Pest Management (AW-IPM): principles, practice and prospects. In: Vreysen, M.J.B., Robinson, A.S. and Hendrichs, J. (eds) *Area-Wide Control of Insect Pests: From Research to Field Implementation*. Springer, Dordrecht, The Netherlands, pp. 3–33.
Waage, J.K. (1993) Making IPM work: developing country experience and prospects. In: Srivastava, J. P. and Alderman, H. (eds) *Agriculture and Environmental Challenges: Proceedings of the Thirteenth Agricultural Sector Symposium*. World Bank, Washington, DC, USA, pp. 119–134.
Waage, J.K. (1998) The future development of IPM. *Entomologia Sinica* **5**: 257–271.
Waibel, H. and Zadoks, J.C. (1996) *Institutional Constraints to IPM*. In: XIIIth International Plant Protection Congress (IPPC), The Hague, July 2–7. Pesticide Policy Project Publication Series. No. 3, Institute of Horticultural Economics, Hannover, Germany.
Wightman, J.A. (1993) Towards the rational management of the insect pests of tropical legumes crops in Asia: review and remedy. In: Chadwick, D.J. and Marsh, J. (eds) *Crop*

Protection and Sustainable Agriculture: CIBA Foundation Symposium 177. John Wiley and Sons, Chichester, UK, pp. 233–256.

Wood, B.J. (2002) Pest control in Malaysia's perennial crops: a half century perspective tracking the pathway to integrated pest management. *Integrated Pest Management Reviews* 7: 173–190.

World Bank (2005) *Sustainable Pest Management: Achievements and Challenges.* Report No. 32714-GBL. World Bank, Washington, DC, USA.

Yepsen, R.B. (1984) *The Encyclopedia of Natural Insect and Disease Control: The Most Comprehensive Guide to Protecting Plants, Vegetables, Fruit, Flowers, Trees and Law.* Rodale Press, Emmaus, PA, USA.

Zalom, F.G., Ford, R.E., Frisbie, R.E., Edwards, C.R. and Telle, J.P. (1992) Integrated pest management: addressing the economic and environmental issues of contemporary agriculture. In: Zalom, F.G. and Fry, W.E. (eds) *Food, Crop Pests, and the Environment: The Need and Potential for Biologically Intensive Integrated Pest Management.* APS Press, St Paul, MN, USA. pp. 1–12.

Zhu, Y., Chen, H., Fan, J., *et al.* (2000) Genetic diversity and disease control in rice. *Nature* **406**:718–722.

Part I

General Background

2

Approaches in Plant Protection: Science, Technology, Environment and Society

Deborah K. Letourneau, Margaret I. FitzSimmons and Diego J. Nieto

2.1 Introduction

Our purpose in this chapter is to introduce Integrated Pest Management (IPM) as a desirable approach to plant protection, but one that requires an ecosystem-wide perspective and substantial shifts within the social, political and economic priorities of food production. In the words of van den Bosch and Stern (1962), 'it is the entire ecosystem and its components that are of primary concern and not a particular pest'. The reason for bringing attention to an even broader context is to understand what forces are shaping pest, disease and weed management today, and to argue that desirable changes in plant protection will require interdisciplinary strategies that foster bold transformations in food systems.

Plant protection approaches in agriculture are derived from a complex array of social and scientific factors, including access to capital, labour and technological tools, the agro-ecosystem and surrounding landscape, relationship dynamics along the commodity chain, market volatility, institutional price supports, regulations and loan stipulations. Yet on the ground, they may look like a series of simple decisions: prophylactically drench broccoli (*Brassica oleracera*) with chlorpyrifos to prevent yield losses from cabbage maggot (*Delia radicum*), sow mixed rice (*Oryza sativa*) varieties to optimize spider predation and virus resistance, grind up unharvested sugar cane (*Saccharum officinarum*) infested with stalk borers (*Diatraea saccharalis*), fumigate the soil to extend harvest cycles, etc. The outcomes of each plant protection decision then provide feedback to growers in terms of experience, knowledge, cultural norms, revenue and marketing options, while also extending out to affect less apparent aspects of the environment, such as microbial shifts in the soil, groundwater purity, allele selection and fixation in weed populations, diversity of parasitoids and dominance of suppressive micro-organisms. Environmental effects then come back to growers, farm workers and consumers indirectly in terms of promoting or disrupting agro-ecosystem health, ecosystem services, public health and food security. Examining pest control approaches in their social contexts highlights some of the barriers against, and opportunities for, integrated and sustainable crop protection solutions.

Environmental Pest Management: Challenges for Agronomists, Ecologists, Economists and Policymakers, First Edition. Edited by Moshe Coll and Eric Wajnberg.

2.2 History of Plant Protection Approaches

Humans have brought native plants into production, selected propagules with desirable traits, and dispersed them as crops into new ecological environments. Emerging in most of the world's regions out of local knowledge of plants and animals (Sauer 1952), early crop protection practices developed as adaptations to local ecological circumstances. As human dispersal transferred desired species into new ecological systems (Sauer 1993), cultural practices changed, in part because plants could be resited away from their natural enemies. The development of plantation agriculture in the 18th century benefitted from such relocation, as coffee (*Coffea* spp.), cotton (*Gossypium* spp.), banana (*Musa* spp.), pineapple (*Ananas comosus*) and other valued crops were freed from pest pressure in new colonial locations. New forms of economic management and control of labour were coupled to large-scale production in these new ecological-social complexes. Over time, specialized intensive production set the stage for new ecological and social challenges requiring different scientific and institutional innovations. The beginnings of standardized, broad-spectrum responses to these challenges were developed at research institutions for plant breeding and pest control and were disseminated through educational extension for farmer and public education.

The economic norms developed in plantation agriculture, such as increasing geographic separation of production and markets, creating powerful intermediate business between producers and consumers, standardizing varieties, and simplifying and intensifying farm activities, spread over a century from plantation and grain crops into horticulture (FitzSimmons 1986). We provide an historical overview of crop protection approaches within their social contexts in the changing production of fresh-market strawberry (*Fragaria ananassa*) in coastal California, USA, over the last 100 years. Major pests and crop protection approaches are listed at 20-year intervals from 1915 to 2015 (Table 2.1) for strawberry production in the Central Coast region of California (Figure 2.1). Our experience with this cropping system and location allows us to illustrate historical changes and social forces that determined the contents of a farmer's 'toolbox' and influenced decision making in plant protection strategies.

In 1915, intensive production of specialty crops in small parcels, including leafy greens, cole crops and strawberry, was initiated in the region by Japanese immigrants, bringing cultural norms, knowledge and production techniques from Asia. However, growing political anti-immigration sentiment in the USA (Higgs 1978) resulted in the Alien Land Law of 1913, which denied Asians in California the right to own, lease or otherwise enjoy land. By listing themselves as 'managers' affiliated with sympathetic growers with European heritage, the Isei (Japanese immigrant families) established and tended strawberry plantings that produced high-quality fruit for 4–6 years, and sold them in local markets. Patchy establishment of a new crop allowed for relatively low accumulation of pests in those early years. Although most pests were introduced species, such as strawberry root weevils from Europe and easily dispersed annual weeds (Mensing and Byrne 1998), native species such as oak root rot (*Armillaria mellea*) occasionally infected the crop, in this case after clearing of its woody hosts for berry production (see Table 2.1).

In 1935, producers in the region had diversified ethnically but were still commonly Nisei – sons of Japanese immigrants – tenant farmers who managed small parcels to grow berries and vegetables using family labour. Strawberries were transplanted in the spring into soil that had never before produced berries. Pest and disease pressure increased nonetheless, compared to the previous generation of growers, whose plants

Table 2.1 Changes in pest-related challenges and crop protection approaches over the last century, as illustrated by strawberry production on the central coast of California, USA.

	Major pests	Crop protection tactics
1915		Sources* include Smith and Goldsmith (1936), Darrow (1966), Farmers' Bulletins (e.g. USDA No. 2184)
Insects	Strawberry root weevil (*Otiorhynchus* spp.) (Eighme, L., pers. notes), strawberry leaf-roller (*Ancylis comptana*)	Conservation of natural enemies**; copper acetoarsenite (Paris Green), powdered arsenate of lead (Hinds 1913)
Mites	Possibly two-spotted mite (*Tetranychus urticae*)	Conservation of natural enemies***
Diseases	Gray mould (*Botrytis cinerea*), oak root fungus (*Armillaria mellea*)	Transplant clean nursery stock to land that has never been in strawberry production. These plants remain in production for 4–6 years with conservation of antagonistic microbes
Nematodes	Possibly root-knot nematodes (*Meloidogyne* spp.)	Naturally occurring *Bacillus thuringiensis* (Bt), predatory nematodes
Weeds	Unknown; few exotic species, some native species likely, such as coast wild cucumber (*Marah oreganus*)	Hand cultivation
1935		Sources* include Thomas (1932, 1939), Stevens (1933), Smith and Goldsmith (1936)
Insects	Strawberry rootworm (*Paria canella*), aphids (Aphididae), white grubs (Scarabaeidae)	Conservation of natural enemies**; top plant (remove leaves) before second spring to control mites and aphids; nicotine tannate, *Derris* plant extract (rotenone), pyrethrum (Ginsberg and Schmit 1932; Little 1931); lead arsenate dusts
Mites	Two-spotted spider mite, occasionally cyclamen mite (*Steneotarsonemus pallidus*)	Conserve natural enemies***; replace popular Nich Ohmer variety with Marshall-Banner varieties with resistance to cyclamen mite, but prone to diseases
Diseases	Red-stele root rot (*Phytophthora fragariae* var. *fragariae*) (Darrow 1966), *Verticillium* spp. wilts	Transplant clean nursery stock to levelled land that has never been in strawberry production; avoid land with a history of potato or tomato; these plants remain in production for 2–3 years with conservation of antagonistic microbes; top plant (remove leaves) before second spring to avoid leaf spot
Nematodes	Root-knot nematodes	Naturally occurring enemies such as the bacterium *Bacillus thuringiensis* (Bt) and predatory nematodes
Weeds	For example, red stem filaree (*Erodium cicutarium*), field bindweed (*Convolvulus arvensis*), pigweed (*Amaranthus* spp.), mustard (*Brassica* spp.)	Hand cultivation, pre-plant irrigation then low water in first year

(*Continued*)

Table 2.1 (Continued)

	Major pests	Crop protection tactics
1955		Sources* include Huffaker and Kennett (1956), Allen (1959), *California Agriculture* articles, e.g. Lange *et al.* (1967), and Smith *et al.* (1958)
Insects	Serpentine leaf miner (*Tischeria* sp.), strawberry aphid (*Chaetosiphon fragaefolii*), which vectors strawberry mild yellow-edge virus (SMYEV) (*Potexvirus*); secondary pests, such as the native western flower thrips (*Frankliniella occidentalis*) and western tarnished plant bug (WTPB) (*Lygus hesperus*)	Kelthane, phosdrin, TEPP, thiodan, chlordane, diazinon, malathion, parathion and methoxychlor; with less attention to conservation of natural enemies**
Mites	Two-spotted spider mite; cyclamen mite in second year plantings	Endrin, azobenzene and isodrin for cyclamen due to their persistence; conservation of natural enemies***
Diseases	Crown rot (*Phytophthora* spp.), SMYEV and crinkle virus (*Cytorhabdovirus*)	Sierra variety resistant to *Verticillium*; chlorobromopropene and methyl bromide soil fumigant, chloropicrin, Captan and phenyl mercury acetate
Nematodes	Root-knot nematode; possibly a secondary pest: root-lesion nematode (*Pratylenchus* spp.)	Chlorobromopropene soil fumigant
Weeds	For example, red stem filaree, field bindweed, pigweed, mustard, etc.	Hand cultivation, tractor cultivation, fumigation
1975		Sources include USDA (1972) and various *California Agriculture* articles, such as Welch *et al.* (1989)
Insects	Western flower thrips and WTPB	Organochlorines were mostly replaced by carbamates and organophosphates including carbaryl, methoxychlor, toxaphene and mevinphos for insects and mites
Mites	Two-spotted spider mite; cyclamen mite in second year plantings	Miticides; conservation or augmentation of predatory mites; release of artificially selected miticide-resistant predatory mites (Roush and Hoy 1980)
Diseases	Root rot (*Verticillium dahliae*)	Tioga variety tolerant to yellows and crinkle virus
Nematodes	Root-knot nematodes	Methyl bromide fumigation, allowed via special use permit
Weeds	For example, red stem filaree, field bindweed, pigweed, mustard, etc.	Monitoring, methyl bromide fumigation, hand and tractor cultivation
1995		Sources: Strand (1994), Gabriel (1989), Welch *et al.* (1989), Pickel *et al.* (1995), Gliessman *et al.* (1996)

Table 2.1 (Continued)

	Major pests	Crop protection tactics
Insects	Western flower thrips and WTPB	Malathion, naled; some organophosphates replaced by novaluron, pyrethroids, spinosyns, malathion; augmentation or conservation of natural enemies**; release of WTPB egg parasitoid *Anaphes iole*; removal of nearby host plants for overwintering WTPB, e.g. wild radish, shepherd's purse; monitor for WTPB on new plantings in fall to set biofix date to predict nymphal exposure and time insecticide spray after 242 degree-days accrue
Mites	Two-spotted spider mite; cyclamen mite in first and second year plantings	Bifenazate, abamectin, fenbutatin, oils; rotate active ingredients to retard resistance build-up; spray thresholds of 5–20 mites per leaflet unless predatory mites are half as abundant; augmentative releases of commercially available predatory mites *Phytoseiulus persimilis*, *Amblyseius californicus* and *Galendromus occidentalis*; conservation of natural enemies***
Diseases	Root rot and SMYEV	Methyl bromide fumigation, which controls white grubs and other previously common pests and diseases; rotation with rye or barley recommended
Nematodes	Northern root-knot nematode (*Meloidogyne hapla*)	Methyl bromide fumigation, hand and tractor cultivation
Weeds	Little mallow (*Malva parviflora*), burclover (*Medicago* spp.), common groundsel (*Senecio vulgaris*), sowthistle (*Sonchus* spp.), purslane (*Portulaca oleracea*), chickweed (*Stellaria media*), red stem filaree, burning nettle (*Urtica urens*), annual bluegrass (*Poa annua*); plants that host WTPB	Methyl bromide fumigation, plastic mulch and drip irrigation, except for field bindweed, sweetclovers, little mallow, burclover and common groundsel, the seeds of which survive methyl bromide and chloropicrin fumigation
2015		Sources: Strand (2008) and Koike *et al.* (2012), unless otherwise indicated
Insects	WTPB, western flower thrips and recently introduced exotic pests: light brown apple moth (*Epiphyas postvittana*), greenhouse whitefly (*Trialeurodes vaporariorum*), spotted-winged *Drosophila* (*Drosophila suzukii*)	Conservation, introduction and augmentation of natural enemies**; alfalfa trap crops (Swezey *et al.* 2014) and 'good bug blends' providing nectar and pollen for parasitoids and predators; introduction of WTPB braconid parasitoid *Peristenus relictus* (Pickett *et al.* 2009); rotation of insecticides with different modes of action, limited applications per season, spinosad, imidicloprid, diazinon; manage the 15 types of wild WTPB hosts; tractor-mounted vaccuums for insect removal on alfalfa or strawberry; organic: insecticidal soap, azadirachtin, neem oil, entomopathogenic fungus *Beauveria bassiana* and pyrethrin

(*Continued*)

Table 2.1 (Continued)

	Major pests	Crop protection tactics
Mites	Two-spotted spider mite, Lewis mite (*Eotetranychus lewisi*), cyclamen mite in organic production	Conservation and augmentation of natural enemies***; synthetic miticides in conventional production such as etoxazole, abamectin, acequinocyl
Diseases	Leaf spot (*Ramularia tulasneii*), crown and root rots, especially *V. dahliae*, SMYEV, and *Phytophthera* spp.	Albion variety for tolerance to major soil-borne pathogens; drip fumigation with an application of chloropicrin mixed with 1,3-dichloropropene followed by metam sodium or chloropicrin alone followed by metam sodium, some methyl bromide through critical use exemptions; cover cropping with mustards, anaerobic soil disinfection (Butler *et al.* 2014), mustard seed meal allelopathy or planting into coconut or rice hull
Nematodes	Root-knot nematodes	Conventional: fumigants, cereal rye or barley cover crops with broadleaf herbicides; organic: cereal rye or barley cover crops, anaerobic soil disinfection, cover-cropping or planting into coconut or rice hull
Weeds	Field bindweed (*Convolvulus arvensis*), burclover and yellow nutsedge (*Cyperus esculentus*) are resistant to fumigants; pigweed, filaree, mustards, radish (*Raphanus raphanistrum*), etc.; plants that host WTPB	Fumigants, flumioxazin and oxyfluorfen herbicides (Samtani *et al.* 2012), cereal rye or barley cover crops, solarization, anaerobic soil disinfection; organic: cover cropping, preirrigation, 12 rounds of tillage, drip irrigation and coverage of planting beds with opaque polyethylene mulch

* Determinations of major pests and likely management tactics used against them on the Central Coast of California, USA, between 1915 and 1955 are from Letourneau's readings of historical documents, including an assessment of what was listed, what common names meant, what was written about and left out in more than 50 sources found online in USDA archives, University of California archives, and on shelves at the Agricultural History Project in Watsonville, CA, USA. http://aghistoryproject.org/.
** Generalist predators, such as minute pirate bug (*Orius tristicolor*), big-eyed bug (*Geocoris* spp.), brown lacewing (*Hemerobius* spp.), ladybird beetle (e.g. *Hippodamia convergens*), predaceous fly larvae (Syrphidae), soil-dwelling beetles (e.g. Staphylinidae), spiders (e.g. Thomisidae) and insectivorous birds. Parasitoids, such as *Lysiphlebus testaceipes* (Braconidae), a strawberry aphid parasitoid (Oatman and Platner 1972) and *Trichogramma pretiosum* (Trichogrammatidae), a parasitoid of corn earworm (UC-IPM 2010).
*** *Phytoseiulus persimilis* and generalist predators, such as *Neoseiulus californicus*, rove beetle (*Oligota oviformis*) and six-spotted thrips (*Scolothrips sexmaculatus*) (Dara *et al.* 2012).

produced for up to six seasons. By the 1930s, cyclamen mite (*Phytonemus pallidus*) and soil-borne pathogens caused declines in yields in older plants, prompting growers to transplant into new ground every 2–3 seasons (Smith and Goldsmith 1936), and cosmetic standards would appear later in the decade (Thomas 1939).

By this time, crop protection practices were not usually devised by farmers acting independently. They were instead orchestrated by farmer co-operatives, which also facilitated marketing orders to nearby urban centres. The California Strawberry Advisory Council, later renamed the California Strawberry Commission (which is

Figure 2.1 Strawberry production on the Central Coast of California, USA, in 2014 showing strip cropping of alfalfa for control of the mirid bug *Lygus hesperus* via suction removal from alfalfa trap crop with tractor-mounted vacuums (a, b), hedgerows with native perennials that provide shelter and food resources for natural enemies of strawberry pests (c), comparison of plants protected with anaerobic soil disinfection (d) and growers' standard pre-plant practice (e) in an organic field infested with *Macrophomina* spp. and *Fusarium oxysporum* fungal diseases, and plastic mulch for weed control (f).

incorporated into the present California Department of Food and Agriculture (CDFA)), took an active role in farmer education, encouraging farmers to adopt new varieties and practices in response to new pests and diseases. The co-operative membership funded research at University of California land grant campuses, which then disseminated county farm advisors and corporate salesmen to promote new varieties, cultural practices and recommended chemical treatments. For example, Central Coast growers incorporated 'Banner' strawberry, which was resistant to cyclamen mite. The mechanism for mite resistance involved rapid leaf unfurling in the crown, which promoted access and suppression by naturally occurring predatory mites (Smith and Goldsmith 1936).

The cyclamen mite was not considered a major pest in this region until the mid-1940s, when growers started prioritizing varieties bred for better shipping and rot resistance in favour of mite-resistant varieties, which were often prone to disease (Stevens 1933). Up to this point, growers could escape the build-up of soil-borne pathogens by planting exclusively into new parcels. However, shorter cropping cycles coupled with land access restrictions resulting from discriminatory laws put a strain on growers' ability to find parcels without a history of strawberry production. As a result, Japanese growers began to grow strawberry consecutively in the same parcels. Subsequent plant protection challenges during the first year of production prompted the development of formal certification for clean nursery stock and preventive insecticides.

Plant-based pesticides and toxic metals were long familiar to growers as medicinal products. On the one hand, nicotine extracts and Paris Green (copper acetoarsenite) had been purchased for decades from travelling salesmen as cure-alls or prescribed by doctors for ailments, and were thus considered medicinal and beneficial. Their safety and effectiveness were reinforced in insecticide advertisements that began appearing in magazines by the late 1920s. Promising easy solutions to pest problems, the advertisements illustrated rapid results, some with creative and appealing cartoons by the illustrator who would later become Dr Seuss, famous for his children's books. On the other hand, public health problems associated with pesticides had been described in a best-selling book (Kallet and Schlink 1932), and both pest resistance and honey bee (*Apis mellifera*) mortality had already been demonstrated as side effects of rotenone and pyrethrum (Ginsberg and Schmit 1932).

By 1955, post-World War II farming practices shifted in response to social phenomena, such as the loss of forced farm labour via prisoners of war, adoption of labour-saving machinery produced in factories that had manufactured tanks and planes during the war, an influx of monetary supports instituted by the US government, new federal investments in the development of foreign markets, and the military use of synthetic pesticides. After the war, consumption patterns also changed, as supermarkets began to replace local sellers, sourcing fresh produce from growers at greater distances.

In strawberries, some of these postwar changes were particularly significant. Many Nisei strawberry growers had lost their land to either European and American investors or government seizure while they were interned with over 100 000 Japanese Americans in World War II concentration camps under President Roosevelt's Executive Order 9066. Newly mechanized cultivation, increased investments in irrigation, and a labour force of Mexican nationals – *braceros* – admitted during the war enabled an expansion of strawberry production. In addition, field testing at a University of California field station demonstrated the effectiveness of the fumigants methyl bromide and chloropicrin against insects, diseases and weeds, which further intensified production through even shorter rotations. Regional strawberry acreage consequently increased five-fold from 1935, with ca. 8000 ha of strawberries in production (Wilhelm and Sagen 1974).

Investments in refrigerated transport enabled production for much larger markets. Local 'truck farm' producers in eastern and central cities began to lose markets to producers in Florida and California, facilitated by the rapid growth of dominant shippers. Technological innovations included by-products of government investment in war industries that turned to agriculture to find new markets.

With the advent of new synthetic pesticides, such as the insecticide DDT and the herbicide 2, 4 D, the Federal Insecticide, Fungicide and Rodenticide Act (FIFRA 1947) required manufacturers to register pesticides, include labels, directions for use and antidotes to reduce the danger from human ingestion. Almost 10 000 new pesticide products were registered with the United States Department of Agriculture. While growers were familiar with toxic substances specific to certain pests, new commercial products were sufficiently toxic across a broad spectrum of pests to justify registration costs. Cyclamen mite, two-spotted spider mite (*Tetranychus urticae*) and strawberry aphid (*Chaetosiphon fragaefolii*), which vectors yellow virus disease, were in almost every field, due to the expansion and greater concentration of strawberry production sites, susceptibility of the prominent 'Shasta' variety and the elimination of various predatory mites and insects through the increased use of broad-spectrum organochlorines such as DDT and toxaphene (Allen 1959). Secondary pests, such as the native western flower thrip (*Frankiniella occidentalis*) and the western tarnished plant bug (*Lygus hesperus*) or WTPB, were also emerging in strawberry fields.

By 1975, strawberry production had taken on an effectively industrial form (FitzSimmons 1986). Land would probably be owned by a family corporation and leased to a former strawberry picker who may have entered California through the *bracero* programme. The short-handled hoe, which allowed field labour managers to easily identify who was standing upright, rather than weeding, was banned in response to organized labour demands. In a counterresponse, physical labour was replaced by tractors and fumigants for weed seed suppression (see Table 2.1). Marketing became increasingly consolidated, as shippers contracted with a concentrated retail sector to provide strawberries on a calendar basis to meet anticipated consumer demand. Some smaller growers found themselves marginalized in these pathways, and began to turn to alternatives like farm stands, farmers' markets and other direct sales strategies. Field packing had become the norm, as shippers closed packing sheds to forestall union organization. Farm workers were widely available and often preferred to work in strawberries because its extended season allowed employment and family settlement over several months. After fumigation, pesticide applications were performed by the harvest workers themselves, creating a pick and spray rotation that maintained a relatively constant level of crop protection (and worker exposure).

Although pesticide regulations (FIFRA 1972) were not substantially different from those in 1947, the climate for crop protection had changed radically after the publication of Rachel Carson's (1962) book *Silent Spring*, with rising public concern about environmental contamination and non-target effects, leading to more active regulation and the banning of DDT in 1972. The notion of IPM was gaining ground as a plant protection approach that used pest monitoring and reserved pesticides for use only when pest numbers reached a level that threatened economic losses. A major IPM research programme was ongoing (the Huffaker Project) at the University of California,

Berkeley, with funding by the National Science Foundation and the Environmental Protection Agency. However, strawberry growers still relied primarily on pesticides, including fumigants for disease, nematode, mite and insect control, along with new resistant or tolerant varieties. Organophosphate insecticides, derived from compounds synthesized as nerve gases during World War II, replaced organochlorines with an array of materials more acutely toxic to farm workers. 'Tioga' strawberries were high yielding and preferred for the firmness of its fruit, which aided field packing and transport in crates. As 'Tioga' plants were also smaller, plant density could be increased by 150%. Installation of drip irrigation reduced both fruit rot and irrigation costs. By using 'Tioga' plants, growers did not need to spray new beds to reduce winged aphids, or treat new plants with parathion, or follow with demeton 3 weeks later and again just before fruit set. However, after fruit set, a typical grower may have applied diazinon for aphids, and parathion after harvest. Beyond specific research on pest control approaches for strawberry, the University of California extension personnel offered much of their advice based on general guides published by pesticide companies (Smith 1982) and the Rodale Guide to Organic Gardening.

By 1995, fall planting was common. Some large shippers were both organizing smaller producers through forward contracting and directly suggesting methods of crop protection. Forward contracting allowed strawberry growers to manage their own labour recruitment, often through farm labour contractors, thus avoiding unionization attempts by the United Farm Workers. Many strawberry producers remained relatively small, but the shippers co-ordinated the process of production and marketing (Wells 1996). Growers had access to state-of-the-art information devoted to crop protection practices, including strawberry IPM, through the comprehensive University of California-IPM publication series. Challenges included widespread pest resistance in the region, proposed restrictions on some commonly used materials, and a phase-out via the Clean Air Act 1990 of methyl bromide, which was recognized as an ozone-depleting compound. Alternative crop protection strategies were needed for conventional and newly established organic strawberry growers. The California Certified Organic Farming organization strictly prohibited the application of synthetic fertilizers and pesticides, with annual certification granted to organic growers only after a 3-year transition period. Land tenure and separation from conventional pesticide drift were critical needs for certified organic strawberry production. University researchers examined organic strawberry production (Gliessman *et al.* 1996), expanded the list of wild plants that harbour the WTPB over winter, and experimented with alfalfa trap-cropping, vacuum removal of pests from plants at vulnerable stages, and incorporating beds devoted to flowering buckwheat and yarrow as resources for parasitoids and predatory insects.

By 2015, most US strawberries are produced in California, valuing US$2 billion (USDA-NASS 2014). California strawberries are also responsible for more than 90 000 kg of applied synthetic insecticides and miticides, as well as 23 000 kg of applied biorational materials, such as *Bacillus thuringiensis* products for lepidopteran pests (active ingredient, CDPR 2015). Apart from regular releases of predatory mites against two-spotted spider mites, pest management in strawberries is mainly dependent on chemical pesticides, and IPM is generally limited to the rotation of pesticides in different modes of action groups. *Verticillium* resistance breeding is ongoing, and although chemical alternatives to methyl bromide are in use, many regional growers are testing or adopting biologically based alternatives (see Table 2.1) such as anaerobic soil disinfection (ASD) (Fennimore *et al.* 2013). ASD relies on heavy additions of a carbon source, such as rice bran or grape pomace, followed by flooding to create high temperatures and anaerobic conditions that, over time, shift the soil biota to favour beneficial microbes and directly control soil-dwelling pests and weeds.

New pest introductions during the past decade have created crop protection challenges. The polyphagous Australian tortricid light brown apple moth (LBAM) (*Epiphyas postvittana*) has increased exponentially in California over the last decade (Suckling *et al.* 2014), causing a rise in both insecticide applications and wholesale losses of exportable berries resulting from LBAM-related quarantines. Upon its discovery, attempts to eradicate LBAM using sex pheromone spraying for mating confusion (Brockerhoff *et al.* 2012) failed, not because of scientific uncertainty but because of intense public reaction spurred by misinformation and ambiguity. For instance, local media reported aerial deployment of 'pesticides' over residential areas at night, government spokespeople came to public meetings unprepared to address citizens' concerns, and University of California scientists publicly debated the critical value of eradication efforts. Ironically, affluent citizens in urban centres filed legal suits against the CDFA to halt public exposure of non-toxic pheromones, which facilitated the pest's establishment and dispersal, and subsequently increased exposure rates of less affluent residents in rural areas to more toxic insecticides.

With a rising demand for pesticide-free produce, certified organic berry farming in the region expanded from 1.6 ha to 891 ha over three decades (California Strawberry Commission (CSC) 2014). Recent use of unionized labour contracts for pickers, training for farm workers in scouting pests, and employee ownership opportunities through stock options have fundamentally challenged the migrant labour model for strawberry. In fact, current California strawberry grower demographics reflect both this upward mobility and the industry's socioeconomic and political history: 65% of growers are of Mexican descent; 20% are of Japanese descent and 15% are of European descent (CSC 2014).

Both organic and conventional strawberries sold at supermarkets are packed in vented clamshells, which display the fruit, protect it from handlers, extend the product's shelf-life, reduce berry shrinkage and decrease supermarket labour costs. Increasingly, clamshells are made from corn and other bio-based sources and exposed to modified gases for preservation against postharvest fruit rot (Caner and Aday 2009). Organic growers struggle with some of the insects and weeds that are effectively controlled in conventional strawberry (see Table 2.1), having no dedicated breeding for organic management conditions and restricted use of chemical treatments. Extra labour and higher losses are offset by price premiums for the sale of organic strawberries.

2.3 Integrated Pest Management: What Does it Take?

Permanent agriculture must be in adjustment with the environment.

(Herbert C. Hanson, 1939)

Integrated Pest Management was intended to be radically different from standardized, schedule-based crop protection approaches focused on broadly effective suppression tactics used against a particular antagonist. IPM ideally orchestrates an emergent, harmonized strategy that uses localized information and thorough pest monitoring (including weeds, pests and plant diseases) to combine specific tactics that will prevent, mitigate and, when necessary, disrupt these organisms before their numbers reach an economic injury level (EIL). The EIL was originally conceived by Stern *et al.* (1959) as the density of insect pests (e.g. number of eggs per leaf, number of egg clusters per plant, number of infested plants per sample) that causes sufficient yield reduction to justify the cost of pest suppression. Thus, insecticides were intended to be applied 'only when damage is imminent', thereby allowing growers 'to avoid dependence upon insurance and prophylactic treatments' (Stern *et al.* 1959). Instead, IPM would aim to protect the crop by first implementing ecologically based prophylactic measures such as soil tillage, pest- and disease-resistant cultivars, temporal or spatial asynchrony between crop and pest, crop rotation, strip-cropping and biological control.

In this section, we discuss the core actions needed for an IPM strategy, assess IPM progress in California strawberry, and suggest needed changes.

2.3.1 Core Actions Needed for an IPM Strategy

2.3.1.1 Monitoring Pests to Minimize Usage and Non-target Effects of Pesticides

Within IPM-based, ecologically resilient cropping systems, monitoring and predicting current and future pest levels are core IPM activities, for which many cost-effective sampling and detection systems have been designed. For example, pheromone trapping and degree-day models used in apple orchards allow the synchronization of codling moth larval emergence with an application of a specific granulosis virus to maximize larval mortality without disrupting complementary biological control (Lacey *et al.* 2008; Witzgall *et al.* 2008). Monitoring of disease, weeds and arthropods is relatively common in strawberry on the central coast of California, with, for example, counts of two-spotted spider mites on strawberry leaves (see Table 2.1) determining the timing of augmentative releases of predatory mites. Theoretically, an increase in monitoring results in decreased pesticide application and a better environment for biological control of pests. However, the effect of IPM developments over time on California pesticide usage and environmental quality is a matter of interpretation, as we will illustrate with an examination of the California Department of Pesticide Regulation's Pesticide Use Reports database (CDPR 2015).

To assess trends in California's annual insecticide and miticide use, we calculated the total application rate of 377 active ingredients from 1974 to 2012. Comparing the 1970s to the 1990s, a period in which California made a concerted effort to promote IPM, the amount of active ingredient (AI) applied to crops increased from an average of 18.1 million kg AI year^{-1} in the 1970s to an average of 29.5 million kg AI year^{-1} in the 1990s (see also Chapter 13). Similarly, pesticide expenditures by growers in California, as a mean percentage of their total expenditures, increased from 5.46% in the 1970s to 7.46% in the 1990s (Mullen *et al.* 2003). That trajectory, however, halted and reversed in the 2000s, with the amount of AI applied to control insects and mites on California crops falling to about 22.7 million kg AI year^{-1}. Was this drop in pesticide usage due to better monitoring of pests, replacement of chemical with biological pest control approaches and a general increase in IPM implementation?

Our interpretation of these data suggests that the answer is no. A decrease in the quantity of active ingredients applied does not necessarily imply a reduction in the number of applications per hectare or any increase in IPM practices (Mullen *et al.* 2003;

Sharma *et al.* 2015; Wilhoit *et al.* 1998). Instead, the quantity of AI applied depends on the dosage rate, which differs substantially among different chemical classes. Specifically, older chemicals, which are applied in relatively large doses, are being used less often, while more recently introduced chemical classes, which are more potent and therefore used at much smaller dosage rates, are being used more often (Epstein and Bassein 2003; Peshin and Zhang 2014). For instance, organophosphates are applied at ~1 $kg\,ha^{-1}$ AI and were largely replaced by pyrethroids which are applied at ~0.1 $kg\,ha^{-1}$ AI. Newer materials that are replacing organophosphates and pyrethroids are often applied at even lower concentrations (e.g. 0.06 $kg\,ha^{-1}$ AI for the neonicotinoids). Therefore, we would expect a sharp decrease in the amount of AI applied by weight over time with the advent of more potent materials, but the 'pesticide pressure' in the field should be measured independently. Only if pesticide pressure and environmental impact decrease will IPM goals be met.

In California strawberries, usage rates of insecticides and miticides, combining 17 commonly used materials, showed no significant change in the quantity applied ($kg\,ha^{-1}$) from 2004 to 2013 using the Pesticide Use Report database (CDPR 2015) (Figure 2.2a). However, the environmental impact of this pesticide usage increased, in part due to potency and to environmental persistence, as shown by incorporating Cornell University's ecological environmental impact quotient (EIQ) for each AI (Eshenaur *et al.* 2015; Kovach *et al.* 1992; Sharma *et al.* 2015). Their ecological component is $[(F \times R) + (D \times ((S + P)/2) \times 3) + (Z \times P \times 3) + (B \times P \times 5)]$, where F = fish toxicity, R = surface loss potential, D = bird toxicity, S = soil half-life, Z = bee toxicity, B = beneficial arthropod toxicity, P = plant surface half-life. Our estimate was calculated by first multiplying the ecological EIQ for each AI by an estimate of annual application frequency for that AI: (applied AI ($kg\,ha^{-1}$))/(application rate ($kg\,ha^{-1}$)). These values were then summed and divided by the sum of all application frequencies to calculate the mean annual ecological EIQ for the quantities and types of AI applied. From 2004 to 2013, mean ecological EIQ increased from 68.1 to 73.1 (Figure 2.2b). Cornell's ecological EIQ equation takes into account both soil and foliage persistence. We isolated the potential effect of relative persistence by obtaining values in days for each AI half-life from aerobic soil metabolism studies. Multiplying the number of half-life days for a given AI by its application frequency and summing those values over all AI showed an upward trajectory for the persistence of these materials in strawberry over time (Figure 2.2c).

Although there are specific examples of reduced frequency and increased specificity of pesticide applications in IPM programmes around the world, the overall trend of increased pesticide use has been evident in many countries. In Thailand, annual pesticide use increased 9.1% during 2000–2009 (Schreinemachers and Tipraqsa 2012). In Bangladesh, pesticide usage increased from 2.8 million kg AI in 1977 to 40.9 million kg AI in 2009, without a corresponding yield improvement (Rahman 2013). In India, pesticide use increased by 39% from 2005 to 2012 (Peshin and Zhang 2014) and in China, pesticide use almost doubled from 1991 to 2005 (Zhang *et al.* 2011). Clearly, IPM policy has not been effective in decreasing farmers' overall reliance on pesticides (Epstein and Zhang 2014).

2.3.1.2 Integrate Strategies for Pest Management

Key concepts in addition to monitoring for pesticide reduction in IPM are diversity, compatibility and extensionality. Diversity can apply to crop and vegetation management and modes of action among tactics, and is fundamental to durable crop

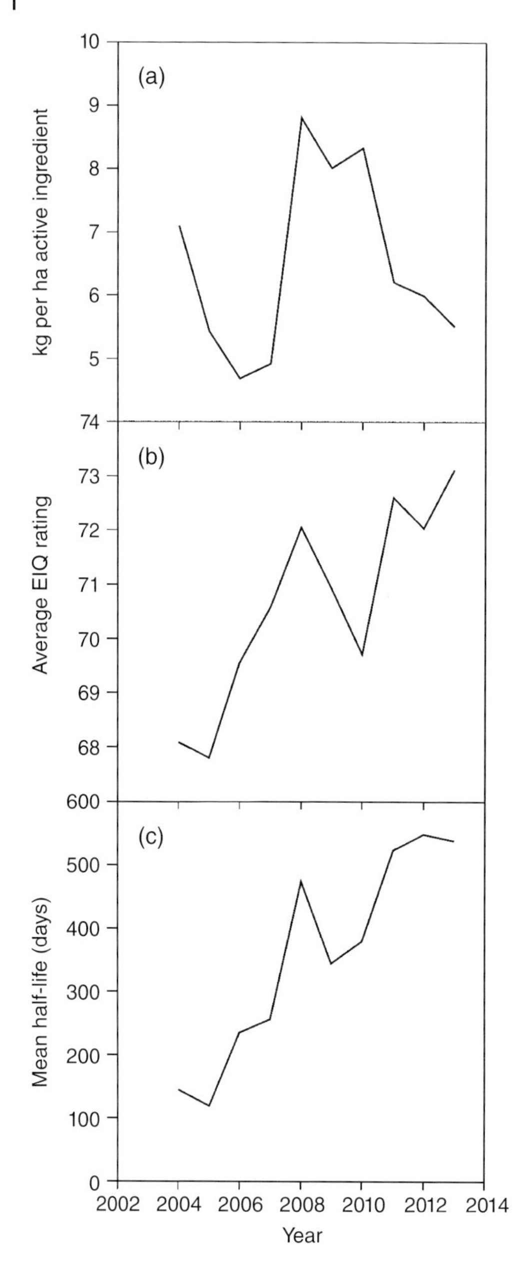

Figure 2.2 Trend in (a) kg of insecticides and miticides active ingredients (AI) per ha per year applied to strawberry on the Central Coast of California 2004–2013, (b) average environment impact quotient (EIQ) rating times (kg AI per ha per year), and (c) average persistence of the same materials as determined by their half-life (persistence in days).

protection. First, genetic diversity within the crop itself provides protection, as shown by widespread losses due to fungal pathogens in genetically similar monocultures of potato in Europe, corn in the Midwestern USA and banana in tropical America (Ordonez *et al.* 2015). Second, a diverse array of tactics with wide-ranging sources of mortality is needed to create sustainable strategies for which the development of pest resistance is evolutionarily difficult (Alyokhin *et al.* 2015). For instance, given the fitness trade-off between insecticide resistance and parasitism susceptibility in the aphid

Myzus persicae, conservation biological control can actually promote insecticide efficacy and preservation (Foster *et al.* 2005, 2007). Combining physical treatments or vegetation management with plant extracts or suppressive organisms is another effective way to diversify tactics, for example, to prevent the introduction of seed-borne fungal pathogens (Mancini and Romanazzi 2014), encourage predatory mites (Ottaviano *et al.* 2015) or protect against *Fusarium* wilt in strawberry (Cha *et al.* 2016). Third, diversifying pesticides can increase durability in IPM programmes. However, the reliance on pesticides alone, even with attention to alternating or combining different modes of action, is not a sufficiently diversified crop protection approach. At the advent of IPM, Stern *et al.* (1959) noted 70 arthropod pests resistant to at least one active ingredient. With continued isolation of pesticide-induced mortality, that figure has grown to more than 500 (Bass *et al.* 2015). Clearly, the fundamental tenets of evolution are too often forgotten or ignored.

In California strawberry, newly integrated management tactics and farmscape diversification have made pest management approaches increasingly robust. Ubiquitous use of tractor-mounted vacuums for WTPB control (Hornick 2015) now helps mitigate widespread resistance within these pest populations (Shimat and Bolda 2014; Thomas 2014) resulting from a historical management approach that relied almost exclusively on WTPB-directed usage of older classes of insecticides. Integration of non-crop vegetation schemes currently diversifies management approaches by increasing field level complexities. For example, barley (*Hordeum vulgare*) is planted around field perimeters to reduce dust, which would otherwise exacerbate mite pest pressure. Also, insectary plants are used on bed-ends to provide refugia and nutritional resources to generalist predators, and alfalfa trap crops are implemented in organic acreage to manage WTPB (Ottaviano *et al.* 2015; Swezey *et al.* 2007). The diversity and richness of biological mortality sources have also been improved by utilizing the entomopathogenic fungus *Beauveria bassiana*, introducing the braconid parasitoid *Peristenus relictus*, and augmentatively releasing predatory mite species, such as *Phytoseiulus persimilis* and *Neoseiulus californicus* (Dara *et al.* 2012; Pickett *et al.* 2009).

Truly integrative pest management may ultimately depend upon the future structural redesign of agro-ecosystems to increase biodiversity on multiple scales, from genetic diversity among cultivars in a crop field to habitat diversity in the landscape (Birch *et al.* 2011), and design pest control approaches as a diversity of 'little hammers' (*sensu* Liebman and Gallandt 1997). A diverse and subtle array of suppression tactics is needed to counter the fitness advantages of high genetic diversity within natural populations of pathogens, weedy plants and arthropod pests. A consideration of arthropod pest diversity must also include the evolutionary capacity of their symbionts, known to protect insect pests from insecticides (Kikuchi *et al.* 2012), resistant crop varieties (Ferrater *et al.* 2013) and parasitoids (Vorburger 2014).

Compatibility becomes important in diversified strategies of crop protection, so that one tactic does not interfere with another, the classic example being broad-spectrum insecticides disrupting biological control (Ruberson *et al.* 1998). IPM also requires compatibility among the tactics used against different pests so that actions taken to prevent one outbreak do not create conditions conducive to outbreaks of another. Such is the mechanism by which secondary pests emerge in cropping systems when pesticides disrupt natural biological control of a non-pest organism. Whitefly management in Arizona cotton (Naranjo and Ellsworth 2009) offers an insightful example of such compatibility: previously used broad-spectrum organophosphates and pyrethroids severely compromised beneficial arthropod communities, required more frequent applications and eventually elicited resistance in whitefly populations. By transitioning to insect growth regulators that targeted whitefly, biological control was restored and total insecticide applications were greatly reduced.

In California strawberry, management practices vary with respect to compatibility. Anaerobic soil disinfestation, which is used as a fumigant alternative to control soil-borne pathogens, may also improve microbial community composition (Mazzola *et al.* 2012) and would therefore be compatible with overall disease suppression efforts. Utilizing LBAM mating disruption pheromone twist ties is probably neutral in that it does not affect other intraspecies (e.g. Bt) or interspecies (e.g. weed control) treatments. Employing tractor-mounted vacuums for WTPB control, on the other hand, is probably not compatible with simultaneous augmentative mite biological control.

Extensionality can apply to all the previously mentioned tenets of IPM, including monitoring, EILs, diversity and compatibility. Extensionality counteracts a singular focus on one pest at one time with one tactic, and is essential for a sustainable pest control approach. IPM strategies are more durable and sustainable when designed comprehensively. Comprehensive designs extend ecologically to include arthropod community interactions throughout an agro-ecosystem, such as the build-up of predators feeding on detritivores and filter-feeders in rice (Settle *et al.* 1996). Approaches extend temporally when they include long-term effectiveness, that is, they are not susceptible to pest resistance. Spatial extensions include landscape-scale mediation of within-field dynamics, such as planting tropical rice asynchronously to better manage pest metapopulations (Ives and Settle 1997) or maintaining natural vegetation that promotes natural enemy diversity for improved biological control in short-rotation crops (Letourneau *et al.* 2012). Economic extensionality broadens economic thresholds by factoring in the marginal costs associated with consumer, worker and environmental health (see Chapter 3 and Chapter 16).

As in a number of other crops in the USA, the version of IPM that is commonly practised today in strawberry is constrained, in part because structural and technological 'path-dependent' or 'lock-in' processes deny much-needed flexibility and resources for implementing agro-ecology (Vanloqueren and Baret 2008). Deep sociohistorical pathways tend to define modern crop protection approaches, leading to what Parsa *et al.* (2014) referred to as a lock-in version of IPM, which relies on 'pesticide management' rather than the 'agro-ecosystem management.' IPM has been criticized as a malleable concept that, when operationalized, does not prioritize pesticide reduction (Ehler 2006) or integrate compatible control tactics for multiple pests (Norris and Kogan 2000). Mainstream IPM is compromised for a number of reasons, most of which are unrelated to science or technology. We do not expect growers to change their practices appreciably unless provided with incentives for implementing agro-ecosystem-based IPM and reducing pesticide use. Change may not happen unless growers or their suppliers are required to incorporate externalities (i.e. societal and environmental costs) associated with pesticide use.

2.3.2 Changes Needed to Incentivize IPM

Here, we consider some of the social, political and economic contributions to the failure of IPM in the USA, and outline several progressive actions to increase the diversity, compatibility and extensionality components of IPM implementation.

2.3.2.1 Strategically Redefine Clinton-era IPM Criteria in the USA and Erect New Policy to Provide Alternative Technology Packages and Rewards for Compliant Growers

The US National IPM initiative was launched in 1994 by the Clinton administration with the goal of achieving a 75% adoption rate of growers qualified as IPM practitioners by 2000 (Ehler 2006). The criterion used to signal adoption of IPM was implementation of at least three of the following four crop protection tactics: prevention, suppression, monitoring and avoidance (Ehler and Bottrell 2000). For practicality and broader buy-in, the criterion was built around practices already used in conventional agriculture. Consequently, an estimated 70% of US growers were in compliance by 2000 (Epstein and Bassein 2003). Guidelines did not require that growers use multiple tactics for each pest, conduct periodic monitoring for the management of multiple pests, perform natural enemy assessments or adhere to economic thresholds before applying pesticides (Ehler 2005). Also excluded were recommendations for reduced pesticide reliance or use of pesticides as a last resort (Ehler and Bottrell 2000). Under the criteria in use, for example, a grower would qualify as implementing IPM if s/he did as little integration as having sprayed herbicide on weeds near the field edge (prevention), planted the field to a different crop 3 years before (avoidance), and used different pesticides with more than one mode of action (suppression). This lax set of IPM criteria may explain why no insecticide reductions were evident between 1994 and 2000 (Coble and Ortman 2009). In fact, Benbrook *et al.* (1996) estimated that only about 6% of US land in production at that time qualified for the more expansive, multi-tactic definitions of IPM. Unfortunately, these trends and their consequences continue (Furlong *et al.* 2014).

The criteria for practising IPM should be redefined to require the use of multiple, compatible tactics to reduce farmer reliance on pesticides and create a more durable and sustainable approach to pest control. Explicitly including pesticide reduction (e.g. application frequency or toxicity) as a central goal of IPM is necessary (Moss 2010; Peshin and Zhang 2014) in order to remove ambiguity otherwise present in some definitions (e.g. University of California) and rebut conflicting definitions that reject insecticide reductions, such as those put forward by the American Crop Protection Association (Epstein and Bassein 2003), half of responding US state-level IPM co-ordinators (GAO 2001) and (to a lesser extent) the Entomological Society of America (ESA 2012). The inclusion of clearly defined pesticide reduction goals has been prevalent in Denmark (see Chapter 13), Sweden and the Netherlands, where ambitious thresholds have been successfully met (Peshin and Zhang 2014).

If the criteria for practising IPM are redefined, then corresponding technical assistance and government support packages will be needed to aid farmers in meeting them. Integrating ecologically based practices can be challenging, and meaningful co-operation is needed for area-wide and landscape-level crop protection strategy implementation. Such technical assistance successfully facilitated IPM adoption in the western USA by changing farmer-perceived impediments to IPM (e.g. increased costs, increased risk and reduced pest control) through the introduction of best practice guidelines and educational programmes between 1996 and 2009 (Farrar *et al.* 2015). Growers adopting an array of harmonious, not antagonistic control tactics for individual pests or groups of different pests will need crop protection packages with compatible management information, subsidized systematic monitoring, provision of technical assistance, contingent crop insurance and governmental incentives for pesticide reduction (Leite *et al.* 2014).

2.3.2.2 Prioritize Integrative, Innovative Research for Crop Protection

Integrated Pest Management is not integrated when it consists of a menu of tactics produced in isolation from one another by natural scientists from different academic disciplines, each with a distinct culture and paradigm. Such tactics do not allow for compatibility between approaches when, for example, an action aimed at insects is irreconcilable with those used to suppress weeds. Unfortunately, land-grant universities have not changed appreciably from the institutional structure that seemed efficient and rigorous in the 19th century, and therefore often lack the interdisciplinary cross-over needed to develop ecologically based strategies against communities of pests (Barfield and Swisher 1994).

Furthermore, the inclusion of social scientists in research teams can improve IPM-based crop protection, as many of the barriers to grower implementation lie within the realm of political economic, policy and social institutions. Research efforts, for example, can be misguided without any socioeconomic context of crop protection needs. Given that the artichoke plume moth (*Platyptilia carduidactyla*) can reduce artichoke (*Cynara scolymus*) yields by half, it may seem obvious to an entomologist that organic fruit and vegetable farmers would benefit from a concerted effort to investigate and improve biological control of that pest. However, such a research programme would be deprioritized with the knowledge that local growers count on tax deductions from artichoke losses to allow for experimentation with, and adoption of, innovative non-chemical alternatives in their strawberry fields (W. Friedland, personal communication). While collaborative, multi-institution, interdisciplinary research projects aimed at addressing broader impacts are encouraged in competitive grants from governmental sources, fundamental barriers still persist with respect to familiarity, reward systems, review processes and funding designations, which can favour disciplinary over interdisciplinary innovation. New reward systems are needed that recognize the flexibility and mutual respect required for working successfully within multidisciplinary teams on comprehensive crop protection plans that achieve synergistic interactions among strategic components of IPM (Birch *et al.* 2011).

2.3.2.3 Transform University Extension Philosophy and Funding Options to Support On-farm, Collaborative Exchanges Focused on Grower-generated Research Priorities

Effective and trustworthy information dissemination from scientists to farmers was pioneered in US agriculture by Seaman Knapp, who promoted culturally appropriate communication strategies and co-operative demonstration farms (Cruzado 2012). However, over time this model has become increasingly 'top-down' and rigid, leading to 'entrenched' pesticide use and failed extension efforts in the developing world (Matteson 2000; Settle *et al.* 2014; Waddington *et al.* 2014). Participatory action research (PAR) or farmer participatory research (FPR) represents an alternative or complementary model for university extension in the USA, in which growers are empowered through participatory and discovery-based education models that address community-generated research priorities (Waddington *et al.* 2014). PAR/FAR programmes have empowered small-scale farmers in developing countries, framed appropriate research for improving farming practices, and operationalized decision support systems (Carberry *et al.* 2002; Mendez *et al.* 2010; Snapp *et al.* 2002).

More specifically, PAR-based IPM models, such as farmer field schools (FFS), are typically successful at reducing pesticide use, which is often a FFS programme goal.

Waddington *et al.* (2014) conducted a meta-analysis of FFS programmes and found that participants reduced pesticide use by an average of 23%, while simultaneously increasing yields and profits. More specifically, cotton FFS participants in Mali reduced synthetic insecticide use by 92.5% over 8 years, while largely maintaining yields comparable with the grower control group (Settle *et al.* 2014). In South-East Asia, rice IPM programmes in 13 countries supported by the United Nations Food and Agriculture Organization implemented FFS and 'strategic extension campaigns' to improve growing practices (Bottrell and Schoenly 2012; Matteson 2000). In Indonesia, 50 000 FFS graduates reduced pesticide use by two-thirds (Matteson 2000). In Vietnam, growers were successfully dissuaded from using counterproductive insecticides during the first 40 days after transplant, resulting in a 50% reduction in their overall use (Escalada *et al.* 2009; Matteson 2000). However, many of these gains were eventually eroded. In Vietnam, for instance, growers' perceptions regarding the value of early-season insecticide treatments was restored to pre-IPM levels, as public funding sources to support IPM programmes diminished and pesticide industry-funded marketing campaigns became increasingly prominent (Escalada *et al.* 2009). The unfortunate lesson learned is that publically funded IPM education initiatives must continually introduce and reinforce relevant multidisciplinary concepts in order to maintain previously achieved benchmarks and prevent undue industry influence (Bottrell and Schoenly 2012).

2.3.2.4 Incorporate Alternative Scenarios, Marginal Costs and Biological Control Into Updated Ecologically Comprehensive EIL Models

New-generation EILs include probabilistic EILs (PEIL) that allow growers to select among risk levels and environmental EILs (EEIL) that include environmental cost estimates into the manage cost parameter, though these have not replaced traditional EILs in the field (Higley and Peterson 2009) (see Chapter 3). To modernize decision-making tools, more comprehensive EILs must be developed, promoted and supported to become standard procedures in crop protection. In diversified IPM strategies that rely on natural enemies of pests, weeds and pathogens, EILs must integrate biological control assessments into economic threshold determinations. Estimates of pest mortality (e.g. due to predators, weed-targeted herbivores or pathogens) indicate if enemy augmentation is needed and refine population growth predictions for determining if pesticide applications are warranted.

Creating ecologically comprehensive EILs may require accelerated research investments, for example, to quantify predator–prey relationships using molecular techniques (Furlong 2015; Hagler 2011), evaluate weed biological control via exclusion methods (Dhileepan 2003) and characterize pollinator deliveries of microbial biocontrol agents against fungal pests (Shafir *et al.* 2006). However, public institutions in the USA, such as the University of California and the California Department of Food and Agriculture, have diminished their capacity for biological control activities (Warner *et al.* 2011). In contrast, federal investments in biological control in Mexico have greatly increased the scale of research and implementation over the last 20 years (Rodríguez *et al.* 2015). Specifically, Mexico's National Reference Center for Biological Control co-ordinates comprehensive multistate biological control projects, such as for the Asian citrus psyllid (*Diaphorina citri*) (Sánchez *et al.* 2015), that serve as a model for federally facilitated biological control efforts.

2.3.2.5 Apply Critical Assessments of New Technologies for Their Advancement of IPM, Sustainable Agriculture and Food Security

Some technological innovations for crop protection are more compatible with ecosystem-based IPM than others. Given site-specific characteristics and variability in agroecosystems, fine-tuning and customizing crop protection through precision agriculture technology can improve the efficiency, accuracy and compatibility of IPM tactics (Tey and Brindal 2012). Mechanical tools for precision IPM include small, manoeuvrable aerial and ground robots with intelligent sensors that deliver flames for targeted weed control or spot insecticide applications (Perez-Ruiz *et al.* 2015). Information and modelling systems can optimize the frequency and timing of biological or chemical pesticide applications. For example, web-based models using interactive precision-directed fungicide applications treating anthracnose and *Botrytis* in strawberry compared favourably against weekly (calendar) treatment records with respect to farmer costs, number of applications and disease incidence (Vorotnikova *et al.* 2014). Alternatively, Lux (2014) offers a customizable simulation for precision IPM that allows for 'virtual preassessments' of site-specific choices of crop combinations, spatial patterns and other tactics designed to prevent pest build-up. DNA barcoding, which is intended to provide fast, accurate and accessible species identification, can improve farmer education and monitoring reliability, particularly in remote areas without a farm extension presence (Hebert and Gregory 2005). Smartphones and agriculturally relevant application software can improve information gathering, interpretation and dissemination (Teacher *et al.* 2013). For instance, field-based polymerase chain reaction is now possible in real time by using just a portable solar-powered thermal cycler and a smartphone (Jiang *et al.* 2014). Ideally, technological information tools for on-farm operations will create opportunities for cost savings, environmental protection and farm worker training and job security.

Methods of genetic modification and manipulation have advanced along with precision agriculture technology, and allow for rapid development of crop resistance to pests and pathogens. Although crop resistance is a cornerstone of IPM, these simple gene alterations may elicit counterresistance adaptations by their target organisms more rapidly than multigene host resistance developed through traditional horizontal transfers (see Chapter 12). Also, simple gene modifications for pest or pathogen resistance can be expressed in crop-wild hybrids and lead to introgression into weed populations. In the latter case, the attempt to protect a crop from a pathogen could increase the fitness of a weed that is suppressed by that pathogen, making these advanced technologies complicated to assess in terms of their compatibility with ecosystem-based IPM. Regulatory requirements of trangenic plants include testing for effects on non-target organisms and require consideration of unintended consequences, such as counterresistance and gene transfer. However, recent attention to gene editing or genome engineering has increased, in part because the regulatory barriers (costs, testing, time) that are in place for transgenic crops can be avoided when only 'native genes' are manipulated (Voytas and Gao 2014). Therefore, new and promising resistant varieties against blight (*Xanthomonas oryzae*) in rice (Jiang *et al.* 2013), powdery mildew (*Blumeria graminis*) in wheat (*Triticum aestivum*) (Wang *et al.* 2014) and rust (*Puccinia* spp.) in cereal crops (Rajendran *et al.* 2015) should be scrutinized for incompatibility within IPM systems.

Risks and benefits of new technological approaches to crop protection often do not distribute themselves equally among corporations, farmers/farm workers, consumers and the environment. For example, glyphosate-tolerant or 2,4 D-tolerant crops increase herbicide use, select for resistant weeds, spread resistance traits into wild relatives, lead to genetic abnormalities in aquatic animals and create public health hazards (see Chapter 4), while benefitting corporations with new patents and technology fees. The rationale often used for patenting technological crop protection tools that clearly benefit the profit margin of biotechnology companies, while risk is borne by consumers, growers and the environment, is that the burgeoning world population must be fed, and regulation is an impediment that stands in the way of global food security (Voytas and Gao 2014). Although it may be rational from a scientific perspective to conclude that an increase in crop productivity leads to the alleviation of food scarcity, in fact poverty, lack of access and powerlessness are the causes of food insecurity, not the amount of food that is produced (Sen 1981). In addition, crop losses due to pests can ultimately provide economic gains for farmers when demand exceeds supply and price points rise.

The real challenge for IPM is to transparently target the means by which contributions to food security, environmental and public health can be made, and to push for the political changes that will increase the adoption and spread of these beneficial approaches.

2.4 Transforming Agriculture Systems for IPM

Problems cannot be solved at the same level of awareness that created them.

(Albert Einstein)

Two decades ago, a National Research Council report (NRC 1996) called for 'a paradigm shift in pest-management theory [...] that examines processes, flows, and relationships among organisms'. Zorner (2000) similarly urged that crop protection treat the systemic causes rather than the symptoms of pest outbreaks. Still, even simple vegetation diversification schemes are rare phenomena in US agriculture. In Kenya, push-pull systems in maize (*Zea mays*) and sorghum (*Sorghum bicolor*) have more than doubled yields (Khan *et al.* 2014) and serve as a pest management exemplar for understanding and incorporating interspecies relationships and behaviours into a successful diversification strategy.

In lieu of innovations like push-pull, crop protection tactics tend to be adopted only if they require few changes in conventional practices and fit well within its associated infrastructure, such as monocultures of transgenic, herbicide-tolerant crops or encapsulated nematodes (Vemmer and Patel 2013) delivered with standard spray equipment. IPM has consequently narrowed its focus to managing insect, pathogen and weed resistance by varying the strength and type of selective pressures imposed on pest populations with different pesticides. Unless newer IPM tools such as stacked transgenic insect-resistant crops and selective insecticides that preserve natural enemies are integrated into a preventive, ecosystem-based IPM strategy, they will be destined for the same path of overexposure and pest resistance selection – the pesticide treadmill (van den Bosch 1978) or the paradox of pesticides (Li and Yang 2015) – that IPM was meant to avoid.

Plant protection in the context of transformational agriculture (*sensu* Hill 1998; Karp *et al.* 2015) arises from the optimization of two ecosystem services: production and pest control (see Chapter 7), and requires radical changes rather than fine adjustments to current intensified, large-scale monoculture systems. Wezel *et al.* (2014) described eight categories of ecologically based practices, whose adoption is predicated on redesigning conventional cropping systems. Categories that are particularly relevant to crop protection are:

- disruption, antibiosis or tolerance to pests via crop varietal choices, mixtures and rotations
- associational resistance to pests via intercropping or diversified agro-forestry
- weed and pathogen control using allelopathic plants and microbial antagonists
- weed control by direct seeding into living cover crops or mulch
- support of natural enemies by integrating semi-natural vegetation at the field or farm scale
- increase source pools and refugia for natural enemies through conservation or restoration of non-crop vegetation at landscape scale (see Chapter 7).

Indeed, research by insect ecologists on the pest-suppressive effects of plant diversification schemes and habitat management for crop protection (Pimentel 1961; van Emden 1964) arose around the same time as the formulation of the IPM concepts (Stern *et al.* 1959; van den Bosch and Stern 1962) and shares a history of research and application in crop protection (Altieri and Letourneau 1984; Andow 1991; Landis *et al.* 2000; Letourneau *et al.* 2011; Nicholls and Altieri 2007; Pimentel and Goodman 1978). The agro-ecosystem-based strategies needed for crop protection through IPM require functioning ecosystems (Wood *et al.* 2015) and social support systems (Altieri *et al.* 1983) to challenge the status of monocultures, which are seen as a 'locked-in' condition of conventional agriculture (*sensu* Vanloqueren and Baret 2009).

Restoring ecosystem structure and function for crop protection emphasizes active integration and reliance on the foundational prevention of insect outbreaks using compatible tactics based on ecological knowledge (Letourneau 2012). Karp *et al.* (2015) showed how simplifying an agro-ecosystem by clearing wildlife habitat around fields of leafy greens in an attempt to address food safety concerns backfired, as levels of harmful microbes are actually positively associated with riparian habitat removal. Letourneau *et al.* (2015) demonstrated positive relationships between the extent of these habitats in the landscape, parasitoid richness and biological control of certain insect pests.

Such landscape perspective was introduced early on within the IPM paradigm for preventing WTPB from reaching the EIL in cotton by strip-cutting adjacent alfalfa fields (Stern *et al.* 1964) and later demonstrated movement of marked WTPB between alfalfa and cotton fields (Sevacherian and Stern 1975). Goodell (2009) adapted this technique to cotton growing near safflower, a potent source of this pest, applying metapopulation theory to avoid insecticide use and conserve natural enemies in cotton. Deguine *et al.* (2008) examined the history of insect pest management in cotton to show how biologically based IPM programmes reduced the frequency of insecticide treatments, secondary pests and insect resistance through a combination of multiple complementary tactics: pheromone trapping and mating disruption, cotton engineered with stacked antifeedant and *Bt* genes accompanied with susceptible refugia, releases of sterilized pests and beneficial insects, conservation biological control through intercropping and food sprays, locally adapted varieties, cover crops, living mulches, altered cropping geometries and sequences, and selective biopesticides (Ikeda *et al.* 2015) applied to associated trap crops.

Ongoing research is addressing localized management of vegetation at various scales to achieve the 'right kind of biodiversity' to support the goals of production agriculture (Landis *et al.* 2000; Lu *et al.* 2014; Macfadyen *et al.* 2009; Straub *et al.* 2008). As organic farming methods improve species richness by 30% (Tuck *et al.* 2014), investigating diversification strategies in this growing agricultural sector may be particularly useful. Critical questions remain, including those relevant to:

- specific soil biota that promote suppressive conditions for pathogens and nematodes (van Bruggen and Semenov 2000)
- vegetative resources that support natural enemies or host pests, pathogens or disease vectors (Barberi *et al.* 2010; Broatch *et al.* 2010; Lu *et al.* 2014; Schellhorn *et al.* 2010)
- the importance of natural enemy species richness *per se* versus evenness, diversity, relative abundance, species identity or species interaction networks for effective biological control services (Anjum-Zubair *et al.* 2010; Cardinale *et al.* 2006; Crowder *et al.* 2010; Moreno *et al.* 2010; Tylianakis *et al.* 2010)
- the effects of landscape heterogeneity and specific landscape features (O'Rourke *et al.* 2011; Tscharntke *et al.* 2005; Werling and Gratton 2010)
- the 'many little hammers' that interact to reduce soil seed banks and suppress weed populations (Shirtliffe and Benaragama 2014; Westerman *et al.* 2005)
- and, most comprehensively, how best to apply knowledge from multiple disciplines to redesign agro-ecosystems to be more resistant and resilient to pest outbreaks (Sigsgaard *et al.* 2014; Steingrover *et al.* 2010; Wood *et al.* 2015), while potentially obtaining many other benefits such as carbon credits for these practices (Gurr and Kvedaras 2010).

Acknowledgments

We thank Larry Wilhoit, California Department of Pesticide Regulation, for guidance on the Pesticide Usage Reports database, Michael Seagraves for clarifying crop protection approaches in strawberry, volunteers at the Agricultural History Project in Watsonville, CA, USA, and greatly appreciate the comments made by the editors and a fantastic reviewer on an earlier version of this manuscript.

References

Allen, W.W. (1959) *Strawberry Pests in California: a Guide for Commercial Growers.* Division of Agricultural Sciences, University of California. Berkeley, CA, USA.

Altieri, M.A. and Letourneau, D.K. (1984) Vegetation diversity and insect pest outbreaks. *Critical Reviews in Plant Sciences* **2**: 131–169.

Altieri, M.A., Letourneau, D.K. and Davis, J.R. (1983) Developing sustainable agroecosystems. *Bioscience* **33**: 45–49.

Alyokhin, A., Mota-Sanchez, D., Baker, M., *et al.* (2015) The Red Queen in a potato field: integrated pest management versus chemical dependency in Colorado potato beetle control. *Pest Management Science* **71**: 343–356.

Andow, D.A. (1991) Vegetational diversity and arthropod population response. *Annual Review of Entomology* **36**: 561–586.

Anjum-Zubair, M., Schmidt-Entling, M.H., Querner, P. and Frank, T. (2010) Influence of within-field position and adjoining habitat on carabid beetle assemblages in winter wheat. *Agricultural and Forest Entomology* **12**: 301–306.
Barberi, P., Burgio, G., Dinelli, G., *et al.* (2010) Functional biodiversity in the agricultural landscape: relationships between weeds and arthropod fauna. *Weed Research* **50**: 388–401.
Barfield, C.S. and Swisher, M. E. (1994) Integrated pest-management – ready for export: historical context and internationalization of IPM. *Food Reviews International* **10**: 215–267.
Bass, C., Denholm, I., Williamson, M.S. and Nauen, R. (2015) The global status of insect resistance to neonicotinoid insecticides. *Pesticide Biochemistry and Physiology* **121**: 78–87.
Benbrook, C.M., Groth, E., Hansen, M., Halloran, J. and Marquardt, S. (1996) *Pest Management at the Crossroads*. Consumers Union of US, New York, USA.
Birch, A.N.E., Begg, G.S. and Squire, G.R. (2011) How agro-ecological research helps to address food security issues under new IPM and pesticide reduction policies for global crop production systems. *Journal of Experimental Botany* **62**: 3251–3261.
Bottrell, D.G. and Schoenly, K.G. (2012) Resurrecting the ghost of green revolutions past: the brown planthopper as a recurring threat to high-yielding rice production in tropical Asia. *Journal of Asia-Pacific Entomology* **15**: 122–140.
Broatch, J.S., Dosdall, L.M., O'Donovan, J.T., Harker, K.N. and Clayton, G.W. (2010) Responses of the specialist biological control agent, *Aleochara bilineata*, to vegetational diversity in canola agroecosystems. *Biological Control* **52**: 58–67.
Brockerhoff, E.G. Suckling, D.M., Kimberley, M., *et al.* (2012) Aerial application of pheromones for mating disruption of an invasive moth as a potential eradication tool. *PloS ONE* 7**(8)**: e433767.
Butler, D.M., Kokalis-Burelle, N., Albano, J.P., *et al.* (2014) Anaerobic soil disinfestation (ASD) combined with soil solarization as a methyl bromide alternative: vegetable crop performance and soil nutrient dynamics. *Plant and Soil* **378**: 365–381.
California Strawberry Commission (CSC) (2014) *Growing the American Dream: California Strawberry Farming's Rich History of Immigrants and Opportunity*. Watsonville, CA, USA.
Caner, C. and Aday, M.S. (2009) Maintaining quality of fresh strawberries through various modified atmosphere packaging. *Packaging Technology and Science* **22**: 115–122.
Carberry, P.S., Hochman, Z., McCown, R.L., *et al.* (2002) The FARMSCAPE approach to decision support: farmers', advisers', researchers' monitoring, simulation, communication and performance evaluation. *Agricultural Systems* **74**: 141–177.
Cardinale, B.J., Srivastava, D.S., Duffey, J.E., *et al.* (2006) Effects of biodiversity on the functioning of trophic groups and ecosystems. *Nature* **443**: 989–992.
Carson, R. (1962) *Silent Spring*. Houghton Mifflin. Boston, MA, USA.
CDPR (California Department of Pesticide Regulation) (2015) Summary of pesticide use report data, 2013: Indexed by chemical. Available at: www.cdpr.ca.gov/docs/pur/pur13rep/chmrpt13.pdf (accessed 28 February 2017).
Cha, J.Y., Han, S., Hong, H.J., *et al.* (2016) Microbial and biochemical basis of a Fusarium wilt-suppressive soil. *Isme Journal* **10**: 119–129.
Coble, H.D. and Ortman, E.E. (2009) The USA national IPM roadmap. In: Radcliffe, E.B., Hutchison, W.D. and Cancelado, R.E. (eds) *Integrated Pest Management*. Cambridge University Press, Cambridge, UK, pp. 471–478.

Crowder, D.W., Northfield, T.D., Strand, M.R. and Snyder, W.E. (2010) Organic agriculture promotes evenness and natural pest control. *Nature* **466**: 109–112.

Cruzado, W. (2012) *Seaman A. Knapp Lecture 'Who Needs Extension, Anyway?' The Relevance and Values for our Next 100 Years of Engagement.* Association of Public Land Grant Universities Annual Meeting, 11 November 2012, Denver, CO, USA.

Dara, S., Bi, J. and Bolda, M. (2012) Arthropod management in organic strawberries. In: Koike, S.T., Bull, C., Bolda, M. and Daugovish, O. (eds) *Organic Strawberry Production Manual.* Publication 3531. University of California Agriculture and Natural Resources, Oakland, CA, USA, pp. 97–110.

Darrow, G.M. (1966) *The Strawberry: History, Breeding and Physiology.* Holt, Rinehart and Winston, San Francisco, CA, USA.

Deguine, J.P., Ferron, P. and Russell, D. (2008) Sustainable pest management for cotton production. A review. *Agronomy for Sustainable Development* **28**: 113–137.

Dhileepan, K. (2003) Evaluating the effectiveness of weed biocontrol at the local scale. In: Spafford Jacob, H. and Briese, D.T. (eds) *Improving the Selection, Testing and Evaluation of Weed Biological Control Agents.* Proceedings of the CRC for Australian Weed Management, Biological Control of Weeds Symposium and Workshop 2002, Perth, Australia, pp. 51–60.

Ehler, L.E. (2005) Integrated pest management: a national goal? *Issues in Science and Technology* **22**: 25–26.

Ehler, L.E. (2006) Integrated pest management (IPM): definition, historical development and implementation, and the other IPM. *Pest Management Science* **62**: 787–789.

Ehler, L.E. and Bottrell, D.G. (2000) The illusion of integrated pest management. *Issues in Science and Technology* **16**: 61–64.

Epstein, L. and Bassein, S. (2003) Patterns of pesticide use in California and the implications for strategies for reduction of pesticides. *Annual Review of Phytopathology* **41**: 351–375.

Epstein, L. and Zhang, M. (2014) The impact of integrated pest management programs on pesticide use in California, USA. In: Peshin, R. and Pimentel, D. (eds) *Integrated Pest Management.* Springer Science, Dordrecht, The Netherlands, pp. 173–200.

ESA (Entomological Society of America) (2012) Issues with 'Least Toxic Pesticides' Applied as 'Last Resort'. Available at: www.entsoc.org/press-releases/issues-associated-least-toxic-pesticides-applied-last-resort (accessed 28 February 2017).

Escalada, M.M., Heong, K.L., Huan, N.H. and Chien, H.V. (2009) Changes in rice farmers' pest management beliefs and practices in Vietnam: an analytical review of survey data from 1992 to 2007. In: Heong, K.L. and Hardy, B. (eds) *Planthoppers: New Threats to the Sustainability of Intensive Rice Production Systems in Asia.* International Rice Research Institute, Los Baños, Philippines, pp. 447–456.

Eshenaur, B., Grant, J., Kovach, J., Petzoldt, C., Degni, J. and Tette, J. (2015) *Environmental Impact Quotient: A Method to Measure the Environmental Impact of Pesticides.* New York State Integrated Pest Management Program, Cornell Co-operative Extension, Cornell University, USA.

Farrar, J.J., Baur, M.E. and Elliott, S. (2015) *Adoption and Impacts of Integrated Pest Management in Agriculture in the western United States.* Western IPM Center, Davis, CA, USA.

Fennimore, S.A., Serohijos, R., Samtani, J.B., *et al.* (2013) TIF film, substrates and nonfumigant soil disinfestation maintain fruit yields. *California Agriculture* **67**: 139–146.

Ferrater, J., de Jong, P., Dicke, M., Chen, Y. and Horgan, F. (2013) Symbiont-mediated adaptation by planthoppers and leafhoppers to resistant rice varieties. *Arthropod-Plant Interactions* 7: 591–605.

FIFRA (1947) Federal Insecticide, Fungicide, and Rodenticide Act of 1947 (P.L. 80-102, June 25, 1947, 61 Stat. 163).

FIFRA (1972) Federal Environmental Pesticide Control Act of 1972 (P.L. 92-516, October 21, 1972, 86 Stat. 973).

FitzSimmons, M.I. (1986) The new industrial agriculture. *Economic Geography* **62**: 334–353.

Foster, S.P., Denholm, I., Thompson, R., Poppy, G.M. and Powell, W. (2005) Reduced response of insecticide-resistant aphids and attraction of parasitoids to aphid alarm pheromone; a potential fitness trade-off. *Bulletin of Entomological Research* **95**: 37–46.

Foster, S.P., Tomiczek, M., Thompson, R., *et al.* (2007) Behavioural side-effects of insecticide resistance in aphids increase their vulnerability to parasitoid attack. *Animal Behaviour* **74**: 621–632.

Furlong, M.J. (2015) Knowing your enemies: integrating molecular and ecological methods to assess the impact of arthropod predators on crop pests. *Insect Science* **22**: 6–19.

Furlong, M.J., Wright, D.J., and Dosdall, L.M. (2014) Diamondback moth ecology and management: problems, progress, and prospects. *Annual Review of Entomology* **58**: 517–541.

Gabriel, A.D. (1989) Ecological Factors Affecting Strawberry Leaf Roller, *Ancylis comptana* (Froelich), Biology. MS Entomology. Iowa State University, Ames, IA, USA.

GAO (Government Accountability Office) (2001) *Agricultural Pesticides: Management Improvements Needed to Further Promote Integrated Pest Management.* GAO-01-815. Government Accountability Office, Washington, DC, USA.

Ginsberg, J.M. and Schmit, J.B. (1932) A comparison between rotenone and pyrethrins as contact insecticides. *Journal of Economic Entomology* **25**: 918–922.

Gliessman, S., Werner, M., Swezey, S., Caswell, E., Cochran, J. and Rosado-May, F. (1996) Conversion to organic strawberry management changes ecological processes. *California Agriculture* **50**: 24–31.

Goodell, P.B. (2009) Fifty years of the integrated control concept: the role of landscape ecology in IPM in San Joaquin valley cotton. *Pest Management Science* **65**: 1293–1297.

Gurr, G.M. and Kvedaras, O.L. (2010) Synergizing biological control: scope for sterile insect technique, induced plant defences and cultural techniques to enhance natural enemy impact. *Biological Control* **52**: 198–207.

Hagler, J.R. (2011) An immunological approach to quantify consumption of protein-tagged *Lygus hesperus* by the entire cotton predator assemblage. *Biological Control* **58**: 337–345.

Hebert, P.D. and Gregory, T.R. (2005) The promise of DNA barcoding for taxonomy. *Systematic Biology* **54**: 852–859.

Higgs, R. (1978) Landless by law – Japanese immigrants in California agriculture to 1941. *Journal of Economic History* **38**: 205–225.

Higley, L.G. and Peterson, R.K.D. (2009) Economic decision rules for IPM. In: Radcliffe, E.B., Hutchinson, W.D. and Cancelado, R.E. (eds) *Integrated Pest Management: Concepts, Tactics, Strategies, and Case Studies.* Cambridge University Press, Cambridge, UK, pp. 14–24.

Hill, S.B. (1998) Redesigning agroecosystems for environmental sustainability: a deep systems approach. *Systems Research and Behavioral Science* **15**: 391–402.

Hinds, W.E. (1913) Powdered arsenate of lead as an insecticide. *Journal of Economic Entomology* **6**: 477–479.
Hornick, M. (2015) Strawberry growers rev up bug vacs. The Packer, 2 April 2015. Available at: www.thepacker.com/shipping-profiles/california-strawberries/strawberry-growers-rev-bug-vacs (accessed 28 February 2017).
Huffaker, C.B. and Kennett, C.E. (1956) Experimental studies on predation: predation and cyclamen-mite studies on strawberry in California. *Hilgardia* **26**: 191–222.
Ikeda, M., Hamajima, R. and Kobayashi, M. (2015) Baculoviruses: diversity, evolution and manipulation of insects. *Entomological Science* **18**: 1–20.
Ives, A.R. and Settle, W.H. (1997) Metapopulation dynamics and pest control in agricultural systems. *American Naturalist* **149**: 220–246.
Jiang, L., Mancuso, M., Lu, Z., Akar, G., Cesarman, E. and Erickson, D. (2014) Solar thermal polymerase chain reaction for smartphone-assisted molecular diagnostics. *Scientific Reports* **4**: 1–5.
Jiang, W., Zhou, H., Bi, H., Fromm, M., Yang, B. and Weeks, D.P. (2013) Demonstration of CRISPR/Cas9/sgRNA-mediated targeted gene modification in Arabidopsis, tobacco, sorghum and rice. *Nucleic Acids Research* **41(21)**: e188.
Kallet, A. and Schlink, F.J. (1932) *100,000,000 Guinea Pigs, Dangers in Everyday Foods, Drugs, and Cosmetics.* Vanguard Press, New York, USA.
Karp, D.S., Gennet, S., Kilonzo, C., *et al.* (2015) Comanaging fresh produce for nature conservation and food safety. *Proceedings of the National Academy of Sciences* **112**: 11126–11131.
Khan, Z.R., Midega, C.A.O., Pittchar, J.O., *et al.* (2014) Achieving food security for one million sub-Saharan African poor through push-pull innovation by 2020. *Philosophical Transactions of the Royal Society B* **369**: 20120284.
Kikuchi, Y., Hayatsu, M., Hosokawa, T., Nagayama, A., Tago, K. and Fukatsu, T. (2012) Symbiont-mediated insecticide resistance. *Proceedings of the National Academy of Sciences* **109**: 8618–8622.
Koike, S.T., Bull, C.T., Bolda, M. and Daugovish, O. (2012) *Organic Strawberry Production Manual.* Publication 3531. University of California, Agriculture and Natural Resources. Oakland, CA, USA.
Kovach, J., Petzoldt, C., Degni, J. and Tette, J. (1992) A method to measure the environmental impact of pesticides. *New York's Food and Life Sciences Bulletin* **139**: 1–8.
Lacey, L.A., Thomson, D., Vincent, C. and Arthurs, S.P. (2008) Codling moth granulovirus: a comprehensive review. *Biocontrol Science and Technology* **18**: 639–663.
Landis, D.A., Wratten, S.D. and Gurr, G.M. (2000) Habitat management to conserve natural enemies of arthropod pests in agriculture. *Annual Review of Entomology* **45**: 175–201.
Lange, A.H., Agamalian, A.K., Humphrey, W.A. and Voth, V. (1967) Weed control studies in strawberry. *California Agriculture* December: 8–9.
Leite, A.E., de Castro, R., Chiappetta Jabbour, C.J., Batalha, M.O. and Govindan, K. (2014) Agricultural production and sustainable development in a Brazilian region (Southwest, Sao Paulo State): motivations and barriers to adopting sustainable and ecologically friendly practices. *International Journal of Sustainable Development and World Ecology* **21**: 422–429.
Letourneau, D.K. (2012) Integrated pest management – outbreaks prevented, delayed, or facilitated? In: Barbosa, P. Letourneau, D. and Agriwal, A. (eds) *Insect Outbreaks Revisited.* John Wiley and Sons, Chichester, UK, pp. 371–394.

Letourneau, D.K., Armbrecht, I., Rivera, B.S., *et al.* (2011) Does plant diversity benefit agroecosystems? A synthetic review. *Ecological Applications* **21**: 9–21.
Letourneau, D.K., Bothwell Allen, S.G. and Stireman III, J.O. (2012) Perennial habitat fragments, parasitoid diversity and parasitism in ephemeral crops. *Journal of Applied Ecology* **49**: 1405–1416.
Letourneau, D.K., Bothwell Allen, S.G., Kula, R.R., Sharkey, M.J. and Stireman III, J.O. (2015) Habitat eradication and cropland intensification may reduce parasitoid diversity and natural pest control services in annual crop fields. *Elementa: Science of the Anthropocene* **3**: 1–13.
Li, Y.C. and Yang, Y.P. (2015) On the paradox of pesticides. *Communications in Nonlinear Science and Numerical Simulation* **29**: 179–187.
Liebman, M. and Gallandt, E.R. (1997) Many little hammers: ecological management of crop-weed interactions. In: Jackson, L. (ed.) *Ecological Agriculture*. Academic Press, New York, USA, pp. 291–343.
Little, V.A. (1931) Devil's shoe-string as an insecticide. *Science* **73**: 315–316.
Lu, Z.X., Zhu, P.Y., Gurr, G.M., *et al.* (2014) Mechanisms for flowering plants to benefit arthropod natural enemies of insect pests: prospects for enhanced use in agriculture. *Insect Science* **21**: 1–12.
Lux, S.A. (2014) PESTonFARM – stochastic model of on-farm insect behaviour and their response to IPM interventions. *Journal of Applied Entomology* **138**: 458–467.
Macfadyen, S., Gibson, R., Polaszek, A., *et al.* (2009) Do differences in food web structure between organic and conventional farms affect the ecosystem service of pest control? *Ecology Letters* **12**: 229–238.
Mancini, V. and Romanazzi, G. (2014) Seed treatments to control seedborne fungal pathogens of vegetable crops. *Pest Management Science* **70**: 860–868.
Matteson, P.C. (2000) Insect pest management in tropical Asian irrigated rice. *Annual Review of Entomology* **45**: 549–574.
Mazzola, M., Muramoto, J. and Shennan, C. (2012) Transformation of soil microbial community structure in response to anaerobic soil disinfestation for soilborne disease control in strawberry. In: Sundin, G. (ed.) *Annual Meeting of the American Phytopathological Society*. Providence, RI, USA, pp. 77–78.
Mendez, V.E., Bacon, C.M., Olson, M., Morris, K.S. and Shattuck, A. (2010) Agrobiodiversity and shade coffee smallholder livelihoods: a review and synthesis of ten years of research in Central America. *Professional Geographer* **62**: 357–376.
Mensing, S. and Byrne, R. (1998) Pre-mission invasion of *Erodium cicutarium* in California. *Journal of Biogeography* **25**: 757–762.
Moreno, C.R., Lewins, S.A. and Barbosa, P. (2010) Influence of relative abundance and taxonomic identity on the effectiveness of generalist predators as biological control agents. *Biological Control* **52**: 96–103.
Moss, S.R. (2010). Non-chemical methods of weed control: benefits and limitations. In: Zydenbos, S.M. (ed.) *Seventeenth Australasian Weeds Conference*. Christchurch, New Zealand, pp. 14–19.
Mullen, J.D., Alston, J.M., Sumner, D.A., Kreith, M.T. and Kuminoff, N.V. (2003) *Returns to University of California Pest Management Research and Extension*. University of California, Agricultural Issues Center. University of California, Agriculture and Natural Resources Communication Services, Oakland, CA, USA.

Naranjo, S.E. and Ellsworth, P.C. (2009) Fifty years of the integrated control concept: moving the model and implementation forward in Arizona. *Pest Management Science* **65**: 1267–1286.
Nicholls, C.I. and Altieri, M.A. (2007) *Agroecology: Contributions Towards a Renewed Ecological Foundation for Pest Management.* Cambridge University Press, New York, USA.
Norris, R.F. and Kogan, M. (2000) Interactions between weeds, arthropod pests, and their natural enemies in managed ecosystems. *Weed Science* **48**: 94–158.
NRC (National Resource Council) (1996) *Ecologically Based Pest Management: New Solutions for a New Century.* National Research Council, National Academy of Sciences, Washington, DC, USA.
Oatman, E.R. and Platner, G.R. (1972) An ecological study of aphids on strawberry in southern California. *Environmental Entomology* **1**: 339–343.
Ordonez, N., Seidl, M.F., Waalwijk, C., *et al.* (2015) Worse comes to worst: bananas and Panama disease – when plant and pathogen clones meet. *PLoS Pathogens* **11(11)**: e1005197.
O'Rourke, M E., Rienzo-Stack, K. and Power, A.G. (2011) A multi-scale, landscape approach to predicting insect populations in agroecosystems. *Ecological Applications* **21**: 1782–1791.
Ottaviano, M.F.G., Cedola, C.V., Sanchez, N E. and Greco, N.M. (2015) Conservation biological control in strawberry: effect of different pollen on development, survival, and reproduction of *Neoseiulus californicus* (Acari: Phytoseiidae). *Experimental and Applied Acarology* **67**: 507–521.
Parsa, S., Morse, S., Bonifacio, A., *et al.* (2014) Obstacles to integrated pest management adoption in developing countries. *Proceedings of the National Academy of Sciences* **111**: 3889–3894.
Perez-Ruiz, M., Gonzalez-de-Santos, P., Ribeiro, A., *et al.* (2015) Highlights and preliminary results for autonomous crop protection. *Computers and Electronics in Agriculture* **110**: 150–161.
Peshin, R., and Zhang, W. (2014) Integrated pest management and pesticide use. In: Pimentel, D. and Peshin, R. (eds) *Integrated Pest Management: Pesticide Problems, Vol. 3.* Springer, Dordrecht, The Netherlands, pp. 1–46.
Pickel, C., Zalom, F., Walsh, D. and Welch, N. (1995) Vacuums provide limited *Lygus* control in strawberries. *California Agriculture* **49**: 19–22.
Pickett, C.H., Swezey, S.L., Nieto, D.J., *et al.* (2009) Colonization and establishment of *Peristenus relictus* (Hymenoptera: Braconidae) for control of *Lygus* spp. (Hemiptera: Miridae) in strawberries on the California Central Coast. *Biological Control* **49**: 27–37.
Pimentel, D. (1961) Species diversity and insect population outbreaks. *Annals of the Entomological Society of America* **54**: 76–86.
Pimentel, D. and Goodman, N. (1978) Ecological basis for management of insect populations. *Oikos* **30**: 422–437.
Rahman, S. (2013). Pesticide consumption and productivity and the potential of IPM in Bangladesh. *Science of the Total Environment* **445**: 48–56.
Rajendran, S.R.C.K., Yau, Y.Y., Pandey, D. and Kumar, A. (2015) CRISPR-Cas9 based genome engineering: opportunities in agri-food-nutrition and healthcare. *Omics: a Journal of Integrative Biology* **19**: 261–275.
Rodriguez, L.A., Arredondo, H.C., Williams, T. and Barrera, F. (2015) Pasado, presente y perspectivas del control biológico en México. In: Rodriguez, L.A. and Arredondo, H.C. (eds) *Casos de Control Biológico en México.* Biblioteca Basica de Agricultura, Guadalajara, Mexico, pp. 17–28.

Roush, R. and Hoy, M. (1980) Selection improves sevin resistance in spider mite predator. *California Agriculture* **34**: 11–14.

Ruberson, J.R., Nemoto, H. and Hirose, Y. (1998) Pesticides and conservation of natural enemies in pest management. In: Barbosa, P. (ed.) *Conservation Biological Control*. Academic Press, San Diego, CA, USA, pp. 207–220.

Samtani, J.B., Ben Weber, J. and Fennimore, S.A. (2012) Tolerance of strawberry cultivars to oxyfluorfen and flumioxazin herbicides. *Hortscience* **47**: 848–851.

Sánchez, J.A., Mellín, M.A., Arredondo, H.C., Vizcara, N.I., González, A. and Montesinos, R. (2015) Psílido asiático de los cítricos, *Diaphornia citri* (Hemiptera: Psyllidae). In: Rodriguez, L.A and Arredondo, H.C. (eds) *Casos de Control Biológico en México*. Biblioteca Basica de Agricultura, Guadalajara, Mexico, pp. 339–372.

Sauer, C.O. (1952) *Agriculture Origins and Dispersals*. American Geographical Society, New York, USA.

Sauer, J.D. (1993) *The Historical Geography of Crop Plants*. CRC Press, Boca Raton, FL, USA.

Schellhorn, N.A., Glatz, R.V. and Wood, G.M. (2010) The risk of exotic and native plants as hosts for four pest thrips (Thysanoptera: Thripinae). *Bulletin of Entomological Research* **100**: 501–510.

Schreinemachers, P. and Tipraqsa, P. (2012) Agricultural pesticides and land use intensification in high, middle and low income countries. *Food Policy* **37**: 616–626.

Sen, A. (1981) *Poverty and Famines: An Essay on Entitlement and Deprivation*. Clarendon Press, Oxford, UK.

Settle, W.H., Ariawan, H., Astuti, E.T., *et al.* (1996) Managing tropical rice pests through conservation of generalist natural enemies and alternative prey. *Ecology* **77**: 1975–1988.

Settle, W., Soumaré, M., Sarr, M., Garba, M.H. and Poisot, A.S. (2014) Reducing pesticide risks to farming communities: cotton farmer field schools in Mali. *Philosophical Transactions of the Royal Society of London B: Biological Sciences* **369**: 20120277.

Sevacherian, V. and Stern, V.M. (1975) Movements of *Lygus* bugs between alfalfa and cotton. *Environmental Entomology* **4**: 163–165.

Shafir, S., Dag, A., Bilu, A., Abu-Toamy, M. and Elad, Y. (2006) Honey bee dispersal of the biocontrol agent *Trichoderma harzianum* T39: effectiveness in suppressing *Botrytis cinerea* on strawberry under field conditions. *European Journal of Plant Pathology* **116**: 119–128.

Sharma, R., Peshin, R.,Shankar, U., Kaul, V. and Sharma, S. (2015) Impact evaluation indicators of an integrated pest management program in vegetable crops in the subtropical region of Jammu and Kashmir, India. *Crop Protection* **67**: 191–199.

Shimat, V.J. and Bolda, M. (2014) Updates on insecticide trials targeting *Lygus* bug in strawberry. Presented at the UCCE Annual Strawberry Meeting, Watsonville, CA, USA.

Shirtliffe, S.J. and Benaragama, D. (2014) Sometimes you need a big hammer: evaluating and appraising selected non-herbicidal weed control methods in an integrated weed management system. In: Martin, R.C. and MacRae, R. (eds) *Managing Energy, Nutients, and Pests in Organic Field Crops*. CCRC Press, Boca Raton, FL, USA, pp. 150–168.

Sigsgaard, L., Naulin, C., Haukeland, S., *et al.* (2014) The effects of strawberry cropping practices on the strawberry tortricid (Lepidoptera: Tortricidae), its natural enemies, and the presence of nematodes. *Journal of Insect Science* **14**: 122.

Smith, L., Allen, W. and Lange, W. (1958) Strawberry leaf miner damage: effectiveness of natural enemies usually holds pest damage below levels of economic importance in commercial plantings. *California Agriculture* **12**: 12–13.

Smith, L.M. and Goldsmith, E.V. (1936) The cyclamen mite, *Tarsonemus pallidus*, and its natural enemies on field strawberries. *Hilgardia* **10**: 53–94.
Smith, M.D. (1982) *The Ortho Problem Solver*. Ortho, Chevron Chemical Co., San Francisco, CA, USA.
Snapp, S., Kanyama-Phiri, G., Kamanga, B., Gilbert, R. and Wellard, K. (2002) Farmer and researcher partnerships in Malawi: developing soil fertility technologies for the near-term and far-term. *Experimental Agriculture* **38**: 411–431.
Steingrover, E.G. Geertsema, W. and van Wingerden, W. (2010) Designing agricultural landscapes for natural pest control: a transdisciplinary approach in the Hoeksche Waard (The Netherlands). *Landscape Ecology* **25**: 825–838.
Stern, V.M., Smith, R.F., van den Bosch, R. and Hagen, K.S. (1959) The integrated control concept. *Hilgardia* **29**: 81–101.
Stern, V.M., van den Bosch, R. and Leigh, T.F. (1964) Strip cutting alfalfa for lygus bug control. *California Agriculture* **18**: 4–6.
Stevens, N.E. (1933) *Strawberry Diseases*. Farmers' Bulletin No. 1458. USDA-Agricultural Research Service, Washington, DC, USA.
Strand, L. (1994) *Integrated Pest Management for Strawberries*. Publication 3351. University of California Division of Agriculture and Natural Resources, Oakland, CA, USA.
Strand, L.L. (2008) *Integrated Pest Management for Strawberries*, 2nd edn. Publication 3351. University of California, Agriculture and Natural Resources, Oakland, CA, USA.
Straub, C.S., Finke, D.L. and Snyder, W.E. (2008) Are the conservation of natural enemy biodiversity and biological control compatible goals? *Biological Control* **45**: 225–237.
Suckling, D.M., Stringer, L.D., Baird, D.B., *et al*. (2014) Light brown apple moth (*Epiphyas postvittana*) (Lepidoptera: Tortricidae) colonization of California. *Biological Invasions* **16**: 1851–1863.
Swezey, S.L., Nieto, D.J. and Bryer, J.A. (2007) Control of western tarnished plant bug *Lygus hesperus* Knight (Hemiptera: Miridae) in California organic strawberries using alfalfa trap crops and tractor-mounted vacuums. *Environmental Entomology* **36**: 1457–1465.
Swezey, S.L., Nieto, D.J., Pickett, C.H., Hagler, J.R., Bryer, J.A. and Machtley, S.A. (2014) Spatial density and movement of the *Lygus* spp. parasitoid *Peristenus relictus* (Hymenoptera: Braconidae) in organic strawberries with alfalfa trap crops. *Environmental Entomology* **43**: 363–369.
Teacher, A.G., Griffiths, D.J., Hodgson, D.J. and Inger, R. (2013) Smartphones in ecology and evolution: a guide for the apprehensive. *Ecology and Evolution* **3**: 5268–5278.
Tey, Y.S. and Brindal M. (2012) Factors influencing the adoption of precision agricultural technologies: a review for policy implications. *Precision Agriculture* **13**: 713–730.
Thomas, H.E. (1932) Verticillium Wilt of Strawberries. Bulletin No. 530. University of California Agriculture Experiment Station, USA.
Thomas, H.E. (1939) *The Production of Strawberries in California*. California Agricultural Extension Service Circular No. 113, USA.
Thomas, H.Q. (2014) Developing an effective integrated pest management (IPM) program for lygus bug. UCCE Annual Strawberry Meeting, Watsonville, CA, USA.
Tscharntke, T., Klein, A.M., Kruess, A., Steffan-Dewenter, I. and Thies, C. (2005) Landscape perspectives on agricultural intensification and biodiversity – ecosystem service management. *Ecology Letters* **8**: 857–874.

Tuck, S.L., Winqvist, C., Mota, F., Ahnström, J., Turnbull, L.A. and Bengtsson, J. (2014) Land-use intensity and the effects of organic farming on biodiversity: a hierarchical meta-analysis. *Journal of Applied Ecology* **51**: 746–755.

Tylianakis, J.M., Laliberte, E., Nielsen, A. and Bascompte, J. (2010) Conservation of species interaction networks. *Biological Conservation* **143**: 2270–2279.

UC-IPM (2010) *UC IPM Pest Management Guidelines: Strawberry*. University of California, Agriculture and Natural Resources, USA.

USDA (1972) *Strawberry Insects... How to Control Them*. Farmers' Bulletin No. 2184. Agricultural Research Service, USDA, Washington, DC, USA.

USDA-NASS (2014) *California Historic Commodity Data: Strawberries 1918–2012*. California Field Office, USA.

Van Bruggen, A.H.C. and Semenov, A.M. (2000) In search of biological indicators for soil health and disease suppression. *Applied Soil Ecology* **15**: 13–24.

Van den Bosch, R. (1978) *The Pesticide Conspiracy*. Doubleday, New York, USA.

Van den Bosch, R. and Stern, V.M. (1962) Integration of chemical and biological control of arthropod pests. *Annual Review of Entomology* **7**: 367–386.

Van Emden, H.F. (1964) The role of uncultivated land in the biology of crop pests and beneficial insects. *Scientific Horticulture* **17**: 121–136.

Vanloqueren, G. and Baret, P.V. (2008) Why are ecological, low-input, multi-resistant wheat cultivars slow to develop commercially? A Belgian agricultural 'lock-in' case study. *Ecological Economics* **66**: 436–446.

Vanloqueren, G. and Baret, P.V. (2009) How agricultural research systems shape a technological regime that develops genetic engineering but locks out agroecological innovations. *Research Policy* **38**: 971–983.

Vemmer, M. and Patel, A.V. (2013) Review of encapsulation methods suitable for microbial biological control agents. *Biological Control* **67**: 380–389.

Vorburger, C. (2014) The evolutionary ecology of symbiont-conferred resistance to parasitoids in aphids. *Insect Science* **21**: 251–264.

Vorotnikova, E., Borisova, T. and van Sickle, J.J. (2014) Evaluation of the profitability of a new precision fungicide application system for strawberry production. *Agricultural Systems* **130**: 77–88.

Voytas, D.F. and Gao, C. (2014) Precision genome engineering and agriculture: opportunities and regulatory challenges. *PLoS Biol* **12(6)**: e1001877.

Waddington, H., Snilstveit, B., Hombrados, J.G., Vojtkova, M., Anderson, J. and White, H. (2014) Farmer field schools for improving farming practices and farmer outcomes: a systematic review. *Campbell Systematic Reviews* **10**: issue 6.

Wang, Y., Cheng, X., Shan, Q., *et al.* (2014) Simultaneous editing of three homoeoalleles in hexaploid bread wheat confers heritable resistance to powdery mildew. *Nature Biotechnology* **32**: 947–951.

Warner, K.D., Daane, K.M., Getz, C.M., Maurano, S.P., Calderon, S. and Powers, K.A. (2011) The decline of public interest agricultural science and the dubious future of crop biological control in California. *Agriculture and Human Values* **28**: 483–496.

Welch, N.C., Pickel, C., Walsh, D. and van Nouhuys, S. (1989) Cyclamen mite control in strawberries. *California Agriculture* **43**: 14–15.

Wells, M.J. (1996) *Strawberry Fields*. Cornell University Press, New York, USA.

Werling, B.P. and Gratton, C. (2010) Local and broadscale landscape structure differentially impact predation of two potato pests. *Ecological Applications* **20**: 1114–1125.

Westerman, P., Liebman, M., Menalled, F.D., Heggenstaller, A.H., Hartzler, R.G. and Dixon, P.M. (2005) Are many little hammers effective? Velvetleaf (*Abutilon theophrasti*) population dynamics in two- and four-year crop rotation systems. *Weed Science* **53**: 382–392.
Wezel, A., Casagrande, M., Celette, F., Vian, J.F., Ferrer, A. and Peigne, J. (2014) Agroecological practices for sustainable agriculture. A review. *Agronomy for Sustainable Development* **34**: 1–20.
Wilhelm, S. and Sagen, J.E. (1974) *A History of the Strawberry, from Ancient Gardens to Modern Markets*. University of California, Division of Agricultural Sciences, Berkeley, CA, USA.
Wilhoit, L., Supkoff, D., Steggall, J., *et al.* (1998) *An Analysis of Pesticide Use in California, 1991–1995*. California Environmental Protection Agency, USA, PM 98-01.
Witzgall, P., Stelinski, L., Gut, L. and Thomson, D. (2008) Codling moth management and chemical ecology. *Annual Review of Entomology* **53**: 503–522.
Wood, S.A., Karp, D.S., DeClerck, F., Kremen, C., Naeem, S. and Palm, C.A. (2015) Functional traits in agriculture: agrobiodiversity and ecosystem services. *Trends in Ecology and Evolution* **30**: 531–539.
Zhang, W., Jiang, F. and Ou, J. (2011) Global pesticide consumption and pollution: with China as a focus. *Proceedings of the International Academy of Ecology and Environmental Sciences* **1**: 125–144.
Zorner, P.S. (2000) Shifting agricultural and ecological context for IPM. In: Kennedy, G.S. and Sutton, T.B. (eds) *Emerging Technologies for Integrated Pest Management: Concepts, Research, and Implementation*. American Phytopathological Society Press, St Paul, MN, USA, pp. 32–41.

3

The Economics of Alternative Pest Management Strategies: Basic Assessment

Clement A. Tisdell, David Adamson and Bruce Auld

3.1 Introduction

The management of pest populations is an important part of human control of the environment. Overall, it increases productivity of bio-industries, improves the quality of their production and adds to human welfare. However, some production methods can have negative consequences for human well-being and the environment. In most contemporary economies, decisions about the management of pests (which adversely affect bio-industries) are driven by the desire of the proprietors of production units (e.g. farms) to maximize their profit. Determining pest management strategies to achieve this objective can be very challenging, given that the optimal choice usually depends on several uncertain variables. In part, this is due to uncertainty about relevant environmental factors. A further complication is that optimal private profit-maximizing decisions about pest control may not result in socially optimal choices, necessitating state intervention in pest control practices.

The focus in this chapter is on the economics of pest control at the farm (mostly field) level. It does not take account of the off-farm environmental and related consequences of decisions about pest control. These are considered in other chapters in this book. This chapter is relevant to the environmental theme of this book in two ways.

- Choices about pest management strategies at the farm level are influenced by environmental factors. For example, environmental factors influence the presence and density of a pest and the effectiveness of control methods.
- The magnitude and nature of off-farm environmental (and related) effects of pest control depend on choices made at the farm level – for example, whether or not a pest is controlled and the method used. Typically, individual farmers are unlikely to take off-farm effects into account in their decision making.

Methods for pest control in agriculture and livestock production are extremely numerous and diverse. Consequently, the economics of pest management is a very complex subject. This analysis continues to evolve and has to be regularly updated to take account of new methods of control, for example, the development of genetically modified (GM) crops to combat pests. Here it is only possible to consider the on-farm economics of adopting a few of these methods. This is done using basic economic

Environmental Pest Management: Challenges for Agronomists, Ecologists, Economists and Policymakers,
First Edition. Edited by Moshe Coll and Eric Wajnberg.

models, mostly economic threshold models. Using these basic models, the following matters are considered.

- Whether or not (and to what extent) it is profitable for a farmer to undertake pest control in the absence of uncertainty.
- How do uncertainties (environmental or otherwise) alter a farmer's decision to engage in pest control?
- How can scale (and some other economic factors) influence on-farm pest control, for example, the technique used for applying a control agent?
- What economic factors influence on-farm choices for choosing which pest management solution to adopt? For example, what conditions determine the adoption of an insect-resistant GM crop or the continued planting of a conventional crop and engaging in traditional pest control? The way in which uncertainty affects the relative profitability of decisions about this is given particular consideration.

All these on-farm choices are influenced by environmental factors and they have differing off-farm environmental consequences.

The following topics are considered in turn.

- Economic decisions at farm level based on threshold models assuming use of a given pest control technique and certainty.
- The same as the above but allowing for uncertainties.
- Choice of alternative means of distributing a pest control agent, particularly changes in this with the area and density of the pest infestation.
- The economics of the timing of pest control and some of the factors which influence farm-level choices when the density of pest infestation is uncertain and alternative methods for control are available. Here the focus is on whether or not to plant GM crops with in-built attributes relevant to pest control, for example, herbicide-resistant crops or Bt (*Bacillus thuringiensis*)-modified ones rather than conventional crops.
- Some aspects of biological control.

3.2 Economic Decisions at Farm Level Based on Threshold Models Assuming Use of a Given Pest Control Technique and Certainty

3.2.1 A Simple Standard Model

A simple economic model, available in the early literature on this subject (Headley 1968; Stern 1966), assumes that farmers know about or anticipate a particular level of infestation of a crop by a pest and have a specific technique available for eliminating the pest. Use of this technique involves a given level of cost per hectare (ha) and the economic benefit achieved by a farmer is the avoidance of the loss in profit which would occur without use of this technique; that is, profit loss avoided by eliminating the pest (Auld and Tisdell 1986; Carlson 1970; Headley 1972; Naranjo *et al.* 2015). This benefit usually increases with the level of pest infestation avoided by the pest control treatment but can be quite variable (Falkenberg *et al.* 2012).

Note that in this analysis, all values are to be taken as per ha values unless otherwise specified. Let us consider the first basic model for determining the economics of pest

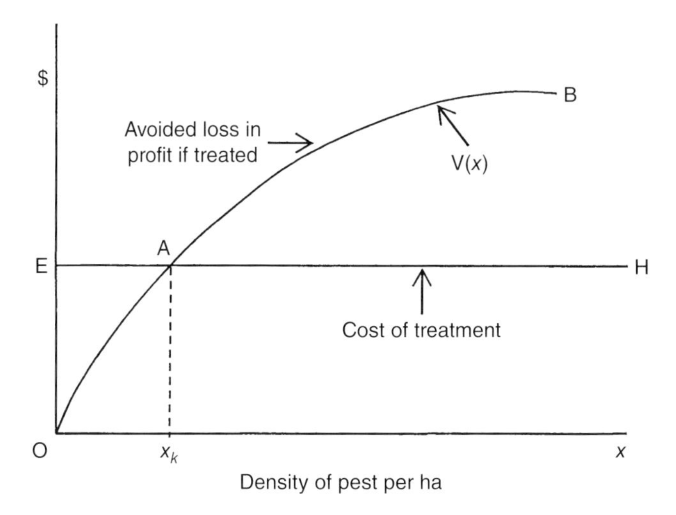

Figure 3.1 A typical representation of the threshold economic model for deciding whether or not it is economic to control a pest. The curve OAB represents the loss in value of crop production experienced by a farmer as a function of pest density and line EH represents the cost of eliminating the pest. See text for additional explanations.

control. Where x is the level of infestation of a crop by a pest (that is, the density of the pest per ha), V is the increase in profit per ha obtained by eliminating it, and C is the cost per ha of doing so, the net benefit per ha of eliminating the pest is:

$$G = V(x) - C \tag{3.1}$$

If this expression is positive, it pays to undertake the pest control. If this expression is negative, it provides a justification for the farmer not to engage in pest control. An example is provided in Figure 3.1. The curve OAB represents the loss in value of crop production experienced by a farmer as a function of the density of a pest and line EH represents the cost of eliminating the pest. If $x < x_k$, the rational producer will not control the pest but if $x > x_k$, the producer will do so. Consequently, x_k is the threshold value where the management costs equal the benefit derived from that management action (Headley 1972).

This type of economic threshold model is sometimes presented differently. An alternative formulation focuses on the cost–benefit ratio of pest control (Brown 1997). If this ratio exceeds unity, it is uneconomic to control the pest but if it is less than this, control is economic. In the relevant literature, this ratio is usually referred to as the economic injury level (EIL) (Brown 1997; Peterson and Hunt 2003).

It is sometimes assumed that the reduction in crop yield due to the presence of pest times the price per unit of crop output represents the extra economic benefit the farmer will obtain by eliminating the pest. This is actually the extra revenue, R, generated by controlling the pest. Therefore, if $f(x)$ represents the extra yield obtained by making sure the pest is eliminated and if p is the price per unit of the output of the crop, then:

$$R(x) = pf(x) \tag{3.2}$$

Elimination of the pest is assumed in the initial basic models. However, many controls only result in reducing the density of the pest instead. Adjustments to the basic model can be made to allow for this and this is done later in this chapter. Usually, p is assumed to be a constant, probably on the assumption that individual farmers are unable to influence the price they receive for their products. However, this assumes that increased yield does not involve any increase in production costs, such as extra harvesting costs. The possibility that it does should be allowed for. Therefore, if $\lambda(x)$ represents the extra cost of processing the higher yield, the economic value of the extra yield if a level of pest infestation of x is avoided is:

$$V(x) = pf(x) - \lambda(x) \tag{3.3}$$

This basic economic model of pest control can be extended further to allow for additional possibilities.

3.2.2 Extensions of the Basic Model

First, the presence of a pest in many cases not only affects the crop yield but also the quality of the produce (e.g. weed seeds contaminating harvested grain). Poorer quality produce will fetch a lower price and if its quality is too poor, it may be unsaleable. So, for some products, p is likely to be a function of x. Therefore, for greater generality, equation (2.2) can be re-expressed as:

$$R(x) = p(x)f(x) \tag{3.4}$$

where, as a rule, $dp/dx < 0$, that is, the price per unit received for produce falls with an increase in the level of pest infestation. Note that it is still appropriate to assume that an individual farmer is unable to influence the price received for his/her produce of a given quality. The type of relationship shown in Figure 3.2 may be common. For produce unaffected by a pest, a price of OH is received but the price falls with greater pest

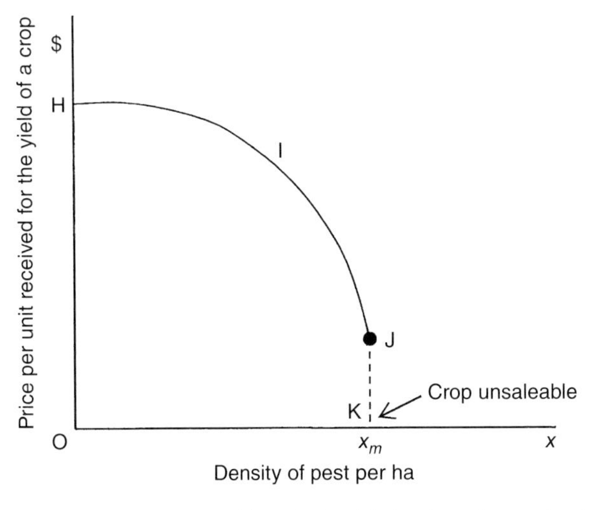

Figure 3.2 Illustration of the relationship for the price received per unit yield of some crops as a function of the level of pest infestation. See text for additional explanations.

infestation. For damages occurring to the produce when the density of a pest is $x \geq x_m$, the output from the crop is unsaleable. However, it may be possible to sort pest-affected produce from the unaffected but this will add to costs (e.g. in the case of harvested fruit). This case is not analysed. Note that because of the adverse quality impacts of a pest, it is possible for V(x) to increase at an increasing rate rather than at a decreasing rate, as is often assumed.

A second aspect is that the pest treatment may not be fully effective in eliminating the pest, perhaps due to the timing of the application or environmental factors (Carlson *et al.* 2011; Myers *et al.* 2005). This can easily be allowed for in this threshold type of analysis. Let x_r represent the density of the pest prior to treatment and x_t represent its density after treatment. Then, the net economic gain from treatment is:

$$G = V(x_r) - V(x_t) - C \tag{3.5}$$

If this expression is positive, it pays to treat the pest but not if it is negative.

A third possibility is that the cost of controlling a pest may not be independent of its density. For example, costs can rise as density increases; for example, labour costs increase in response to density levels – as density increases, more labour (time) is required to deal with the infestation if the management solution involves hand weeding or cutting and painting the stumps of woody weeds with herbicides. Consequently, the density of the pest can result in a 'double threshold'. It may not pay to control the pest if it is present at a low density, nor if it occurs at a high density. An example is shown in Figure 3.3.

A fourth important aspect of the economics of pest control is that the cost per ha of pest control often depends on the size of the area which needs to be treated. This aspect is covered below when the economics of alternative techniques for controlling pests is discussed.

If a number of alternative techniques (methods) are available to control a pest then the one adopted should be the cheapest one for the level of pest infestation experienced. Therefore, in Figure 3.3, this least cost relationship, C(x), corresponds to the

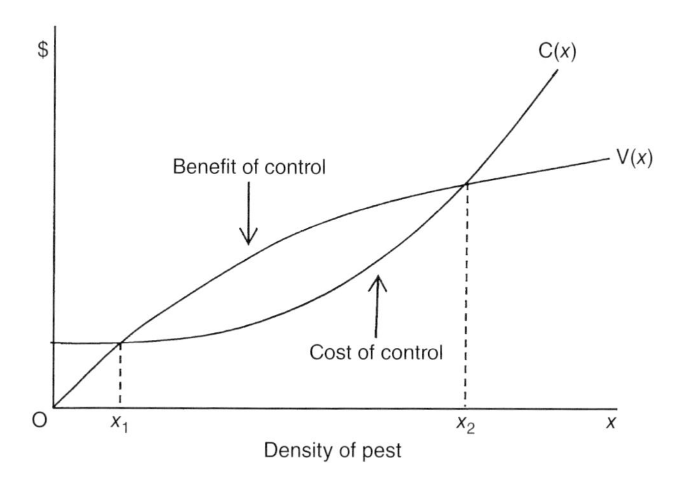

Figure 3.3 Illustration of a case in which two pest control thresholds exist. In this case, it does not pay to control the pest if its density is less than x_1 or greater than x_2.

lower envelope (boundary) of the cost curves of all the available techniques (methods) for controlling the pest. Another possible response to the increased prevalence of a pest is for a farmer to adopt a new production system which is less attractive to the pest. For example, due to increased numbers of feral pigs (Tisdell 1982) and wild dogs, some Australian graziers have switched from running sheep to grazing cattle because cattle are more successful than sheep in protecting themselves and their young against these predators, thus increasing profit.

Note that the standard economic threshold model does not allow for variability in the level of pest density in a field. However, the model can be adjusted to allow for this (Auld and Tisdell 1988). Potentially, also, it can be adjusted to allow for advances in precision agricultural technology which make it possible to adjust pest treatments with site variations in the presence of pests in a field (Liu *et al.* 2014). Furthermore, these simple economic threshold models do not take account of the dynamics of pest reproduction on farms, and the possibility of off-farm immigration of pests. Tisdell (1982, pp. 361–378) specifically discusses both these issues in relation to Headley's (1972) threshold models. Yokomizo *et al.* (2009) take some account of relevant population reproduction issues.

Basic economic threshold models also do not take account of the management of multiple pests and the impact of controls on economically beneficial organisms. In addition, some pest control techniques (although effective in controlling a pest) may have some negative effect on the level of crop yield and its quality. The latter can be expected to influence the price received for the output.

3.2.3 Summarized Implications of the Above Basic Models

Despite these limitations, several inferences can be drawn from the above basic models. If other things are held constant and if $V'(x) > 0$ throughout, pest control is more likely to be profitable when:

- a higher price per unit is obtained for the yield of a crop
- a larger amount of yield is saved as a result of treatment
- the cost of control is lower
- a larger reduction in the price of the product is avoided as a result of pest control
- a larger level of pest infestation is avoided as a result of pest control.

3.3 Uncertainties and Economic Decisions at Farm Level About Pest Control: Assumes a Given Pest Control Technique and Applies the Threshold Approach

Most variables of relevance for decisions about whether or not to undertake pest control are subject to uncertainty. These variables include the effectiveness of the control (the kill rate), the price of the product, and the increase in yield attributable to pest treatment. Both the kill rate and the increase in yield may be influenced by environmental factors. Furthermore, in many situations, the decision about whether or not to institute a pest control measure is made before pest density is known. This applies, for example, to the common practice of prophylactic applications of fungicides for disease management in vineyards and orchards, before the disease and/or symptoms are present. In these circumstances, optimal decisions at farm level will depend on attitudes of farmers to risk bearing. Nevertheless, in most cases, we can narrow optimal choices down to a restricted set, no matter what the attitude of a farmer to risk bearing.

Let us consider the consequences of uncertainty for decisions about whether or not to control a pest for two different types of situations. In the first case, it is assumed that the benefit of controlling the pest is uncertain. In the second case, it is supposed that the (anticipated) level of pest infestation is uncertain. Bear in mind that both uncertainties can occur together.

3.3.1 The Benefit Function for Pest Control is Uncertain

In Figure 3.4, the benefit function is uncertain. Yokomizo *et al.* (2009) have explored some of the economic consequences from incorrectly specifying the mathematical form of the density-impact curve, but that is not covered here. Instead, this form is assumed to be known but the position of the density-impact curve is supposed to be uncertain. Consequently, there is a band of possible benefits from pest control for each given level of pest infestation. As a result, a range of pest densities now exists for which it is uncertain whether it would be economically rational to control the pest. However, as showed in Figure 3.4, outside this uncertain range one can be certain about whether or not it is economic to eliminate the pest. If the anticipated level of pest infestation is uncertain, then management decisions are dependent upon the attitudes of farmers to risk taking. Consider the consequences of a few of the possible different attitudes to this. In the case shown in Figure 3.4, it is believed that the benefit function may be as low as represented by the relationship OCD, $V_1(x)$, or as high as OAB, $V_2(x)$, and that it may assume any value in between. Consequently, in these circumstances (given the goal of profit

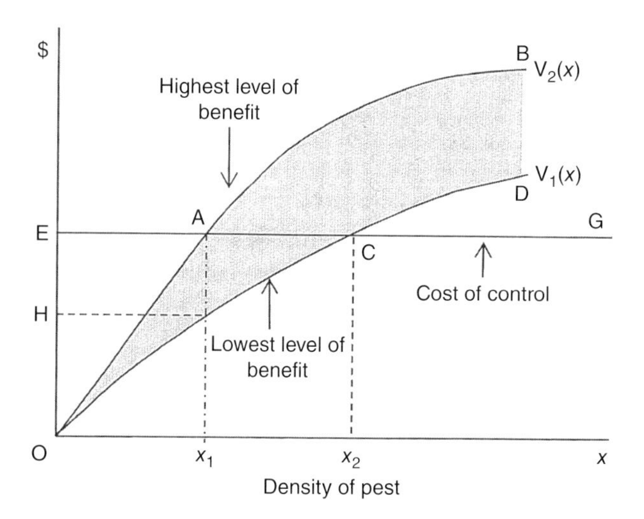

Figure 3.4 A case in which the benefit function of pest control is uncertain. In the case shown, it is believed that the benefit function may be as low as shown by the relationship OCD, $V_1(x)$, or as high as OAB, $V_2(x)$, and that it may assume any value in between. See text for additional explanations.

maximization), it will always be rational for a farmer to undertake pest treatment if $x > x_2$ and not to do so if $x < x_1$. For values of x in the range $x_1 < x < x_2$, the choice of whether or not to undertake pest control depends on the attitude of the farmer to risk taking.

If the farmers' prime objective is to minimize risk (i.e. risk-averse strategy), then he or she should adopt the 'minimax loss rule'. This requires the maximum possible loss to be minimized. In the case illustrated, the maximum possible loss occurs when the benefit function is $V_1(x)$ and the extent of this loss is at its highest level when $x = x_1$ (it is then equivalent to the distance EH) and tapers off as x approaches x_2. In these circumstances, the adoption of this risk-averse strategy reduces the willingness of farmers to control the pest. Therefore, it is not always the case that a high preference for income security favours the control of a pest. At the other end of the spectrum, farmers who like to gamble (i.e. risk-prone strategy) may adopt a 'maximax strategy', that is, a strategy which maximizes their profit in the most favourable circumstances. They will assume (in the case shown in Figure 3.4) that the benefit function for control of the pest is $V_2(x)$. Therefore, if the range of possible levels of pest infestation is $x_1 \leq x \leq x_2$, they will control the pest.

In some cases, the farmer may weigh the likelihood of different benefit functions occurring by their subjective probabilities and maximize expected net benefit of pest control on that basis. Depending on the distribution of probabilities, this will result in a value of x between x_1 and x_2 becoming the critical value for determining whether or not to undertake pest control. If the probabilities are skewed towards $V_2(x)$, control will occur at a lower pest density than if they are skewed towards $V_1(x)$. Most farmers operate within this framework of uncertainty. Note that the case illustrated in Figure 3.4 represents 1 year or growing season. If the current year's control impacts on future pest levels, the threshold density x_1 may be significantly lower if one is considering longer term benefits. This particularly applies to annual weeds in annual crops such as wheat. In practice, threshold levels for weed density can be very low (Trezzi *et al.* 2015).

3.3.2 The Level of Pest Infestation is Uncertain

Another important case (and an associated economic problem) is to find the most economic method of assessing the population of a pest, for example, thrips (Sutherland and Parrella 2011). This arises when the density of the pest is uncertain at the time pest control is undertaken. This is so, for example, for pest controls adopted before the emergence of the pest, and is effectively the case for the use of Bt (*Bacillus thuringiensis*)-modified plants (see also Chapter 12). Bt-modified crops are those which have undergone transgenic change to express toxins present in the bacterium *Bacillus thuringiensis*. These toxins are fatal to some species of caterpillars.

Pre-emergence uncertainty about pest populations is common in many situations. It can, for example, be difficult to predict from the density of their eggs the subsequent levels of infestation by caterpillars of lepidoptera species (Paula-Moraes *et al.* 2013). Risk aversion strongly favours the adoption of pest control in these cases. Some of the effects on the decision of whether or not to control a pest can be illustrated by Figure 3.5 if there is uncertainty about the level of pest infestation.

In Figure 3.5, the same basic assumptions are made as those relating to Figure 3.1. However, the economic benefit function (disregarding control costs) of eliminating a pest is in this illustration assumed to be incremental at an increasing rate; that is, it is

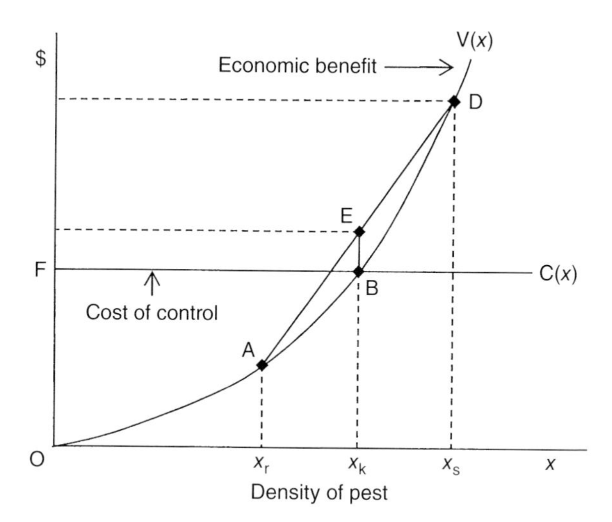

Figure 3.5 Diagram to illustrate influences on decisions to undertake pest control if there is uncertainty about the level of a pest infestation when pest control is undertaken.

supposed that $V''(x) > 0$. Nevertheless, depending on circumstances, $V''(x)$ may be positive, negative or zero. The likelihood is high that $V''(x)$ is positive if the price received for a pest-damaged product falls rapidly with the extent of its damage by the pest.

Given the relationships illustrated in Figure 3.5, first note that if all the possible densities of the pest are less than x_k, it does not rationally pay to control the pest. Similarly, if the set of possible densities of the pest all exceed x_k, it is always rational to undertake pest control irrespective of attitudes of the pest manager to taking risks and the distribution of probable values of x. These are strong results. They hold for all possible forms of $V(x)$ for which $V(x)$ is less than $C(x)$ for $x < x_k$, and for which $V(x)$ exceeds $C(x)$ for $x > x_k$. However, the optimal decision about whether or not to control a pest is sensitive to attitudes to risk taking (and some other factors) and the density of the pest when $x_r \leq x \leq x_s$. In these circumstances, a highly risk-averse approach to decision making will result in a decision to control the pest. For example, if the minimax loss approach is adopted, it results in the pest manager deciding to undertake pest control. This decision prevents the largest possible reduction in profit, $V(x_s)$, occurring.

For a decision maker with the objective of minimizing the expected reduction in profit (i.e. the average loss in profit avoided) by deciding whether or not to engage in pest control, the situation is more complicated. In this case, the optimal decision is sensitive to the nature of the probability distribution of x and to the sign and size of the second derivative of $V(x)$. As the range of uncertain values of x increases (and if the expected value of x remains constant), this tends to increase the likelihood that pest control minimizes the expected loss in profit, if $V''(x) > 0$. This is known as Jensen's inequality (Jensen 1906). The opposite is the case if $V''(x) < 0$ (Hardy *et al.* 1934).

A simple illustration of this is provided in Figure 3.5. If the value of x is certain and equal to x_k, there is no net benefit to be had by engaging in pest control because $V(x_k) = C(x_k)$. However, assuming that x_r and x_s are equidistant from x_k and that each has a probability of 0.5, the expected net benefit from controlling the pest is equal to the distance EB. It pays to control the pest in this case. The further apart are x_r and x_s the greater is the net benefit to be obtained by controlling the pest. The opposite relationships occur if $V''(x) < 0$.

Taking another example, suppose that:

$$V(x) = ax \pm bx^2 \tag{3.6}$$

then, if E[V] represents the expected value of V(x):

$$E[V] = aE[x] \pm bE[x]^2 \pm b\text{var}(x) \tag{3.7}$$

In this expression, var(x) represents the variance of x and is a measure of the extent of uncertainty about its value. Hence, given E[x], the loss in profit if pest control is not undertaken will increase with b and with the value of var(x) if b is positive. Consequently, with E[x] constant, the likelihood that pest control is optimal rises with x. If b is negative, the opposite relationship occurs. In some cases, a quadratic function is a close approximation to V(x). Note that only the branches of parabolas in the positive quadrant of Cartesian space are relevant. In all these cases, given that $V(x) > 0$ for all x, E[V] increases with E[x], and var(x) remains constant. Consequently, it is also true that as the expected level of pest infestation rises (other things being held constant) and if the aim of the decision maker is to maximize his or her expected profit, the likelihood of pest control being optimal increases.

In general, when the economic penalties imposed by a pest infestation tend to escalate rapidly with that level of infestation, increased uncertainty about the level of infestation increases the likelihood that pest control is a farmer's superior economic choice compared to no control. Research on the likelihood of pest outbreaks and their probable magnitude (Guillemin *et al.* 2013; Izquierdo *et al.* 2013) reduces the level of uncertainty about these. Increased information about the magnitude of pest outbreaks not only improves the profit-maximizing decisions of farmers (because they are less likely to undertake pest control when they know that the level of pest infestation will be lower than they would have otherwise thought possible) but it also has social benefits if pest controls have negative environmental spillovers or health risks. Furthermore, the most economic control of some pests requires the collective gathering of information and in some instances, collective action for instance, by state bodies, for example, in the case of highly mobile pests such as locusts or *Heliothis* species.

3.4 Choice of Alternative Pest Control Techniques at Farm Level Assuming Certainty

Methods of controlling pests can be classified in several different ways. For example, this can be done according to:

- the means used to kill a pest or limit its population, e.g. destruction of the pest by hand, machinery, chemical pesticide use or biologically based controls
- in relevant cases, the method used to distribute the control agent
- according to the effectiveness of the method adopted for controlling the pest.

Consider situations in which the optimal choice of a method for distributing a control agent varies with the size of the area to be treated. This analysis enables the modelling considered in section 3.2.2 to be extended.

3.4.1 Cost Minimization

Assume that a control agent is to be applied to a crop, and suppose that no matter what distribution method is used, it is equally effective in controlling a pest. The least cost per ha method of applying it to the crop should be chosen in order to maximize the profitability of pest control. For example, the lowest cost per ha of treating a small area may be by hand but if a large area is to be treated, the least cost method per ha may be by the use of a tractor or, if the area is quite large, by a plane or drone.

The following indicates (for a simple case) how this matter can be analysed. Suppose that two techniques, I and II, are available for applying a control agent and that in each case, the cost per ha of applying it declines with the size of the area to be treated. Using technique II to treat a small area results in greater cost per ha than using technique I but the position is reversed when a larger area needs to be treated. This relationship may exist because using technique II results in higher overhead costs (fixed costs) than does using technique I but lower variable costs.

Figure 3.6 illustrates this choice problem. Let function $C_1(z)$ represent the cost per ha of the use of the control agent if technique I is adopted. The variable z indicates the size of the area to be treated. For example, the relationship $C_1(z)$ might be as shown by the curve KLM in Figure 3.6. Similarly, let $C_2(z)$ represent the cost per ha of controlling a pest when technique II is used. This is represented in Figure 3.6 by curve HLJ.

In the case illustrated in Figure 3.6, the cost per ha of controlling a pest is minimized when $z < z_h$ by adopting technique I and, if $z > z_h$, by adopting technique II. The lower boundaries of the cost curves shown (that is, their envelope), KLJ, designate the least cost

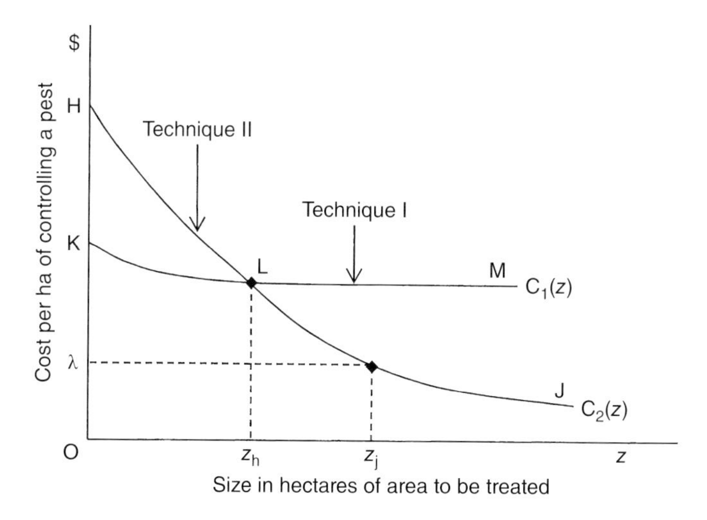

Figure 3.6 A case in which two optimal techniques of applying a control agent depend on the size of the area to be treated.

per ha application method of controlling a pest. Represent this relationship (envelope) by the function by C(z) in order to extend the analysis.

3.4.2 Extensions of Previous Threshold Models

The previous models in which the cost of controlling a pest were assumed to be independent of its density can be given greater generality because C(z) can be substituted for C in equation (3.1). In the illustration shown in Figure 3.1, the line EAH will be lower the larger z is. It follows, then, that the larger the area to be treated to control a pest, the lower is the level of infestation at which it is economic to control it. This is so, provided V(x) is monotonically increasing, for instance, provided V(x)' > 0 for all values of x. This relationship is, however, reversed if C(z) increases with z. It seems likely that economies of scale for controlling pests exist in many cases at farm level. Consequently, those with larger farms are more likely to find pest control more economical than smaller landholders. This implies that those with larger landholdings would be more likely to undertake pest control than those with smaller holdings as a matter of routine.

Further extension of this type of analysis is possible. For example, the optimal choice of technique to control a pest may in some cases depend on its density and the area to be treated. Then the cost minimizing technique depends on both x and z. Hence, equation (3.1) in this case becomes:

$$G = V(x) - C(x,z) \tag{3.8}$$

and the choice of the pest control technique which minimizes costs is sensitive to both x and z.

An additional important extension has to do with the effectiveness of alternative techniques in reducing the density of a pest. Often a pest is not entirely eliminated by a control method. Consequently, the following decision rules can be applied. Does the gain in gross economic return from using a particular technique exceed its costs, taking into account its effectiveness in reducing the density of the targeted pest? If yes, its use is profitable and otherwise not. If several alternative techniques are available, compare their additions to profit taking in relation to the factors just mentioned and select the one making the greatest addition to profit. Note that the economically optimal technique may not be the one resulting in the greatest reduction in pest density, because the private benefit-cost ratio may be highest for a technique which does not result in the maximum achievable pest reduction. The mathematical analysis of this can be formalized, but this will not be done here. It should, however, be kept in mind that private decisions about the choices of a pest control technique may not be socially optimal.

3.5 The Economics of the Timing of Pest Control and the Optimal Choice of Techniques Given Uncertainty

Uncertainty can influence the optimal choice of pest control techniques as well as the optimal timing of pest control. First, let us briefly consider some of the factors that may influence the timing of pest control and, subsequently, how uncertainty about the level of pest infestation can influence decisions to adopt the use of GM herbicide-resistant crops rather than the non-resistant ones, and about whether to use Bt-modified seeds rather than unmodified seeds.

3.5.1 Timing of Pest Control

The timing of pest controls often influences ecological effectiveness and the level of economic benefits (Keller *et al.* 2014). For some techniques, flexibility exists about the time at which pest control can be undertaken. If the likely level of the pest infestation is uncertain, delay will increase information about its distribution and density within the landscape. As a result of delaying a control, it may, for example, become clear that the level of the infestation is going to be too low to warrant pest control. However, the benefit of this information needs to be weighed against possible economic penalties from delaying the decision. For example, the longer the delay, the lower can be the yield of the crop because the pest may have already damaged the produce. Furthermore, applying a pest control at a later stage may add to application costs, damage to the crop or create problems if there is a withholding period before marketing. Therefore, the extra benefits from delaying the control of a pest need to be compared with any loss in the economic value of the crop caused by the delay and any extra cost involved in applying the control.

In the case of mung beans and other legumes, the control of insects needs to occur before the seed pod is compromised. While insects may only cause cosmetic damage to the seedpod, the weakening of this protective layer allows fungus, pathogens and moisture into the seedpod, ruining the grain.

3.5.2 Choice of GM Herbicide-resistant Crops versus Non-resistant Ones

The choice of planting a GM herbicide-resistant crop rather than a non-resistant one can be influenced by uncertainty about the density of weeds in the crop. Initially, a decision maker has to pay more to purchase the herbicide-resistant rather than a non-resistant seed. However, the economic benefit for farmers is that they gain flexibility in their decision-making process and now can use herbicides (if rational) without damaging the crop.

Assuming that the decision maker aims to maximize profit, a simple model can be used to highlight the value that the flexibility of herbicide-resistant GM crops offers to producers. This model assumes that both the GM and non-GM crop have an identical relationship between the loss in economic value and weed density, $V(x)$; GM seed is more expensive than non-modified seed, and that it is impossible to control the level of weed infestation in the non-GM crop once it is planted. In this case, the extra cost of GM seed is the price paid for increased flexibility of weed control. Once this cost is incurred, it is a sunk cost but it has an economic benefit because it keeps options open. If GM seed is planted, it allows the subsequent use of herbicides when it is economic to do so.

Figure 3.7 illustrates the economic outcome from choosing to adopt a GM herbicide-resistant crop or plant a conventional variety (i.e. non-GM). Although the benefit of weed control per ha is represented by the straight line OABD, the following argument is applicable provided that the function is upward sloping. OF represents the herbicide cost of producing a conventional crop and EF is the additional cost of using the GM

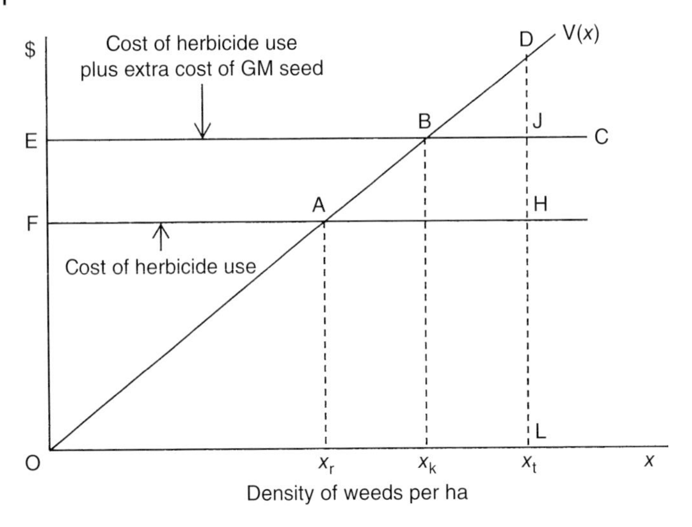

Figure 3.7 A diagram to illustrate the economics of choosing between a GM herbicide-resistant crop and one that is not.

seeds (e.g. the seed costs and the licence fee). In this case, if prior to planting all predicted values of x exceed x_k, it pays to plant GM seeds but if all predicted values are less than x_k, it is more profitable to plant non-modified seeds. It does not matter whether these values of x are uncertain or not. The weed density $x = x_k$ is the critical value in this case. If these predictions are correct, no spraying of herbicide will occur if $x < x_k$. However, if GM seeds are wrongly selected (for any reason), in which case it is not the most profitable choice, then spraying will occur at a lower weed density, namely at any density in the range $x_r \leq x \leq x_k$, because the extra cost of GM seeds is a sunk cost.

The use of GM herbicide-resistant crops may itself produce new weed problems (Kniss *et al.* 2011) as well as accelerating the evolution of herbicide-resistant weed species (Vencill *et al.* 2012), thus producing a range of externality issues. Various economic aspects of pest resistance to controls (including the use of GM crops) are discussed in Tisdell (2015).

If the predicted levels of weed density straddle x_k, the decision about whether it pays to buy GM seeds or not is more complicated. If $x < x_r$, the comparative loss in profit of GM seeds rather than non-GM is equal to the extra cost of GM seeds, EF. If $x_r < x < x_k$, this loss is partially offset by a net gain from herbicide use. If $x > x_k$, there is a net gain in avoided loss of profit. For example, if $x = x_t$, it is equivalent to DJ. If expected profit is to be maximized, net values of losses and benefits (times their probability) in these ranges should be computed and summed. If the result is positive, the decision maker maximizes expected profit by planting a GM crop. If it is negative, the decision maker would maximize profit by planting a traditional crop.

A producer who has an overriding desire for income security will favour planting a herbicide-resistant crop, unless all predicted values of $x < x_k$. If $x > x_k$ and a herbicide-resistant crop is planted, the minimum possible reduction in profit is C. If a non-herbicide resistant crop is planted, it is $V(x) > C$. So the maximum possible loss in profit is minimized when the herbicide-resistant crop is chosen. For example, if the highest predicted possible level of $x = x_t$, the loss in profit, if a non-herbicide resistant crop is planted, is

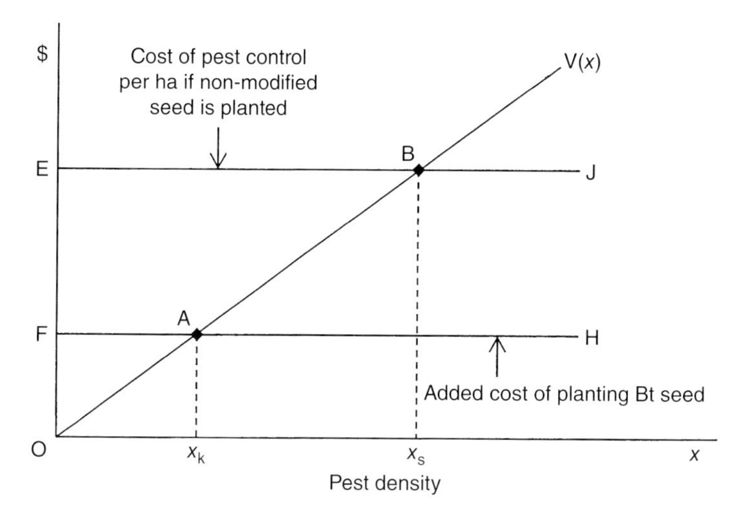

Figure 3.8 An illustration of the economics at farm level of choosing between Bt seeds and conventional seeds when a pest is to be managed. Planting of Bt seeds involves inescapable upfront costs once it is decided to plant it whereas the option of controlling the pest remains open if conventional seeds are planted. See text for additional explanations.

equal to an amount equivalent to the distance DL. However, the planting of a herbicide-resistant crop reduces this by an amount equivalent to the distance JL. Consequently, this will lower the loss in profit by an amount equivalent to the distance DJ.

3.5.3 Choice of Bt-modified Crops versus Non-modified Ones

Consider economic factors which can be expected to influence the choice of Bt crops versus non-modified crops. An important influence on the optimal economic decision will be the uncertainty about possible levels of pest infestations. These levels are influenced by environmental conditions. Figure 3.8 can be used to illustrate the relevant choice problem assuming that if a non-modified crop is planted, alternative means of killing the pest (such as spraying insecticide) are available. The adoption of the latter alternative is assumed to be more costly than control which is achieved by planting a Bt-modified crop. However, it permits greater flexibility in deciding whether controlling the pest is economic.

In Figure 3.8, the extra cost of relying on a Bt crop (such as extra cost of seeds or provision of a refuge crop) compared to planting the same conventional crop is indicated by OF. If the conventional crop is planted and pest control is undertaken, the cost of controlling the pest is assumed to be equal to OE. The economic benefit from controlling the pest is shown by the line OB. For simplicity, the alternative means of pest control are assumed to be equally effective in eliminating the pest. Therefore, in the case illustrated, if it is certain that $x < x_k$, it is less profitable to plant Bt-modified seeds rather than conventional seeds. If $x > x_k$, this relationship is reversed.

If uncertainty about the level of pest infestation exists, this complicates the optimal decision, unless of course all uncertain values of x are less than x_k or they are larger than x_k. Suppose that possible values of x may be less than x_k or greater than x_k. If the farmer places a very high emphasis on income security, the minimax loss rule may be adopted. In that case, if there is any possibility that $x > x_k$, the farmer should plant Bt seeds because this minimizes the largest possible reduction in profits, taking into account pest management options. However, if the (mathematically) expected value of the reduction in profit is to be minimized, the optimal choice is not so clearcut. Suppose that at the time when conventional pest control may be undertaken, the level of pest infestation will be known. Then the reduction in profit (as a result of pest management for $x < x_k$ for the planting of conventional seed) will be $V(x)$ where $x < x_k$. This is a smaller reduction than if Bt seeds are planted because then this loss is equal to $V(x_k)$. However, if $x > x_k$, the use of conventional seeds results in a greater reduction in profit than if Bt seeds are planted. In this case, the reduction in profit is $V(x)$ for $x_k < x < x_s$ and $V(x_s)$ for $x > x_s$ compared to $OF = V(x_k)$ if the Bt crop is planted. Consequently, the greater the skew of probable x values towards larger values of x above x_k, the more likely is the expected reduction in profit to be minimized by the adoption of a Bt crop. On the other hand, the more marked the skew in the opposite direction, the greater is the likelihood that the planting of conventional seeds will minimize the expected reduction in the profit of the farmer.

3.6 A Note on Biological Pest Control

It is worthwhile specifically mentioning biological means of managing pests. Mackauer *et al.* (1990) provide a useful introduction to this subject. Biological control usually refers to the release of living organisms to manage pests and the subject is normally divided into a consideration of classic and non-classic methods of control. However, there are also additional possible categories which are listed by van Driesche and Abell (2008). In the case of classic biological control, the agent, once released (at one or more sites), spreads of its own accord to target the pest, which is its host or prey. Its effect on the pest population is usually long term (if the control is a success) by reducing both the distribution and density of its host/prey.

In introducing an exotic organism to a region to control a pest, there is always the risk that it will also target beneficial organisms (see contributions in Mackauer *et al.* 1990, especially Harris 1990; see also Chapter 5, this volume). This is why decisions about such controls should not be left to individual farmers. Moreover, finding and selecting appropriate organisms to act as a means of classic pest control can be very costly and beyond the financial capacity of individual farmers. There is also effectively no market for this type of control and this means that government initiatives are usually required in this field. Economic aspects of classic biological control are summarized by Hoddle and van Driesche (2009) and Tisdell (1990). More recent studies include Letourneau *et al.* (2015).

On the other hand, sometimes non-classic biological means of managing some pests are available and can be one of the strategies chosen by individual farmers to control pests. For example, in some cases beneficial insects can be purchased and distributed for pest management or the Bt bacterium can be purchased, mixed with water and sprayed onto a crop.

Often non-classic forms of biological pest control are divided into inundative and augmentative (Tisdell *et al.* 1984), but this classification is not exhaustive and the dividing line between the two is not always clear. In the case of inundative control, the biological control organism is distributed in a heavy concentration but its population usually has a short life-span. Augmentative biological pest control involves the adoption of methods that supplement or support the population of biological control organisms already present.

The type of threshold models outlined in this chapter are relevant to considering whether several types of non-classic biological controls will be adopted by farmers in preference to alternative pest management strategies. In some cases, but not all, a price premium might be paid for agricultural produce which has been protected from pests by non-classic biological methods rather than by the use of chemical pesticides or GM crops. However, as stated by van Driesen and Abell (2008): 'Augmentative biological control may be either more or less expensive than other approaches depending on details such as the cost of natural enemy production by commercial insectaries that sell beneficial organisms, and efficacy of other control tactics.' The farmer's choice will in most cases be determined by the relative profitability of the available options. As a result, the social value of relying on non-classic biological pest management (when this is possible) can be expected to exceed its private value because this form of pest management is unlikely to have negative environmental and health consequences. However, that does not necessarily mean that it is always socially optimal to adopt this type of control. In some cases, society may find that the aggregate net economic benefits from this choice are less than for an alternative means of control. Individual cases have to be evaluated.

3.7 Discussion of the Modelling of the Economics of Pest Management at the Farm Level

The above models only cover a limited set of possible influences on the economics of pest control at the farm level. Other factors that can be relevant at the farm level include:

- the possibility that multiple pest control measures are needed serially or even simultaneously
- the speed with which a pest population recovers from a control
- the likelihood of immigration of a pest population occurring when its on-farm population is reduced. This may depend on the extent to which other farmers and agencies control the pest
- the likelihood that populations of secondary pests will increase if populations of primary pests are controlled
- the costs and benefits of alterations in the 'intensity' of the control agent or method
- the relevance of the evolution of pest resistance to controls.

From the above modelling, it is clear that farmers face many environmental uncertainties which affect the type of pest controls they adopt and whether or not they undertake pest control at all. Changing environmental conditions lead to uncertainty about the effectiveness of different types of pest control and to uncertainty about crop yields (Jones *et al.* 2006). Furthermore, uncertain environmental conditions influence the prices received for agricultural produce because there are major factors changing the (aggregate) market supplies of this produce. The pest control strategies adopted by farmers to respond to these uncertainties depend on their attitudes to risk and their economic returns.

The farm-level models outlined above assume that the basic objective of farmers is to maximize their profit. Under conditions of uncertainty, this narrows the range of pest control measures which it is rational to adopt, and in some circumstances (as was demonstrated using the above models), the presence of uncertainty is irrelevant for making optimal choices. However, in 'straddle-type' cases (which may be common), this is not so. In these cases, the attitudes of individual farmers to risk need to be taken into account and usually attention needs to be paid to the nature of the probability distribution of relevant uncertain events. In these circumstances (but not in all, as was shown), increased risk aversion tends to increase the likelihood of pest control measures being adopted, and probably favours pest control techniques which show greater reliability in controlling pests than other methods. However, the decision-making process is made more complicated by the need to take account of the flexibility which different techniques allow in responding to changes in environmental conditions which, among other things, includes changes in estimates of likely levels of pest infestation. Although an economic premium is usually placed on flexibility (for example, if with the passage of time, knowledge improves about the variables which influence the profitability of pest controls), there are often extra costs associated with adopting techniques that permit greater flexibility in pest control as relevant conditions change. The economics of flexibility in decision making generally is discussed and modelled in Tisdell (1968, 1996) (see also Chapter 16). The costs and benefits therefore need to be compared, as was demonstrated by considering two types of GM-based strategies for pest control, namely the planting of herbicide-resistant crops (which permit significant flexibility in weed control) and the growing of Bt-modified crops which result in less flexibility in pest control than possible alternatives.

Genetically modified crops can provide additional benefit for producers, including a reduction in time allocated to crop management, transaction costs associated with regulations designed to internalize externalities, improving relationships with neighbours from using fewer chemical pesticides and a reduction in stress associated with worry about crop management (Back and Beasley 2007). However, the planting of GM crops can also add to social conflict, for example, between growers of GM crops and GM-free crops.

As mentioned above, economic behaviour depends on motives. In some cases, farmers may aim for a satisfactory level of profit rather than for profit maximization. This is liable to alter their choice of pest control strategies (Doohan *et al.* 2010). These choices can be quite different to those based on profit maximization. For example, suppose that a farmer seeks a particular level of profit and no more. Then, if the price of a relevant product rises, this increases the likelihood that a profit-maximizing farmer will undertake pest control but it reduces the likelihood that the profit-satisficing farmer will do so. The latter case is believed to occur in some less developed countries (LDCs).

Note also that the most economic choice of pest control techniques is liable to differ between countries. In LDCs, where labour is abundant and capital is scarce, labour-intensive pest control techniques are likely to be more economic than in developed countries. Lack of availability of finance for smallholders may further reinforce this effect. It is also possible that in some LDCs, more weight will be placed on a greater quantity of food supplies than on the negative environmental and health effects associated with the use of some pesticides.

3.8 Concluding Comments

Most references in this chapter have been to the economics of control of pests in crops. Nevertheless, the economic threshold models outlined in this chapter can also be applied in several situations involving the control of pests in livestock (Davis and Tisdell 2002).

Pest management situations are very diverse in relation to the type of pests to be controlled, the various techniques available for their management and prevailing economic and environmental conditions. Moreover, several techniques may be employed in an integrated pest management approach to individual or multiple problems serially or simultaneously (Harker and O'Donovan 2012). In addition, pest management is frequently the source of social conflict, subject to communal constraints, and is further complicated by uncertainties. Consequently, a variety of economic and ecological models are needed to effectively analyse the optimality of decisions about pest control. It has only been possible to introduce a few of these in this chapter. In later chapters, attention will be given to several pest management issues which involve market failure (see Part VI). These include the importance of various types of environmental externalities or spillovers and the consequences of pest control for the supply of public (non-marketed) goods, for example, the conservation of wildlife. Another issue considered is the degree of consumer awareness about the extent to which their purchases have been subjected to pest controls and their consequences, for instance, for human health. Economic analyses have been developed that do take some account of these issues.

A major constraint on economically optimal decision making is controlling pests in the bounded rationality (Tisdell 1996) of all parties with an interest in it. For example, farmers often have limited knowledge about the effects and cost–benefits ratio of alternative methods of pest control. They are therefore likely to be heavily influenced in their decisions by information provided by suppliers of saleable pesticides and pest control products. This information naturally tends to be one-sided. Some studies in China revealed that farmers were quite ignorant about the economic benefits of the pest controls which they had adopted (Zhao *et al.* 2011).

References

Auld, B.A. and Tisdell, C.A. (1986) Economic threshold/critical density models in weed control. *Economic Weed Control, Proceedings, EWRS Symposium*, pp. 261–268.

Auld, B.A. and Tisdell, C.A. (1988) Influence of spatial distribution on weeds on crop yield loss. *Plant Protection Quarterly* **3**: 81.

Back, W. and Beasley, S. (2007) Case study analysis of the benefits of genetically modified cotton. In: O'Reilly, S., Keane, M. and Enright, P. (eds) *Proceedings of 16th International Farm Management Association Congress*. University College Cork, Cork, Ireland, pp. 247–266.

Brown, G.C. (1997) Simple models of natural enemy action and economic thresholds. *American Entomologist* **43**: 117–124.

Carlson, G.A. (1970) A decision theoretic approach to crop disease prediction and control. *American Journal of Agricultural Economics* **52**: 216–223.

Carlson, T.P., Webster, E.P., Salassi, M.E., Bond, J.A. Hensley, J.B. and Blouin, D.C. (2011) Economic evaluations of *imazethapyr* rates and timings on rice. *Weed Technology*, **26**: 24–28.

Davis, R. and Tisdell, C.A. (2002) Alternative specifications and extensions of the economic threshold concept and the control of livestock pests. In: Hall, D.C. and Moffitt, L.J. (eds) *Economics of Pesticides, Sustainable Food Production and Organic Food Markets.* Elsevier Science, Oxford, UK, pp. 55–79.

Doohan, D., Wilson, R. Canales, E. and Parker, J. (2010) Investigating the human dimension of weed management: new tools of the trade. *Weed Science*, **58**: 503–510.

Falkenberg, N.R., Cogdill, T.J. Rister, M.E. and Chandler, J.M. (2012) Economic evaluation of common sunflower (*Helianthus annuus*) competition in field corn. *Weed Technology* **26**: 137–144.

Guillemin, J.P., Gardarin, A., Granger, S., Reibel, C., Munier-Jolain, N. and Colbach, N. (2013) Assessing potential germination period of weeds with base temperatures and base water potentials. *Weed Research* **53**: 76–87.

Hardy, G.H., Littlewood, J.E. and Polya, G. (1934) *Inequalities.* Cambridge University Press, Cambridge, UK.

Harker, K.N. and O'Donovan, J.T. (2012) Recent weed control, weed management, and integrated weed management. *Weed Technology* **27**: 1–11.

Harris, P. (1990) Environmental impact of introduced biological control agents. In: Mackauer, M., Ehler, L.E. and Roland, J. (eds) *Critical Issues in Biological Control.* Intercept, Andover, UK, pp. 289–300.

Headley, J.C. (1968) Estimating the productivity of agricultural pesticides. *American Journal of Agricultural Economics* **50**: 13–23.

Headley, J.C. (1972) Defining the economic threshold. In: National Research Council (ed.) *Pest Control Strategies for the Future.* National Academy of Sciences, Washington, DC, USA, pp. 100–108.

Hoddle, M.S. and van Driesche, R.G. (2009) Biological control of insect pests. In: Cardé, R.T. and Resch, V.H. (eds) *Encyclopedia of Insects*, 2nd edn. Academic Press, Burlington, MA, USA, pp. 91–101.

Izquierdo, J., Bastida, F., Lezaún, J.M., Sánchez del Arco, M.J. and Gonzalez-Andujar, J.L. (2013) Development and evaluation of a model for predicting *Lolium rigidum* emergence in winter cereal crops in the Mediterranean area. *Weed Research*, **53**: 269–278.

Jensen, J.L.W.V. (1906) Sur les fonctions convexes et les inégalités entre valeurs moyennes. *Acta Mathematica* **30**: 175–193.

Jones, R., Cacho, O. and Sinden, J. (2006) The importance of seasonal variability and tactical responses to risk on estimating the economic benefits of integrated weed management. *Agricultural Economics* **35**: 245–256.

Keller, M., Gantoli, G., Möhring, J., Gutjahr, C., Gerhards, R. and Rueda-Ayala, V. (2014) Integrating economics in the critical period for weed control concept in corn. *Weed Science* **62**: 608–618.

Kniss, A.R., Sbatella, G.M. and Wilson, R.G. (2011) Volunteer glyphosate-resistant corn interference and control in glyphosate-resistant sugarbeet. *Weed Technology* **26**: 348–355.

Letourneau, D.K., Ando, A.W., Jedlicka, J.A., Narwani, A. and Barbier, E. (2015) Simple-but-sound methods for estimating the value of changes in biodiversity for biological pest control in agriculture. *Ecological Economics* **120**: 215–255.
Liu, L., Griffin, T. and Kirkpatrick, T.L. (2014) Statistical and economic techniques for site-specific nematode management. *Journal of Nematology* **46**: 12–17.
Mackauer, M., Ehler, L.E. and Roland, J. (1990) *Critical Issues in Biological Control.* Intercept, Andover, UK.
Myers, M.W., Curran, W.S., Vangessel, M.J., *et al.* (2005) The effect of weed density and application timing on weed control and corn grain yield. *Weed Technology* **19**: 102–107.
Naranjo, S.E., Ellsworth, P.C. and Friswold, G.B. (2015) Economic value of biological control in integrated pest management of managed plant systems. *Annual Review of Entomology* **16**: 621–645.
Paula-Moraes, S., Hunt, T.E., Wright, R.J., Hein, G.L. and Blankenship, E.E. (2013) Western bean cutworm survival and the development of economic injury levels and economic thresholds in field corn. *Journal of Economic Entomology* **106**: 1274–1285.
Peterson, R.K.D. and Hunt, T.E. (2003) The probabilistic economic injury level: incorporating uncertainty into pest management decision-making. *Journal of Economic Entomology* **96**: 536–542.
Stern, V.M. (1966) Significance of the economic threshold in Integrated Pest Control. In: Food and Agriculture Organisation of the United Nations (ed.) *Proceedings of FAO Symposium on Integrated Pest Control*, pp. 41–56.
Sutherland, A.M. and Parrella, M.P. (2011) Accuracy, precision, and economic efficiency for three methods of thrips (Thysanoptera: Thripidae) population density assessment. *Journal of Economic Entomology* **104**: 1323–1328.
Tisdell, C.A. (1968) *Price Uncertainty, Production and Profit.* Princeton University Press, Princeton, NJ, USA.
Tisdell, C.A. (1982) *Wild Pigs: Environmental Pest or Economic Resource?* Pergamon Press, Sydney, Australia.
Tisdell, C.A. (1990) Economic impact of biological control of weeds and insects. In: Mackauer, M., Ehler, L.E. and Roland, J. (eds) *Critical Issues in Biological Control.* Intercept, Andover, UK, pp. 301–316.
Tisdell, C.A. (1996) *Bounded Rationality and Economic Evolution.* Edward Elgar, Cheltenham, UK.
Tisdell, C.A. (2015) *Sustaining Biodiversity and Ecosystem Functions: Economic Issues.* Edward Elgar, Cheltenham, UK.
Tisdell, C.A., Auld, B.A. and Menz, K.M. (1984) On assessing the value of the biological control of weeds. *Protection Ecology* **6**: 169–179.
Trezzi, M.M., Vidal, R.A., Patel, F., *et al.* (2015) Impact of *Conyza bonariensis* density and establishment period on soyabean grain yield, yield components and economic threshold. *Weed Research* **55**: 34–41.
Van Driesche, R.G. and Abell, K. (2008) Classical and augmentative biological control. In: Jorgensen, S.E. and Fath, B. (eds) *Encyclopedia of Ecology*. Elsevier, Amsterdam, The Netherlands, pp. 578–582.
Vencill, W.K., Nichols, R.L., Webster, T.M., *et al.* (2012) Herbicide resistance: toward an understanding of resistance development and the impact of herbicide-resistant crops. *Weed Science* **60 (sp1)**: 2–30.

Yokomizo, H., Possingham, H.P., Thomas, M.B. and Buckley, Y.M. (2009) Managing the impact of invasive species: the value of knowing the density–impact curve. *Ecological Applications* **19**: 376–386.
Zhao, J.H., Ho, P. and Azadi, H. (2011) Benefits of Bt cotton counterbalanced by secondary pests? Perceptions of ecological change in China. *Environmental Monitoring and Assessment* **173**: 985–994.

Part II

Impact of Pest Management Practices on the Environment

4

Effects of Chemical Control on the Environment

Francisco Sánchez-Bayo

4.1 Introduction

Since Neolithic times, humanity has had to deal with the recurrent problem of pests, weeds and diseases in agriculture. Numerous devices and ingenious solutions have been used to overcome these problems and so reduce crop losses. Among them are organic and inorganic chemicals, whether natural compounds or artificial products, which can be very effective weapons in the fight against the causal agents that diminish our agricultural productivity. From the natural pyrethrum and nicotine extracts to their current artificial derivatives, such as pyrethroids and neonicotinoids, a large array of man-made chemical compounds have been used to tackle these issues. It is interesting to note that the vast majority of them have been produced in the last 70 years. Large-scale use of chemicals in agriculture started in the 1940s, and was fostered by the implementation of the 'Green Revolution' in most developed countries during the 1950s and 1960s. Currently, global pesticide usage is estimated at 4 million tons per year (Sánchez-Bayo 2011), and is increasing in developing countries of Africa, Asia and South America. The majority of pesticides are used in farming and storage of farm products.

The beneficial effects of agricultural chemicals were estimated by Pimentel *et al.* (1993) as 37% increases in yield due to a reduction in combined losses by insect pests (13%), competing weeds (12%) and diseases (12%). Pests and diseases cause such losses not only in the crop fields but also during storage of grain and other produce, particularly in tropical countries where humid conditions tend to spoil the harvested crop. This chapter, however, does not deal with the efficiency of this method of crop protection, but rather with its side effects on the environment. Unlike other methods of pest and weed control, chemicals kill indiscriminately the bad and the good, thus causing a series of problems for both the crops and the environment at large.

4.2 Pesticides in Agriculture

4.2.1 When Should Pesticides be Applied?

Awareness of the potential problems that pesticides can bring matters, because only thus can we minimize their use to strictly necessary circumstances, that is, as a final weapon in the fight against pests, weeds and diseases. By pests, we refer mostly to

Environmental Pest Management: Challenges for Agronomists, Ecologists, Economists and Policymakers, First Edition. Edited by Moshe Coll and Eric Wajnberg.

insects, but mammals (e.g. rodents) and birds are also included. In this context, the principles of Integrated Pest Management (IPM), derived from dozens of years of field experiments and scientific research (Baur *et al.* 2011), provide reliable guidance to agronomists and farmers, as discussed in other chapters (see Chapter 2). All too often, these common-sense principles are not heeded, despite being made compulsory in the European Union by Directive 2009/128/CE. As a result, pesticides can be used unnecessarily or misused to the extent of causing more harm than good. For example, the recent trend of applying systemic insecticides such as neonicotinoids and fipronil as seed dressings in the prophylactic treatment of maize, sunflower, rape and cotton crops is against IPM principles because these insecticides are applied regardless of whether the target pests (e.g. aphids or borers) pose a problem or not. There is also evidence that crop yields do not increase significantly under these treatments, while the cost of the chemicals counterbalances any yield benefits, and so their use is not justified (Furlan and Kreutzweiser 2015). Moreover, the environmental harm of seed dressing includes:

- elimination of beneficial predatory insects and other non-target organisms (He *et al.* 2012)
- contamination of pollen and nectar that affects pollinators (Krupke and Long 2015)
- build-up of residues in soil, which eventually causes contamination of surface and ground waters (Hladik *et al.* 2014)
- fostering resistance in the target pests by continuously applying the same products (Alyokhin *et al.* 2007).

Current practices in IPM tend to utilize selective chemical insecticides only for management of recalcitrant and resistant insect pests in combination with biocontrol agents. But chemicals must be scrutinized first for their effects on non-target organisms, particularly natural enemies and pollinators, so as to avoid unnecessary risks to the environment.

4.2.2 Fate and Transport of Pesticides

Pesticides are applied in agriculture in various ways, depending on the product formulation and the target they control. Typically, the concentrated active ingredients are diluted in water and/or surfactants before application. Aqueous or oily solutions of insecticides are then sprayed directly on to the crop plants using aeroplanes, ground-rig machinery or manual sprayers. The tiny droplets of insecticide fall mostly on the leaves and other parts of the plants but a fraction, which could be large or small depending on the type of crop and its stage of development, falls on the soil beneath or around the field edges. Herbicide and fungicide solutions are commonly sprayed onto the soil before a crop is planted, around the trees of orchards, groves and vineyards, or even when the crop is at the early stages of development. A substantial amount of the applied solution goes into the soil profile through the pores and cracks of the surface. Whatever pesticides are sprayed, some 15% of the airborne droplets are carried away by the wind and, depending on weather conditions, may drift even kilometres from the target area (Woods *et al.* 2001).

Many herbicide formulations, as well as some insecticides and biocides, are granular, in which case they are incorporated into soil using planters. Seedlings of rice and other plants grown in nurseries may be treated with granules a day or two before they are transplanted. Granular formulations allow a slow but steady release of pesticide active ingredient into the soil, being very effective in controlling weeds, grubs, nematodes and other nuisance pests that have life stages in the soil. Unfortunately, granivorous birds and rodents tend to consume these granules scattered around the treated fields, often with disastrous consequences (Wilson *et al.* 2002).

Seed dressing, the coating of seeds with pesticides, is commonly used in crops of large acreage such as maize, cotton, rape and sunflower. Most fungicides and systemic insecticides are applied this way, usually together, so that from germination to flowering, the pesticides are taken up by the roots and distributed to the entire plant tissues. When fungicides are applied this way, they protect the seedling against pathogenic fungi from the beginning of its life whereas, as mentioned above, seeds coated with systemic insecticides are not justified on the grounds of IPM. Moreover, prophylactic treatments foster pest or fungal resistance, while the soil acts as a constant reservoir of chemical residues that contaminate surface and ground waters. As with granular formulations, seeds coated with pesticides also fall by the side of the fields, where they pose a lethal risk to birds and other animals (Greig-Smith 1987).

Once applied, residues in the soil will spread around the environment through various mechanisms of dispersion: attached to wind-blown dust particles that are carried far away from the fields, in run-off after storm and irrigation events, which discharge soluble residues and sediment particles into drains, streams, ponds, rivers and eventually the oceans, or through percolation in the soil profile, thus contaminating groundwater aquifers (Figure 4.1). Volatile pesticides will also evaporate from the plant and soil surfaces on which they fell, or directly from the airborne droplets, dissipating and

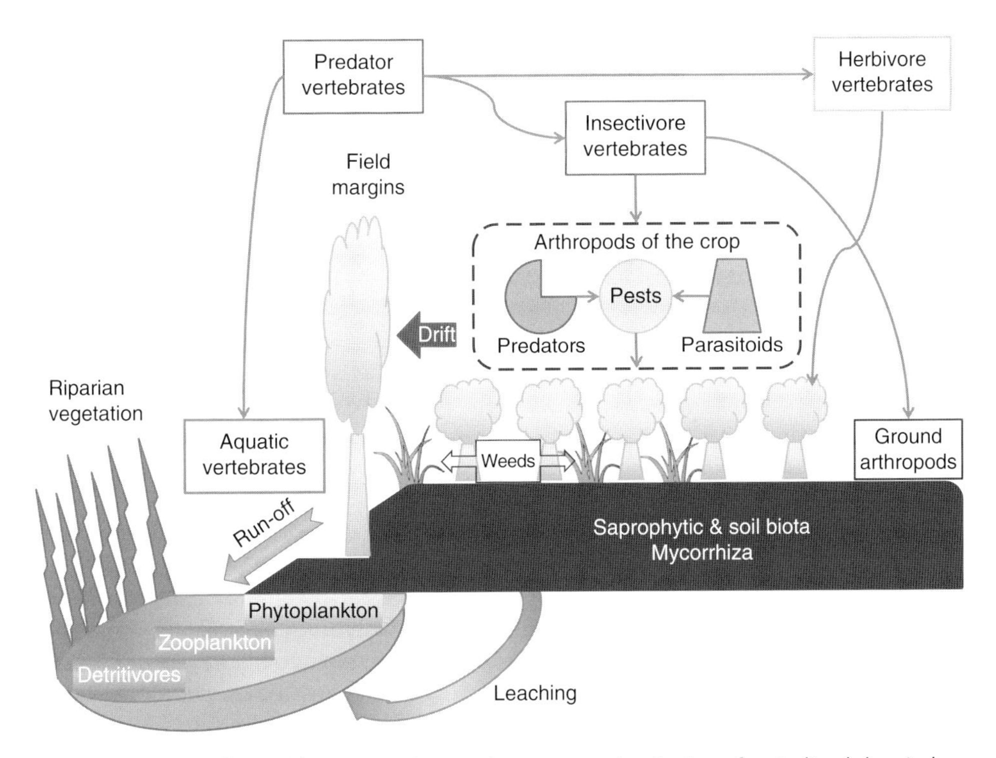

Figure 4.1 Impact of pesticides on organisms and ecosystems. Application of agricultural chemicals affects all animals and plants in and around the crop, including soil biota. Aquatic organisms may also be affected by residues in run-off and ground water.

contaminating the surrounding air with their vapours. Most organochlorines are notorious for that – they are included among the persistent organic pollutants (POPs), and have been found hundreds and even thousands of kilometres away from the applied fields, in the Himalayas (Gong *et al.* 2014), the mountains of Costa Rica (Daly *et al.* 2007) and even the polar caps (Muir *et al.* 1988), where they undergo biomagnification (i.e. residue levels in the predators are much higher than in their prey) through the marine and terrestrial food chain (Goerke *et al.* 2004). These are now banned for use in agriculture, but are still illegally used in some developing countries (Yadav *et al.* 2015).

Flowable herbicide formulations can also be poured into irrigation channels, either for clearing the aquatic plants that may obstruct the ditches or for distribution to the fields prior to planting. Such channels are hazardous to a diverse fauna of invertebrates (Suárez-Serrano *et al.* 2010) as well as vertebrates that may live there or drink from the contaminated waters (Venturino *et al.* 2007).

4.2.3 The Problem of Persistence

Ideally, once a pesticide has done its job, it should disappear from the environment. Modern insecticides tend to break down within a few days, thus minimizing the impacts on non-target organisms and the environment. Foliar applications of insecticides are quite effective, as populations of the target pest are reduced below the non-injury levels in a matter of hours or a few days. Pesticides that are applied to the soil (e.g. many herbicides and fungicides) may take longer to be effective because they must reach the target after being taken up from the soil. This means the chemical must be stable in soil for a while, or else it would be ineffective. Such pesticides are usually persistent and stay in the field for months or even longer – the so-called residual pesticides. A pesticide is considered to be persistent when its field half-life ($t_{1/2}$), that is, the time for a substance to be reduced by half, is longer than 90 days, as more than 5% of the original amount of chemical applied will remain in the field after 1 year.

Persistent pesticide residues can build up in the environment year after year, thus causing problems to organisms. Examples of persistent compounds in soil are the insecticides bifenthrin ($t_{1/2}$ = 136 days), clothianidin ($t_{1/2}$ = 160 days) and fipronil ($t_{1/2}$ = 190 days), the herbicides diuron ($t_{1/2}$ = 126 days), pendimethalin ($t_{1/2}$ = 107 days) and monosodium methyl-arsonate (MSMA; $t_{1/2}$ = 200 days), and the fungicides azoxystrobin ($t_{1/2}$ = 190 days) and fludioxonil ($t_{1/2}$ = 130 days).

Microbial and fungal degradation is one of the most effective natural processes that remove pesticide residues from the soil. Given the right conditions, these microbes and fungi can metabolize most organic pesticides, including recalcitrant chlorinated compounds (Maule *et al.* 1987), especially in warm and humid climates.

Pesticide residues will accumulate in plant and animal tissues only if the degradation rate of a chemical is slower than its rate of uptake. If persistent, such residues may be transferred through the food chain, as contaminated plants or animals are eaten by predators (and also parasitoids) that stand higher in the food pyramid, causing biomagnification. Persistent organochlorine pesticides such as dichloro-diphenyl-trichloroethane (DDT), heptachlor and others, that are nowadays banned in agriculture, accumulate in the fatty tissues of all organisms and reach their highest concentrations in raptors like the peregrine falcon (*Falco peregrinus*), eagles, fish-eating birds (Henny *et al.* 2009) and mammal predators, in particular cetaceans (Tanabe *et al.* 1994), and also humans (Ogbeide *et al.* 2015) (see also Chapter 11).

4.3 Impacts of Pesticides on the Environment

4.3.1 Insecticides, Acaricides and Rodenticides

Chemicals in this group can be very toxic to animals because many of the compounds are neurotoxic or act upon the cellular respiration system, which is common to all animals – the so-called broad-spectrum insecticides. Vertebrates usually require larger doses of these chemicals to reach lethality than insects and other invertebrates. In general, vertebrates are quite tolerant of synthetic pyrethroids (e.g. cypermethrin, fenvalerate), neonicotinoids (e.g. imidacloprid, thiamethoxam) and organochlorine insecticides (e.g. lindane), but are very susceptible to cholinesterase inhibitors (organophosphorus such as diazinon or phosmet, and carbamates such as carbaryl or pirimicarb), in particular birds and small mammals such as shrews, moles and rodents, due to their high feeding and metabolic rates.

Among vertebrates, birds are the most susceptible to neurotoxic compounds because of their fast metabolism and deficient detoxification system. Mammals are more tolerant of certain pesticide groups than other taxa because they possess active detoxification mechanisms. Reptiles appear to have either less or similar sensitivities to mammals in regard to neurotoxic compounds. Amphibians are generally very sensitive to pyrethroids but more tolerant of cholinesterase inhibitors than birds and mammals. Fish, crustaceans and many aquatic invertebrate taxa are very sensitive to all kinds of insecticides and acaricides because they lack the enzymatic mechanisms to detoxify complex molecules, while their mono-oxygenase system (i.e. cytochrome P450) is primitive and inefficient (Walker *et al.* 2001). Rodenticides are poisonous to mammals and birds (Christensen *et al.* 2012), most of them acting as inhibitors of vitamin K, which results in internal bleeding and haemorrhaging.

4.3.1.1 Direct Effects

These are toxic effects caused by the specific mode of action of the chemical, which kills the target pest and some non-target organisms as well. Most beneficial insects undergo large losses when crops are sprayed with organochlorine, organophosphorus, carbamate and pyrethroid insecticides, as these neurotoxic compounds are deadly to all insects and mites (Figure 4.2a). Populations of ladybirds, earwigs, lacewings, mantis, spiders, predatory mites and parasitoids, for example, crash immediately after a single insecticide spray, but this effect is temporary as they recover in the weeks following the sprays, usually at a slower pace than the recovery of the target pests they feed on. Hymenoptera parasitoids appear to be particularly sensitive to insecticide treatments: a review of 39 ecosystems found that agro-chemicals negatively affect these parasitoids in 46% of cases (Butler *et al.* 2009).

Of course, the overall diversity and abundance of insect communities may change after repeated application of insecticides (Theiling and Croft 1988), and this often results in unbalanced ecosystems that may do more harm than good to the treated crops because they prompt pest resurgence. Typically, problems caused by these chemicals are due to a combination of two factors: the lack of predators and insecticide resistance developed within the pest species. For example, insecticide sprays on rice crops of Indonesia and the Philippines eliminated the predators of brown plant hoppers (*Nilaparvata lugens*) and created a heavy selection pressure for

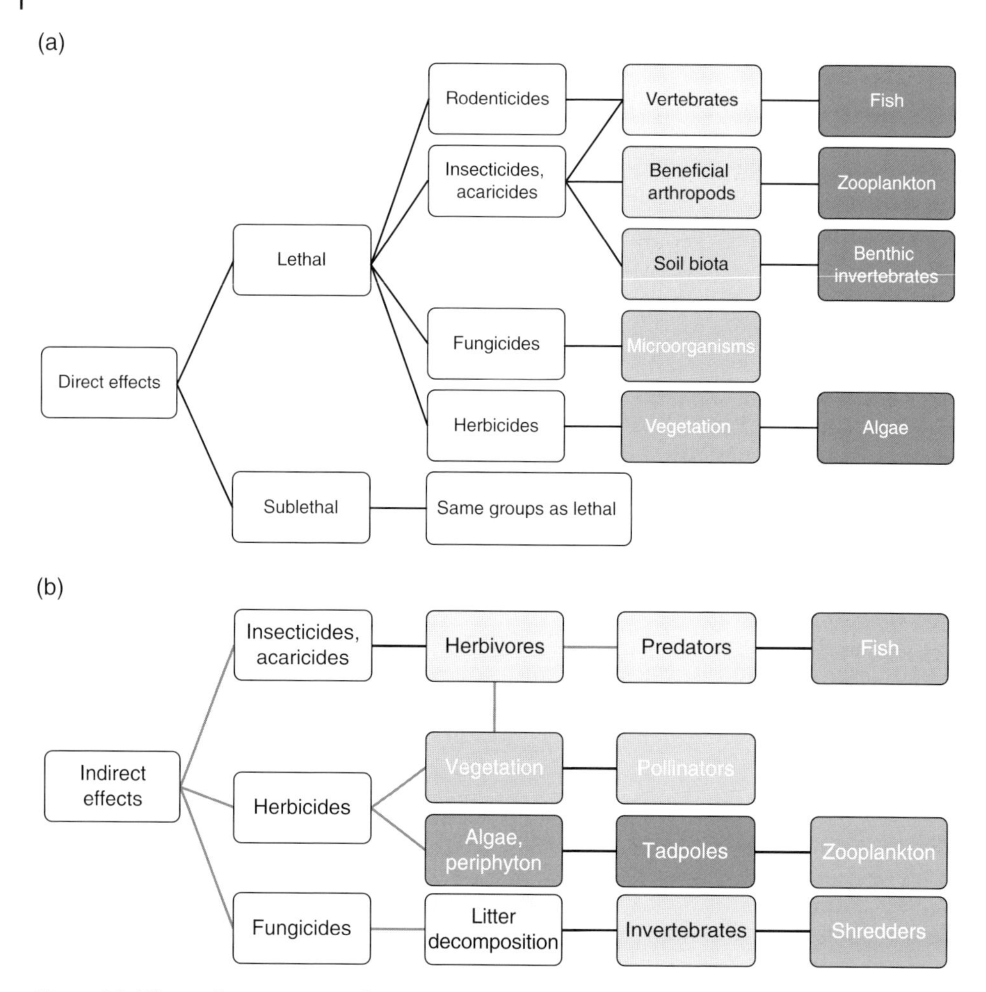

Figure 4.2 Effects of various types of pesticides on organisms. (a) Direct effects. (b) Indirect effects. Different grey tones refer to different groups of organisms.

resistant strains of this pest until entire crops collapsed in the early 1970s (Daryanto 1998). Impaired biological control led to outbreaks of secondary rice pests such as stem borers, while applications of gamma-BHC (lindane), parathion, imidacloprid and fipronil significantly decreased the dragonflies, spiders and parasitoids that control other pests of rice (Way and Heong 1994). The ban on DDT and the implementation of IPM practices allowed the return of those communities in subsequent years in the Philippines, Indonesia and other countries (see more examples in Brown 1978, Sánchez-Bayo 2011).

Droplets of insecticide drift may fall on flying insects and birds that happen to be nearby at the time of application. Insects such as bees would be killed almost immediately, while birds, which are particularly sensitive to cholinesterase inhibitor insecticides, would be negatively affected as well. The collateral damage on insects due to spray drift is impossible to quantify, whereas the death toll of birds around treated fields can be estimated to some extent but never accurately, as many casualties will pass unnoticed.

The crops responsible for most potential bird mortality in the USA are corn and cotton, followed more distantly by alfalfa, wheat, potato, peanut, sugar beet, sorghum, tobacco and citrus. Risks can be as high as 50–75% for some organophosphorus insecticides (Mineau and Whiteside 2006). Equally, a significant proportion of bees exposed to neonicotinoid dust arising from the planting of corn seeds usually do not return to their hive, as they are caught in the drift of particles and die on the spot (Tremolada *et al.* 2010). Butterflies and other pollinators may suffer similar losses (Hahn *et al.* 2015).

A diverse community of arthropods and other soil biota suffer as a result of chemical applications early in the cropping season, when a large proportion of the spray falls on the soil. Because of their tiny size, predatory and saprophytic mites are especially sensitive to all insecticides and fumigants (Koehler 1992), and their populations can take years to recover. Springtails are also very susceptible to fumigants, carbamates and many organophosphorus insecticides. Elimination of springtails slows down the recycling of nutrients in agricultural fields. For example, applications of 0.5 kg ha^{-1} lindane to corn crops in Africa reduced springtail numbers by 80% and consequently reduced the breakdown of organic matter by 45% (Wiktelius *et al.* 1999). Residues of organochlorines, still present in soil of many countries, reduce the abundance of most species of springtails (Collembola), saprophagous mites, symphylids and pauropods (Myriapoda). They also kill a higher proportion of predatory mites than other insecticides, but had little or no effect on earthworms, enchytraeid worms and nematodes (Edwards and Thompson 1973). Among Myriapoda, pauropods seem to be most susceptible to all kinds of insecticides, and some populations are completely eliminated by organophosphorus compounds. Millipedes are more tolerant, and so are symphylids, which do not suffer as much because they feed on plant rootlets deep in the soil layers.

Larvae of beetles, cicadas and flies play an important role in breaking down dead plant or animal matter, so the repeated application of insecticides leads to significant losses of insect larvae and potential accumulation of organic material in the soil. In this regard, systemic and persistent neonicotinoids such as imidacloprid have the highest toll on non-target insects when applied for control of white grubs (Scarabeidae) in turfgrass (Peck 2009).

Vertebrate poisoning by ingesting granular insecticides is very common: fatal incidents have been reported for a variety of crops, with songbirds, quails, waterfowl and voles being the typical casualties. Average killing rates have been measured as 1.0–1.5 birds ha^{-1} in rapeseed crops in Canada (James 1995). Seeds coated with systemic neonicotinoids or fipronil cause similar problems: a single grain of maize contains enough insecticide to kill a small bird (Mineau and Palmer 2013), and red-legged partridges (*Alectoris rufa*) feeding on imidacloprid-coated grain at recommended rates for cereals died in less than a month (López-Antia *et al.* 2015). Most of the mortality is due to direct or primary poisoning of the individual animals but quite often, especially among raptors, is due to secondary poisoning, that is by ingesting prey that was already contaminated with toxic residues.

Secondary pesticide poisoning is a regular feature among wildlife that feed on earthworms, grubs, locusts and other invertebrates that contain residues of cholinesterase inhibitor insecticides (Story *et al.* 2013), as well as rodents killed by rodenticides. For example, secondary poisoning killed some bald eagles (*Haliaetus leucocephalus*) and red-tailed hawks (*Buteo jamaicensis*) in Canada after scavenging on dead waterfowl that had eaten granular phorate, an organophosphorus insecticide, used a few months earlier for controlling root grubs in potato fields (Elliott *et al.* 1997). Aldicarb, carbofuran, fonofos, terbufos and phorate are the main insecticides involved in this kind of fatal incident, but modern systemic neonicotinoids and fipronil are now taking their place (López-Antia *et al.* 2015). Among the rodenticides, the second generation of anticoagulant coumarins are very persistent and their residues, ingested with the carcasses of rodents, accumulate in the predators' bodies, causing internal and external bleeding until they die. Some 70% of the owls collected in Canada between 1988 and 2003 had residues of at least one rodenticide at levels up to 0.93 $mg\,kg^{-1}$ (brodifacoum) or 1.01 $mg\,kg^{-1}$ (bromadiolone) in their liver (Albert *et al.* 2010).

Insecticide and acaricide residues in soil of the treated crops are eventually removed from the field by run-off water, particularly during storm events or by regular field irrigation. In addition, pesticides adhered to the plant leaves and other surfaces end up in the soil due to wash-off during rainfall events. Water erosion removes soil particles containing all those residues and the contaminated run-off is eventually discharged into surface waters. For lipophilic pesticides, some 0.5–2% of the applied amounts are lost in run-off (Wauchope 1978). In agricultural regions, storms and irrigation events produce pulses of residues that are highly concentrated for a few hours until further dilution reduces their levels. Zooplankton and epibenthic crustaceans are very susceptible to insecticides and most species disappear during those pulses, as residue concentrations are typically above their toxic thresholds, with recolonization occurring only after the residue levels drop (van den Brink *et al.* 1996). Ostracods are drastically reduced or disappear from rice fields treated with carbufuran, endosulfan, imidacloprid or fipronil. In streams of the Argentine pampa, water contaminated with chlorpyrifos, endosulfan and alpha-cypermethrin from agricultural fields produced high mortality of the macrocrustaceans *Hyalella curvispina* and *Macrobrachium borelli* during peak pulses, but mortality declined further downstream due to migration (Jergentz *et al.* 2004).

Mayflies, stoneflies, caddisflies and midges are the most sensitive taxa among aquatic invertebrates and do not tolerate insecticide residues even in minimal concentrations (Beketov *et al.* 2009), whereas dragonflies, aquatic beetles, water striders, amphipods and snails are more tolerant. Neonicotinoid, fipronil and pyrethroid insecticides are particularly toxic to these taxa, but whereas population effects of the latter chemicals are temporary, chronic exposure to the former tends to eliminate populations for several months (Hayasaka *et al.* 2012b).

When several insecticides are present, additive toxicity can be expected, while synergistic interactions have been observed when other stressor factors (i.e. predators) are present (Relyea 2004). Mixtures of insecticides and herbicides usually have a synergistic toxic effect on midge larvae (Chironomidae) in the sediments (Lydy and Austin 2004). Apart from being an essential food resource for fish and water birds, aquatic insect larvae are key organisms in the recycling of nutrients in aquatic ecosystems, as they break down leaves and other plant material, so their elimination greatly disturbs the ecosystem and results in poor water quality.

Due to their naked skins, frogs are directly exposed to insecticide residues in agricultural ditches and ponds, and their populations can be reduced dramatically under peak pulses of malathion, carbaryl, endosulfan and herbicides (Rohr and Crumrine 2005). Fish kills have been reported after insecticide spray operations for mosquito, malaria and locust control in many countries, but are more rare when the insecticides are applied to agricultural fields. Nevertheless, during peak contamination events in rivers of cotton-growing areas in Australia, residues of profenofos were sufficiently high as to affect populations of European carp (*Cyprinus carpio*), bony bream (*Nematalosa erebi*) and mosquitofish (*Gambusia holbrooki*) (Kumar and Chapman 2001). Despite constant exposure to water-borne insecticides, many fish escape death as residues tend to be adsorbed onto organic sediments and aquatic plants (Carriquiriborde *et al.* 2007).

Worldwide, waterborne residues of pesticides that have their source in agricultural fields are causing loss of biodiversity in the adjacent aquatic ecosystems (Beketov *et al.* 2013). Eventually, such residues end up in the oceans, whether associated with sediments in the estuaries, in particles that are filtered by clams, mussels and oysters along the coasts, or further down. They have even been detected in abyssal fish (Mormede and Davies 2001).

4.3.1.2 Indirect Effects on Pests and Non-target Organisms

These derive from the ecological relationships that organisms have in natural environments (Figure 4.2b), not from the toxicity of the chemical. For example, the decline in bird populations in the Netherlands is not due to the toxicity of imidacloprid or other chemicals to bird species, but rather to the elimination of their food resources (i.e. aquatic invertebrates) by the chemicals (Hallmann *et al.* 2014).

The unintended consequences of scattering insecticides in the environment were discovered in the early days of chemical pest control with organochlorines, as pest resurgence and the appearance of secondary pests were obvious. The latter pests resulted from indirect effects of insecticides on the disturbed arthropod communities, as ecosystems lost their natural enemies (Sánchez-Bayo 2011). Indirect effects, however, are also apparent among non-target organisms, including beneficial arthropods as well as vertebrates (see Figure 4.2b). For example, some predatory and scavenger arthropods such as rove beetles, carabids, ants and spiders may increase in numbers in the aftermath of a chemical treatment, as they feed on the dead insects killed by the chemical, even though this is only a temporary effect. Also, springtails usually increase in numbers when fields are treated with normal doses of insecticides, as these chemicals often kill the predatory mites that prey on them (Badji *et al.* 2007).

The best-known indirect effects of insecticides are on birds. Many bird species feed on insects and invertebrates of the soil, and even granivorous species use this resource while rearing their young. It is not surprising, therefore, that bird populations decline in agricultural environments as a consequence of both direct mortality caused by insecticide sprays and lack of food (Mineau and Whiteside 2013). For example, nest success of chestnut-collared longspurs (*Calcarius ornatus*) was reduced when their staple food, grasshoppers, were decimated as a result of pyrethroid locust control in Canada (Martin *et al.* 1998). Starvation was also attributed as the main cause of nestling mortality among lesser kestrels (*Falco naumanni*) in agricultural areas treated with insecticides in Spain (Negro 1993).

Reduced availability of invertebrates, seeds and weeds has been recognized as the key factor that explains the disappearance of birds under agricultural intensification (Wilson *et al.* 1999). These effects of pesticide usage on food resources eventually affect the breeding performance of most birds (Boatman *et al.* 2004), as has happened to partridges (*Perdix perdix*) in England and other European countries since the 1950s (Potts 1986). Starvation effects, however, take a long time to be noticed, since experimental field trials conducted over one or two years are unable to statistically prove the cause–effect relationship that insecticides have on bird food resources and their breeding success, as the impacts are marginal on a yearly basis (Howe *et al.* 1996). Thus, long-term studies are required to establish causality. For example, it took 5 years of monitoring to demonstrate the significant reduction caused by *Bacillus thuringiensis* (Bt) on the abundance of reed-dwelling invertebrates that constitute the food of passerines, and more specifically its negative effect on the breeding success of house martins (*Delichon urbicum*) in rural areas of the Camargue in France (Poulin 2012). In the past 20 years, populations of at least 15 species of passerines in the Netherlands have declined at an average rate of 3.5% annually, due mainly to the indirect effects of imidacloprid on food resources. However, this was statistically demonstrated in only 40% of the species studied (Hallmann *et al.* 2014), probably because other confounding pollutants also intervened.

Indirect effects are sometimes beneficial to species that take advantage of the vacuum of predators, graziers or competitors that result from insecticide usage. In aquatic environments, the abundance of copepods and rotifers can increase as their competitors (waterfleas and ostracods) are eliminated by organophosphorus insecticides such as malathion. Similarly, tadpoles of various species of frogs can become more abundant as there is more growth of periphyton and algae in the absence of zooplankton grazers (Relyea 2005). In an experiment with terrestrial rodents, populations of feral house mouse (*Mus musculus*) increased when those of meadow voles (*Microtus pennsylvanicus*) were reduced after application of a carbamate insecticide to oat fields (Barrett 1988). In all cases, population increases are the result of competition among species of similar ecological niche or of the dynamics of prey and predator relationships.

4.3.1.3 Sublethal Effects

Bioaccumulation and bioconcentration of pesticide residues are a regular feature in filtering organisms (clams, mussels, barnacles, etc.) and fish that live in contaminated waters. The residues are for the most part passed on to the predators, and move up the food chain to contaminate other species (Katagi 2010). Earthworms and insect larvae that live in the soil also accumulate residues and pass them on to birds and other animals that feed on them. Some individuals may die after accumulating such residues, but commonly organisms develop side effects or chronic conditions that are not lethal but may affect their health – the so-called sublethal effects (see Figure 4.2a).

For example, consumption of invertebrates contaminated with organochlorine insecticides may kill insectivorous birds and bats (Guillén *et al.* 1994), but rarely causes death in larger birds and other animals. Instead, contamination of prey with DDT and cyclodienes (e.g. dieldrin, endosulfan, chlordane) results in bioaccumulation of their residues in the tissues of predatory animals. In birds of prey, including raptors and fish-eating birds, these residues and their metabolites (i.e. dichloro-diphenyl-dichloroethylene, DDE) interfere with the deposition of calcium in the eggshells, resulting in eggshell thinning, which in turn causes the breakage of eggs and thus fewer offspring (Ratcliffe 1970). Although this physiological disturbance is caused by sublethal doses of the insecticides and/or metabolites, it affects the viability of the species as a whole when the reproductive success rate drops by 10% (Sibly *et al.* 2005). Thus, entire populations of peregrine falcons and other birds of prey were wiped out in past decades due to this side effect of organochlorine insecticides, and they only recovered after a ban was imposed in many countries (Kirk and Hyslop 1998).

Sublethal effects are also common among organophosphorus and neonicotinoid insecticides, often manifested as neuropathies that lead to abnormal behaviour. For example, thyroid impairment has been reported for songbirds exposed to mixtures of pesticides since the 1960s, and more recently for imidacloprid, so the affected birds neglect the rearing of their nestlings (Bishop *et al.* 2000). Bees exposed to neonicotinoids experience olfactory, learning and memory impairment that affect their feeding, performance in foraging, disorientation and other tasks in the hives (Desneux *et al.* 2007). Imidacloprid also inhibits feeding in bumble bees, while imidacloprid and clothianidin disable the bee immune system by promoting the repressors of this defence system (Di Prisco *et al.* 2013). Consequently, bees feeding on pollen and nectar contaminated with neonicotinoids often experience high rates of infection by *Varroa* mites and their associated viruses, which result in unhealthy colonies that eventually die (Sánchez-Bayo *et al.* 2016).

Numerous sublethal effects of insecticides have been reported for aquatic organisms. For example, growth impairment in midges (*Chironomus tentants*) exposed to pyrethroid residues in sediments (Maul *et al.* 2008), and in medaka fish (*Oryzias latipes*) exposed to fipronil and neonicotinoids (Hayasaka *et al.* 2012a); malformations in caddisfly larvae (*Hydropsyche slossonae*) exposed to sublethal residues of malathion (Tessier *et al.* 2000); impaired spermatogenesis in the testis of bluegill (*Lepomis macrochirus*) by diazinon (Dutta and Meijer 2003); and feeding inhibition of the freshwater amphipod *Gammarus pulex* in waters contaminated with imidacloprid (Nyman *et al.* 2013). Cholinesterase inhibitors, pyrethroids, neonicotinoids and probably other insecticides usually cause stress in fish, alter their hematocrit and metabolism and make them more susceptible to fungal and parasite infections by reducing the production of lymphocytes (Barry *et al.* 1995; Gül *et al.* 2012). Removal of the pesticides from water eventually improves the aquatic ecosystem health (Gagliardi and Pettigrove 2013).

4.3.2 Herbicides

Toxicity of these chemicals affects mainly plant species, with few compounds being selective for either monocots or dicots. Effects on animals are at sublethal levels, and usually the impacts are indirectly through the food chain (see Figure 4.2b).

4.3.2.1 Direct Effects

Herbicides applied to crops reduce the abundance and diversity of arable weeds so as to increase agricultural yields. However, herbicide drift also reduces the plant biodiversity in nearby hedgerows (Aude *et al.* 2003), field margins and riparian strips that harbour numerous animals, mostly birds, rodents and their predators. Farmers often blame such habitats as refuges of crop-damaging pests, so they use herbicides to control them. Yet herbicide sprayings may not bring any benefit in that regard, as treated areas do not appear to reduce the abundance of the most harmful bird pests (Deschenes *et al.* 2003). In experimental trials conducted over 3 years, the impact of herbicides on field margins was enhanced when fertilizers were used: both agro-chemicals led to shifts in plant community compositions over time, causing significantly lower species diversities in the treated areas than in the controls (Schmitz *et al.* 2014).

The effectiveness of herbicides in reducing plant biomass is often underestimated; many annual plants are excluded from being established, and although vegetation communities may recover after the cropping season, widespread application of herbicides year after year leads to the depletion of soil seed banks. For example, after many years of intensive agricultural practices using a range of herbicides, the Hilly Country of Saxony has lost many landscapes and their associated flora diversity (Schlüter *et al.* 1990). Some individual herbicides may have minimal impacts. A review of the broad-spectrum herbicide glyphosate found the shifts in species floral composition and structure of habitats on a variety of forest and agro-ecosystems to be within the normal range of variation in natural ecosystems. Nevertheless, 'reductions in plant biomass and related moose (*Alces alces* L.) forage and habitat use generally occur for 1–5 years after treatment' of forests with this herbicide (Sullivan and Sullivan 2003).

Few herbicides can directly affect invertebrates. An exception is pendimethalin, which applied at 0.75–1.0 $kg\,ha^{-1}$ reduces abundance of soil nematodes by 35–60% and negatively affects microbial biota, specifically plant–*Rhizobium* symbiosis (Strandberg and Scott-Fordsmand 2004). Acrolein, a contact herbicide used to control submerged and floating weeds in irrigation channels, is unusually toxic to many fish and aquatic invertebrates (e.g. tadpoles, snails, some crustaceans and insects), causing lethal exposures in more than 70–90 % of the species up to 20 km downstream from the application point (Venturino *et al.* 2007).

As herbicides are more hydrophilic than insecticides, their residues are more likely to appear in run-off and even ground waters. In fact, they constitute the most common residues found in surveys of rivers in the USA and other countries, usually at few ppb ($\mu g\,L^{-1}$) levels (Ensminger *et al.* 2013; Hermosin *et al.* 2013; Kreuger 1998). Although water-borne residues are transient, restricted to crop growing seasons, sediments at the bottom of rivers, wetlands and estuaries capture and retain longer residual herbicides, affecting aquatic macrophytes and algae to various degrees. For atrazine, the most widespread contaminant of surface waters, ecological effects on algae and macrophytes may occur when residue levels are above 50 $\mu g\,L^{-1}$ (Solomon *et al.* 1996). For tebuthiuron, the impacts on algae and *Chironomus* larvae appear when residues in water are above 200 $\mu g\,L^{-1}$ (Temple *et al.* 1991), whereas for hexazinone, the threshold is 1 $mg\,L^{-1}$ (Thompson *et al.* 1993). Usually, a residue of herbicide concentration equivalent to 0.3 effective concentration for 50% of plants (EC50) triggers ecological effects in the aquatic communities, mainly due to reduction of phytoplankton and periphyton as well as macrophytes (Hartgers *et al.* 1998). For sulfonylureas, this threshold may be as low as 0.03 $\mu g\,L^{-1}$ (de Lafontaine *et al.* 2014). Waterborne residues of urea herbicides (e.g. tebuthiuron) last quite a long time and can thus affect aquatic plants a few months after they enter the streams and rivers (Dam *et al.* 2004), slowing their subsequent recovery and that of dependent animal communities.

4.3.2.2 Indirect Effects

Since herbicides inhibit growth of plants and algae, less primary productivity and food resources for the animal species that depend on them can be expected. Studies in managed ecosystems distinguish two types of interactions that can be affected by herbicide usage: direct trophic interactions, which occur when pest or beneficial arthropods feed directly on weeds; and indirect trophic interactions, which occur when damage to crops by feeding arthropods impacts weeds through alteration of ecosystem resource availability, or through weeds serving as hosts for alternative prey for beneficial arthropods (Norris and Kogan 2000).

Long-term studies carried out over several years in vegetable crops have revealed that the soil arthropod community structure is positively correlated with the weed community biomass, which varies with the use of specific herbicides and other management practices (Wardle *et al.* 1999). For example, while glyphosate and conservation tillage may help increase density of soil arthropods compared to conventional ploughing with no herbicide (House 1989), high rates of glyphosate consistently reduce the total number of web-spinner spiders in arable field margins, as a result of changes in vegetation structure caused by this herbicide (Haughton *et al.* 1999). Healthy arthropod communities are essential for controlling pests in agro-ecosystems. For example, cane and sugar yields averaged 19% higher in weedy sugarcane plantations than in the weed-free plantations in Lousiana, USA, because broadleaf weeds enhanced the populations of beneficial carabids, ants and spiders that control the sugarcane borer (*Diatraea saccharalis*) (Ali and Reagan 1985). By contrast, in tomato crops, paraquat and trifluralin indirectly reduced the density of ground beetle predators (Carabidae) as their herbivorous prey decreased due to starvation (Yardim and Edwards 2002). A meta-analysis of 23 experimental studies also revealed that herbicide usage reduced populations of most arthropods (except Heteroptera) and invertebrates in arable crop edges (Frampton and Dorne 2007).

Such indirect impacts on arthropods are illustrated by the reduction of weeds in orange groves in Spain: many years of herbicide applications have reduced the abundance and biodiversity of consumer ants to the point that fewer ant colonies have made the soil progressively less porous and more compacted, thus enhancing rainfall erosion and slowly depleting the orchard's soil fertility (Cerdà and Jurgensen 2008). The introduction of recent transgenic herbicide-tolerant (TGHT) crops may encourage no-tillage practices, which are beneficial for soil fertility, but there is concern that such crops may lead to a more intensive use of herbicides and the removal of many weeds that support populations of pollinators.

Pollination by bees is a very important ecological service provided to agriculture, as 25% of tropical crops and possibly up to 84% of temperate crops depend on insect pollination (Garibaldi *et al.* 2013). Thus, management and protection of pollinator populations and habitats of nectar-producing plants can be essential for some crops and for the environment at large. Agricultural intensification and habitat loss due to herbicides are the most frequent causes of pollinator impoverishment (64% of cases), although direct bee mortality by insecticides is evident and cannot be ignored either (Greig-Smith *et al.* 1994).

The demise of the grey partridge (*Perdix perdix*) in England is due to a decrease in chick survival as a result mainly of herbicide usage; as treatment of crops with these chemicals became widespread, chick survival dropped from 49% to 32% in the period 1952–1993, while additional nest predation caused the collapse of partridge populations (Potts 1986). Declining bird species (e.g. skylark, corn bunting, etc.) are not associated with particular foods, but with overall reductions in abundance and diversity of plants, seeds and insects resulting from intensive agriculture. Granivorous species feed on cereal grain and seeds of many 'weeds' like knotgrasses (Polygonaceae), chickweeds (*Stellaria* spp.), goosefoots (*Chenopodium* spp.), and others, so their decline has been driven primarily by herbicide use and the switch from spring-sown to autumn-sown cereals, both of which have massively reduced the food supplies of these birds (Newton 2004). In intensively managed grasslands, loss of grasshoppers and lepidopteran larvae due to weed removal deprives chicks of a wide range of bird species of an important food resource (Wilson *et al.* 1999).

Residual herbicides also indirectly affect the invertebrate communities of freshwater systems, and their mixtures have additive effects. For example, reduction of phytoplankton and periphyton by atrazine, tebuthiuron and other herbicides decreases or temporarily eliminates the populations of waterfleas, chironomid larvae, snails and tadpoles, as a result of which copepods tend to increase (Juttner *et al.* 1995). Glyphosate suppresses periphyton and diatoms, producing a long-term shift in the typology of waterbodies as cyanobacteria tend to fill a gap in the absence of algae (Vera *et al.* 2012). Field surveys in the Philippines have shown that herbicides had significant effects on photosynthetic activity in flood water and on populations of benthic aquatic oligochaetes, although the impact of nitrogen fertilizers on algal and invertebrate populations was even larger (Roger *et al.* 1994). In Japanese rice fields, the impact of chlormethoxynil, oxadiazon or a mixture of thiobencarb and simetryne resulted invariably in a rapid decrease of arthropods as their algae food source disappeared. Concurrently, plant-parasitic nematodes and snails increased compared to hand-weeded control plots (Ishibashi *et al.* 1983). Mixtures of herbicides and insecticides produce unbalanced communities in which outcomes are difficult to predict. While frogs are sensitive to insecticides, their tadpoles feed on periphyton and are more affected by the indirect impacts of herbicides, so the combined effect of both types of pesticides typically results in the collapse of amphibian populations (Relyea 2005).

Waterborne herbicide residues have wider indirect impacts as they eventually reach the sea. For example, there is evidence that the declining populations of fish and invertebrates in San Joaquin Delta (California, USA) are related to the decreasing primary productivity of the estuary; phytoplankton is the most important food source of these organisms, and residual herbicides have reduced microalgal populations since the late 1960s to the point of affecting the food chain at the top (Jassby *et al.* 2003). Concerns about the impacts of sulfonylurea and PSII-inhibitor herbicides (e.g. atrazine, diuron) on coral reefs around Australia and the Caribbean are justified based on the constant inputs of residues from agricultural sources as well as biofouling paints from ships and recreational boats; a detailed account of their impacts is given in van Dam *et al.* (2011).

4.3.2.3 Sublethal Effects on Animals

Endocrine disruption of herbicides has been reported mainly in amphibians under laboratory conditions. Effects include hermaphroditism, male feminization and low testosterone levels in African clawed frogs (*Xenopus laevis*) and leopard frogs (*Rana pipiens*) exposed to environmental levels of 25 and 0.1 $\mu g L^{-1}$ atrazine, respectively (Hayes *et al.* 2010). Atrazine is also genotoxic to oysters (*Crassostrea gigas*) at concentrations of 10–100 $\mu g L^{-1}$ (Bouilly *et al.* 2004).

4.3.3 Fungicides

Many fungicides disrupt essential cellular processes such as respiration, so they are considered broad-spectrum biocides. Selective compounds, however, target biochemical pathways specific for fungi such as ergosterol biosynthesis. They are very effective in controlling plant fungal diseases, with negligible effects on plants.

4.3.3.1 Direct Effects

Ecological impacts of fungicides affect mostly soil microbial communities that include fungi and bacteria alike (see Figure 4.2a). For example, benomyl and captan significantly reduce soil microbial processes such as substrate-induced respiration, soil enzyme activities, microbial biomass nitrogen and dissolved organic nitrogen concentrations, resulting in slower than normal decomposition of wheat straw. However, urease activity, ammonium and nitrate concentrations as well as initial nitrogen mineralization increase, particularly in captan-treated microcosms (Chen *et al.* 2001). Ecosystem functions that depend on soil biota (e.g. organic matter decomposition) can also be negatively affected by carbendazim, which reduces the abundance of earthworms and millipedes in tropical soils (Förster *et al.* 2006). Omnivorous nematodes and enchytraeids of the genus *Fridericia* appear to be most sensitive to this fungicide, with EC50s for biomass reduction in the range 0.9–24 mg kg^{-1} of soil (Moser *et al.* 2004), but carbendazim does not appear to have significant impacts on springtails nor mites in controlled microcosms (Koolhaas *et al.* 2004). Most fungicides seem not to affect terrestrial invertebrates such as Collembola, Arachnida, Formicidae or Nematoda (Alves *et al.* 2014; Jaensch *et al.* 2006).

Fungicides are the most common pesticide residues found in water (63%) and sediments (44%) of agricultural areas in Australia, albeit at concentrations that may pose negligible direct risk to aquatic organisms (Wightwick *et al.* 2012). By contrast, in Spanish orange groves some 85% of aquatic ecotoxicity is due to fungicides applied before harvest (Juraske and Sanjuan 2011). In Spanish rice fields, mixtures of four herbicides and three fungicides reduced algae growth and *Daphnia magna* populations at the same time (Suárez-Serrano *et al.* 2010). Herbicides, fungicides and insecticides mixtures caused substantial reductions in phytoplankton and zooplankton in ditches near vegetable fields when the drift amounts after spraying were 5% of the application rate, but smaller and transient effects were observed if the drift was 1% or less of the rate (Arts *et al.* 2006). Glochidia and juveniles of freshwater mussels such as *Lampsilis siliquoidea* are highly sensitive to the fungicides chlorothalonil, propiconazole and pyraclostrobin (Bringolf *et al.* 2007), and chlorothalonil can accumulate in mussels' tissues up to 4 days (Ernst *et al.* 1991). Aquatic worms of the Oligochaeta, Turbellaria and Hirudinea taxa as well as waterfleas, cyclopoid copepods and rotifers are very sensitive to carbendazim and azoxystrobin residues in water, which are very persistent in this medium, and can alter the zooplankton and macroinvertebrate communities whenever their residues are above 1 μg L^{-1} and 0.1 mg L^{-1}, respectively (Daam *et al.* 2010; van Wijngaarden *et al.* 2014).

4.3.3.2 Indirect Effects

The best known indirect effect is the suppresion by ergosterol-inhibiting fungicides (azoles) of the detoxification mechanism mediated by cytochrome P450 in animals. This enhances the toxicity of neonicotinoid and pyrethroid insecticides to honeybees with factors up to 1000-fold for thiacloprid and 10-fold for alpha-cyhalothrin (Iwasa *et al.* 2004; Pilling *et al.* 1995). In aquatic organisms, recent studies with *Daphnia magna* have shown that azole fungicides act as synergists of pyrethroids, with toxicities increasing during the recovery period by factors of up to six-fold (epoxiconazole), 7–13-fold (propiconazole) and 61-fold (prochloraz) compared to that without fungicide exposure (Kretschmann *et al.* 2015; Norgaard and Cedergreen 2010). While tebuconazole, imazalil and kasugamycin significantly reduce decomposition rates in river sediments contaminated with these residues (Artigas *et al.* 2012; Flores *et al.* 2014; Huang *et al.* 2010), the combination of azole fungicides and pyrethroid insecticides in surface waters significantly reduces macro-invertebrate shredding activity due to the synergistic effect of the fungicide (Rasmussen *et al.* 2012). Such residues are often found in surface waters and sediments of banana plantations in tropical countries of Central America (Castillo *et al.* 2000).

4.3.3.3 Sublethal Effects

Fungicides such as quinoxyfen, cyprodinil, carbendazim, azoxystrobin and tebuconazole and their mixtures inhibit the feeding activity of aquatic shredders like *Asellus aquaticus*, posing a risk to the fundamental ecosystem function of litter decomposition (Zubrod *et al.* 2014). Ketoconazole exhibits endocrine disruption properties, as it consistently depresses gonadal synthesis of testosterone in both sexes of fathead minnow (*Pimephales promelas*). This fungicide inhibits the activity of two cytochrome P450s (CYP11a and CYP17) that are key to sex steroid production in vertebrates (Ankley *et al.* 2012).

Many aquatic environments have been contaminated with mercury due to the widespread use of mercurial fungicides in agriculture and the pulp industry during the 1950s and 1960s, especially in Sweden and Japan. Bacteria in soil and sediments convert it to methyl mercury, which accumulates in the wild fauna, grain-eating birds and their predators, causing neurological problems. A ban on such products has reduced levels of mercury in terrestrial animals since the 1970s, but unfortunately mercury is still being transferred to the rivers and estuaries, where it accumulates in fish (Ackefors 1971).

4.4 Concluding Remarks

Insecticides are often not needed and may not always contribute effectively to yield gain, as is the case with systemic insecticides (Balconi *et al.* 2011). However, chemicals are still, and will continue to be, the last weapon used for controlling the pest outbreaks that inevitably happen from time to time in any agricultural crop. Also, in most developing countries insecticides are needed to control grain insect pests in storage facilities, as no other practical systems are available.

Apart from the impacts on the environment described above, the constant use of pesticides leads inevitably to increased resistance in pests, weeds and fungi. Resistance can be overcome by introducing new products as the old ones become ineffective or banned, but this solution in the long run does not benefit the environment, which becomes polluted with more and more toxic chemicals as time goes by.

Alternative strategies for insect pest control include:

- pheromone traps and other attractants, that use insecticides in the trap instead of scattering them around
- agronomic practices that disrupt the life cycle of key pests (e.g. tillage timing, crop rotation, fallow) or attract beneficial natural enemies (i.e. intercropping)
- biological control, i.e. the use of beneficial organisms such as predators, parasitoids and viral, nematodes or other diseases
- transgenics if available.

For the control of rodent pests, there are other solutions, such as python snakes or other friendly predators. For weed control, the traditional practice of hand weeding is a viable solution for small plots, but not for large acreage. For control of diseases, however, no other choice is available but to rely on specific chemical products.

References

Ackefors, H. (1971) Effects of particular pollutants - 3. Mercury pollution in Sweden with special reference to conditions in the water habitat. *Proceedings of the Royal Society of London (B)* **177**: 365–387.

Albert, C.A., Wilson, L.K., Mineau, P., Trudeau, S. and Elliott, J.E. (2010) Anticoagulant rodenticides in three owl species from Western Canada, 1988–2003. *Archives of Environmental Contamination and Toxicology* **58**: 451–459.

Ali, A.D. and Reagan, T.E. (1985) Vegetation manipulation impact on predator and prey populations in Louisiana (USA) sugarcane ecosystems. *Journal of Economic Entomology* **78**: 1409–1414.

Alves, P.R.L., Cardoso, E.J.B.N., Martines, A.M., Sousa, J.P. and Pasini, A. (2014) Seed dressing pesticides on springtails in two ecotoxicological laboratory tests. *Ecotoxicology and Environmental Safety* **105**: 65–71.

Alyokhin, A., Dively, G., Patterson, M., *et al.* (2007) Resistance and cross-resistance to imidacloprid and thiamethoxam in the Colorado potato beetle *Leptinotarsa decemlineata. Pest Management Science* **63**: 32–41.

Ankley, G.T., Cavallin, J.E., Durhan, E.J., *et al.* (2012) A time-course analysis of effects of the steroidogenesis inhibitor ketoconazole on components of the hypothalamic-pituitary-gonadal axis of fathead minnows. *Aquatic Toxicology* **114–115**: 88–95.

Artigas, J., Majerholc, J., Foulquier, A., *et al.* (2012) Effects of the fungicide tebuconazole on microbial capacities for litter breakdown in streams. *Aquatic Toxicology* **122–123**: 197–205.

Arts, G.H.P., Buijse-Bogdan, L.L., Belgers, J.D.M., *et al.* (2006) Ecological impact in ditch mesocosms of simulated spray drift from a crop protection program for potatoes. *Integrated Environmental Assessment and Management* **2**: 105–125.

Aude, E., Tybirk, K. and Pedersen, M.B. (2003) Vegetation diversity of conventional and organic hedgerows in Denmark. *Agriculture, Ecosystems and Environment* **99**: 135–147.

Badji, C.A., Guedes, R.N.C., Silva, A.A., Correa, A.S., Queiroz, M.E.L.R. and Michereff -Filho, M. (2007) Non-target impact of deltamethrin on soil arthropods of maize fields under conventional and no-tillage cultivation. *Journal of Applied Entomology* **131**: 50–58.

Balconi, C., Mazzinelli, G. and Motto, M. (2011) Neonicotinoid insecticide seed coatings for the protection of corn kernels and seedlings, and for plant yield. *Maize Genetics Cooperation Newsletter* **3**.

Barrett, G.W. (1988) Effects of sevin on small mammal populations in agricultural and oil field ecosystems. *Journal of Mammalogy* **69**: 731–739.

Barry, M.J., O'Halloran, K., Logan, D.C., Ahokas, J.T. and Holdway, D.A. (1995) Sublethal effects of esfenvalerate pulse exposure on spawning and non-spawning Australian crimson-spotted rainbow fish (*Melanotaenia fluviatilis*). *Archives of Environmental Contamination and Toxicology* **28**: 459–463.

Baur, R., Wijnands, F. and Malavolta, C. (2011) Integrated production – objectives, principles and technical guidelines. *IOBC/WPRS Bulletin Special Issue.*

Beketov, M.A., Foit, K., Schäfer, R.B., *et al.* (2009) SPEAR indicates pesticide effects in streams – comparative use of species- and family-level biomonitoring data. *Environmental Pollution* **157**: 1841–1848.

Beketov, M.A., Kefford, B.J., Schäfer, R.B. and Liess, M. (2013) Pesticides reduce regional biodiversity of stream invertebrates. *Proceedings of the National Academy of Sciences* **110**: 11039–11043.

Bishop, C., Ng, P., Mineau, P., Quinn, J. and Struger, J. (2000) Effects of pesticide spraying on chick growth, behavior, and parental care in tree swallows (*Tachycineta bicolor*) nesting in an apple orchard in Ontario, Canada. *Environmental Toxicology and Chemistry* **19**: 2286–2297.

Boatman, N.D., Brickle, N.W., Hart, J.D., *et al.* (2004) Evidence for the indirect effects of pesticides on farmland birds. *Ibis* **146**: 131–143.

Bouilly, K., McCombie, H., Leitao, A. and Lapegue, S. (2004) Persistence of atrazine impact on aneuploidy in Pacific oysters, *Crassostrea gigas. Marine Biology* **145**: 699–705.

Bringolf, R.B., Cope, W.G., Eads, C.B., Lazaro, P.R., Barnhart, M.C. and Shea, D. (2007) Acute and chronic toxicity of technical-grade pesticides to glochidia and juveniles of freshwater mussels (Unionidae). *Environmental Toxicology and Chemistry* **26**: 2094–2100.

Brown, A.W.A. (1978) *Ecology of Pesticides.* John Wiley and Sons, New York, USA.

Butler, C.D., Beckage, N.E. and Trumble, J.T. (2009) Effects of terrestrial pollutants on insect parasitoids. *Environmental Toxicology and Chemistry* **28**: 1111–1119.

Carriquiriborde, P., Díaz, J., Mugni, H., Bonetto, C. and Ronco, A.E. (2007) Impact of cypermethrin on stream fish populations under field-use in biotech-soybean production. *Chemosphere* **68**: 613–621.

Castillo, L.E., Ruepert, C. and Solís, E. (2000) Pesticide residues in the aquatic environment of banana plantation areas in the North Atlantic zone of Costa Rica. *Environmental Toxicology and Chemistry* **19**: 1942–1950.

Cerdà, A. and Jurgensen, M.F. (2008) The influence of ants on soil and water losses from an orange orchard in eastern Spain. *Journal of Applied Entomology* **132**: 306–314.

Chen, S.K., Edwards, C. and Subler, S. (2001) Effects of the fungicides benomyl, captan and chlorothalonil on soil microbial activity and nitrogen dynamics in laboratory incubations. *Soil Biology and Biochemistry* **33**: 1971–1980.

Christensen, T.K., Lassen, P. and Elmeros, M. (2012) High exposure rates of anticoagulant rodenticides in predatory bird species in intensively managed landscapes in Denmark. *Archives of Environmental Contamination and Toxicology* **63**: 437–444.

Daam, M.A., Satapornvanit, K., van den Brink, P.J. and Nogueira, A.J.A. (2010) Direct and indirect effects of the fungicide carbendazim in tropical freshwater microcosms. *Archives of Environmental Contamination and Toxicology* **58**: 315–324.

Daly, G.L., Lei, Y.D., Teixeira, C., Muir, D.C.G., Castillo, L.E. and Wania, F. (2007) Accumulation of current-use pesticides in neotropical montane forests. *Environmental Science and Technology* **41**: 1118–1123.

Dam, R.A., Camilleri, C., Bayliss, P. and Markich, S.J. (2004) Ecological risk assessment of tebuthiuron following application on tropical Australian wetlands. *Human and Ecological Risk Assessment* **10**: 1069–1097.

Daryanto, I. (1998) Pesticide management policy in Indonesia. In: Kennedy, I.R., Skerritt, J.H., Johnson, G.I. and Highley, E. (eds) *Seeking Agricultural Produce Free of Pesticide Residues.* Australian Centre for International Agricultural Research, Canberra, Australia, pp. 31–36.

De Lafontaine, Y., Beauvais, C., Cessna, A.J., Gagnon, P., Hudon, C. and Poissant, L. (2014) Sulfonylurea herbicides in an agricultural catchment basin and its adjacent wetland in the St. Lawrence River basin. *Science of the Total Environment* **479–480**: 1–10.

Deschenes, M., Bélanger, L. and Giroux, J.F. (2003) Use of farmland riparian strips by declining and crop damaging birds. *Agriculture, Ecosystems and Environment* **95**: 567–577.

Desneux, N., Decourtye, A. and Delpuech, J.M. (2007) The sublethal effects of pesticides on beneficial arthropods. *Annual Review of Entomology* **52**: 81–106.

Di Prisco, G., Cavaliere, V., Annoscia, D., *et al.* (2013) Neonicotinoid clothianidin adversely affects insect immunity and promotes replication of a viral pathogen in honey bees. *Proceedings of the National Academy of Sciences* **110**: 18466–18471.

Dutta, H.M. and Meijer, H.J.M. (2003) Sublethal effects of diazinon on the structure of the testis of bluegill, *Lepomis macrochirus*: a microscopic analysis. *Environmental Pollution* **125**: 355–360.

Edwards, C.A. and Thompson, A.R. (1973) Pesticides and the soil fauna. *Residue Reviews* **45**: 1–79.

Elliott, J.E., Wilson, L.K., Langelier, K.M., Mineau, P. and Sinclair, P.H. (1997) Secondary poisoning of birds of prey by the organophosphorus insecticide, phorate. *Ecotoxicology* **6**: 219–231.

Ensminger, M.P., Budd, R., Kelley, K.C. and Goh, K.S. (2013) Pesticide occurrence and aquatic benchmark exceedances in urban surface waters and sediments in three urban areas of California, USA, 2008–2011. *Environmental Monitoring and Assessment* **185**: 3697–3710.

Ernst, W., Doe, K., Jonah, P., Young, J., Julien, G. and Henningar, P. (1991) The toxicity of chlorothalonil to aquatic fauna and the impacts of its operational use on a pond ecosystem. *Archives of Environmental Contamination and Toxicology* **21**: 1–9.

Flores, L., Banjac, Z., Farré, M., *et al.* (2014) Effects of a fungicide (imazalil) and an insecticide (diazinon) on stream fungi and invertebrates associated with litter breakdown. *Science of the Total Environment* **476–477**: 532–541.

Förster, B., García, M., Francimari, O. and Römbke, J. (2006) Effects of carbendazim and lambda-cyhalothrin on soil invertebrates and leaf litter decomposition in semi-field and field tests under tropical conditions (Amazonia, Brazil). *European Journal of Soil Biology* **42**: S171–S179.

Frampton, G.K. and Dorne, J.L.C.M. (2007) The effects on terrestrial invertebrates of reducing pesticide inputs in arable crop edges: a meta-analysis. *Journal of Applied Ecology* **44**: 362–373.

Furlan, L. and Kreutzweiser, D. (2015) Alternatives to neonicotinoid insecticides for pest control: case studies in agriculture and forestry. *Environmental Science and Pollution Research* **22**: 135–147.

Gagliardi, B. and Pettigrove, V. (2013) Removal of intensive agriculture from the landscape improves aquatic ecosystem health. *Agriculture, Ecosystems and Environment* **176**: 1–8.

Garibaldi, L.A., Steffan-Dewenter, I., Winfree, R., *et al.* (2013) Wild pollinators enhance fruit set of crops regardless of honey bee abundance. *Science* **339**: 1608–1611.

Goerke, H., Weber, K., Bornemann, H., Ramdohr, S. and Plötz, J. (2004) Increasing levels and biomagnification of persistent organic pollutants (POPs) in Antarctic biota. *Marine Pollution Bulletin* **48**: 295–302.

Gong, P., Wang, X.P., Li, S.H., *et al.* (2014) Atmospheric transport and accumulation of organochlorine compounds on the southern slopes of the Himalayas, Nepal. *Environmental Pollution* **192**: 44–51.

Greig-Smith, P.W. (1987) Hazards to wildlife from pesticide seed treatments. In: Martin T. (ed.) *Application to Seeds and Soil.* Monograph No. **39**. British Crop Protection Council, Croydon, UK, pp. 127–133.

Greig-Smith, P.W., Thompson, H.M., Hardy, A.R., Bew, M.H., Findlay, E. and Stevenson, J.H. (1994) Incidents of poisoning of honeybees (*Apis mellifera*) by agricultural pesticides in Great Britain 1981–1991. *Crop Protection* **13**: 567–581.

Guillén, A., Ibáñez, C., Pérez, J.L., *et al.* (1994) Organochlorine residues in Spanish common pipistrelle bats (*Pipistrellus pipistrellus*). *Bulletin of Environmental Contamination and Toxicology* **52**: 231–237.

Gül, A., Benli, A.Ç.K., Ayhan, A., *et al.* (2012) Sublethal propoxur toxicity to juvenile common carp (*Cyprinus carpio* L., 1758): biochemical, hematological, histopathological, and genotoxicity effects. *Environmental Toxicology and Chemistry* **31**: 2085–2092.

Hahn, M., Schotthöfer, A., Schmitz, J., Franke, L.A. and Brühl, C.A. (2015) The effects of agrochemicals on Lepidoptera, with a focus on moths, and their pollination service in field margin habitats. *Agriculture, Ecosystems and Environment* **207**: 153–162.

Hallmann, C.A., Foppen, R.P.B., van Turnhout, C.A.M., de Kroon, H. and Jongejans, E. (2014) Declines in insectivorous birds are associated with high neonicotinoid concentrations. *Nature* **511**: 341–343.

Hartgers, E.M., Aalderink, G.H.R., van den Brink, P.J., Gylstra, R., Wiegman, J.W.F. and Brock, T.C.M. (1998) Ecotoxicological threshold levels of a mixture of herbicides (atrazine, diuron and metolachlor) in freshwater microcosms. *Aquatic Ecology* **32**: 135–152.

Haughton, A.J., Bell, J.R., Boatman, N.D. and Wilcox, A. (1999) The effects of different rates of the herbicide glyphosate on spiders in arable field margins. *Journal of Arachnology* **27**: 249–254.

Hayasaka, D., Korenaga, T., Sánchez-Bayo, F. and Goka, K. (2012a) Differences in ecological impacts of systemic insecticides with different physicochemical properties on biocenosis of experimental paddy fields. *Ecotoxicology* **21**: 191–201.

Hayasaka, D., Korenaga, T., Suzuki, K., Saito, F., Sánchez-Bayo, F. and Goka, K. (2012b) Cumulative ecological impacts of two successive annual treatments of imidacloprid and fipronil on aquatic communities of paddy mesocosms. *Ecotoxicology and Environmental Safety* **80**: 355–362.

Hayes, T.B., Khoury, V., Narayan, A., *et al.* (2010) Atrazine induces complete feminization and chemical castration in male African clawed frogs (*Xenopus laevis*). *Proceedings of the National Academy of Sciences* **107**: 4612–4617.

He, Y., Zhao, J., Zheng, Y., Desneux, N. and Wu, K. (2012) Lethal effect of imidacloprid on the coccinellid predator *Serangium japonicum* and sublethal effects on predator voracity and on functional response to the whitefly *Bemisia tabaci*. *Ecotoxicology* **21**: 1291–1300.

Henny, C.J., Kaiser, J.L. and Grove, R.A. (2009) PCDDs, PCDFs, PCBs, OC pesticides and mercury in fish and osprey eggs from Willamette River, Oregon (1993, 2001 and 2006) with calculated biomagnification factors. *Ecotoxicology* **18**: 151–173.

Hermosin, M.C., Calderón, M.J., Real, M. and Cornejo, J. (2013) Impact of herbicides used in olive groves on waters of the Guadalquivir river basin (southern Spain). *Agriculture, Ecosystems and Environment* **164**: 229–243.
Hladik, M.L., Kolpin, D.W. and Kuivila, K.M. (2014) Widespread occurrence of neonicotinoid insecticides in streams in a high corn and soybean producing region, USA. *Environmental Pollution* **193**: 189–196.
House, G.J. (1989) Soil arthropods from weed and crop roots of an agroecosystem in a wheat-soybean-corn rotation: impact of tillage and herbicides. *Agriculture, Ecosystems and Environment* **25**: 233–244.
Howe, F., Knight, R., McEwen, L. and George, T. (1996) Direct and indirect effects of insecticide applications on growth and survival of nestling passerines. *Ecological Applications* **6**: 1314–1324.
Huang, C.Y., Ho, C.H., Lin, C.J. and Lo, C.C. (2010) Exposure effect of fungicide kasugamycin on bacterial community in natural river sediment. *Journal of Environmental Science and Health, Part B* **45**: 485–491.
Ishibashi, N., Kondo, E. and Ito, S. (1983) Effects of application of certain herbicides on soil nematodes and aquatic invertebrates in rice paddy fields in Japan. *Crop Protection* **2**: 289–304.
Iwasa, T., Motoyama, N., Ambrose, J.T. and Roe, R.M. (2004) Mechanism for the differential toxicity of neonicotinoid insecticides in the honey bee, *Apis mellifera. Crop Protection* **23**: 371–378.
Jaensch, S., Frampton, G.K., Römbke, J., van den Brink, P.J. and Scott-Fordsmand, J.J. (2006) Effects of pesticides on soil invertebrates in model ecosystem and field studies: a review and comparison with laboratory toxicity data. *Environmental Toxicology and Chemistry* **25**: 2490–2501.
James, P.C. (1995) Internalizing externalities: granular carbofuran use on rapeseed in Canada. *Ecological Entomology* **13**: 181–184.
Jassby, A.D., Cloern, J.E. and Muller-Solger, A.B. (2003) Phytoplankton fuels delta food web. *Californian Agriculture* **57**: 104–109.
Jergentz, S., Pessacq, P., Mugni, H., Bonetto, C. and Schulz, R. (2004) Linking *in situ* bioassays and population dynamics of macroinvertebrates to assess agricultural contamination in streams of the Argentine pampa. *Ecotoxicology and Environmental Safety* **59**: 133–141.
Juraske, R. and Sanjuan, N. (2011) Life cycle toxicity assessment of pesticides used in integrated and organic production of oranges in the Comunidad Valenciana, Spain. *Chemosphere* **82**: 956–962.
Juttner, I., Peither, A., Lay, J.P., Kettrup, A. and Ormerod, S.J. (1995) An outdoor mesocosm study to assess ecotoxicological effects of atrazine on a natural plankton community. *Archives of Environmental Contamination and Toxicology* **29**: 435–441.
Katagi, T. (2010) Bioconcentration, bioaccumulation, and metabolism of pesticides in aquatic organisms. *Reviews of Environmental Contamination and Toxicology* **204**: 1–132.
Kirk, D.A. and Hyslop, C. (1998) Population status and recent trends in Canadian raptors: a review. *Biological Conservation* **83**: 91–118.
Koehler, H.H. (1992) The use of soil mesofauna for the judgement of chemical impact on ecosystems. *Agriculture, Ecosystems and Environment* **40**: 193–205.
Koolhaas, J.E., van Gestel, C.A.M., Römbke, J., Soares, A.M.V.M. and Jones, S.E. (2004) Ring-testing and field-validation of a Terrestrial Model Ecosystem (TME) – an

instrument for testing potentially harmful substances: effects of carbendazim on soil microarthropod communities. *Ecotoxicology* **13**: 75–88.

Kretschmann, A., Gottardi, M., Dalhoff, K. and Cedergreen, N. (2015) The synergistic potential of the azole fungicides prochloraz and propiconazole toward a short α-cypermethrin pulse increases over time in *Daphnia magna*. *Aquatic Toxicology* **162**: 94–101.

Kreuger, J. (1998) Pesticides in stream water within an agricultural catchment in southern Sweden, 1990–1996. *Science of the Total Environment* **216**: 227–251.

Krupke, C.H. and Long, E.Y. (2015) Intersections between neonicotinoid seed treatments and honey bees. *Current Opinion in Insect Science* **10**: 8–13.

Kumar, A. and Chapman, J.C. (2001) Profenofos residues in wild fish from cotton-growing areas of New South Wales. *Journal of Environmental Quality* **30**: 740–750.

López-Antia, A., Ortiz-Santaliestra, M.E., Mougeot, F. and Mateo, R. (2015) Imidacloprid-treated seed ingestion has lethal effect on adult partridges and reduces both breeding investment and offspring immunity. *Environmental Research* **136**: 97–107.

Lydy, M.J. and Austin, K.R. (2004) Toxicity assessment of pesticide mixtures typical of the Sacramento–San Joaquin Delta using *Chironomus tentans*. *Archives of Environmental Contamination and Toxicology* **48**: 49–55.

Martin, P., Johnson, D., Forsyth, D. and Hill, B. (1998) Indirect effects of the pyrethroid insecticide deltamethrin on reproductive success of chestnut-collared longspurs. *Ecotoxicology* **7**: 89–97.

Maul, J.D., Brennan, A.A., Harwood, A.D. and Lydy, M.J. (2008) Effect of sediment-asociated pyrethroids, fipronil, and metabolites on *Chironomus tentans* growth rate, body mass, condition index, immobilization, and survival. *Environmental Toxicology and Chemistry* **27**: 2582–2590.

Maule, A., Plyte, S. and Quirk, A.V. (1987) Dehalogenation of organochlorine insecticides by mixed anaerobic microbial populations. *Pesticide Biochemistry and Physiology* **27**: 229–236.

Mineau, P. and Palmer, C. (2013) *The Impact of the Nation's Most Widely Used Insecticides on Birds*. American Bird Conservancy, Virginia, USA.

Mineau, P. and Whiteside, M. (2006) Lethal risk to birds from insecticide use in the United States – a spatial and temporal analysis. *Environmental Toxicology and Chemistry* **25**: 1214–1222.

Mineau, P. and Whiteside, M. (2013) Pesticide acute toxicity is a better correlate of U.S. grassland bird declines than agricultural intensification. *PLoS ONE* **8**(2): e57457.

Mormede, S. and Davies, I.M. (2001) Polychlorobiphenyl and pesticide residues in monkfish *Lophius piscatorius* and black scabbard *Aphanopus carbo* from the Rockall Trough. *Journal of Marine Science* **58**: 725–736.

Moser, T., Schallnass, H.J., Jones, S.E., *et al.* (2004) Ring-testing and field-validation of a Terrestrial Model Ecosystem (TME) – an instrument for testing potentially harmful substances: effects of carbendazim on nematodes. *Ecotoxicology* **13**: 61–74.

Muir, D.C.G., Norstrom, R.J. and Simon, M. (1988) Organochlorine contaminants in Arctic marine food chains: accumulation of specific polychlorinated biphenyls and chlordane-related compounds. *Environmental Science and Technology* **22**: 1071–1079.

Negro, J.J. (1993) Organochlorine and heavy metal contamination in non-viable eggs and its relation to breeding success in a Spanish population of Lesser Kestrels (*Falco naumanni*). *Environmental Pollution* **82**: 201–205.

Newton, I. (2004) The recent declines of farmland bird populations in Britain: an appraisal of causal factors and conservation actions. *Ibis* **146**: 579–600.
Norgaard, K.B. and Cedergreen, N. (2010) Pesticide cocktails can interact synergistically on aquatic crustaceans. *Environmental Science and Pollution Research* **17**: 957–967.
Norris, R.F. and Kogan, M. (2000) Interactions between weeds, arthropod pests, and their natural enemies in managed ecosystems. *Weed Science* **48**: 94–158.
Nyman, A.M., Hintermeister, A., Schirmer, K. and Ashauer, R. (2013) The insecticide imidacloprid causes mortality of the freshwater amphipod *Gammarus pulex* by interfering with feeding behavior. *PLoS ONE* **8(5)**: e62472.
Ogbeide, O., Tongo, I., Enuneku, A., Ogbomida, E. and Ezemonye, L. (2015) Human health risk associated with dietary and non-dietary intake of organochlorine pesticide residues from rice fields in Edo State Nigeria. *Exposure and Health* **8**: 53–66.
Peck, D.C. (2009) Long-term effects of imidacloprid on the abundance of surface- and soil-active nontarget fauna in turf. *Agricultural and Forest Entomology* **11**: 405–419.
Pilling, E.D., Bromley-Challenor, K.A.C., Walker, C.H. and Jepson, P.C. (1995) Mechanism of synergism between the pyrethroid insecticide λ-cyhalothrin and the imidazole fungicide prochloraz, in the honeybee (*Apis mellifera* L.). *Pesticide Biochemistry and Physiology* **51**: 1–11.
Pimentel, D., McLaughlin, L., Zepp, A., *et al.* (1993) Environmental and economic effects of reducing pesticide use in agriculture. *Agriculture, Ecosystems and Environment* **46**: 273–288.
Potts, G.R. (1986) *The Partridge – Pesticides, Predation and Conservation*. Collins, London, UK.
Poulin, B. (2012) Indirect effects of bioinsecticides on the nontarget fauna: the Camargue experiment calls for future research. *Acta Oecologica* **44**: 28–32.
Rasmussen, J.J., Monberg, R.J., Baattrup-Pedersen, A., *et al.* (2012) Effects of a triazole fungicide and a pyrethroid insecticide on the decomposition of leaves in the presence or absence of macroinvertebrate shredders. *Aquatic Toxicology* **118–119**: 54–61.
Ratcliffe, D.A. (1970) Changes attributable to pesticides in egg breakage frequency and shell thickness in some British birds. *Journal of Applied Ecology* **7**: 67–115.
Relyea, R.A. (2004) Synergistic impacts of malathion and predatory stress on six species of North American tadpoles. *Environmental Toxicology and Chemistry* **23**: 1080–1084.
Relyea, R.A. (2005) The impact of insecticides and herbicides on the biodiversity and productivity of aquatic communities. *Ecological Applications* **15**: 618–627.
Roger, P.A., Simpson, I., Oficial, R., Ardales, S. and Jiménez, R. (1994) Impact of pesticides on soil and water microflora and fauna in wetland ricefields. *Australian Journal of Experimental Agriculture* **34**: 1057–1068.
Rohr, J.R. and Crumrine, P.W. (2005) Effects of an herbicide and an insecticide on pond community structure and processes. *Ecological Applications* **15**: 1135–1147.
Sánchez-Bayo, F. (2011) Impacts of agricultural pesticides on terrestrial ecosystems. In: Sánchez-Bayo, F., van den Brink, P.J. and Mann, R. (eds) *Ecological Impacts of Toxic Chemicals*. Bentham Science Publishers, Sharjah, UAE, pp. 63–87.
Sánchez-Bayo, F., Goulson, D., Pennacchio, F., Nazzi, F., Goka, K. and Desneux, N. (2016) Are bee diseases linked to pesticides? A brief review. *Environment International* **89–90**: 7–11.
Schlüter, H., Böttcher, W. and Bastian, O. (1990) Vegetation change caused by land-use intensification – examples from the Hilly Country of Saxony. *GeoJournal* **22**: 167–174.

Schmitz, J., Hahn, M. and Brühl, C.A. (2014) Agrochemicals in field margins – an experimental field study to assess the impacts of pesticides and fertilizers on a natural plant community. *Agriculture, Ecosystems and Environment* **193**: 60–69.
Sibly, R.M., Akçakaya, H.R., Topping, C.J. and O'Connor, R.J. (2005) Population-level assessment of risks of pesticides to birds and mammals in the UK. *Ecotoxicology* **14**: 863–876.
Solomon, K.R., Baker, D.B., Richards, R.P., *et al.* (1996) Ecological risk assessment of atrazine in North American surface waters. *Environmental Toxicology and Chemistry* **15**: 31–76.
Story, P.G., Mineau, P. and Mullié, W.C. (2013) Insecticide residues in Australian plague locusts (*Chortoicetes terminifera* Walker) after ultra-low volume aerial application of the organophosphorus insecticide fenitrothion. *Environmental Toxicology and Chemistry* **32**: 2792–2799.
Strandberg, M. and Scott-Fordsmand, J.J. (2004) Effects of pendimethalin at lower trophic levels – a review. *Ecotoxicology and Environmental Safety* **57**: 190–201.
Suárez-Serrano, A., Ibáñez, C., Lacorte, S. and Barata, C. (2010) Ecotoxicological effects of rice field waters on selected planktonic species: comparison between conventional and organic farming. *Ecotoxicology* **19**: 1523–1535.
Sullivan, T.P. and Sullivan, D.S. (2003) Vegetation management and ecosystem disturbance: impact of glyphosate herbicide on plant and animal diversity in terrestrial systems. *Environmental Reviews* **11**: 37–59.
Tanabe, S., Iwata, H. and Tatsukawa, R. (1994) Global contamination by persistent organochlorines and their ecotoxicological impact on marine mammals. *Science of the Total Environment* **154**: 163–177.
Temple, A.J., Murphy, B.R. and Cheslak, E.F. (1991) Effects of tebuthiuron on aquatic productivity. *Hydrobiologia* **224**: 117–127.
Tessier, L., Boisvert, J.L., Vought, L.B.M. and Lacoursière, J.O. (2000) Anomalies on capture nets of *Hydropsyche slossonae* larvae (Trichoptera; Hydropsychidae), a potential indicator of chronic toxicity of malathion (organophosphate insecticide). *Aquatic Toxicology* **50**: 125–139.
Theiling, K.M. and Croft, B.A. (1988) Pesticide side-effects on arthropod natural enemies: a database summary. *Agriculture, Ecosystems and Environment* **21**: 191–218.
Thompson, D.G., Holmes, S.B., Wainio-Keizer, K., MacDonald, L. and Solomon, K.R. (1993) Impact of hexazinone and metsulfuron methyl on the zooplankton community of a boreal forest lake. *Environmental Toxicology and Chemistry* **12**: 1709–1717.
Tremolada, P., Mazzoleni, M., Saliu, F., Colombo, M. and Vighi, M. (2010) Field trial for evaluating the effects on honeybees of corn sown using cruiser® and Celest® treated seeds. *Bulletin of Environmental Contamination and Toxicology* **85**: 229–234.
Van Dam, J.W., Negri, A.P., Uthicke, S. and Mueller, J.F. (2011). Chemical pollution on coral reefs: exposure and ecological effects. In: Sánchez-Bayo, F., van den Brink, P.J. and Mann, R. (eds) *Ecological Impacts of Toxic Chemicals*. Bentham Science Publishers, Sharjah, UAE, pp. 187–211.
Van den Brink, P.J., Wijngaarden, R.P.A.V., Lucassen, W.G.H., Brock, T.C.M. and Leeuwangh, P. (1996) Effects of the insecticide Dursban 4E (active ingredient chlorpyrifos) in outdoor experimental ditches: II. Invertebrate community responses and recovery. *Environmental Toxicology and Chemistry* **15**: 1143–1153.

van Wijngaarden, R.P.A., Belgers, D.J.M., Zafar, M.I., Matser, A.M., Boerwinkel, M.C. and Arts, G.H.P. (2014) Chronic aquatic effect assessment for the fungicide azoxystrobin. *Environmental Toxicology and Chemistry* **33**: 2775–2785.

Venturino, A., Montagna, C.M. and Pechen de D'Angelo, A.M. (2007) Risk assessment of magnicide® herbicide at rio Colorado irrigation channels (Argentina). Tier 3: studies on native species. *Environmental Toxicology and Chemistry* **26**: 177–182.

Vera, M., di Fiori, E., Lagomarsino, L., *et al.* (2012) Direct and indirect effects of the glyphosate formulation Glifosato Atanor® on freshwater microbial communities. *Ecotoxicology* **21**: 1805–1816.

Walker, C.H., Hopkin, S.P., Sibly, R.M. and Peakall, D.B. (2001) *Principles of Ecotoxicology*, 2nd edn. Taylor and Francis, Glasgow, UK.

Wardle, D., Nicholson, K., Bonner, K. and Yeates, G. (1999) Effects of agricultural intensification on soil-associated arthropod population dynamics, community structure, diversity and temporal variability over a seven-year period. *Soil Biology and Biochemistry* **31**: 1691–1706.

Wauchope, R.D. (1978) The pesticide content of surface water draining from agricultural fields: a review. *Journal of Environmental Quality* **7**: 459–472.

Way, M.J. and Heong, K.L. (1994) The role of biodiversity in the dynamics and management of insect pests of tropical irrigated rice – a review. *Bulletin of Entomological Research* **84**: 567–587.

Wightwick, A.M., Bui, A.D., Zhang, P., *et al.* (2012) Environmental fate of fungicides in surface waters of a horticultural-production catchment in Southeastern Australia. *Archives of Environmental Contamination and Toxicology* **62**: 380–390.

Wiktelius, S., Chiverton, P.A., Meguenni, H., *et al.* (1999) Effects of insecticides on non-target organisms in African agroecosystems: a case for establishing regional testing programmes. *Agriculture, Ecosystems and Environment* **75**: 121–131.

Wilson, J., Morris, A., Arroyo, B., Clark, S. and Bradbury, R. (1999) A review of the abundance and diversity of invertebrate and plant foods of granivorous birds in northern Europe in relation to agricultural change. *Agriculture, Ecosystems and Environment* **75**: 13–30.

Wilson, L.K., Elliott, J.E., Vernon, R.S., Smith, B.D. and Szeto, S.Y. (2002) Persistence and retention of active ingredients in four granular cholinesterase-inhibiting insecticides in agricultural soils of the lower Fraser River valley, British Columbia, Canada, with implications for wildlife poisoning. *Environmental Toxicology and Chemistry* **21**: 260–268.

Woods, N., Craig, I.P., Dorr, G. and Young, B. (2001) Spray drift of pesticides arising from aerial application in cotton. *Journal of Environmental Quality* **30**: 697–701.

Yadav, I.C., Devi, N.L., Syed, J.H., *et al.* (2015) Current status of persistent organic pesticides residues in air, water, and soil, and their possible effect on neighboring countries: a comprehensive review of India. *Science of the Total Environment* **511**: 123–137.

Yardim, E.N. and Edwards, C.A. (2002) Effects of weed control practices on surface-dwelling arthropod predators in tomato agroecosystems. *Phytoparasitica* **30**: 379–386.

Zubrod, J.P., Baudy, P., Schulz, R. and Bundschuh, M. (2014) Effects of current-use fungicides and their mixtures on the feeding and survival of the key shredder *Gammarus fossarum*. *Aquatic Toxicology* **150**: 133–143.

5

Environmental Impacts of Arthropod Biological Control: An Ecological Perspective

David E. Jennings, Jian J. Duan and Peter A. Follett

5.1 Introduction

Arthropod biological control (hereafter 'biocontrol') is the use of natural enemies such as herbivores, parasitoids and predators that act as a 'top-down' ecological force to regulate or suppress an arthropod or weed pest, with the goal of preventing populations from reaching sizes that would result in significant economic and/or ecological damage. The 'top-down' effect of biocontrol results from the dynamics of community-level processes acting on populations, in which the natural enemy (upper trophic level) lowers the number of arthropod pests that feed on agricultural crops, forest trees or desirable native plants, or, in the case of weed pests, the natural enemy reduces the number of individual plants or biomass in affected ecosystems.

Biocontrol using conservation, augmentation and/or indundative release of native (or local) natural enemies (primarily insect herbivores, parasitoids and predators) has long been employed for controlling or managing agricultural and forest pests, with few documented cases of significant adverse environmental impact, and in general is environmentally benign compared with use of chemical pesticides (Follett and Duan 2000; van Driesche *et al.* 2008, 2010; van Lenteren *et al.* 2006) (see Chapter 14). Native or local natural enemies have a long co-evolutionary history with their hosts or prey in their co-inhabited ecosystems, and manipulation or augmentation of populations of the native natural enemies is perhaps unlikely to cause serious interruption of ecosystem functions or services. Therefore, these types of biocontrol should almost always be favoured in the development of sustainable pest management programmes.

However, serious ecological impacts can be associated with the introduction of non-native natural enemies for classic biocontrol of non-native pests in agriculture, forests and natural ecosystems (Follett and Duan 2000; Howarth 1991; Lynch and Thomas 2000; Wajnberg *et al.* 2001). This is largely because those non-native species may themselves become 'invasive' (e.g. spread beyond their intended range of introduction and cause negative environmental impacts) and the invasiveness of non-native species in a new environment or ecosystem has been difficult to predict (see Chapter 10). Thus, modern classical biocontrol programmes are governed by strict regulations and extensive safety (host specificity) testing.

In this chapter, we examine the 'invasion' process of establishing non-native natural enemies after release into the target ecosystem, and the consequences associated with

Environmental Pest Management: Challenges for Agronomists, Ecologists, Economists and Policymakers, First Edition. Edited by Moshe Coll and Eric Wajnberg.

each phase of the invasion in terms of both target and non-target impacts. Risk assessment for different forms of arthropod biocontrol is evaluated in the context of sustainable pest management, and the risks and benefits of biocontrol relative to other pest control methods, such as no action, and chemical or mechanical control, are assessed. Finally, two recent case studies are presented to illustrate in more detail how different ecological processes can be incorporated into biocontrol programme decision making and risk assessment.

5.2 The 'Invasion' Process of Establishing Non-native Biocontrol Agents

The process of establishing populations of classic biocontrol agents shares many aspects with biological invasions (Ehler 1998) (see Chapter 10). The introduction of a non-native arthropod for biocontrol is a deliberate targeted invasion, with six main stages (Figure 5.1) that are similar to the four described by Lockwood *et al.* (2007) for biological invasions (transport, introduction, spread and impact). Indeed, much of what is known regarding propagule pressure in invasion ecology stems from work conducted in the field of biocontrol (Grevstad 1999).

Using knowledge of the ecological processes that comprise much of invasion biology can help to plan the release strategy for biocontrol programmes. For instance, research on biological invasions shows that propagule pressure can be an important factor determining whether or not populations successfully establish (Hayes and Barry 2008; Simberloff 2009). Similarly, in the early stages of invasion, population densities may be regulated by Allee effects and stochastic processes (Liebhold and Tobin 2008). These processes are equally applicable to the successful establishment of biocontrol agents, where sufficient numbers need to be released for populations to establish.

Different ecological processes are important at various points throughout the stages of an invasion, and likewise exploiting aspects of a species' ecology can inform the

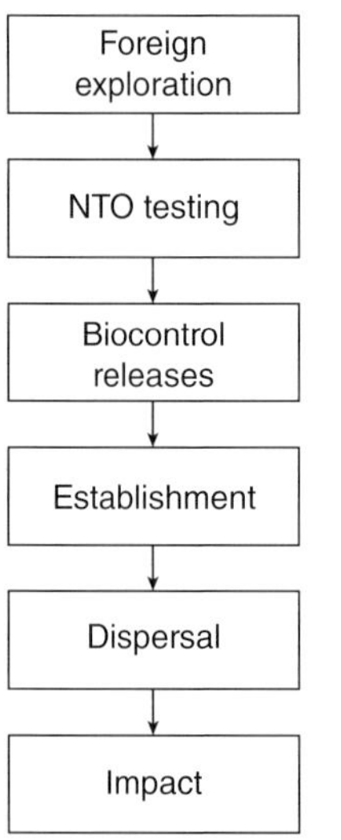

Figure 5.1 Conceptual model of the classic biocontrol methodology. The process begins with foreign exploration for suitable natural enemies in the home range of the target invasive pest, followed by non-target organism (NTO) testing to assess the risk to native species from the selected biocontrol agents at the target area. If the biocontrol agents are judged to pose little to no risk, they are released, before ideally establishing, dispersing and suppressing pest populations.

choice of biocontrol agents and greatly enhance the success of a biocontrol release. For instance, polyphagy or oligophagy can facilitate establishment of invasive species such as aphids (Messing *et al.* 2007) or Asian and citrus longhorned beetles (Haack *et al.* 2010). Likewise, when conducting foreign exploration and non-target organism testing, locating host-specific natural enemies is especially important as it can not only minimize environmental impacts, but also help to make biocontrol more effective and economical than alternative management strategies. Invasive species can competitively displace native species (Holway 1999; Snyder and Evans 2006), and similarly, after biocontrol agents are released, their intra- and interspecific interactions may determine how effective they are at suppressing pest populations, as well as their effects on other species in the community (Snyder *et al.* 2004). Additionally, spatial and temporal processes influence how far (and fast) invasive species and biocontrol agents disperse and establish (or fail to do so) (Johnson *et al.* 2006; Muirhead *et al.* 2006; Wilson *et al.* 2009). Finally, it is important to understand how these ecological processes will be affected by predicted changes in climate (Crowder and Harwood 2014; Dukes and Mooney 1999; Hellmann *et al.* 2008), possibly altering the success of biocontrol agents and their potential impacts on non-target organisms and the environment.

5.3 Ecological Processes Underlying the Environmental Impact of Biocontrol

Biocontrol involves exploiting the spatial and temporal community- and population-level processes between natural enemies and their targeted pests (e.g. prey or hosts). Such community- and population-level processes can occur with or without the action of humans. However, success in biocontrol relies on releasing the agent at an optimal location and time when the appropriate pest life stage is present and vulnerable so that its population is reduced to an acceptable threshold level. For example, when evaluating the efficacy of predatory phytoseiid mites released for biocontrol of *Oligonychus perseae* (Acari: Tetranychidae) on avocados, Hoddle *et al.* (1999) found that more of the predators were recovered when releases were conducted early in the season when *O. perseae* population densities were increasing and of sufficient size to sustain the predators.

Therefore, biocontrol is an ecologically based pest control technology and its effectiveness is influenced by many ecological factors such as host specificity, spatial and temporal dynamics of spread, and species interactions, all of which could be affected by climatic change. Below, we discuss some of these key ecological processes and explore how they can determine the risk of non-target impacts.

5.3.1 Host Specificity

The number of host species that a biocontrol agent is capable of utilizing for its survival and reproduction determines its host specificity. Host specificity is a primary consideration for biocontrol agents before their introduction against an arthropod or weed pest, as it affects not only the efficacy in controlling the target pest populations but also whether attacks on non-targets are likely to occur in the targeted ecosystem (Bigler *et al.* 2006; Follett and Duan 2000; Howarth 1983; Lockwood 1993). In general, greater host specificity results in a lower probability of non-target impacts. However, some debate remains over whether or not specialists actually make for more effective biocontrol agents than generalists (Kaser and Heimpel 2015; Stiling and Cornelissen 2005; Symondson *et al.* 2002).

Differences in the agent's taxonomy, biology, evolutionary history and population genetics, and the species present in the receiving ecological system can influence host attack and utilization spectra. Koinobionts (parasitoids that develop in/on hosts that continue to grow) have evolved delicate physiological and/or biochemical pathways that evade the host's immune system, and host range is usually determined by host physiology. In contrast, idiobionts (parasitoids that develop in/on hosts whose development has been halted) do not need to suppress or evade the host's immune system, and habitat or host plant may play a more important role in determining host range than host physiology (van Driesche and Murray 2004).

Increased concern surrounding potential non-target effects from biocontrol programmes has led to stricter regulations and rigorously developed methodologies for assessing the risks posed by candidate arthropod natural enemies prior to introduction. However, even after laboratory or even field testing, host ranges of potential biocontrol agents can still exceed expectations. Along with various other indirect ecological impacts (Bigler *et al.* 2006), host range expansion can be difficult to predict. Even without any host range expansion, on average, arthropods released for biocontrol may attack between two and three different host species (Stiling and Simberloff 2000).

Introduced natural enemies are now usually required to undergo pre-release non-target testing with regard to host specificity. Release of biocontrol agents with low host specificity has resulted in some of the most notorious examples of the negative effects of biocontrol. For instance, the parasitoid fly *Compsilura concinnata* Meigen (Diptera: Tachinidae) is a generalist that was introduced into North America throughout much of the 20th century to control a variety of pests, including the gypsy moth (*Lymantria dispar dispar* L., Lepidoptera: Erebidae) and browntail moth (*Euproctis chrysorrhoea* L., Lepidoptera: Erebidae) (Boettner *et al.* 2000; Louda *et al.* 2003). This species subsequently has been implicated in the decline of several native moth species (Boettner *et al.* 2000). However, it is worth noting that *C. concinnata* also appears to have been relatively successful in suppressing populations of an intended target, the browntail moth (Elkinton and Boettner 2012).

5.3.2 Spatial and Temporal Dynamics of Spread

The spatial and temporal dynamics of the spread of biocontrol agents are important factors when considering their ecological impacts, as they can result in natural enemies being exposed to new hosts against which they were not previously tested. However, these dynamics are not always accounted for when examining potential impacts of biocontrol agents before or after releases (Fagan *et al.* 2002; Pratt and Center 2012; Strong and Pemberton 2000). An invasive pest with a well established range and an invasive pest that is actively expanding its range may present different challenges for biocontrol. For new invasive pests, a generalist natural enemy may be better able to persist in the environment than a specialist, as at the edge of the invasion front there are likely to be fewer hosts for a highly host-specialized biocontrol agent to utilize (Fagan *et al.* 2002). Such a scenario with generalist natural enemies can lead to use of non-target hosts given the low density of the target species.

Biocontrol of prickly pear cacti (*Opuntia* spp.) provides an example of an introduced natural enemy spreading beyond its intended range and negatively affecting non-target organisms. The moth *Cactoblastis cactorum* Berg (Lepidoptera: Pyralidae) had been used successfully for biocontrol of *Opuntia stricta* in Australia in the 1920s. Based on this success, *C. cactorum* was subsequently introduced to other areas where *Opuntia* was invasive, such as the Caribbean and South Africa (Stiling and Simberloff 2000). However, after being introduced to the Caribbean in the 1950s, *C. cactorum* was subsequently discovered in the Florida Keys in 1989. It has since spread throughout much of the south-eastern United States where it threatens many native *Opuntia* spp. (Jezorek *et al.* 2010, 2012). In addition to the risk it presents to rare cacti (Stiling 2010), *C. cactorum* is also a threat to opuntoid crop production in Mexico which has an annual value of around $85 million (Soberon *et al.* 2001).

The issues with *C. cactorum* are not isolated, and there are examples of other biocontrol agents spreading far from their original release locations over relatively short temporal scales. For instance, a biocontrol programme was initiated against the invasive tree *Melaleuca quinquenervia* in Florida, USA, using the curculionid *Oxyops vitiosa* Pascoe (Coleoptera: Curculionidae) and the psyllid *Boreioglycaspis melaluecae* Moore (Hemiptera: Psyllidae) (Center *et al.* 2000, 2006). Seven years after being released in Florida, *B. melaluecae* was detected on *M. quinquenervia* in Los Angeles, California, USA, approximately 3500^km from its original release location (Pratt and Center 2012). Introduction of the ladybird *Harmonia axyridis* (Pallas) (Coleoptera: Coccinellidae) into Europe provides another example of unanticipated spread. This ladybird, native to parts of Asia, was introduced to at least 12 European countries between the 1960s and early 2000s, but has subsequently been found in a further eight countries where it was not deliberately introduced (Brown *et al.* 2008). This is especially concerning as a risk assessment suggested that *H. axyridis* is highly polyphagous and is known to feed on many other insects, as well as various plant materials (van Lenteren *et al.* 2008). Based on their risk assessment, van Lenteren *et al.* (2008) concluded that *H. axyridis* should never have been released in northern Europe.

As indicated by these examples, the dispersal ability of biocontrol agents (both natural and human assisted) can play an important role in their risk to non-target organisms, and thorough risk assessments should be conducted before any introduction. Good dispersal ability of candidate biocontrol agents is often a trait required for effective control of their targeted pests, and there may be a trade-off between risks and benefits. Where possible, potential spread of biocontrol agents should be modelled and host specificity testing should include candidate taxa within the predicted range of spread. These examples also emphasize the necessity for international collaboration on biocontrol programmes, and the need to consider biocontrol introductions on a continental scale (Pratt and Center 2012).

5.3.3 Species Interactions

Both target and non-target impacts of biocontrol agents can be influenced by various biotic factors such as interspecific competition between introduced and native natural enemies, apparent competition and hyperparasitism. Interspecific competition between natural enemies can lead to declines in populations of some species or even competitive exclusion (Schellhorn *et al.* 2002). In addition to effects on native natural enemies, interspecific competition between introduced biocontrol agents can also be a concern for management programmes, as it can lead to reduced parasitism/predation pressure on the target pest. Pre-release testing in the laboratory can be conducted to assess the likely strength and outcomes of such interactions between biocontrol agents (Ulyshen *et al.* 2010; Wang *et al.* 2015). However, the outcomes of these interactions typically depend on multiple factors such as host and parasitoid densities and spatial scale, and laboratory experiments may be limited in their predictive ability (Wang *et al.* 2015).

Apparent competition can occur when the target and non-target species have a common natural enemy, and it typically has been studied among biocontrol agents of weeds (Carvalheiro *et al.* 2008; Willis and Memmott 2005). Also, a considerable amount of laboratory work has examined apparent competition in insect parasitoids (Bonsall and Hassell 1997, 1998), increasingly utilizing field research too (Morris *et al.* 2001, 2004). Models suggest that, with moderate risk to non-target species, apparent competition can be beneficial for biocontrol of target pests (Kaser and Heimpel 2015), and manipulating systems where apparent competition is an important factor could allow for biocontrol enhancement (Chailleux *et al.* 2014). However, apparent competition between arthropod pests involves non-specific natural enemies, an undesirable trait in classic biocontrol. Using available knowledge, potentially important interspecific interactions between biocontrol agents and competitors, hosts, predators and prey should be highlighted and investigated prior to releases to assess the risks to other species in the target ecosystems.

Although empirical examples are relatively rare, there is some evidence that hyperparasitism can interfere with biocontrol programmes (Gomez-Marco *et al.* 2015; Pérez-Lachaud *et al.* 2004; Schooler *et al.* 2011). For example, Schooler *et al.* (2011) examined interactions between the hyperparasitoid *Asaphes suspensus* (Nees) (Pteromalidae: Asaphinae) and the primary aphid parasitoid *Aphidius ervi* Haliday (Hymenoptera: Braconidae) using glasshouse experiments and field surveys. In the glasshouse experiments, the primary parasitoids were driven to extinction by hyperparasitism, but field surveys did not support the glasshouse findings. While providing another example of how the strength of interactions can be inflated when using laboratory or glasshouse experiments, the results obtained by Schooler *et al.* (2011) provide some indication of the potential for hyperparasitism to influence biocontrol. In terms of non-target impacts caused by hyperparasitism, empirical evidence is also scarce. However, Wang and Messing (2004) investigated the effects of the facultative hyperparasitoid *Pachycrepoideus vindemmiae* Rondani (Hymenoptera: Pteromalidae) on four species of primary tephritid fruit fly parasitoids. They found that the hyperparasitoid was able to complete development from hosts previously parasitized by all four of the primary parasitoids, and that the hyperparasitoid exhibited no preference for unparasitized hosts over hosts already parasitized by one of the primary parasitoids (Wang and Messing 2004). Thus, there is potential for hyperparasitoids with low host specificity to negatively affect non-target organisms.

5.3.4 Implications of Climatic Change

Global climate change is predicted to increase mean temperatures and the frequency of extreme temperature events, and such changes could have important implications for biocontrol (Gerard *et al.* 2013; Hellmann *et al.* 2008; Tylianakis and Binzer 2014) (see Chapter 9). Many species are adjusting their distributions, phenologies and behaviours to adapt to their new environmental conditions (Parmesan 2006). Consequently, the effects of climate change should be considered during biocontrol programmes in the context of both target and non-target impact assessment (Schuldiner-Harpaz and Coll 2013). For instance, host specificity testing may need to be broader to consider non-target species outside the current range of the target species (Gerard *et al.* 2013). Some biocontrol agents may be able to disperse to new areas beyond the range of their target and track shifting climates whereas others may not. Mismatches in synchrony between natural enemies and their hosts, plants or prey resulting from climatic change could also affect biocontrol (Hance *et al.* 2007; Jeffs and Lewis 2013; Thomson *et al.* 2010). For example, changes in mean temperatures and frequencies of extreme temperatures could result in more instances of phenological disparities between parasitoids and their hosts, possibly leading to reduced efficacy on target pests and increased chance of host switching to non-targets (Gerard *et al.* 2013).

Climate change will also likely affect various aspects of the biology of different biocontrol agents. For parasitoids, temperature shifts may have particularly important consequences for diapause (Li *et al.* 2008; Polgár and Hardie 2000), attack rates (Baffoe *et al.* 2012; Duan *et al.* 2014; Romo and Tylianakis 2013) and fertility (Cascone *et al.* 2015). Climate change could also affect biocontrol by altering the nutritional quality of plants (Coll and Hughes 2008; Thomson *et al.* 2010) or affecting interspecific interactions such as competition (Guzmán *et al.* 2016).

Although climatic change clearly has broad implications for the environmental impacts of biocontrol, to date empirical evidence demonstrating these effects is rare. However, Lu *et al.* (2015) provide an example linking climate change to non-target effects. These authors studied the non-target effects of *Agasicles hygrophila* Selman and Vogt (Coleoptera: Chrysomelidae) in relation to various temperatures. *A. hygrophila* was introduced to China for the control of *Alternanthera philoxeroides*, an invasive freshwater weed. Field surveys and experiments indicated that increased damage from *A. hygrophila* on the non-target *A. sessilis* coincided with higher temperatures (Lu *et al.* 2015). Thus, a general warming trend or greater incidence of temperature extremes could lead this beetle to expand its range and overwinter on non-target plants. This research and others suggest that climate should be carefully considered when planning releases and assessing the risk to non-targets, for example by modelling potential spread (Mukherjee *et al.* 2012) or thermal tolerance of biocontrol agents (Allen *et al.* 2014), or by examining the response of biocontrol agents and their intended targets using experimental manipulation of climate variables (Baffoe *et al.* 2012; Benzemer *et al.* 1998; Chen *et al.* 2007).

5.4 Ecological Impact Assessment and Cost–benefit Analysis

The general ecological impact assessment framework for other pest control technologies, such as transgenic insect-control plants (e.g. Environmental Protection Agency 1998) (see Chapter 6), may be applicable also to biocontrol. This impact or risk assessment framework includes three primary phases: problem formulation, analysis and risk characterization. In problem formulation, researchers identify and evaluate ecological concerns so that appropriate goals (e.g. protection of a specific group or taxon of non-target organism) and assessment endpoints (e.g. number of non-target organisms selected for host specificity testing) can be selected and conceptual models developed. During the analysis phase, researchers evaluate the exposure of a list of potential non-target organisms to the biocontrol agent and define the relationship between the exposure levels and ecological effects. During risk characterization, researchers estimate risk of biocontrol agents through integration of exposure and agent-response profiles, describe risk by discussing lines of evidence and determine ecological adversity. The interface among researchers (or risk assessors), risk managers and stakeholders during the planning stage and communication of risk as well as at the end of the risk assessment is critical to ensure that the results of the assessment can be used to support a management decision. Because of the diverse expertise required (especially in complex ecological risk assessments), risk assessors and risk managers frequently work in multi-disciplinary teams.

In recent years, there have been more detailed examinations of the ecological risks associated with biocontrol in an effort to develop more predictive guidelines (Louda *et al.* 2003). In this line, host range testing and identifying potential indirect non-target effects have been highlighted as areas where ecology can help to improve the safety of biocontrol (Fowler *et al.* 2012). Although various negative environmental impacts (e.g. attacks on non-targets) have been alleged to be associated with biocontrol in general (Howarth 1991), there appear to be relatively few examples of significant ecological damage caused by arthropod natural enemies used in modern biocontrol programmes.

The economic costs and benefits of biocontrol have been studied on numerous occasions (see Chapters 3 and 16), but the ecological costs and benefits to the use of arthropods as biocontrol agents are much more difficult to quantify. This is mainly because of the difficulty in assigning a monetary value to a non-target organism itself (e.g. an arthropod or plant species), particularly in association with its ecological services, which are often not well documented or understood (de Lange and van Wilgen 2010) (see also Chapters 7 and 8). The risk (or cost) of biocontrol has to be considered relative to its potential benefit in developing a sustainable pest management programme, particularly in comparison with other pest control options such as the use of synthetic insecticides or no action. Below, we discuss the environmental impacts of arthropod biocontrol using two recent case studies from North America, and highlight some of the ecological processes and cost–benefit analyses involved.

5.5 Case Study I: Biocontrol of Emerald Ash Borer (*Agrilus planipennis*)

5.5.1 Background

Emerald ash borer (EAB), *Agrilus planipennis* Fairmaire (Coleoptera: Buprestidae), is an invasive forest insect native to north-eastern Asia (China, the Korean Peninsula and the Russian Far East), that is thought to have been accidentally introduced into North America in the 1990s (Siegert *et al.* 2014). It is a serious pest of North American ash trees (*Fraxinus* spp.), where in sufficient numbers larval galleries can effectively girdle and kill host trees in 2–3 years after the initial infestation. In its native range, where Asian ash species have co-evolved with this beetle, trees appear to have effective host defences against it (Rebek *et al.* 2008), and a complex of parasitoids supplement host defences to help suppress populations of EAB (Duan *et al.* 2012b; Liu *et al.* 2007).

The ecological impacts of the EAB invasion in North America are already widespread, and EAB has the potential to functionally extirpate ash trees from the continent (Klooster *et al.* 2014). An estimated 282 arthropod species are dependent on ash to varying extents (Gandhi and Herms 2010), including iconic insects such as the western Hercules beetle, *Dynastes granti* Horn (Coleoptera: Scarabaeidae) (Wagner and Todd 2016). Additionally, the loss of ash has already affected taxa such as amphibians (Stephens *et al.* 2013), beetles (Gandhi *et al.* 2014) and earthworms (Ulyshen *et al.* 2011), and caused wider effects on ecosystem processes such as nutrient cycling (Flower *et al.* 2013). Economic impacts of EAB are also significant, with management costs running into hundreds of millions of dollars each year (Aukema *et al.* 2011).

Management options include chemical treatments or physical removal and replacement of ash trees (McCullough and Mercader 2012; McCullough *et al.* 2011), as well as biocontrol (Bauer *et al.* 2015a).

5.5.2 Biocontrol Agents and Their Ecology

The EAB biocontrol programme was initiated in 2003 with foreign exploration in China and, by 2004, three promising natural enemies of EAB had been discovered: the larval parasitoids *Tetrastichus planipennisi* (Hymenoptera: Eulophidae) (Figure 5.2) and *Spathius agrili* (Hymenoptera: Braconidae), and the egg parasitoid *Oobius agrili* (Hymenoptera: Encyrtidae). Although other generalist parasitoids such as *Sclerodermus pupariae* (Hymenoptera: Bethylidae) have been found parasitizing EAB, these were not considered for biocontrol in North America because of their generalist nature (Yang *et al.* 2012).

Figure 5.2 Larvae and adults of the parasitoid wasp *Tetrastichus planipennisi* (Hymenoptera: Eulophidae) emerging from a gallery of their host, the emerald ash borer larva, *Agrilus planipennis* Fairmaire (Coleoptera: Buprestidae). Photo credit: Jian J. Duan, United States Department of Agriculture.

After collection, extensive host range specificity testing was conducted in quarantine for all three parasitoid species (Bauer *et al.* 2015a). *Tetrastichus planipennisi* was tested in no-choice assays against eight buprestid species, five cerambycids and one sawfly (Liu and Bauer 2007). Results from these tests indicated that *T. planipennisi* attacked only EAB larvae. The other two parasitoid species were also tested for host specificity but the results were not quite as conclusive. *Spathius agrili* parasitized six of the other nine *Agrilus* species it was tested against, although the rates of parasitism were significantly lower than for EAB (Yang *et al.* 2008). *Oobius agrili* was tested against eggs from six species of *Agrilus*, two cerambycids and four lepidopterans (Bauer and Liu 2007). In no-choice tests, *O. agrili* was found to parasitize three species of *Agrilus* with large eggs similar in size to those of EAB, but in paired choice tests *O. agrili* consistently demonstrated a preference for EAB eggs. All three species (*T. planipennisi*, *S. agrili*, and *O. agrili*) were nonetheless approved for release by the North American Plant Protection Organization (NAPPO) in 2006, and releases started in 2007.

More recently, another braconid parasitoid of EAB larvae, *Spathius galinae*, was discovered in the Russian Far East and underwent host specificity testing. Some concerns regarding the EAB biocontrol programme had been raised by Simberloff (2012), particularly regarding the risks to native buprestids. Consequently, even more extensive host testing was conducted for *S. galinae*. Using choice and no-choice tests with 15 different species of host larvae (13 wood-boring beetles, one clearwing moth and one sawfly) infesting host plants such as ash, birch, maple and/or red oak, Duan *et al.* (2015b) found that *S. galinae* only attacked one non-target species of North American woodborer, the goldspotted oak borer (*Agrilus coxalis*), which itself is an invasive pest in California. Based on these results, *S. galinae* was approved by NAPPO in 2013 and releases began in 2015.

The interactions among several of the EAB biocontrol agents have been investigated in both laboratory and field experiments. For example, experimental work with *T. planipennisi* and *S. agrili* suggested that *S. agrili* might outcompete *T. planipennisi* in the field (Ulyshen *et al.* 2010). However, multiple years of release and recovery data have recovered far fewer *S. agrili* than *T. planipennisi* (Duan *et al.* 2012a, 2015a; Jennings *et al.* 2015). Although the cause of this trend might be more related to the phenology of *S. agrili* and EAB in North America, this example serves as an indicator that results from laboratory experiments might not always mirror the outcomes of field releases.

More recently, competition between *T. planipennisi* and *S. galinae* was examined under various different experimental conditions (Wang *et al.* 2015), including different host densities, parasitoid densities, host plant sizes and host–parasitoid ratios. Ultimately, little evidence for competition between these two species was found. Thus, it appears that any negative effects on biocontrol from releasing multiple parasitoids will be minimal.

Future climate change in North America could have considerable effects on the EAB biocontrol programme, and some research has been conducted in this area for both EAB and some of its biocontrol agents (DeSantis *et al.* 2013; Duan *et al.* 2014; Liang and Fei 2014). At present, winter temperatures in most of the USA and many parts of Canada are rarely low enough to kill overwintering EAB larvae (DeSantis *et al.* 2013). However, predicted increases in temperature could cause the climatic niche overlap between EAB and their North American ash tree hosts to diverge (Liang and Fei 2014).

Temperature is also known to strongly affect how long EAB eggs are susceptible to parasitism, as well as the attack and diapause rates of the egg parasitoid *O. agrili* (Duan *et al.* 2014). It remains to be seen how effective *S. galinae* will be at establishing populations and suppressing EAB, but this parasitoid may be better suited to colder climates than *S. agrili* (Bauer *et al.* 2015b). However, the climate in many of the areas where *S. agrili* does not appear to have established (e.g. Michigan, Maryland, USA) could become better suited to this species in the future.

5.5.3 Ecological Impact Assessment and Cost–benefit Analysis

Ultimately, the economic losses and cascading ecological effects from wide-scale mortality of ash trees caused by EAB were judged to outweigh the possible risk of introducing parasitoids that might also attack native *Agrilus* beetles. To date, this decision appears to have been justified, as at least one of the parasitoids released (*T. planipennisi*) has successfully established at several sites and is helping to suppress populations of EAB (Duan *et al.* 2015a), without any documented evidence of attacks on native beetles. Notably, *T. planipennisi* is also the most host-specific of the biocontrol agents released. *Oobius agrili* has established populations in Michigan but does not disperse quickly, and while *S. agrili* has occasionally been recovered in a few locations, populations of this parasitoid do not appear to have established (Bauer *et al.* 2015a). Because releases only began in 2015, it is too early to tell if *S. galinae* has established populations anywhere.

Thus, the two biocontrol agents with the lowest host specificity have been the least successful at establishing in North America and, nine years after initial releases, no non-target effects from the EAB biocontrol parasitoids have been detected. Although there is undoubtedly a lack of field data regarding populations of non-target species, it appears as though the strategy of incorporating various aspects of ecology into the impact assessment helped to produce a thorough evaluation.

It seems most likely that an integrated approach to managing EAB will be required to help maintain viable populations of North American ash trees, particularly in urban ecosystems. While biocontrol has shown some promise, the parasitoids have not established fast enough to protect many trees. Consequently, chemical and mechanical methods will also need to be used, particularly in urban areas. In combination, the relatively minimal non-target risks from EAB biocontrol or trunk-injected pesticides are a far superior alternative to taking no action at all. Leaving ash trees to succumb to EAB would almost certainly result in secondary extinctions of arthropods as well as considerable changes to the structure of many forests throughout the continent (Wagner and Todd 2016).

5.6 Case Study II: Biocontrol of Tamarisk (*Tamarix* spp.)

5.6.1 Background

Tamarisk (*Tamarix* spp.) is a genus of shrubs or trees native to parts of Asia, Europe and Africa (Di Tomaso 1998; Everitt 1980, 1998; Hultine and Dudley 2013). Several species of tamarisk were introduced to the United States as ornamental or shade plants in the 1800s, but it was not until riparian systems were modified in the 20th century that these plants dramatically expanded their range in arid and semi-arid areas of the south-western USA (Di Tomaso 1998). This range expansion was expedited by the ability of tamarisk to colonize open substrates adjacent to streams and reservoirs (Lesica and Miles 2004). The two most common introduced species are *T. chinensis* and *T. ramosissima*, and notably most of the invasion is composed of hybrids of these species (Gaskin and Schaal 2002).

The spread of tamarisk has had many adverse effects on native taxa and ecosystem processes. For instance, tamarisk invasion has altered plant communities, displacing native species such as willow (*Salix* spp.) and cottonwood (*Populus* spp.) (Busch and Smith 1995). Additionally, removing tamarisk from riparian areas can increase the abundance of native fish (Kennedy *et al.* 2005). Leaf litter from tamarisk decomposes at a faster rate than leaf litter from the native plants it replaced, which can negatively affect aquatic macroinvertebrate communities (Bailey *et al.* 2001). Tamarisk invasion also has far-reaching ecosystem effects, such as increasing the severity of flooding and the frequency of fires (Drus 2013). Additionally, annual economic losses caused by tamarisk (from changes to water supplies, hydroelectric power generation and flood control) have been estimated to reach $285 million (Zavaleta 2000). Given the scale of the invasion, a range of management techniques have been utilized including the use of herbicides, mechanical removal of plants and biocontrol.

5.6.2 Biocontrol Agents and Their Ecology

The biocontrol programme for tamarisk was initiated in the 1970s (DeLoach *et al.* 1996; Hultine and Dudley 2013). One of the main challenges for tamarisk biocontrol was that a widely planted congener, the evergreen athel (*T. aphylla*), was considered a non-target species. One advantage, however, was that there were no native members of the Tamaricaceae family in the USA, and very few members of the allied Frankeniaceae. Foreign exploration in the native range of tamarisk subsequently discovered more than 300 candidate species for use in the programme, and a small sample was selected for more extensive host specificity testing (DeLoach *et al.* 1996). Among this group were the leaf beetle *Diorhabda carinulata* Debroschers (Coleoptera: Chrysomelidae) (originally considered as *D. elongata* Brullé) (Figure 5.3), the mealy bug *Trabutina mannipara* (Hemprich and Ehrenberg) (Hemiptera: Pseudococcidae) and the weevil *Coniatus tamarisci* F. (Coleoptera: Curculionidae) (Dudley and Bean 2012; Tracy and Robbins 2009).

Host specificity testing for some of these candidates was conducted in the 1990s (prior to any releases in the USA). DeLoach *et al.* (1996) reported on such trials for 15 species of insect, including *D. carinulata* and *T. mannipara*, among others. *Diorhabda carinulata* was tested against plants in the Tamaricaceae and was found to feed almost exclusively on those in the *Tamarix* genus (DeLoach *et al.* 1996). Additionally, no larvae reared on *Frankenia* spp. were able to successfully develop into reproductively functional adults. *Trabutina mannipara* was tested on 14 genera of Caryophyllales as well as willow and cottonwood, and larvae were able to survive only on *Tamarix* spp. (DeLoach *et al.* 1996). Several other species tested (including moths and psyllids) indicated that they might have suitable host ranges to be deployed in future biocontrol efforts with minimal risk to non-target species.

In addition to pre-release testing, laboratory and field assessments to further examine the risk to non-targets were conducted after *D. carinulata* was approved for caged field

Figure 5.3 Tamarisk beetle, *Diorhabda elongate* (Coleoptera: Chrysomelidae). Photo credit: Eric Coombs, Oregon Department of Agriculture, www.bugwood.org.

releases in 1998–1999 and open field releases in 2001 (Lewis *et al.* 2003a). For instance, Dudley and Kazmer (2005) examined the effects of *D. carinulata* populations in the field on two species of *Frankenia* (plants in the same suborder as *Tamarix* spp.). Their study indicated that, under field conditions, *D. carinulata* posed very little threat to *Frankenia* spp., in slight contrast to earlier laboratory research (Lewis *et al.* 2003b). Thomas *et al.* (2010) compared host plant preference between *D. carinulata* beetles that had become established in the field with those from the original laboratory colony, and found that the former had a lower threshold for host acceptance. More specifically, the source colony exhibited a preference for *T. ramosissima* over *T. parviflora*, while the beetles collected from the field had no significant feeding preference between the two species of tamarisk (Thomas *et al.* 2010). Similar results were found by Dudley *et al.* (2012), who showed that *D. carinulata* had a general preference for using *T. ramosissima* ahead of *T. parviflora*. Herr *et al.* (2014) also investigated host plant selection and oviposition preference of *D. carinulata* on *Tamarix* spp. and *Frankenia* spp. using field cages and open field tests. Under field conditions, *Tamarix* spp. were significantly preferred over *Frankenia* spp., suggesting that some laboratory experiments may have overestimated the risk to non-target plants (Herr *et al.* 2014). These examples further highlight how the findings from laboratory experiments should be interpreted cautiously before extrapolating them to outcomes in the field.

Since being released, *D. carinulata* populations have successfully established and dispersed widely in a relatively short period of time (Bean *et al.* 2013). Specifically, 3 years after releases were conducted in Nevada, USA, *D. carinulata* numbers had expanded to defoliate an area of around 200^ha and more than an order of magnitude more the following year (Dudley 2005; Pattison *et al.* 2011). Establishment was also found at release locations in Wyoming, Utah and Colorado and, consistent with the results from much of the non-target organism testing, few direct effects on non-targets have been detected. Populations of *D. carinulata* did not successfully establish in California, probably because a day length diapause cue meant that no establishment was detected at any sites at a latitude below 37° (Bean *et al.* 2007; Dudley *et al.* 2012). A further possible cause for the lack of establishment is that a different species of tamarisk (*T. parviflora*) is more dominant there, compared with *T. ramosissima* at the other release locations.

5.6.3 Ecological impact Assessment and Cost–benefit Analysis

In terms of its objective, biocontrol of tamarisk has shown considerable success, with natural enemies establishing, dispersing and defoliating populations of these invasive plants (Dudley and Bean 2012). However, the expansion of the tamarisk biocontrol programme has been hindered because of concerns over the potential risk that biocontrol may pose to an endangered subspecies of willow flycatcher (*Empidonax traillii extimus* Phillips), a bird which sometimes uses tamarisk for nesting sites. In some areas now dominated by tamarisk, the willow flycatcher nests in these trees because native species (e.g. willows) have been displaced, which provides an example of how biocontrol can indirectly affect ecosystems through the process of ecological replacement (when invasive species physically or functionally replace native species) (Pearson and Callaway 2003). If tamarisk is removed from the environment, there is concern that, without accompanying restoration efforts, less nesting habitat will be available for willow flycatchers, and that even in tamarisk that remain, defoliation by biocontrol agents will lead to nests being exposed to heat and desiccation. Even though some data suggest that willow flycatcher populations have responded well to the unassisted restoration of ecosystems where tamarisk densities have been greatly reduced (Ahlers and Moore 2009), concerns regarding willow flycatcher conservation still led to a lawsuit being filed against the agency that approved *D. carinulata* releases (the United States Department of Agriculture – Animal and Plant Health Inspection Service) and caused lengthy delays to the implementation and expansion of the biocontrol programme (Dudley and Bean 2012).

Removing such a widespread and established invasive plant could also impact communities of small mammals (Longland 2014) and amphibians and reptiles (Bateman *et al.* 2015), as well as having potentially effects on ecosystem processes such as leaf litter decomposition (Uselman *et al.* 2011) and nutrient cycling (Hultine *et al.* 2010). Without concurrent plans to restore areas where tamarisk has been removed, there is some concern that ecosystems could further be vulnerable to other invasive species such as Russian knapweed (*Acroptilon repens*) (Dudley 2005).

Alternative approaches to biocontrol such as chemical and mechanical treatments have been utilized for tamarisk management. For example, chemical management of tamarisk has been attempted with herbicides such as imazapyr (Duncan and McDaniel 1998). Imazapyr alone or in combination with glyphosate can reduce tamarisk density by over 90% (Duncan and McDaniel 1998). However, the widespread use of herbicides such as imazapyr presents some degree of risk to non-target organisms (Douglass *et al.* 2016; Kaeser and Kirkman 2010), necessitates frequent reapplication of chemicals and typically is combined with mechanical control measures. Mechanical control (e.g. bulldozing, excavating) can also be costly and require repeated treatments, but this management strategy has been used for tamarisk (Ostoja *et al.* 2014).

Although these methods can be quite successful at suppressing populations of tamarisk, they may still fall short of complete eradication. Mechanical removal of tamarisk may provide small improvements to native ecosystems in terms of species diversity, although it remains unclear how much this improves habitat for wildlife (Ostoja *et al.* 2014).

Numerous studies have shown that tamarisks have negative effects on ecosystems, and that ultimately removing these species is better for native ecosystems in the long term. Biocontrol has been highly effective at defoliating tamarisk and conserving water resources (Pattison *et al.* 2011), and generally with fewer negative effects on ecosystems and at a lower cost than other methods such as chemical and mechanical control. Indeed, one of the advantages of tamarisk biocontrol is that because it is a slower process compared with other forms of control, there is more chance for native plants and other species to respond as the target plants decline. However, this case study highlights the importance of engaging relevant stakeholders in management decisions, and also the need to consider ecological restoration in weed biocontrol in general. More generally, it provides an example of why it is important for all stakeholders to consider the long-term impacts of biocontrol. Ultimately, biocontrol aims to ameliorate the environmental effects of invasive or pest species by suppressing their populations, and consequently biocontrol agents will almost certainly indirectly affect other species in target ecosystems in some way. Although short-term effects on ecosystems (e.g. loss of potential nesting habitat) can sometimes be perceived as negative, it is important to maintain a holistic view of the biocontrol process and utilize the best available knowledge when developing management strategies. In the case of tamarisk biocontrol, most of the putative adverse effects on the willow flycatcher and some other taxa have thus far failed to emerge.

5.7 Concluding Remarks

Discussions regarding the use of biocontrol fundamentally involve trade-offs. In some cases, the invasive species targeted by biocontrol agents can be severely damaging to the environment and jeopardize the conservation of rare species (see Chapter 10). For instance, the invasive gypsy moth has been demonstrated as having negative effects on native species of Lepidoptera (Wagner and van Driesche 2010), EAB threatens obligate feeders on ash (Wagner and Todd 2016), and invasive weeds like *M. quinquenervia* and tamarisk have dramatically changed native ecosystems (Bailey *et al.* 2001; Drus 2013; Gordon 1998; Rayamajhi *et al.* 2009).

Biological invasions have been increasing as global trade increases (see Chapter 10), and consequently there is even greater need for effective and safe biocontrol programmes. For example, invasive wood-boring arthropods, in particular, are being detected more frequently (Aukema *et al.* 2010), and in natural forests biocontrol often represents the best (potentially even the only) option for long-term suppression of pest populations. However, even though there are examples of arthropod biocontrol agents attacking non-target organisms, larger scale negative effects on ecosystems are rare. Indeed, for weed biocontrol, these agents may have a biosafety record of more than 99% (Suckling and Sforza 2014). Therefore, when viewed in the context of sustainability, the benefits of biocontrol using arthropods often appear to far outweigh the risks (Bale *et al.* 2008).

To ensure that risks to the environment remain acceptably low, adequate consideration must be given to the numerous ecological processes affecting arthropod biocontrol agents. This includes all three of the main phases of the risk assessment process. In the problem formulation and analysis phases, assessment endpoints should include extensive host specificity testing in quarantine accounting for potential spread beyond the targeted system over time. In the risk characterization phase, effects of interactions among biocontrol agents, and also between biocontrol agents and other native natural enemies, should also be included.

Finally, throughout all phases of risk assessment, the influence of these ecological processes ought to be considered under predicted scenarios of global climatic change. Including these different factors in risk assessments should help to ensure that negative environmental impacts of arthropod biocontrol are minimized.

Acknowledgements

We thank the editors, Tom Dudley (University of California, Santa Barbara, CA, USA), Doug Luster (United States Department of Agriculture, MD, USA), Kim Hoelmer (United States Department of Agriculture, DE, USA) and one anonymous reviewer for providing comments that greatly improved this manuscript.

References

Ahlers, D. and Moore, D. (2009) *A Review of Vegetation and Hydrologic Parameters Associated With the Southwestern Willow Flycatcher – 2002 to 2008 Elephant Butte Reservoir Delta, NM*. USDI-BOR Technical Service Center, Denver, CO, USA.

Allen, J.L., Clusella-Trulls, S. and Chown, S.L. (2014) Thermal tolerance of *Cyrtobagous salviniae*: a biocontrol agent in a changing world. *BioControl* **59**: 357–366.

Aukema, J.E., McCullough, D.G., Von Holle, B., Liebhold, A.M., Britton, K. and Frankel, S.J. (2010) Historical accumulation of nonindigenous forest pests in the continental United States. *BioScience* **60**: 886–897.

Aukema, J.E., Leung, B., Kovacs, K., *et al.* (2011) Economic impacts of non-native forest insect in the continental United States. *PLoS ONE* **6**(9): e24587.

Baffoe, K.O., Dalin, P., Nordlander, G. and Stenberg, J.A. (2012) Importance of temperature for the performance and biocontrol efficiency of the parasitoid *Perilitus brevicollis* (Hymenoptera: Braconidae) on *Salix*. *BioControl* **57**: 611–618.

Bailey, J.K., Schweitzer, J.A. and Whitham, T.G. (2001) Salt cedar negatively affects biodiversity of aquatic macroinvertebrates. *Wetlands* **21**: 442–447.

Bale, J.S., van Lenteren, J.C. and Bigler, F. (2008) Biological control and sustainable food production. *Philosophical Transactions of the Royal Society B – Biological Sciences* **363**: 761–776.

Bateman, H.L., Merritt, D.M., Glenn, E.P. and Nagler, P.L. (2015) Indirect effects of biocontrol of an invasive riparian plant (*Tamarix*) alters habitat and reduces herpetofauna abundance. *Biological Invasions* **17**: 87–97.

Bauer, L.S. and Liu, H. (2007) *Oobius agrili* (Hymenoptera: Encyrtidae), a solitary egg parasitoid of emerald ash borer from China. In: Mastro, V., Lance, D., Reardon, R. and Parra, G. (eds) *Proceedings of the 2006 Emerald Ash Borer and Asian Longhorned Beetle Research and Technology Development Meeting*. USDA Forest Service, Morgantown, WV, USA, pp. 63–64.

Bauer, L.S., Duan, J.J., Gould, J.R. and van Driesche, R. (2015a) Progress in the classical biological control of *Agrilus planipennis* Fairmaire (Coleoptera: Buprestidae) in North America. *Canadian Entomologist* **147**: 300–317.

Bauer, L.S., Duan, J.J., Lelito, J.P., Liu, H. and Gould, J.R. (2015b) Biology of emerald ash borer parasitoids. In: van Driesche, R.G. and Reardon, R.C. (eds) *Biology and Control of Emerald Ash Borer*. USDA Forest Service, Morgantown, WV, USA, pp. 97–112.

Bean, D.W., Dudley, T.L. and Keller, J.C. (2007) Seasonal timing of diapause induction limits the effective range of *Diorhabda elongata deserticol*a (Coleoptera: Chrysomelidae) as a biological control agent for tamarisk (*Tamarix* spp.). *Environmental Entomology* **36**: 15–25.

Bean, D.W., Dudley, T. and Hultine, K. (2013) Bring on the beetles! The history and impact of tamarisk biological control. In: Sher, A. and Quigley, M.F. (eds) *Tamarix: A Case Study of Ecological Change in the American West*. Oxford University Press, New York, USA, pp. 377–403.

Benzemer, T.M., Jones, T.H. and Knight, K.J. (1998) Long-term effects of elevated CO_2 and temperature on populations of the peach potato aphid *Myzus persicae* and its parasitoid *Aphidius matricariae*. *Oecologia* **116**: 128–135.

Bigler, F., Babendreier, D. and Kuhlmann, U. (2006) *Environmental Impact of Invertebrates for Biological Control of Arthropods: Methods and Risk Assessment*. CABI, Wallingford, UK.

Boettner, G.H., Elkinton, J.S. and Boettner, C.J. (2000) Effects of a biological control introduction on three nontarget native species of Saturniid moths. *Conservation Biology* **14**: 1798–1806.

Bonsall, M.B. and Hassell, M.P. (1997) Apparent competition structures ecological assemblages. *Nature* **388**: 371–373.

Bonsall, M.B. and Hassell, M.P. (1998) Population dynamics of apparent competition in a host- parasitoid assemblage. *Journal of Animal Ecology* **67**: 918–929.

Brown, P.M.J., Adriaens, T., Bathon, H., *et al.* (2008) *Harmonia axyridis* in Europe: spread and distribution of a non-native coccinellid. *BioControl* **53**: 5–21.

Busch, D.E. and Smith, S.D. (1995) Mechanisms associated with decline of woody species in riparian ecosystems of the southwestern U.S. *Ecological Monographs* **65**: 347–370.

Carvalheiro, L.G., Buckley, Y.M., Ventim, R., Fowler, S.V. and Memmott, J. (2008) Apparent competition can compromise the safety of highly specific biocontrol agents. *Ecology Letters* **11**: 690–700.

Cascone, P., Carpenito, S., Slotsbo, S., *et al.* (2015) Improving the efficiency of *Trichogramma achaeae* to control *Tuta absoluta*. *BioControl* **60**: 761–771.

Center, T.D., Van, T.K., Rayachhetry, M., *et al.* (2000) Field colonization of the melaleuca snout beetle (*Oxyops vitiosa*) in south Florida. *Biological Control* **19**: 112–123.

Center, T.D., Pratt, P.D., Tipping, P.W., *et al.* (2006) Field colonization, population growth, and dispersal of *Boreioglycaspis melaleucae* Moore, a biological control agent of the invasive tree *Melaleuca quinquenervia* (Cav.) Blake. *Biological Control* **39**: 363–374.

Chailleux, A., Mohl, E.K., Teixeira Alves, M., Messelink, G.J. and Desneux, N. (2014) Natural enemy-mediated indirect interactions among prey species: potential for enhancing biocontrol services in agroecosystems. *Pest Management Science* **70**: 1769–1779.

Chen, F.J., Wu, G., Parajulee, M.N. and Ge, F. (2007) Impact of elevated CO_2 on the third trophic level: a predator *Harmonia axyridis* and a parasitoid *Aphidius picipes*. *Biocontrol Science and Technology* **17**: 313–324.

Coll, M. and Hughes, L. (2008) Effects of elevated CO_2 on an insect omnivore: a test for nutritional effects mediated by host plants and prey. *Agriculture, Ecosystems and Environment* **123**: 271–279.

Crowder, D.W. and Harwood, J.D. (2014) Promoting biological control in a rapidly changing world. *Biological Control* **75**: 1–7.

de Lange, W.J. and van Wilgen, B.W. (2010) An economic assessment of the contribution of biological control to the management of invasive alien plants and to the protection of ecosystem services in South Africa. *Biological Invasions* **12**: 4113–4124.

DeLoach, C.J., Gerling, D., Fornasari, L., *et al.* (1996) Biological control programme against saltcedar (*Tamarix* spp.) in the United States of America: progress and problems. In: Moran, V.C. and Hoffmann, J.H. (eds) *Proceedings of the IX International Symposium on Biological Control of Weeds*. University of Cape Town, Cape Town, South Africa, pp. 253–260.

DeSantis, R.D., Moser, W.M., Gormanson, D.D., Bartlett, M.G. and Vermunt, B. (2013) Effects of climate on emerald ash borer mortality and the potential for ash survival in North America. *Agricultural and Forest Meteorology* **178–179**: 120–128.

Di Tomaso, J.M. (1998) Impact, biology, and ecology of saltcedar (*Tamarix* spp.) in the southwestern United States. *Weed Technology* **12**: 326–336.

Douglass, C.H., Nissen, S.J. and Knis, A.R. (2016) Efficacy and environmental fate of imazapyr from directed helicopter applications targeting *Tamarix* species infestations in Colorado. *Pest Management Science* **72**: 379–387.

Drus, G.M. (2013) Fire ecology of *Tamarix*. In: Sher, A. and Quigley, M.F. (eds) *Tamarix: A Case Study of Ecological Change in the American West*. Oxford University Press, New York, USA, pp. 240–255.

Duan, J.J., Bauer, L.S., Abell, K.J. and van Driesche, R.G. (2012a) Population responses of hymenopteran parasitoids to the emerald ash borer (Coleoptera: Buprestidae) in recently invaded areas in the north central United States. *BioControl* **57**:199–209.

Duan, J.J., Yurchenko, G. and Fuester, R.W. (2012b) Occurrence of emerald ash borer (Coleoptera: Buprestidae) and biotic factors affecting its immature stages in the Russian Far East. *Environmental Entomology* **41**: 245–254.

Duan, J.J., Jennings, D.E., Williams, D.C. and Larson, K.M. (2014) Patterns of parasitoid host utilization and development across a range of temperatures: implications for biological control of an invasive forest pest. *BioControl* **59**: 659–669.

Duan, J.J., Bauer, L.S., Abell, K.J., Ulyshen, M.D. and van Driesche, R.G. (2015a) Population dynamics of an invasive forest insect and associated natural enemies in the aftermath of invasion: implications for biological control. *Journal of Applied Ecology* **52**: 1246–1254.

Duan, J.J., Gould, J.R. and Fuester, R.W. (2015b) Evaluation of the host specificity of *Spathius galinae* (Hymenoptera: Braconidae), a larval parasitoid of the emerald ash borer (Coleoptera: Buprestidae) in Northeast Asia. *Biological Control* **89**: 91–97.

Dudley, T.L. (2005) Progress and pitfalls in the biological control of saltcedar (*Tamarix* spp.) in North America. In: Gottschalk, K.W. (ed.) *Proceedings of the 16th US Department of Agriculture Interagency Research Forum on Gypsy Moth and Other Invasive Species*. USDA Forest Service, Morgantown, WV, USA, pp 12–15.

Dudley, T.L. and Bean, D.W. (2012) Tamarisk biocontrol, endangered species risk and resolution of conflict through riparian restoration. *BioControl* **57**: 331–347.

Dudley, T.L. and Kazmer, D.J. (2005) Field assessment of the risk posed by *Diorhabda elongata*, a biocontrol agent for control of saltcedar (*Tamarix* spp.), to a nontarget plant, *Frankenia salina*. *Biological Control* **35**: 265–275.

Dudley, T.L., Bean, D.W., Pattison, R.R. and Caires, A. (2012) Selectivity of a biological control agent, *Diorhabda carinulata* Desbrochers, 1870 (Coleoptera: Chrysomelidae) for host species within the genus *Tamarix* Linneaus, 1753. *Pan-Pacific Entomologist* **88**: 319–341.

Dukes, J.S. and Mooney, H.A. (1999) Does global change increase the success of biological invaders? *Trends in Ecology and Evolution* **14**: 135–139.

Duncan, K.W. and McDaniel, K.C. (1998) Saltcedar (*Tamarix* spp.) management with imazapyr. *Weed Technology* **12**: 337–344.

Ehler, L.E. (1998) Invasion biology and biological control. *Biological Control* **13**: 127–133.

Elkinton, J.S. and Boettner, G.H. (2012) Benefits and harm caused by the introduced generalist tachinid, *Compsilura concinnata*, in North America. *BioControl* **57**: 277–288.

Environmental Protection Agency (1998) Guidelines for ecological risk assessment. Available at: www.epa.gov/risk/guidelines-ecological-risk-assessment (accessed 2 March 2017).

Everitt, B.L. (1980) Ecology of saltcedar – a plea for research. *Environmental Geology* **3**: 77–84.

Everitt, B.L. (1998) Chronology of the spread of tamarisk in the central Rio Grande. *Wetlands* **18**: 658–668.

Fagan, W.F., Lewis, M.A., Neubert, M.G. and van den Driessche, P. (2002) Invasion theory and biological control. *Ecology Letters* **5**: 148–157.

Flower, C.E., Knight, K.S. and Gonzalez-Meler, M.A.(2013) Impacts of the emerald ash borer (*Agrilus planipennis*) induced ash (*Fraxinus* spp.) mortality on forest carbon cycling and successional dynamics in the eastern United States. *Biological Invasions* **15**: 931–944.

Follett, P.A. and Duan, J.J. (2000) *Nontarget Effects of Biological Control*. Springer, Boston, MA, USA.

Fowler, S.V., Paynter, Q., Dodd, S. and Groenteman, R. (2012) How can ecologists help practitioners minimize non-target effects in weed biocontrol? *Journal of Applied Ecology* **49**: 307–310.

Gandhi, K.J.K. and Herms, D.A. (2010) North American arthropods at risk due to widespread *Fraxinus* mortality caused by the alien emerald ash borer. *Biological Invasions* **12**: 1839–1846.

Gandhi, K.J.K., Smith, A., Hartzler, D.M. and Herms, D.A. (2014) Indirect effects of emerald ash borer-induced ash mortality and canopy gap formation on epigaeic beetles. *Environmental Entomology* **43**: 546–555.

Gaskin, J.F. and Schaal, B.A. (2002) Hybrid *Tamarix* widespread in U.S. invasion and undetected in native Asian range. *Proceedings of the National Academy of Sciences USA* **99**: 11256–11259.

Gerard, P.J., Barringer, J.R.F., Charles, J.G., *et al.* (2013) Potential effects of climate change on biological control systems: case studies from New Zealand. *BioControl* **58**: 149–162.

Gomez-Marco, F., Urbaneja, A., Jaques, J.A., Rugman-Jones, P.F., Stouthamer, R. and Tena, A. (2015) Untangling the aphid-parasitoid food web in citrus: can hyperparasitoids disrupt biological control? *Biological Control* **81**: 111–121.

Gordon, D.R. (1998) Effects of invasive, non-indigenous plant species on ecosystem processes: lessons from Florida. *Ecological Applications* **8**: 975–989.

Grevstad, F.S. (1999) Experimental invasions using biological control introductions: the influence of release size on the chance of population establishment. *Biological Invasions* **1**: 313–323.

Guzmán, C., Aguilar-Fenollosa, E., Sahún, R.M., *et al.* (2016) Temperature-specific competition in predatory mites: implications for biological pest control in a changing climate. *Agriculture, Ecosystems, and Environment* **216**: 89–97.
Haack, R.A., Herard, F., Sun, J.H. and Turgeon, J.J. (2010) Managing invasive populations of Asian longhorned beetle and citrus longhorned beetle: a worldwide perspective. *Annual Review of Entomology* **55**: 521–546.
Hance, T., van Baaren, J., Vernon, P. and Boivin, G. (2007) Impact of extreme temperatures on parasitoids in a climate change perspective. *Annual Review of Entomology* **52**: 107–126.
Hayes, K.R. and Barry, S.C. (2008) Are there any consistent predictors of invasion success? *Biological Invasions* **10**: 483–506.
Hellmann, J.J., Byers, J.E., Bierwagen, B.G. and Dukes, J.S. (2008) Five potential consequences of climate change for invasive species. *Conservation Biology* **22**: 534–543.
Herr, J.C., Herrera-Reddy, A.M. and Carruthers, R.I. (2014) Field testing *Diorhabda elongata* (Coleoptera: Chrysomelidae) from Crete, Greece, to assess potential impact on nontarget native California plants in the genus *Frankenia*. *Environmental Entomology* **43**: 642–653.
Hoddle, M.S., Aponte, O., Kerguelen, V. and Heraty, J. (1999) Biological control of *Oligonychus perseae* (Acari: Tetranychidae) on avocado: I. Evaluating release timings, recovery and efficacy of six commercially available phytoseiids. *International Journal of Acarology* **25**: 211–219.
Holway, D.A. (1999) Competitive mechanisms underlying the displacement of native ants by the invasive Argentine ant. *Ecology* **80**: 238–251.
Howarth, F.G. (1983) Classic biocontrol: Panacea or Pandora's Box? *Proceedings of the Hawaiian Entomological Society* **24**: 239–244.
Howarth, F.G. (1991) Environmental impacts of classic biological control. *Annual Review of Entomology* **36**: 485–509.
Hultine, K.R. and Dudley, T.L. (2013) *Tamarix* from organism to landscape. In: Sher, A. and Quigley, M.F. (eds) *Tamarix: A Case Study of Ecological Change in the American West*. Oxford University Press, New York, USA, pp. 149–167.
Hultine, K.R., Belnap, J., van Riper III, C., *et al.* (2010) Tamarisk biocontrol in the western United States: ecological and societal implications. *Frontiers in Ecology and the Environment* **8**: 467–474.
Jeffs, C.T. and Lewis, O.T. (2013) Effects of climate warming on host-parasitoid interactions. *Ecological Entomology* **38**: 209–218.
Jennings, D.E., Duan, J.J. and Shrewsbury, P.M. (2015) Biotic mortality factors affecting emerald ash borer (*Agrilus planipennis*) are highly dependent on life stage and host tree crown condition. *Bulletin of Entomological Research* **105**: 598–606.
Jezorek, H.A., Stiling, P.D. and Carpenter, J.E. (2010) Targets of an invasive species: oviposition preference and larval performance of *Cactoblastis cactorum* (Lepidoptera: Pyralidae) on 14 North American opuntoid cacti. *Environmental Entomology* **39**: 1884–1892.
Jezorek, H.A., Baker, A.J. and Stiling, P.D. (2012) Effects of *Cactoblastis cactorum* on the survival and growth of North American *Opuntia*. *Biological Invasions* **14**: 2355–2367.
Johnson, D.M., Liebhold, A.M., Tobin, P.C. and Bjornstad, O.N. (2006) Allee effects and pulsed invasion by the gypsy moth. *Nature* **444**: 361–363.
Kaeser, M.J. and Kirkman, L.K. (2010) The effect of pre-and post-emergent herbicides on non-target native plant species of the longleaf pine ecosystem. *Journal of the Torrey Botanical Society* **137**: 420–430.

Kaser, J.M. and Heimpel, G.E. (2015) Linking risk and efficacy in biological control host-parasitoid models. *Biological Control* **90**: 49–60.

Kennedy, T.A., Finlay, J.C. and Hobbie, S.E. (2005) Eradication of invasive *Tamarix ramosissima* along a desert stream increases native fish density. *Ecological Applications* **15**: 2072–2083.

Klooster, W.S., Herms, D.A., Knight, K.S., *et al.* (2014) Ash (*Fraxinus* spp.) mortality, regeneration, and seed bank dynamics in mixed hardwood forests following invasion by emerald ash borer (*Agrilus planipennis*). *Biological Invasions* **16**: 859–873.

Lesica, P. and Miles, S. (2004) Ecological strategies for managing tamarisk on the C. M. Russell National Wildlife Refuge, Montana, USA. *Biological Conservation* **119**: 535–543.

Lewis, P.A., DeLoach, C.J., Knutson, A.E., Tracy, J.L. and Robbins, T.O. (2003a) Biology of *Diorhabda elongata deserticola* (Coleoptera: Chrysomelidae), an Asian leaf beetle for biological control of saltcedars (*Tamarix* spp.) in the United States. *Biological Control* **27**: 101–116.

Lewis, P.A., DeLoach, C.J., Herr, J.C., Dudley, T.L. and Carruthers, R.I. (2003b) Assessment of risk to native *Frankenia* shrubs from an Asian leaf beetle, *Diorhabda elongata deserticola* (Coleoptera: Chrysomelidae), introduced for biological control of saltcedars (*Tamarix* spp.) in the western United States. *Biological Control* **27**: 148–166.

Li, W., Li, J., Coudron, T.A., *et al.* (2008) Role of photoperiod and temperature in diapause induction of endoparasitoid wasp *Microplitis mediator* (Hymenoptera: Braconidae). *Annals of the Entomological Society of America* **101**: 613–618.

Liang, L. and Fei, S. (2014) Divergence of the potential invasion range of emerald ash borer and its host distribution in North America under climate change. *Climatic Change* **122**: 735–746.

Liebhold, A.M. and Tobin, P.C. (2008) Population ecology of insect invasions and their management. *Annual Review of Entomology* **53**: 387–408.

Liu, H. and Bauer, L.S. (2007) *Tetrastichus planipennisi* (Hymenoptera: Eulophidae), a gregarious larval endoparasitoid of emerald ash borer from China. In: Mastro, V., Lance, D., Reardon, R. and Parra, G. (eds) *Proceedings of the 2006 Emerald Ash Borer and Asian Longhorned Beetle Research and Technology Development Meeting*. USDA Forest Service, Morgantown, WV, USA, pp. 61–62.

Liu, H., Bauer, L.S., Miller, D.L., *et al.* (2007) Seasonal abundance of *Agrilus planipennis* (Coleoptera: Buprestidae) and its natural enemies *Oobius agrili* (Hymenoptera: Encyrtidae) and *Tetrastichus planipennisi* (Hymenoptera: Eulophidae) in China. *Biological Control* **42**: 61–71.

Lockwood, J.A. (1993) Environmental issues involved in biological control of rangeland grasshoppers (Orthoptera: Acrididae) with exotic agents. *Environmental Entomology* **22**: 503–518.

Lockwood, J.L., Hoopes, M.F. and Marchetti, M.P. (2007) *Invasion Ecology*. Blackwell, Malden, MA, USA.

Longland, W.S. (2014) Biological control of saltcedar (*Tamarix* spp.) by saltcedar leaf beetles (*Diorhabda* spp.): effects on small mammals. *Western North American Naturalist* **74**: 378–385.

Louda, S.M., Pemberton, R.W., Johnson, M.T. and Follett, P.A. (2003) Nontarget effects – the Achilles' heel of biological control? Retrospective analyses to reduce risk associated with biocontrol introductions. *Annual Review of Entomology* **48**: 365–396.

Lu, X., Siemann, E., He, M., Shao, X. and Ding, J. (2015) Climate warming increases biological control agent impact on a non-target species. *Ecology Letters* **18**: 48–56.

Lynch, L.D. and Thomas, M.B. (2000) Nontarget effects in the biocontrol of insects with insects, nematodes and microbial agents: the evidence. *Biocontrol News and Information* **21**: 117N–130N.

McCullough, D.G. and Mercader, R.J. (2012) SLAM in an urban forest: evaluation of potential strategies to Slow Ash Mortality caused by emerald ash borer (*Agrilus planipennis*). *International Journal of Pest Management* **57**: 9–23.

McCullough, D.G., Poland, T.M., Anulewicz, A.C., Lewis, P. and Cappaert, D. (2011) Evaluation of *Agrilus planipennis* control provided by emamectin benzoate and two neonicotinoid insecticides, one and two seasons after treatment. *Journal of Economic Entomology* **104**: 1599–1612.

Messing, R.H., Tremblay, M.N., Mondor, E.B., Foottit, R.G. and Pike, K.S. (2007) Invasive aphids attack native Hawaiian plants. *Biological Invasions* **9**: 601–607.

Morris, R.J., Müller, C.B. and Godfray, H.C.J. (2001) Field experiments testing for apparent competition between primary parasitoids mediated by secondary parasitoids. *Journal of Animal Ecology* **70**: 301–309.

Morris, R.J., Lewis, O.T. and Godfray, H.C.J. (2004) Experimental evidence for apparent competition in a tropical forest food web. *Nature* **428**: 310–313.

Muirhead, J.R., Leung, B., van Overdijk, C., *et al.* (2006) Modelling local and long-distance dispersal of invasive emerald ash borer *Agrilus planipennis* (Coleoptera) in North America. *Diversity and Distributions* **12**: 71–79.

Mukherjee, A., Diaz, R., Thom, M., Overholt, W.A. and Cuda, J.P. (2012) Niche-based prediction of establishment of biocontrol agents: an example with *Gratiana boliviana* and tropical soda apple. *Biocontrol Science and Technology* **22**: 447–461.

Ostoja, S.M., Brooks, M.L., Dudley, T. and Lee, S.R. (2014) Short-term vegetation response following mechanical control of saltcedar (*Tamarix* spp.) on the Virgin River, Nevada, USA. *Invasive Plant Science and Management* **7**: 310–319.

Parmesan, C. (2006) Ecological and evolutionary responses to recent climate change. *Annual Review of Ecology, Evolution, and Systematics* **37**: 637–669.

Pattison, R.R., d'Antonio, C.M., Dudley, T.L., Allander, K.K. and Rice, B. (2011) Early impacts of biological control on canopy cover and water use of the invasive saltcedar tree (*Tamarix* spp.) in western Nevada, USA. *Oecologia* **165**: 605–616.

Pearson, D.E. and Callaway, R.M. (2003) Indirect effects of host-specific biological control agents. *Trends in Ecology and Evolution* **18**: 456–461.

Pérez-Lachaud, G., Batchelor, T.P. and Hardy, I.C.W. (2004) Wasp eat wasp: facultative hyperparasitism and intra-guild predation by bethylid wasps. *Biological Control* **30**: 149–155.

Polgár, L.A. and Hardie, J. (2000) Diapause induction in aphid parasitoids. *Entomologia Experimentalis et Applicata* **97**: 2--27.

Pratt, P.D. and Center, T.D. (2012) Biocontrol without borders: the unintended spread of introduced weed biological control agents. *BioControl* **57**: 319–329.

Rayamajhi, M.B., Pratt, P.D., Center, T.D., Tipping, P.W. and Van, T.K. (2009) Decline in exotic tree density facilitates increased plant diversity: the experience from *Melaleuca quinquenervi*a invaded wetlands. *Wetlands Ecology and Management* **17**: 455–467.

Rebek, E.J., Herms, D.A. and Smitley, D.R. (2008) Interspecific variation in resistance to emerald ash borer (Coleoptera: Buprestidae) among North American and Asian ash (*Fraxinus* spp.). *Environmental Entomology* **37**: 242–246.

Romo, C.M. and Tylianakis, J.M. (2013) Elevated temperature and drought interact to reduce parasitoid effectiveness in suppressing hosts. *PLoS ONE* **8**(3): e58136.

Schellhorn, N.A., Kuhman, T.R., Olson, A.C. and Ives, A.R. (2002) Competition between native and introduced parasitoids of aphids: nontarget effects and biological control. *Ecology* **83**: 2745–2757.

Schooler, S.S., De Barro, P. and Ives, A.R. (2011) The potential for hyperparasitism to compromise biological control: why don't hyperparasitoids drive their primary parasitoid hosts extinct? *Biological Control* **58**: 167–173.

Schuldiner-Harpaz, T. and Coll, M. (2013) Effects of global warming on predatory bugs supported by data across geographic and seasonal climatic gradients. *PLoS ONE* **8**(6): e66622.

Siegert, N.W., McCullough, D.G., Leibhold, A.M. and Telewski, F.W. (2014) Dendrochronological reconstruction of the epicentre and early spread of emerald ash borer in North America. *Diversity and Distributions* **20**: 847–858.

Simberloff, D. (2009) The role of propagule pressure in biological invasions. *Annual Review of Ecology, Evolution, and Systematics* **40**: 81–102.

Simberloff, D. (2012) Risks of biocontrol for conservation purposes. *Biological Invasions* **57**: 263–276.

Snyder, W.E. and Evans, E.W. (2006) Ecological effects of invasive arthropod generalist predators. *Annual Review of Ecology, Evolution, and Systematics* **37**: 95–122.

Snyder, W.E., Clevenger, G.M. and Eigenbrode, S.D. (2004) Intraguild predation and successful invasion by introduced ladybird beetles. *Oecologia* **140**: 559–565.

Soberon, J., Golubov, J. and Sarukhan, J. (2001) The importance of *Opuntia* in Mexico and routes of invasion and impact of *Cactoblastis cactorum* (Lepidoptera: Pyralidae). *Florida Entomologist* **84**: 486–492.

Stephens, J.P., Berven, K.A. and Tiegs, S.D. (2013) Anthropogenic changes to leaf litter input affect the fitness of a larval amphibian. *Freshwater Biology* **58**: 1631–1646.

Stiling, P. (2010) Death and decline of a rare cactus in Florida. *Castanea* **75**: 190–197.

Stiling, P. and Cornelissen, T. (2005) What makes a successful biocontrol agent? A meta-analysis of biological control agent performance. *Biological Control* **34**: 236–246.

Stiling, P. and Simberloff, D. (2000) The frequency and strength of nontarget effects of invertebrate biological control agents of plant pests and weeds. In: Follett, P.A. and Duan, J.J. (eds) *Nontarget Effects of Biological Control*. Springer, Boston, MA, USA, pp. 31–43.

Strong, D.R. and Pemberton, R.W. (2000) Biological control of invading species – risk and reform. *Science* **288**: 1969–1970.

Suckling, D.M. and Sforza, R.F.H. (2014) What magnitude are observed non-target impacts from weed biocontrol? *PLoS ONE* **9**(1): e84847.

Symondson, W.O.C., Sunderland, K.D. and Greenstone, M.H. (2002) Can generalist predators be effective biocontrol agents? *Annual Review of Entomology* **47**: 561–594.

Thomas, H.Q., Zalom, F.G. and Roush, R.T. (2010) Laboratory and field evidence of post-release changes to the ecological host range of *Diorhabda elongata*: has this improved biological control efficiency? *Biological Control* **53**: 353–359.

Thomson, L.J., Macfayden, S. and Hoffmann, A.A. (2010) Predicting the effects of climate change on natural enemies of agricultural pests. *Biological Control* **52**: 296–306.

Tracy, J.L. and Robbins, T.O. (2009) Taxonomic revision and biogeography of the *Tamarix*-feeding *Diorhabda elongata* (Brullé, 1832) species group (Coleoptera: Chrysomelidae: Galerucinae: Galerucini) and analysis of their potential in biological control of tamarisk. *Zootaxa* **2101**: 1–152.

Tylianakis, J.M. and Binzer, A. (2014) Effects of global environmental changes on parasitoid-host food webs and biological control. *Biological Control* **75**: 77–86.

Ulyshen, M.D., Duan, J.J. and Bauer, L.S. (2010) Interactions between *Spathius agrili* (Hymenoptera: Braconidae) and *Tetrastichus planipennisi* (Hymenoptera: Eulophidae), larval parasitoids of *Agrilus planipennis* (Coleoptera: Buprestidae). *Biological Control* **52**: 188–193.

Ulyshen, M.D., Klooster, W.S., Barrington, W.T. and Herms, D.A. (2011) Impacts of emerald ash borer-induced tree mortality on leaf litter arthropods and exotic earthworms. *Pedobiologia* **54**: 261–265.

Uselman, S.M., Snyder, K.A. and Blank, R.R. (2011) Insect biological control accelerates leaf litter decomposition and alters short-term nutrient dynamics in a *Tamarix*-invaded riparian ecosystem. *Oikos* **120**: 409–417.

Van Driesche, R.G. and Murray, T.J. (2004) Overview of testing schemes and designs used to estimate host ranges. In: Van Driesche, R.G., Murray, T.J. and Reardon, R. (eds) *Assessing Host Ranges of Parasitoids and Predators Used for Classic Biological Control: A Guide to Best Practice*. USDA Forest Service, Morgantown, WV, USA, pp. 68–89.

Van Driesche, R.G., Hoddle, M. and Center, T. (2008) *Control of Pests and Weeds by Natural Enemies: An Introduction to Biological Control*. Blackwell, Malden, MA, USA.

Van Driesche, R.G., Carruthers, R.I., Center, T., *et al.* (2010) Classic biological control for the protection of natural ecosystems. *Biological Control* **54**:S2–S33.

Van Lenteren, J.C., Bale, J., Bigler, F., Hokkanen, H.M.T. and Loomans, A.J.M. (2006) Assessing risks of releasing exotic biological control agents of arthropod pests. *Annual Review of Entomology* **51**: 609–634.

Van Lenteren, J.C., Loomans, A.J.M., Babendreier, D. and Bigler, F. (2008) *Harmonia axyridis*: an environmental risk assessment for Northwest Europe. *BioControl* **53**: 37–54.

Wagner, D.L. and Todd, K.J. (2016) New ecological assessment for the emerald ash borer: a cautionary tale about unvetted host-plant literature. *American Entomologist* **62**: 26–35.

Wagner, D.L. and van Driesche, R.G. (2010) Threats posed to rare or endangered insects by invasions of nonnative species. *Annual Review of Entomology* **55**: 547–568.

Wajnberg, E., Scott, J.K. and Quimby, P.C. (2001) *Evaluating Indirect Ecological Effects of Biological Control*. CABI, Wallingford, UK.

Wang, X.G. and Messing, R.H. (2004) The ectoparasitic pupal parasitoid, *Pachycrepoideus vindemmiae* (Hymenoptera: Pteromalidae), attacks other primary tephritid fruit fly parasitoids: host expansion and potential non-target impact. *Biological Control* **31**: 227–236.

Wang, X.Y., Jennings, D.E. and Duan, J.J. (2015) Trade-offs in parasitism efficiency and brood size mediate parasitoid coexistence, with implications for biological control of the invasive emerald ash borer. *Journal of Applied Ecology* **52**: 1255–1263.

Willis, A.J. and Memmott, J. (2005) The potential for indirect effects between a weed, one of its biocontrol agents and native herbivores: a food web approach. *Biological Control* **35**: 299–306.

Wilson, J.R.U., Dormontt, E.E., Prentis, P.J., Lowe, A.J. and Richardson, D.M. (2009) Something in the way you move: dispersal pathways affect invasion success. *Trends in Ecology and Evolution* **24**: 136–144.

Yang, Z.Q., Wang, X.Y., Gould, J.R. and Wu, H. (2008) Host specificity of *Spathius agrili* Yang (Hymenoptera: Braconidae), an important parasitoid of the emerald ash borer. *Biological Control* **47**: 216–221.

Yang, Z.Q., Wang, X.Y., Yao, Y.X., Gould, J.R. and Cao, L.M. (2012) A new species of *Sclerodermus* (Hymenoptera: Bethylidae) parasitizing *Agrilus planipennis* (Coleoptera: Buprestidae) from China, with a key to Chinese species in the genus. *Annals of the Entomological Society of America* **105**: 619–627.

Zavaleta, E. (2000) The economic value of controlling an invasive shrub. *Ambio* **29**: 462–467.

6

Effects of Transgenic Crops on the Environment

Peter B. Woodbury, Antonio DiTommaso, Janice Thies,
Matthew Ryan and John Losey

6.1 Range and Scope of Transgenic Crops

Globally, the use of transgenic crops has increased rapidly during recent decades; they are now grown for food in 31 countries and for feed in 19 counties (Aldemita *et al.* 2015). The most commonly incorporated trait is herbicide tolerance (HT; e.g. crop tolerance to glyphosate and glufosinate), followed by insect resistance (IR; e.g. crops containing genes that produce insecticidal proteins derived from the soil bacterium *Bacillus thuringiensis* (Bt)) (Aldemita *et al.* 2015). Crops with stacked traits, those containing more than one trait, are becoming increasingly common. The crop with the largest number of varieties that contain single or stacked traits is maize, with stacked traits representing 30% of the total trait approvals (Aldemita *et al.* 2015). As of September 2013, for example, the USA Animal and Plant Health Inspection Service (APHIS) had approved 96 petitions for 145 transgenic crop releases to be sold as follows: maize, 30; cotton, 15; tomatoes, 11; soybeans, 12; rapeseed/canola, 8; potatoes, 5; sugar beets, 3; papaya, rice, and squash, 2 each; and alfalfa, plum, rose, tobacco, flax, and chicory, 1 each (Fernandez-Cornejo *et al.* 2014).

In the USA in 2015, a large proportion of the major field crops include some form of either pest or herbicide resistance traits, including soy (94%), maize (89%) and cotton (89%), covering about half of the total cropland (www.ers.usda.gov/data-products/adoption-of-genetically-engineered-crops-in-the-us/recent-trends-in-ge-adoption.aspx), with a majority of these containing stacked traits (Fernandez-Cornejo *et al.* 2014). In 2015, 77% of maize and 79% of cotton in the USA had both herbicide-tolerant and insect-resistant traits. Other traits, such as resistance to bacterial, fungal and viral pathogens, continue to be developed and many new trait combinations have been released in recent years. For example, releases of transgenic cultivars with properties such as drought resistance increased from 1043 in 2005 to 5190 in 2013 (Fernandez-Cornejo *et al.* 2014). Increasing numbers of reports of herbicide resistance in weeds and insect pest resistance to Bt crops make it very clear that more comprehensive assessments of risks over longer temporal and larger spatial scales are required. The increase in the number and types of traits being engineered into crops indicates a need for assessments that account for different types of potential environmental effects beyond those associated with herbicide and pest resistance.

Environmental Pest Management: Challenges for Agronomists, Ecologists, Economists and Policymakers,
First Edition. Edited by Moshe Coll and Eric Wajnberg.

For example, it is important to evaluate changes in use of herbicides and pesticides due to adoption of genetically engineered crops. In the USA, for example, adoption of transgenic crops with herbicide resistance caused an increase of 239 million kg of herbicides to be used but adoption of transgenic crops with pest resistance traits caused a reduction of 56 million kg pesticides used for an overall increase of 183 million kg or approximately 7% from 1996 to 2011 (Benbrook 2012) (see also Chapter 12). These figures may underestimate the increase in pesticide use from 1996 to the present as they do not include the recent rapid rise in the use of neonicotinoid seed treatments (Douglas and Tooker 2015).

Viewing trends from an alternate perspective, Brookes and Barfoot (2015) estimate that global pesticide use with transgenic crops is 8.6% lower than it would have been if the most likely alternative controls were used, presumably with a concomitant reduction in adverse environmental impacts. In the USA, insecticide use on maize declined 10-fold from 1995 to 2010, consistent with the decline in European corn borer populations shown to be a direct result of Bt crop adoption (Fernandez-Cornejo *et al.* 2014). However, pest resistance to Bt proteins has increased substantially as well. For example, out of 27 sets of monitoring data, seven showed severe field-evolved resistance, eight showed less severe field-evolved resistance and 12 showed no evidence of decreased susceptibility to Bt proteins (Tabashnik and Carrière 2015).

Herbicide resistance in weed species has also increased substantially. For example, since glyphosate-tolerant crops were introduced in 1996, 32 glyphosate-resistant weed species have been identified worldwide (Heap 2016). Because of the use of transgenic crops, total herbicide use in the USA has been projected to increase from ~1.5 kg ha^{-1} in 2013 to more than 3.5 kg ha^{-1} in 2025 (Mortensen *et al.* 2012).

6.2 Conceptual Framework

Although, as we detail below, the effects of transgenic crops or any pest management tactic can be compounded (additively, antagonistically or synergistically) over wide spatial and temporal scales, most pest management decisions are made at the scale of a single field in a single season. The major factor determining if a given effect is positive or negative from the perspective of an individual grower is the function or guild of the non-target species affected. We examine primary effects on two non-target guilds, herbivorous insects and non-crop plants, and then contrast these effects with the secondary effects on four guilds: pollinators, decomposer fauna, predators (in a broad sense) and micro-organisms. Finally, we examine the effects on all groups across a broader spatial and temporal scale. As far as possible, we try to draw a distinction between quantified or at least identified effects as distinct from potential effects.

6.3 Primary Effects

6.3.1 Effects on Non-target Herbivorous Insects

Non-target arthropods may be exposed to the insecticidal proteins termed 'plant incorporated protectants' or PIPs (www3.epa.gov/pesticides/chem_search/reg_actions/pip/index.htm) present in genetically modified crops. Exposure can occur by direct

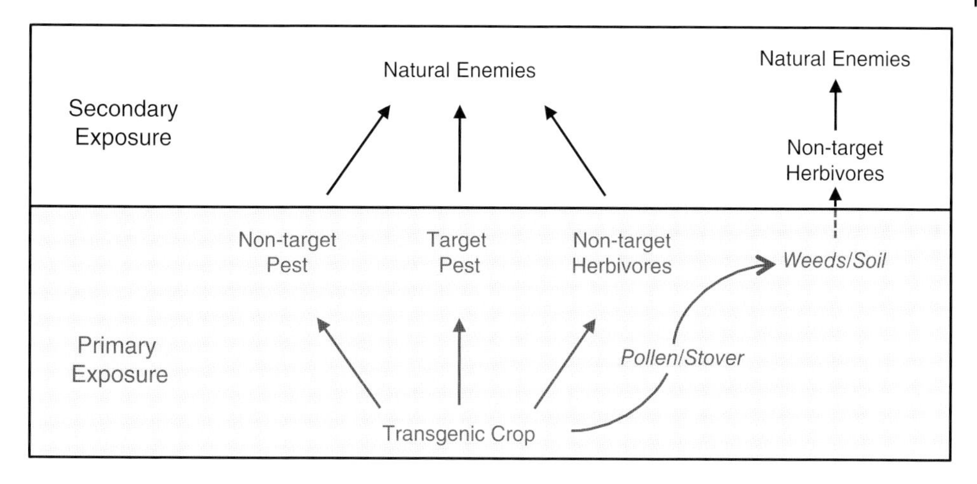

Figure 6.1 Modes of interaction with transgenic crops. Source: Adapted from Obrycki *et al.* (2001).

feeding on plant tissues, ingestion of insecticidal proteins in the soil, feeding on prey that have ingested insecticidal proteins, and indirectly through reductions in prey/host populations (Figure 6.1) (Obrycki *et al.* 2001). We define primary exposure as effects from direct feeding on the focal transgenic crop and this narrows our scope to insects such as those in Lepidoptera (moths and butterflies) and Coleoptera (beetles) that feed on the two primary Bt crops (maize and cotton). Bt toxins are generally more specific than conventional insecticides and these two orders of insects are also the selective targets. However, toxicity to other orders has been demonstrated (Amichot *et al.* 2016).

In general, since any organism feeding on a crop plant can be considered a pest, growers will not alter their pest management strategies to minimize direct effects on species that feed on the crop plant even if these species are not the targeted pest. The monarch butterfly, *Danaus plexippus*, is one example of a species that is not a target pest but is affected by directly consuming a plant tissue (pollen) (Losey *et al.* 1999). Grower behaviour could change if aggregate impacts became severe enough that one or more directly impacted species was listed as endangered or threatened at broader levels. The probability that a directly affected species would be endangered is low since organisms that feed on common crop plants are generally not rare (Losey *et al.* 2003). It is important to note that some growers might choose to alter their use of transgenic crops to conserve iconic species even if such changes were not legally required or necessary to optimize profitability. The 'willingness to pay' or at least forego potential profits has been quantified for certain insects, including the monarch butterfly (Diffendorfer *et al.* 2013). Growers could decide to plant non-transgenic varieties to minimize environmental impacts and, in some cases, to take advantage of broader scale suppression of pests (e.g. *Ostrinia nubilalis* on maize in the USA) (Hutchison *et al.* 2010).

6.3.2 Effects on Non-crop Plants

Plants other than the focal crop (e.g. weeds) in a production field are potential competitors with the crop itself and are thus usually also considered pests. Since these plants can reduce yield and crop quality, any negative impact on them from the use of transgenic crops would be seen as positive and would be facilitated rather than avoided by the grower. The use of herbicides, such as glyphosate, with herbicide-tolerant crops is very effective at removing essentially all weeds that have not developed resistance. In some situations, reduction in non-crop plant density and diversity creates an 'ecological desert' (Obrycki *et al.* 2001) that can lower the density or diversity of the animals that depend on those plant species. Plant diversity is known to enhance and facilitate ecosystem services (Egan and Mortensen 2012; Quijas *et al.* 2012) (see also Chapter 7), especially by supporting pollinators.

Although recent studies did not find significant impact of herbicide-tolerant crops on plant diversity in general (Schwartz *et al.* 2015; Young *et al.* 2013), this effect has been reported for specific herbivore groups including monarch butterflies (Pleasants and Oberhauser 2012) and birds (Taylor *et al.* 2006).

6.4 Secondary Effects

6.4.1 Effects on Beneficial Species

In contrast to the profile of effects of PIP crops on non-target plants or herbivorous insects, beneficial insects (i.e. those that enhance crop production) are presumably less likely to be affected because they are either unaffected by the insecticidal protein or do not ingest it – a prerequisite for the Bt protein to have insecticidal activity. However, any impact on beneficial species is also more likely to have a negative effect on the services they provide and this could lower yield and profitability (see Figure 6.1). Functional groups of beneficial insects (beneficials) that might be affected include pollinators, decomposers and predators (or parasitoids) of insect pests and weeds. Potential effects and consequences for each group of beneficials are described below.

6.4.1.1 Pollinators

The largest and most important group of pollinators is the Hymenoptera, specifically bees. No currently registered product is targeted to affect Hymenoptera so it is not surprising that there are very few effects reported. Malone and Burgess (2009) reviewed 22 studies (some within the same publication) that tested for effects of the Bt toxin on bees and only two found any effects and these were variable. They did find more potential for effects with other types of toxins (e.g. protease inhibitors) that may be more widely incorporated in the future (Malone and Burgess 2009). Loss of plant diversity following herbicide application could have a negative impact on pollinators. Soybeans and maize are not insect pollinated but there is potential for impact on cotton (Han *et al.* 2010) or other crops (e.g. fruits or vegetables) on the same farm.

6.4.1.2 Decomposer Fauna

Most decomposer fauna live in or on the soil and are thus easily overlooked. However, between 60% and 90% of net primary production is processed by decomposer organisms (Brady and Weil 2008). While the bulk of this processing is accomplished by bacteria and fungi (addressed later in this chapter), larger organisms, including worms and arthropods, play an important role by modifying physical and chemical properties of the plant material and greatly enhancing decomposition efficiency (Meyer *et al.* 2011). There are relatively few studies that address the potential effects of transgenic crops on meso- or macroinvertebrates, but one of the few studies found a negative impact on night crawlers (*Lumbricus terrestris*) (Zwahlen *et al.* 2003), while another found no effect on the earthworm *Eisenia fetida* or the springtail *Folsomia candida* (Clark and Coats 2006). These mixed results, coupled with the lengthy retention time of a few Bt proteins in soil (Cry1Ab, Stotzky 2000; Cry1Ac, Sun *et al.* 2007), suggest that further study in this area is warranted.

In a study combining assessment of both diversity and function, Londoño-R *et al.* (2013) examined in-field decomposition rates and diversity of microarthropod decomposers after 5 months in a field trial in which residues from two Cry1Ab Bt corn hybrids were compared to their non-transgenic counterparts in litterbags placed on the soil surface or buried at 10 cm depth. Microarthropod diversity varied by residue placement and by plant part, but there was no significant effect of genotype. Looking across a wide range of organisms, Saxena and Stotzky (2001) incorporated Cry1Ab Bt maize root exudates or plant biomass in soil in a laboratory experiment and found that these did not significantly affect populations of earthworms, nematodes, protozoa, bacteria or fungi 45 days after the soil was amended compared to non-transgenic maize.

6.4.1.3 Predators

Predators, used here in the broad sense of any organism (including parasitoids and parasites) that preys on either insect pests or weeds, suppress the majority of potential pest or weed populations before they reach economically damaging levels (Losey and Vaughan 2006). One unintended effect of the use of pesticides to control pests is that they can interfere with the pest suppression delivered by the natural enemies of pests. Since some of the most important predators of insect pests and weed seeds are beetles, there is the potential for negative impacts especially from crops expressing the Cry3 Bt toxin that is toxic to beetles.

Stephens *et al.* (2012) reviewed 24 laboratory and field studies involving the Cry1 Bt protein (primarily active against lepidopteran insects) and 11 studies involving the Cry3 protein. They reported five and two cases of negative impact for Cry1 and Cry3, respectively (Stephens *et al.* 2012). Based on multi-year large-scale field studies, this same study reports that densities of carabids and the coccinellid *Harmonia axyridis* were significantly lower in Bt maize plots compared to control plots. Using data from the same field study, DiTommaso *et al.* (2014) also report a reduction in the rate of weed seed feeding in Bt maize plots compared to control plots. It is important to note that, while these studies found differences between Bt and control plots, they did not find differences between the Bt and the insecticide plots, implying that the effects of Bt were no worse than the conventional (herbicide) alternative. This comparison is important because risk assessments should evaluate appropriate alternative management strategies in order to provide results that can support improved decision making by both growers and regulators.

6.4.2 Effects on Non-target Micro-organisms

Micro-organisms play a crucial role in decomposition so any negative effect on this group has implications for the system as a whole. Saxena and Stotsky (2001) reported reduced decomposition rates for Cry1Ab Bt maize in a laboratory experiment and attributed this to a higher lignin content in the transgenic maize. Flores *et al.* (2005) also found reduced rates of decomposition for Bt residues of rice (Cry1Ab), tobacco (Cry1Ac), canola (Cry1Ac), cotton (Cry1Ac) and potato (Cry3A) compared to their non-Bt isolines in a laboratory study. These results could not be explained by differences in lignin content and no other potential mechanism for observed results was proposed. These results have not been replicated and a series of multi-year field trials in which different varieties of Cry1Ab rice, Cry3Bb maize and Cry1Ac cotton were grown did not yield similar results. Instead, in field-based litterbag studies where Bt and non-Bt residues were placed either at the soil surface or buried at 10 cm depth, no differences in decomposition between transgenic and non-transgenic varieties were found for two varieties of Cry1Ab maize (Londoño-R *et al.* 2013; Tarkalson *et al.* 2007), one variety of Cry3Bb maize (Xue *et al.* 2011), two varieties of Cry1Ab rice (Lu *et al.* 2010a, 2010b; Wu *et al.* 2009a) and one variety of Cry1Ac cotton (Kumari *et al.* 2014). Instead, in all cases, differences in decomposition were driven by residue placement (surface versus buried) and plant part, with leaves decomposing most quickly and cobs (for maize) and roots (for rice) decomposing more slowly. In no case was genotype a significant factor controlling residue decomposition.

In several of these field trials, bacterial and fungal decomposer community diversity in the litterbag samples was also investigated using molecular fingerprinting methods. For three varieties of Cry1Ab rice (Lu *et al.* 2010a, 2010b; Wu *et al.* 2009b), one variety of Cry3Bb maize (Xue *et al.* 2011) and one variety of Cry1Ab maize (Londoño-R *et al.* 2013), no differences in bacterial or fungal communities colonizing the residues could be attributed to genotype (Bt versus non-Bt), except as found by Lu *et al.* (2010b). Rather, residue placement and plant part were again the major drivers of changes in community composition observed. The singular difference in community composition observed for Cry1Ab rice in a paddy soil was in the fungal community decomposing Bt rice roots compared to the non-Bt rice (Lu *et al.* 2010b). However, no differences in the fungal communities decomposing rice straw were observed.

Overall, these results indicate that plant tissue composition is a very strong driver controlling decomposition rate, regardless of the presence of the Bt protein. Bacteria and fungi produce exoenzymes, such as proteases, to decompose soil proteins, such as the Bt protein. There is no mechanism that has been demonstrated or proposed that would suggest microbial decomposers would be directly affected in their ability to decompose these proteins. The persistence of the Cry1Ab protein from Bt maize reported by Stotzky (2000) is most probably due to adsorption of the protein on clays or soil organic matter such that proteases are unable to access them. For Cry3Bb maize, residue and protein decomposition is rapid, with the protein detectable in the rhizosphere only during active plant growth and undetectable after harvest (Xue *et al.* 2014). Icoz and Stotzky (2008) also found that the Cry3Bb protein does not persist in soil. The Cry3Bb maize is intended to control the corn root worm (*Diabrotica* spp.). Xue *et al.* (2014) found that, among 15 different genotypes of Cry3Bb maize, including several with stacked HT resistance traits, all displayed low expression in the roots and the protein decomposed very readily. These combined characteristics produced a relatively weak, short-lived presence of the toxin. While that profile of traits could represent relatively weak selection pressure, it apparently also posed only a minor barrier that the target pest overcame relatively rapidly. The variety expressing this protein (MON863) is no longer sold commercially as a result.

Devare *et al.* (2004, 2007) studied the effects of Cry3Bb Bt maize and tefluthrin pesticide (Force G, Dow Elanco, St Louis, MO, USA) on soil microbial biomass, nitrogen mineralization potential, short-term nitrification rate and respiration rate in a 3-year field trial. While there was variation across time and significant differences in all variables between the bulk and rhizosphere soil, there was no effect of genotype on nitrogen mineralization potential or short-term nitrification rate. Using molecular fingerprinting, no differences were observed between bacterial or fungal communities colonizing the rhizosphere or bulk soil attributable to maize genotype. However, the authors did find that soils sampled from the Bt maize had increased levels of microbial biomass and microbial respiratory activity. While 'more' is not necessarily 'better', these results suggest that the Bt maize did not have any repressive effects on soil microbial abundance or activity and thus is unlikely to be harmful. For plants treated with the pesticide teflutrin, they found depressed respiratory activity midseason for 2002 samples only.

In only one study (Wu *et al.* 2009b) was the community composition of rhizosphere bacteria found to differ substantially in Cry1Ab rice, compared to its non-transgenic counterpart in a paddy soil. Wu *et al.* (2009b) used phospholipid fatty acid analysis to characterize these bacterial communities. They suggested that potential differences in the content or extent of root exudation between the transgenic and non-transgenic rice could have led to the bacterial community level differences observed. Arbuscular mycorrhizal fungi spore abundance and root colonization were examined in field studies with Cry3Bb maize and non-transgenic isolines over a 5-year period (Zeng *et al.* 2015). Only minor effects on the arbuscular mycorrhizal fungi community were observed over the 5-year span of these trials.

Lastly, Liu *et al.* (2008) studied the effects of Cry1Ab Bt rice and the insecticide triazophos [3-(o,o-diethyl)-1-phenyl thiophosphoryl-1,2,4-triazol] in a paddy soil on microbial activity and community composition. Molecular fingerprinting was used to assess changes in bacterial and fungal communities in the rhizosphere of Bt rice compared to the non-transgenic isoline. Measurements were taken at four stages in the rice developmental cycle over a 2-year period. No significant differences in phosphatase activity, dehydrogenase activity, respiration, methanogenesis or fungal community composition were found in the transgenic compared to the non-transgenic variety.

Fewer studies have been conducted to examine the effects of herbicide-tolerant crops on the soil microbial community. Glyphosate inhibits the enzyme 5-enolpyruvylshikimate-3-phosphate (EPSP) synthase that is required for the biosynthesis of the aromatic amino acids phenylalanine, tyrosine and tryptophan via the shikimate pathway in bacteria, fungi and plants. Roundup Ready crops (herbicide-tolerant crops) contain a gene derived from *Agrobacterium* sp. strain CP4 that encodes a glyphosate-tolerant enzyme (CP4 EPSP synthase) (Funke *et al.* 2006) and are thus relatively unaffected by glyphosate application. Nakatani *et al.* (2014) examined the effects of herbicide-tolerant soybean on soil microbial biomass-carbon and nitrogen and the activities of the enzymes beta-glucosidase and acid phosphatase. They worked at six sites in Brazil across 2 years. Their results show no significant effect of genotype on any of the variables measured.

6.5 Tertiary Effects: Broader Spatial and Temporal Scales

While pest management decisions are made at the field level, effects often accrue across broader spatial and temporal scales and these scales need to be considered as regulations for transgenic crops continue to be developed and refined. Traditionally, data for risk assessments of transgenic crops have usually been collected at the plot or field scale. However, many environmental impacts occur at landscape and larger spatial scales and may not be apparent at small plot or field scales. Assessments of the effects of transgenic crops are performed on a case-by-case basis in order to address the specific characteristics of the trait, crop and environment where the transgenic crop is to be deployed (Andow and Zwahlen 2006; Andow *et al.* 2006; Peterson *et al.* 2000; Wolfenbarger and Phifer 2000). Recent research has highlighted the importance of considering landscape context and extended spatial and temporal scales when evaluating the impacts of land use changes associated with introducing new genetically engineered crops. Since broad-scale effects take longer to accrue and have not been as rigorously assessed, we will address them primarily in terms of risks of future effects.

Given the current widespread use of transgenic crops in several parts of the world and the projected rapid adoption of next-generation transgenic crops with stacked traits (Mortensen *et al.* 2012) (see also Chapter 12), there is an urgent need for improved tools and methods for assessing risks at regional to national spatial scales and multidecadal time scales so that these methods are ready to use on the next generations of transgenic crops.

There has been a large amount of work devoted to developing and standardizing risk assessment methods. The United States Environmental Protection Agency (USEPA) has produced numerous guidance documents for conducting ecological risk assessments. However, a recent compilation indicates that only three of the 38 documents address landscape-scale analyses, and none of these are focused on pesticides (USEPA 2012). In other countries, guidance for risk assessment for other topics such as maintaining biodiversity in Australia (Smith *et al.* 2013) and managing invasive (alien) species in Norway (Sandvik 2013) may be applicable to genetically engineered crops.

For genetically engineered crops, landscape and regional geospatial analysis is critically important because these traits are deployed on a large portion of the landscape and interspersed with non-genetically engineered crops, organic fields and other sensitive, non-target vegetation and ecosystems. Because the benefits and risks of genetically engineered crops are often landscape dependent, it is critically important to represent these landscapes accurately in risk assessments. For example, in some regions, genetically engineered crop fields may dominate the landscape, so effects that are small in a single plot or field may become large as they accumulate in a watershed or region. Conversely, in other regions, genetically engineered crop fields may be interspersed in a patchwork of non-agricultural lands, non-genetically engineered row crop fields including organic systems, and high-value horticultural crops.

The risks of a particular genetically engineered trait may be mitigated or magnified by these spatial arrangements at the landscape to regional scale. Furthermore, if a genetically engineered crop has increased yields compared with the non-genetically engineered alternative, it could contribute to 'land-sparing' such that less area is required for the crop, providing the opportunity for land use change toward other uses. But if a genetically engineered crop allows commercial production to occur on lands that are marginal for alternative crops, including conservation lands, there could be land use change toward more intensive use with the potential to contribute to cumulative risks. For these reasons, a geospatial approach to risk analysis can improve upon current approaches by better representing spatial patterns of genetically engineered crops and other land uses, how these spatial patterns affect risks, and how patterns may change with the introduction of a new genetically engineered crop.

For transgenic crops, increasing attention has been paid to the need to develop geospatial methods to conduct ecological risk assessment and management in Europe. For example, a web-based geographical information system has been designed that incorporates data on the location of all genetically engineered crop production in Germany, facilitating analysis with geospatial data on climate, soil and agricultural patterns (Kleppin *et al.* 2011). Models have also been developed to scale up from plot and field scale data to perform landscape and regional ecological risk assessments for transgenic crops, particularly in Europe (Breckling *et al.* 2011; Reuter *et al.* 2011; Wurbs *et al.* 2012). Unplanned release of genetically engineered crops may occur along transportation corridors such as railways, as has been documented for herbicide-tolerant rapeseed in Switzerland (Schoenenberger and D'Andrea 2012). These types of approaches are required to address both local and regional environmental impacts of transgenic crops.

In addition to accounting for landscape-dependent effects, risk assessments at larger scales may detect effects that are not apparent at plot scales. While genetically engineered traits are generally intended to improve yields, for soybean, small yield decreases have been found at the regional scale in the USA of 0.07 t ha^{-1} in the Central Corn Belt and 0.11 t ha^{-1} elsewhere (Xu *et al.* 2013). Furthermore, these results demonstrate that effects that are important at regional and national scales may be difficult to detect at the field scale. During recent years, USA maize yields have averaged 8–10 t ha^{-1}, so detecting small losses of 0.07–0.11 t ha^{-1} could be quite difficult at the plot or field scale, but not difficult at regional scales due to the very large sample size and concomitant statistical power to detect small differences.

Recent advancements have contributed to the development of the field of probabilistic regional geospatial environmental risk assessment, for example by analysing climate change effects on forest growth (Woodbury *et al.* 1998), improving methods for spatially explicit risk assessments (Woodbury 2003) or identifying promising methods for probabilistic assessment of multiple types of risks to agricultural and forest ecosystems at the regional scale (Woodbury and Weinstein 2010). Careful attention to appropriate temporal and spatial scales and to cumulative impacts has been recommended for all types of ecological risk assessments (Dale *et al.* 2008), as have regional risk assessments that cover multiple types of environmental stressors (Landis and Wiegers 2007). There is a need and an opportunity to apply these approaches of cumulative, probabilistic, regional geospatial risk assessment to transgenic crops at landscape, regional and national scales. For example, there is evidence that widespread deployment of Bt crops has reduced insect pest populations in China at the landscape scale. Specifically, adoption of Bt cotton caused an increase in the abundance of generalist predators in non-Bt crops that increased the biological control of aphid pests beyond the genetically engineered crop fields (Lu *et al.* 2012). In Europe, geospatial modelling suggests that Bt maize pollen could cause mortality to the protected butterfly *Inachis io* in southern Europe where it is multivoltine, but not in northern Europe where it is univoltine (Holst *et al.* 2013).

Increasing the temporal scale will also provide a more comprehensive risk assessment compared to assessments exclusively at shorter time scales. Incorporating multidecadal time scales is important because environmental benefits and impacts may change over time. For example, glyphosate-resistant genetically engineered crops have provided a benefit to farmers and others due to improving the ease of weed management, and replacing older, more toxic herbicides with glyphosate (Green 2012). However, the relatively rapid development of multiple populations of glyphosate-resistant weeds after introduction of glyphosate-resistant crops was an unpleasant surprise to many growers and policy makers (Powles 2008, Shaner *et al.* 2012). Development of glyphosate-resistant weeds has prompted development of genetically engineered crops resistant to synthetic auxin herbicides. A multidecadal time scale is required to more fully evaluate both benefits and risks of herbicide-resistant crops and the effects of management practices.

6.6 Quantifying Risks and Benefits of Transgenic Traits

Although much of the research and regulation for transgenic crops is based on assessment and management of risk, there are also benefits from genetically engineered crops that should be balanced against the risks in order to assess the overall impacts. For example, expressing the Bt protein in crop tissues reduces insect damage and increases yield due to improved efficacy of pest control in treated fields. It also reduces insecticide application and concomitant risks, as discussed earlier. Such co-benefits may also provide economic savings to farmers despite the increased cost of genetically engineered seeds. There is evidence for both yield increases and economic benefits due to deployment of genetically engineered traits in the USA. For example, for maize in the USA, yield increases with full adoption of genetically engineered traits are estimated to be 1.3 t ha^{-1} in the Central Corn Belt and 0.6 t ha^{-1} elsewhere (Xu *et al.* 2013). For European corn borer in five Mid-Western states in the USA (IL, MN, WI, IA, NE), recent analyses suggest economic benefits of Bt maize over 14 years as high as US$6.8 billion, with US$4.3 billion of this total due to indirect effects, specifically reduction of pests in fields without the Bt trait (Hutchison *et al.* 2010). These results also highlight the value of retaining non-Bt maize refugia to slow development of resistance to Bt in maize pests. In China, increases in arthropod predators and decreases in aphid pests were found in Bt cotton, along with potential improvement in biocontrol in neighbouring crops including maize and soybean (Lu *et al.* 2012).

6.6.1 Quantifying Effects on Ecosystem Services at Landscape and Regional Scales

Humans derive an array of services from ecosystems, which can be classified as provisioning, regulating, supporting or cultural (MEA 2005) (see also Chapter 7). Agriculture provides services such as crop production, a provisioning service. But to do so, it relies on supporting and regulating services such as nutrient and water cycling, pollination, pest regulation, and maintenance of soil quality and biodiversity (Power 2010). Such services are extremely valuable; for example, a limited set of ecosystem services provided by wild insects in the USA were found to be worth US$57 billion per year (Losey and Vaughan 2006).

To incorporate the many services provided by and required by agriculture, risk assessments should address a broad suite of ecosystem services. As an example of a screening-level risk assessment of the impacts of future bioenergy crop production on a comprehensive suite of ecosystem services across a 12-state region of the upper Mid-Western USA, Bruins *et al.* (2009) identified a large number of services and endpoints that are relevant because of the focus on both annual and perennial crops (including

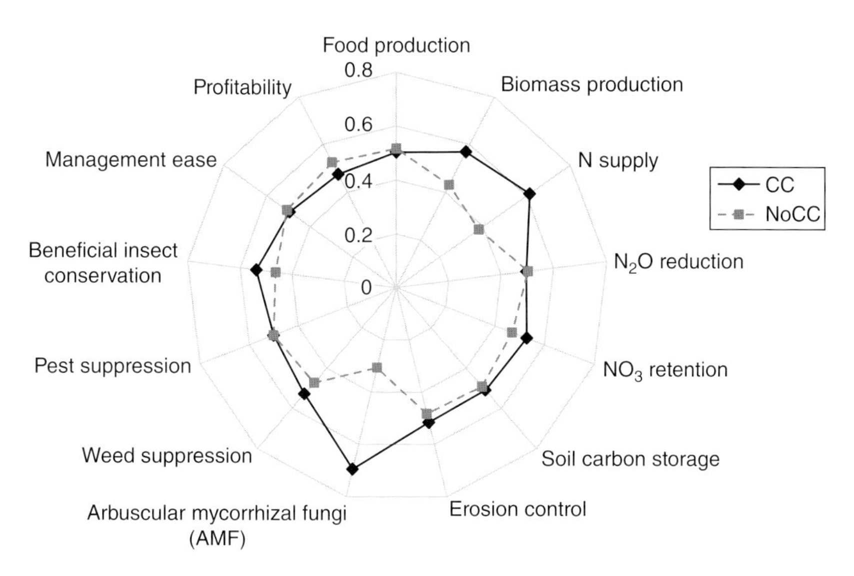

Figure 6.2 Example of the effects of an agricultural management practice (cover cropping, CC) on a suite of indicators of ecosystem services, with higher numbers indicating increased ecosystem service benefit. Source: Aadapted from Schipanski *et al.* (2014). Reproduced with permission from Elsevier.

maize, soybean and switchgrass) at the regional scale. As another example, both quantitative models and semi-quantitative estimates were used to analyse the impact of cover crops on the temporal dynamics of 11 ecosystem services and two economic metrics in a 3-year soybean–wheat–maize rotation in a typical Mid-Atlantic climate (Schipanski *et al.* 2014) (Figure 6.2). These studies illustrate that it is feasible to account for both risks and benefits to ecosystem services in a quantitative regional risk assessment framework for genetically engineered crops. However, it will be an ongoing challenge for risk assessment methodologies and for collecting data needed to conduct risk assessments as the pace of transgenic technology increases.

One example of these types of advances is the recent development of CRISPR (clustered regularly interspaced short palindromic repeats) technology that has the potential to greatly increase the ease and speed of transgenic crop trait development by allowing straightforward 'editing' of genomes. Crops have already been developed using advanced genome editing technologies including oilseed rape (canola), maize and wheat (Ainsworth 2015). If these crops produced with gene editing technology do not include genetic material from other species, they may not be defined as 'transgenic', and in fact the first such crop has already been planted in California (Ainsworth 2015). However, this same crop may be considered transgenic in Europe, thus further complicating risk analysis and management for crops that may be traded internationally.

6.6.2 Risk Management Impacts for Transgenic, Non-transgenic and Best Management Practices

Most crops in the USA are highly managed, and management practices substantially affect risks. For example, for synthetic auxin herbicides such as 2,4-D and dicamba that have the potential for vapour and spray drift, best management practices such as low-vapour formulation, correct timing of application, accounting for wind speed and direction, correct rate, correct type of spray nozzles and thorough cleaning of equipment can greatly reduce or eliminate problems of drift damage (Mueller *et al.* 2013; Parker 2011). Conversely, widespread use of these herbicides without best management practices has the potential to substantially damage sensitive crops (Egan *et al.* 2014; Mueller *et al.* 2013; Parker 2011). Thus, management strongly influences the risk of damage to non-target crops and other vegetation. However, risk assessment guidance and risk assessments rarely focus adequate attention on risk management. For the USA, for example, a recent compilation indicates that only four of 38 USEPA guidance documents directly address risk management, while another 11 'touched upon' this subject (USEPA 2012).

This issue is discussed in the current draft genetically engineered synthetic auxin herbicide Environmental Impact Statement prepared by APHIS (USDA-APHIS 2013). This document notes that the 'Save our Crops Coalition', a group of growers who raise crops sensitive to 2,4-D, was opposed to 2,4-D-tolerant (Enlist™; Dow AgroSciences; www.enlist.com/en) crops due to concerns about off-target drift of 2,4-D. However, once they became familiar with the Stewardship Agreement that adopters of the Enlist™ plants must enter into with the developer, data on the volatility of proposed 2,4-D formulations and requirements for drift reduction planned for the label, they were no longer opposed (www.regulations.gov/document?D=APHIS-2012-0032-0143). Concerning this, it is important to evaluate the impacts of different degrees of compliance with these best management practices, as they will have strong impacts on risks.

In a broader sense, more case studies are needed to support quantification of impacts of different management practices at landscape to regional scales.

6.6.3 Quantitative Uncertainty Analysis

In any risk assessment, there is uncertainty about the magnitude and effects of a particular stressor (any physical, chemical or biological entity that can induce adverse responses). This is particularly true for assessing potential risks of genetically engineered crops, since agricultural systems are complex, with different environments, management practices and landscape contexts in different fields, farms and regions of the world. Furthermore, both uncertainty and variability are important in risk assessments, and it is useful to distinguish between them. Uncertainty may be caused by lack of knowledge or lack of data. Variability is a property of natural systems, including agricultural systems, with variation within and among plants, pests and non-target species, as well as biophysical properties of the environment such as soil characteristics.

Quantitative uncertainty analysis has long been recommended to help understand and quantify the sources and magnitudes of uncertainties, and how they may affect risk management decisions (Hammonds *et al.* 1994; Thompson and Graham 1996; USEPA 2001; Warren-Hicks and Moore 1998). Such techniques have been used to improve previous analysis of environmental risks (Nagle *et al.* 2007, 2012; Woodbury *et al.* 1998) and should also be utilized to improve risk analyses for genetically engineered crops.

6.7 Variation Among Countries in Risk Assessment and Management

Approaches to regulation and concomitant adoption of genetically engineered crops vary substantially among countries, reflecting broad cultural, social and political differences (Heinemann *et al.* 2013; NAS 2016). During the late 1990s, adoption of genetically engineered crops increased 20-fold worldwide, prompting an explosion of concern, particularly in Europe, over the health and environmental impacts of these crops (Peterson *et al.* 2000). Since that time, genetically engineered crop area has grown globally each year such that in 2014, 18 million farmers in 28 countries planted more than 181 million hectares (James 2014). However, very large differences in adoption of genetically engineered crops among countries and global regions have continued, with very rapid adoption in North America, most of South America, especially Brazil, as well as South and East Asia, contrasted with adoption in only a few countries in Europe and in Africa (James 2014). These differences appear to be driven not so much by different data on environmental impacts of genetically engineered crops in different parts of the globe or by use of entirely different risk assessment methods, but rather by differences in the approach to regulating risks and benefits (NAS 2016). For example, in both the USA and Europe, risk assessment and regulation are in theory based on a requirement for review of each crop variety produced by a specific set of genetically engineered techniques, while similar varieties developed through other breeding technologies are not regulated (Bartsch 2014; NAS 2016). However, as discussed earlier, genetically engineered crop adoption is very widespread in the USA, while in Europe a precautionary approach has dominated, and there was adoption in only five European Union countries as of 2014 (James 2014).

In part, this difference may be because the policy goals specified in many countries may be so broad that it is difficult to translate them into specific assessment endpoints for risk assessment (Garcia-Alonso and Raybould 2014). Additionally, regulations in some countries go well beyond food safety and environmental protection to address social goals such as protecting organic production systems and product labelling for consumers (NAS 2016). More generally, there are political, social and psychological factors that affect approval of genetically engineered crops that go beyond scientific assessments of environmental risk (Bartsh 2014; Devos *et al.* 2014; NAS 2016). Therefore, even with increasing amounts of data on various types of environmental impacts of genetically engineered crops, it is likely that there will continue to be large differences in policy, management and adoption among countries and regions.

6.8 Conclusions

Improved assessment tools that comprehensively consider the impacts of genetically engineered crops on non-target organism and ecosystem services are clearly needed. These assessments need to be conducted not just at the field scale, but at longer temporal and larger spatial scales as well.

Effects on soil ecology provide a salient example to consider, especially if lands where transgenic crops are grown are to be transitioned to organic agriculture in the future. Any carryover effects could adversely affect the ability of land owners to obtain and maintain organic certification (Thies 2015). However, when do we consider sufficient data to have been collected to determine whether a particular transgenic crop has no harmful effects on the soil microbial community? In many cases, for the Bt crops, Bt protein constructs are beginning to change as resistance in target pests escalates. Once one protein has been 'retired' and another comes on line, what data will we still need to gather to assure environmental safety of this new trait? And, if a given Bt protein is no longer detectable in soil after a single season, is there a need to continually monitor? Some rationale is needed to target those 'risks' that may persist and continue to measure those while we 'cease and desist' where the weight of the evidence suggests this is prudent to do.

Besides prospective risk assessment, there is a need for ongoing risk management, including analyses of new data collected after release of a transgenic cultivar. Ongoing technological developments such as CRISPR technology will pose a challenge for risk assessments, especially to the extent that genetically engineered crops may not be defined as transgenic and thus could fall out of the infrastructure in place to assess effects and risk for genetically engineered crops. Furthermore, uncertainty should not be ignored, but rather addressed using quantitative uncertainty analysis.

All of these recommendations would improve our knowledge of the environmental impact of genetically engineered crops and provide better decision support for risk management and sustainable crop production.

References

Ainsworth, C. (2015) Agriculture: a new breed of edits. *Nature* **528**: S15–S16.

Aldemita, R.R., Reano, I.M.E., Solis, R.O. and Hautea, R.A. (2015) Trends in global approvals of biotech crops (1992–2014). *GM Crops and Food* **6**: 150–166.

Amichot, M., Curty, C., Benguettat-Magliano, O., Gallet, A. and Wajnberg, E. (2016) Side effects of *Bacillus thuringiensis* var. kurstaki on the hymenopterous parasitic wasp *Trichogramma chilonis*. *Environmental Science and Pollution Research* **23**: 3097–3103.

Andow, D.A. and Zwahlen, C. (2006) Assessing environmental risks of transgenic plants. *Ecology Letters* **9**: 196–214.

Andow, D.A., Lovei, G.L. and Arpaia, S. (2006) Ecological risk assessment for Bt crops. *Nature Biotechnology* **24**: 749–751.

Bartsch, D. (2014) GMO regulatory challenges and science: a European perspective. *Journal Fur Verbraucherschutz und Lebensmittelsicherheit* **9**: S51–S58.

Benbrook, C.M. (2012) Impacts of genetically engineered crops on pesticide use in the US – the first sixteen years. *Environmental Sciences Europe* **24**: 24.

Brady, N.C. and Weil, R.R. (2008) *The Nature and Properties of Soils*, 14th edn. Prentice Hall, Upper Saddle River, New Jersey, USA.

Breckling, B., Reuter, H., Middelhoff, U., *et al.* (2011) Risk indication of genetically modified organisms (GMO): modelling environmental exposure and dispersal across different scales oilseed rape in Northern Germany as an integrated case study. *Ecological Indicators* **11**: 936–941.

Brookes, G. and Barfoot, P. (2015) Environmental impacts of genetically modified (GM) crop use 1996-2013: impacts on pesticide use and carbon emissions. *GM Crops and Food* **6**: 103–133.

Bruins, R.J.F., Franson, S.E., Foster, W.E., Daniel, F.B. and Woodbury, P.B. (2009) *A Methodology for the Preliminary Scoping of Future Changes in Ecosystem Services – with an Illustration from the Future Midwestern Landscapes Study*. EPA/600/R-09/134. Office of Research and Development, US Environmental Protection Agency, USA.

Clark, B.W. and Coats, J.R. (2006) Subacute effects of Cry1Ab Bt corn litter on the earthworm *Eisenia fetida* and the springtail *Folsomia candida*. *Environmental Entomology* **35**: 1121–1129.

Dale, V.H., Biddinger, G.R., Newman, M.C., *et al.* (2008) Enhancing the ecological risk assessment process. *Integrated Environmental Assessment and Management* **4**: 306–313.

Devare, M.H., Jones, C.M. and Thies, J.E. (2004) Effect of Cry3Bb transgenic corn and tefluthrin on the soil microbial community. *Journal of Environmental Quality* **33**: 837–843.

Devare, M., Londoño-R, L.M. and Thies, J.E. (2007) Neither transgenic Bt maize (MON863) nor tefluthrin insecticide adversely affect soil microbial activity or biomass: a 3-year field analysis. *Soil Biology and Biochemistry* **39**: 2038–2047.

Devos, Y., Aguilera, J., Diveki, Z., *et al.* (2014) EFSA's scientific activities and achievements on the risk assessment of genetically modified organisms (GMOs) during its first decade of existence: looking back and ahead. *Transgenic Research* **23**: 1–25.

Diffendorfer, J.E., Loomis, J.B., Ries, L., *et al.* (2013) National valuation of monarch butterflies indicates an untapped potential for incentive-based conservation. *Conservation Letters* **7**: 253–262.

DiTommaso, A., Ryan, M.R., Mohler, C.L., *et al.* (2014) Effect of Cry3Bb Bt corn and tefluthrin on postdispersal weed seed predation. *Weed Science* **62**: 619–624.

Douglas, M.R. and Tooker, J.F. (2015) Large-scale deployment of seed treatments has driven rapid increase in use of neonicotinoid insecticides and preemptive pest management in US field crops. *Environmental Science and Technology* **49**: 5088–5097.

Egan, J.F. and Mortensen, D.A. (2012) A comparison of land-sharing and land-sparing strategies for plant richness conservation in agricultural landscapes. *Ecological Applications* **22**: 459–471.

Egan, J.F., Barlow, K.M. and Mortensen, D.A. (2014) A meta-analysis on the effects of 2,4-D and dicamba drift on soybean and cotton. *Weed Science* **62**: 193–206.

Fernandez-Cornejo, J., Wechsler, S., Livingston, M. and Mitchell, L. (2014) *Genetically Engineered Crops in the United States*. ERR-162. US Department of Agriculture, Economic Research Service, Washington, DC, USA.

Flores, S., Saxena, D. and Stotzky, G. (2005) Transgenic Bt plants decompose less in soil than non-Bt plants. *Soil Biology and Biochemistry* **37**: 1073–1082.

Funke, T., Han, H., Healy-Fried, M.L., Fischer, M. and Schönbrunn, E. (2006) Molecular basis for the herbicide resistance of Roundup Ready crops. *Proceedings of the National Academy of Sciences* **103**: 13010–13015.

Garcia-Alonso, M. and Raybould, A. (2014) Protection goals in environmental risk assessment: a practical approach. *Transgenic Research* **23**: 945–956.

Gilbert, N. (2013) A hard look at GM crops. *Nature* **497**: 24–26.

Green, J.M. (2012) The benefits of herbicide-resistant crops. *Pest Management Science* **68**: 1323–1331.

Hammonds, J.S., Hoffman, F.O. and Bartell, S.M. (1994) *An Introductory Guide to Uncertainty Analysis in Environmental and Health Risk Assessment*. ES/ER/TM-35/R1. Prepared by SENES Oak Ridge, Inc. under direction from the Environmental Restoration Risk Assessment Council for Oak Ridge National Laboratory, Oak Ridge, TN, USA.

Han, P., Niu, C.Y., Lei, C.L., Cui, J.J. and Desneux, N. (2010) Quantification of toxins in a Cry1Ac+CpTI cotton cultivar and its potential effects on the honey bee *Apis mellifera* L. *Ecotoxicology* **19**: 1452–1459.

Heap, I. (2016) International survey of herbicide resistant weeds. Available at: http://weedscience.org/ (accessed 2 March 2017).

Heinemann, J.A., Agapito-Tenfen, S.Z. and Carman, J.A. (2013) A comparative evaluation of the regulation of GM crops or products containing dsRNA and suggested improvements to risk assessments. *Environment International* **55**: 43–55.

Holst, N., Lang, A., Lovei, G. and Otto, M. (2013) Increased mortality is predicted of *Inachis io* larvae caused by Bt-maize pollen in European farmland. *Ecological Modelling* **250**: 126–133.

Hutchison, W.D., Burkness, E.C., Mitchell, P.D., *et al.* (2010) Areawide suppression of European corn borer with Bt maize reaps savings to Non-Bt maize growers. *Science* **330**: 222–225.

Icoz, I. and Stotzky, G. (2008) Cry3Bb1 protein from *Bacillus thuringiensis* in root exudates and biomass of transgenic corn does not persist in soil. *Transgenic Research* **17**: 609–620.

James, C. (2014) *Global Status of Commercialized Biotech/GM Crops: 2014*. ISAAA Brief No. 49. ISAAA, Ithaca, NY, USA. Available at: www.isaaa.org/resources/publications/briefs/49/ (accesssed 2 March 2017).

Kleppin, L., Schmidt, G. and Schroeder, W. (2011) Cultivation of GMO in Germany: support of monitoring and coexistence issues by WebGIS technology. *Environmental Sciences Europe* **23**: 4.

Kumari, S., Rakshit, A. and Beura, K. (2014) Decomposition of Bt cotton and non Bt cotton residues under varied soil types. *Journal of Microbiology, Biotechnology and Food Sciences* **3**: 360–363.

Landis, W.G. and Wiegers, J.K. (2007) Ten years of the relative risk model and regional scale ecological risk assessment. *Human and Ecological Risk Assessment* **13**: 25–38.

Liu, W., Lu, H.H., Wu, W., Wei, Q.K., Chen, Y.X. and Thies, J.E. (2008) Transgenic Bt rice does not affect enzyme activities and microbial composition in the rhizosphere during crop development. *Soil Biology and Biochemistry* **40**: 475–486.

Londoño-R, L.M., Tarkleson, D. and Thies, J.E. (2013) In-field rates of decomposition and microbial communities colonizing residues vary by depth of residue placement and plant part, but not by crop genotype for residues from two Cry1Ab Bt corn hybrids and their non-transgenic isolines. *Soil Biology and Biochemistry* **57**: 349–355.

Losey, J.E. and Vaughan, M. (2006) The economic value of ecological services provided by insects. *BioScience* **56**: 311–323.

Losey, J.E., Rayor, L.S. and Carter, M.E. (1999) Transgenic pollen harms monarch larvae. *Nature* **399**: 214–214.

Losey, J.E., Hufbauer, R.A. and Hartzler, R.G. (2003) Enumerating lepidopteran species associated with maize as a first step in risk assessment in the USA. *Environmental Biosafety Research* **2**: 247–261.

Lu, H.H., Wu, W.X., Chen, Y.X., Zhang, X.J, Devare, M. and Thies, J.E. (2010a) Decomposition of Bt transgenic rice residues and response of soil microbial community in rapeseed-rice cropping system. *Plant and Soil* **336**: 279–290.

Lu, H.H., Wu, W.X., Chen, Y.X., Wang, H.L., Devare, M. and Thies, J.E. (2010b) Soil microbial community responses to Bt transgenic rice residue decomposition in a paddy field. *Journal of Soils and Sediments* **10**: 1598–1605.

Lu, Y.H., Wu, K.M., Jiang, Y.Y., Guo, Y.Y. and Desneux, N. (2012). Widespread adoption of Bt cotton and insecticide decrease promotes biocontrol services. *Nature* **487**: 362–365.

Malone, L.A. and Burgess, E.P.J. (2009) Impact of genetically modified crops on pollinators. In: Ferry, N. and Gatehouse, A.M.R. (eds) *Environmental Impact of Genetically Modified Crops*. CABI, Cambridge, MA, USA, pp. 199–224.

MEA (2005) *Millennium Ecosystem Assessment, 2005. Ecosystems and Human Well-Being: Synthesis*. Island Press, Washington, DC, USA.

Meyer, W.M., Ostertag, R. and Cowie, R.H. (2011) Macro-invertebrates accelerate litter decomposition and nutrient release in a Hawaiian rainforest. *Soil Biology and Biochemistry* **43**: 206–211.

Mortensen, D., Egan, J., Maxwell, B., Ryan, M. and Smith, R. (2012) Navigating a critical juncture for sustainable weed management. *BioScience* **62**: 75–84.

Mueller, T.C., Wright, D.R. and Remund, K.M. (2013) Effect of formulation and application time of day on detecting dicamba in the air under field conditions. *Weed Science* **61**: 586–593.

Nagle, G.N., Fahey, T.J., Ritchie, J.C. and Woodbury, P.B. (2007) Variations in sediment sources and yields in the Finger Lakes and Catskills regions of New York. *Hydrological Processes* **21**: 828–838.

Nagle, G.N., Fahey, T.J., Ritchie, J.C. and Woodbury, P.B. (2012) Bank erosion in fifteen tributaries in the glaciated upper Susquehanna basin of New York and Pennsylvania. *Physical Geography* **33**: 229–251.

Nakatani, A.S., Fernandes, M.F., Aparecida de Souza, R., *et al.* (2014) Effects of the glyphosate-resistance gene and of herbicides applied to the soybean crop on soil microbial biomass and enzyme. *Field Crops Research* **162**: 20–29.

NAS (2016) *Genetically Engineered Crops: Experiences and Prospects*. National Academies Press, Washington, DC, USA.

Obrycki, J.J., Losey, J., Taylor, O. and Jesse, L.C.H. (2001) Transgenic insecticidal corn: beyond insecticidal toxicity to ecological complexity. *BioScience* **51**: 353–361.

Parker, J.S. (2011) *The New 2,4-D and Dicamba-Tolerant Crops: Managing Risks to Farms and Communities*. Ohio State University. Columbus, OH, USA.

Peterson, G., Cunningham, S., Deutsch, L., *et al.* (2000) The risks and benefits of genetically modified crops: a multidisciplinary perspective. *Conservation Ecology* **4**: 13.

Pleasants, J.M. and Oberhauser, K.S. (2012) Milkweed loss in agricultural fields because of herbicide use: effect on the monarch butterfly population. *Insect Conservation and Diversity* **6**: 135–144.

Power, A.G. (2010) Ecosystem services and agriculture: tradeoffs and synergies. *Philosophical Transactions of the Royal Society B – Biological Sciences* **365**: 2959–2971.

Powles, S.B. (2008) Evolved glyphosate-resistant weeds around the world: lessons to be learnt. *Pest Management Science* **64**: 360–365.

Quijas, S., Jackson, L.E., Maass, M., Schmid, B., Raffaelli, D. and Balvanera, P. (2012) Plant diversity and generation of ecosystem services at the landscape scale: expert knowledge assessment. *Journal of Applied Ecology* **49**: 929–940.

Reuter, H., Schmidt, G., Schröder, W., Middelhoff, U., Pehlke, H. and Breckling, B. (2011) Regional distribution of genetically modified organisms (GMOs) – up-scaling the dispersal and persistence potential of herbicide resistant oilseed rape (*Brassica napus*). *Ecological Indicators* **11**: 989–999.

Sandvik, H., Saether, B.E., Holmern, T., Tufto, J., Engen, S. and Roy, H.E. (2013) Generic ecological impact assessments of alien species in Norway: a semi-quantitative set of criteria. *Biodiversity and Conservation* **22**: 37–62.

Saxena, D. and Stotzky, G. (2001) *Bacillus thuringiensis* (Bt) toxin released from root exudates and biomass of Bt corn has no apparent effect on earthworms, nematodes, protozoa, bacteria, and fungi in soil. *Soil Biology and Biochemistry* **33**: 1225–1230.

Schipanski, M.E., Barbercheck, M., Douglas, M.R., *et al.* (2014) A framework for evaluating ecosystem services provided by cover crops in agroecosystems. *Agricultural Systems* **125**: 12–22.

Schoenenberger, N. and D'Andrea, L. (2012) Surveying the occurrence of subspontaneous glyphosate-tolerant genetically engineered *Brassica napus* L. (Brassicaceae) along Swiss railways. *Environmental Sciences Europe* **24**: 23.

Schwartz, L.M., Gibson, D.J., Gage, K.L., *et al.* (2015) Seedbank and field emergence of weeds in glyphosate-resistant cropping systems in the United States. *Weed Science* **63**: 425–439.

Shaner, D.L., Lindenmeyer, R.B. and Ostlie, M.H. (2012) What have the mechanisms of resistance to glyphosate taught us? *Pest Management Science* **68**: 3–9.

Smith, F.P., Prober, S.M., House, A.P.N. and McIntyre, S. (2013) Maximizing retention of native biodiversity in Australian agricultural landscapes – the 10:20:40:30 guidelines. *Agriculture Ecosystems and Environment* **166**: 35–45.

Stephens, E.J., Losey, J.E., Allee, L.L., DiTommaso, A., Bodner, C. and Breyre, A. (2012) The impact of Cry3Bb Bt-maize on two guilds of beneficial beetles. *Agriculture Ecosystems and Environment* **156**: 72–81.

Stotzky, G. (2000) Persistence and biological activity in soil of insecticidal proteins from *Bacillus thuringiensis* and of bacterial DNA bound on clays and humic acids. *Journal of Environmental Quality* **29**: 691–701.

Sun, C.X., Chen, L.J., Wu, J., Zhou, LK. and Shimizu, H. (2007) Soil persistence of *Bacillus thuringiensis* (Bt) toxin from transgenic Bt cotton tissues and its effect on soil enzyme activities. *Biology and Fertility of Soils* **43**: 617–620.

Tabashnik, B.E. and Carrière, Y. (2015) Successes and failures of transgenic Bt crops. In: Soberón, M., Gao, Y. and Bravo, A. (eds) *Bt Resistance: Characterization and Strategies for GM Crops Producing Bacillus thuringiensis Toxins*. CABI International, New York, USA, pp. 1–15.

Tarkalson, D.D., Kachman, S.D., Knops, J.M.N., Thies, J.E. and Wortman, C.S. (2007) Decomposition of Bt and non-Bt corn hybrid residues in the field. *Nutrient Cycling in Agroecosystems* **80**: 211–222.

Taylor, R.L., Maxwell, B.D. and Boik, R.J. (2006) Indirect effects of herbicides on bird food resources and beneficial arthropods. *Agriculture Ecosystems and Environment* **116**: 157–164.

Thies, J.E. (2015) Coexistence in the fields? GM, organic and non-GM. In: Herring, R. (ed.) *The Oxford Handbook of Food, Politics and Society*. Oxford University Press, Oxford, UK, pp. 689–716.

Thies, J.E. and Devare, M.H. (2007) An ecological assessment of transgenic crops. *Journal of Development Studies* **43**: 97–129.

Thompson, K.M. and Graham, J.D. (1996) Going beyond the single number: using probabilistic risk assessment to improve risk management. *Human and Ecological Risk Assessment* **2**: 1008–1034.

USDA-APHIS (2013) *Dow AgroSciences Petitions (09-233-01p, 09-349-01p, and 11-234-01p) for Determinations of Nonregulated Status for 2,4-D-Resistant Corn and Soybean Varieties – Draft Environmental Impact Statement.* US Department of Agriculture, Riverdale, MD, USA.

USEPA (2001) *Risk Assessment Guidance for Superfund (RAGS) Volume III - Part A: Process for Conducting Probabilistic Risk Assessment.* EPA 540-R-02-002. United States Environmental Protection Agency, Washington, DC, USA.

USEPA (2012) *Annotated Library for Guidance on Ecological Assessments at EPA.* United States Environmental Protection Agency, Washington, DC, USA.

Warren-Hicks, W.J. and Moore D.R.J. (1998) *Uncertainty Analysis in Ecological Risk Assessment.* Society of Environmental Toxicology and Chemistry, Pensacola, FL, USA.

Wolfenbarger, LL. and Phifer, P.R. (2000) The ecological risks and benefits of genetically engineered plants. *Science* **290**: 2088–2093.

Woodbury, P.B. (2003) Dos and don'ts of spatially explicit ecological risk assessments. *Environmental Toxicology and Chemistry* **22**: 977–982.

Woodbury, P.B. and Weinstein D.A. (2010) Review of methods for developing regional probabilistic risk assessments, part 2: Modeling invasive plant, insect, and pathogen species. In: Pye, J.M., Rauscher, H.M., Sands, Y. and Lee, D.C. (eds) *Advances in Threat Assessment and Their Application to Forest and Rangeland Management.* Gen. Tech. Rep. PNW-GTR-802. US Department of Agriculture, Forest Service, Pacific Northwest and Southern Research Stations. Portland, OR, USA, pp. 521–538.

Woodbury, P.B., Smith, J.E., Weinstein, D.A. and Laurence, J.A. (1998) Assessing potential climate change effects on loblolly pine growth: a probabilistic regional modeling approach. *Forest Ecology and Management* **107**: 99–116.

Wu, W.X., Liu, W., Lu, H.H., Chen, Y.X., Devare, M. and Thies, J. (2009a) Use of ^{13}C labeling to assess carbon partitioning in transgenic and nontransgenic (parental) rice and their rhizosphere soil microbial communities. *FEMS Microbiology and Ecology* **67**: 93–102.

Wu, W.X., Lu, H.H., Liu, W., Devare, M., Thies, J.E. and Chen, Y.X. (2009b) Decomposition of *Bacillus thuringiensis* (Bt) transgenic rice residues (straw and roots) in paddy fields. *Journal of Soils and Sediments* **9**: 457–467.

Wurbs, A., Bethwell, C. and Stachow, U. (2012) Assessment of regional capabilities for agricultural coexistence with genetically modified maize. *Environmental Sciences Europe* **24**: 17.

Xu, Z., Hennessy, D.A., Sardana, K. and Moschini, G. (2013) The realized yield effect of genetically engineered crops: US maize and soybean. *Crop Science* **53**: 735–745.

Xue, K., Serohijos, R.C., Devare, M. and Thies, J.E. (2011) Decomposition rate and microbial communities colonizing residues do not differ between Cry3Bb Bt and Non-Bt corn hybrids in the field. *Applied and Environmental Microbiology* **77**: 839–846.

Xue, K., Serohijos, R.C., Devare, M., Duxbury, J., Lauren, J. and Thies, J.E. (2012) Short-term carbon allocation and root lignin of Cry3Bb Bt and NonBt corn in the presence of corn rootworm. *Applied Soil Ecology* **57**: 16–22.

Xue, K., Diaz, B.R. and Thies, J.E. (2014) Stability of Cry3Bb1 protein in soils and its degradation in transgenic corn residues. *Soil Biology and Biochemistry* **76**: 119–126.

Young B.G., Gibson, D.J., Gage, K.L., *et al.* (2013) Agricultural weeds in glyphosate-resistant cropping systems in the United States. *Weed Science* **61**: 85–97.

Zeng, H., Tan, F., Shu, Y., Zhang, Y., Feng, Y. and Wang, J. (2015) The Cry1Ab protein has minor effects on the arbuscular mycorrhizal fungal communities after five seasons of continuous Bt maize cultivation. *PloS ONE* **10(12)**: e0146041.
Zwahlen, C., Hilbeck, A., Howald, R. and Nentwig, W. (2003) Effects of transgenic Bt corn litter on the earthworm *Lumbricus terrestris*. *Molecular Ecology* **12**: 1077–1086.

Part III

Influence of Unmanaged Habitats on Pest Management

7

Ecosystem Services Provided by Unmanaged Habitats in Agricultural Landscapes

Stefano Colazza, Morgan W. Shields, Ezio Peri and Antonino Cusumano

7.1 Introduction

Agriculture currently faces global challenges such as sufficient food production and security, insect pest damage and climate change (Culliney 2014; de Schutter 2010; Godfray and Garnett 2014; Gurr *et al.* 2012; Sandhu *et al.* 2015; Sparks and Nauen 2015; Wratten *et al.* 2014). One approach to meeting these challenges is using agro-ecology, which is the management of agricultural systems in an ecologically sound and sustainable way (Pywell *et al.* 2015) by enhancing ecosystem services (ES) provided by beneficial organisms. ES comprise ecosystem functions such as predation, pollination and water purification to which value to humans (usually but not always monetary) has been ascribed (Costanza *et al.* 1997).

The concept of ES that was introduced at the end of the 20th century (Costanza *et al.* 1997; Daily 1997) is now broadly accepted and viewed as a fundamental principle in conservation biology, agro-ecology, sustainability research and implementation (Reid *et al.* 2005). The global monetary value of ES is currently estimated at US$145 trillion per year (Costanza *et al.* 2014). ES can be enhanced by human manipulation of the environment such as the improvement of insect pest management by planting non-crop vegetation that benefits natural enemies. However, many ES have substantial value to humans with no manipulation required, such as earthworms aerating the soil and breaking down plant material in agriculture (Blouin *et al.* 2013) or pollination of orchards by insects (Földesi *et al.* 2016).

The value of ES is not always monetary, such as reduced costs and increased yield (Pywell *et al.* 2015; Tschumi *et al.* 2015). Cultural value, for instance protecting areas of importance to indigenous people and endangered wildlife, as well as aesthetics and human well-being are also important but often neglected (Roberts *et al.* 2015). The ES delivery pathway includes the idea of a 'service-providing unit', which is the smallest unit (such as individuals, a population or a community) at the desired scale that directly provides an ES (Luck *et al.* 2003). An example is a naturally occurring population of parasitoids controlling insect pests (Frank *et al.* 2007; Tschumi *et al.* 2015). Ideally, such a population needs to be managed in some way to maintain or enhance its efficacy. In this context, the idea of an ecosystem service provider (ESP) is sometimes to be invoked. An ESP is a species, food web, habitat or managed system that facilitates and supports the provision of ES (Kremen 2005). For example, an ESP could be a strip of flowering

Environmental Pest Management: Challenges for Agronomists, Ecologists, Economists and Policymakers, First Edition. Edited by Moshe Coll and Eric Wajnberg.

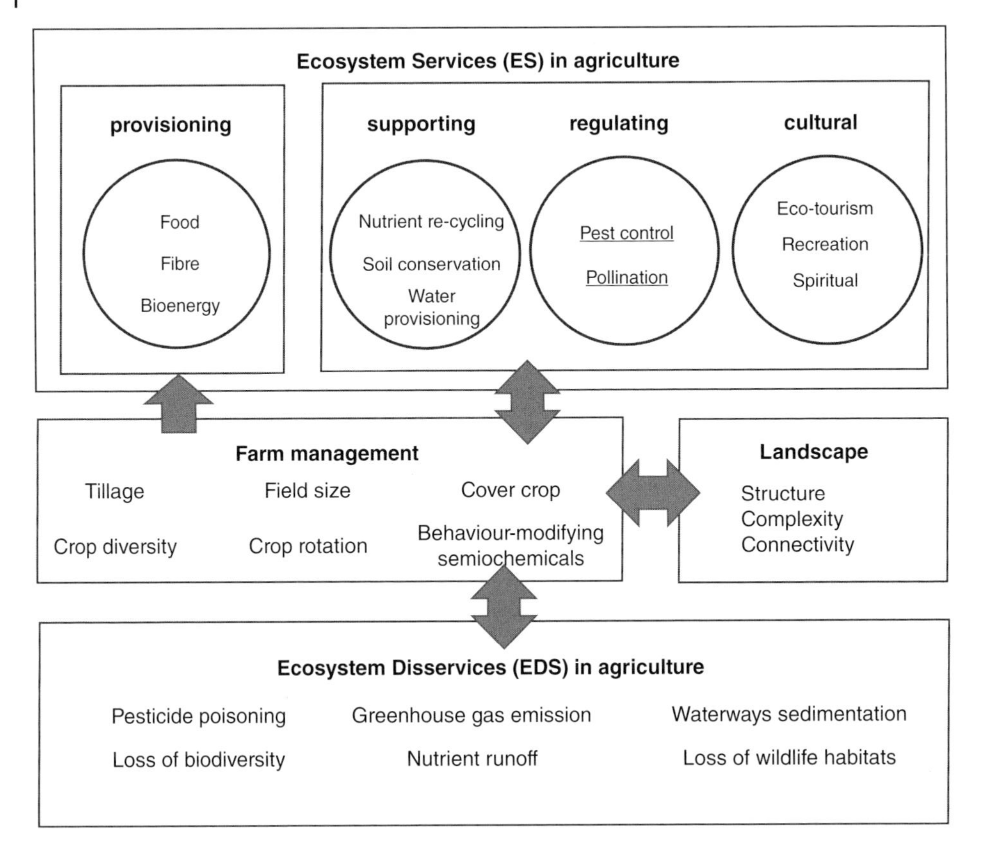

Figure 7.1 Interactions between farm management, landscape features and agricultural ecosystem services (ES) and disservices (EDS). The underlined services are primarily provided by insects.

buckwheat, *Fagopyrum esculentum* Moench, providing nectar and/or pollen for natural enemies (Scarratt *et al.* 2008).

Ecosystem services in agriculture are primarily managed to maximize the provisioning of food, fibre and fuel. However, such manipulations can also lead to ecosystem disservices (EDS), such as loss of biodiversity, increased greenhouse gas emissions, nutrient run-off and loss of natural habitats (see Chapter 8). The balance between ES and EDS depends on how agro-ecosystems are managed at the farm scale as well as the diversity, composition and functioning of the surrounding landscape (Power 2010; Tilman 1999; Zhang *et al.* 2007) (Figure 7.1). This chapter discusses the ES provided by naturally occurring populations of natural enemies in order to control insect pest populations with the aim of reducing crop damage and insecticide usage. Furthermore, we discuss how unmanaged vegetation is used by natural enemies and how landscape complexity affects their efficacy. The potential benefits of natural enemies are well recognized in agro-ecosystems and conservation biological control is appreciated as an important regulating ES (Cardinale *et al.* 2012; Cullen *et al.* 2008; Holland *et al.* 2012; Jonsson *et al.* 2008; Wratten *et al.* 2014).

Ecosystem services can be provided by non-crop vegetation specifically introduced to agricultural habitats or by unmanaged wild plants already found in the landscape. One trait of habitat manipulation is the addition of ecological infrastructures within the agricultural landscape to conserve and enhance the activity of beneficial biological control agents by providing them with needed resources such as shelter, food resources and alternative hosts or prey (Barbosa 1998). For example, the introduction of flowering plant strips in agricultural landscapes can supply beneficial parasitoid insects with suitable sugars, which increase their lifespan and fecundity. This enhances the parasitoids' fitness and biological control efficacy, enabling them to provide the ES of pest suppression (Foti *et al.* 2017; Landis *et al.* 2000). The extent of such positive effects of non-crop vegetation depends on the distance of flowering plants from the crop (Platt *et al.* 1999). Conversely, wild plants present in unmanaged vegetation may provide ES that influence crop protection in positive manners. For example, attractive and repellent plants may reduce pest populations in nearby fields. This may be the result, for example, of unmanaged vegetation serving as sinks that pull pests from crop plants or by deterring pest populations from wild vegetation that otherwise would be a source for crop infestation (Cook *et al.* 2007). Likewise, unmanaged vegetation may harbour susceptible pest genotypes that mix with resistant pests found in crop fields. This would act to slow down the development of resistance in pest populations to defensive plant traits and to pesticides (Fuentes-Contreras *et al.* 2014; Reyes *et al.* 2009). Yet, such effects of unmanaged vegetation on pest management are expected to be much smaller than their effect on biological pest control.

However, it should be noted that plants in unmanaged vegetation may also act as a source of herbivorous pest populations in nearby commercial crops (Wang *et al.* 2016). The potential of providing EDS has limited the use of unmanaged vegetation in biological control. In many cases, it resulted in the removal of non-crop vegetation from the agricultural landscape (Bianchi *et al.* 2006; Littlejohn *et al.* 2015). Nevertheless, unmanaged vegetation may provide a largely unrealized wealth of biological control services if employed in a suitable manner. This chapter explores the potential benefits provided by non-crop vegetation.

First, we briefly discuss the global importance of arthropod natural enemies. We then review the ES provided by natural enemy communities and, later, examine the effect of unmanaged vegetation and landscape complexity on conservation biological control provided by arthropod natural enemies. Finally, we conclude with the role of implementing habitat management practices at a landscape scale to maximize ES provided by arthropod natural enemies. In this chapter, the term 'non-crop vegetation' is considered to be vegetation intentionally planted by humans and managed with anthropogenic inputs such as irrigation, which enhances ES but does not produce a harvestable crop. The term 'unmanaged vegetation' refers to semi-natural vegetation that does not involve any anthropogenic inputs and does not produce a harvestable crop. The term 'landscape complexity' refers to the amount, distribution and 'quality' of non-crop and unmanaged vegetation at the landscape scale.

7.2 Global Importance of Arthropod Natural Enemies in Pest Management

Insect pests constitute a serious threat to agriculture and food security worldwide. Global insect pest damage already costs an estimated US$470 billion per annum (Sparks and Nauen 2015). In the USA alone, about 37% of potential crops are destroyed by insect pests annually, even under intensive insecticide application (Pimentel *et al.* 1992).

Annual use of insecticides in the USA is around 78471 tonnes per 1000 ha (http://faostat3.fao.org/browse/R/RP/E). This excessive insecticide use is likely to rise with further development of unsustainable agricultural intensification and insect pest resistance. The unnecessary and unsustainable use of insecticides has generated concern due to their potential societal and external costs and effects on human health and the environment (Heimpel *et al.* 2013; Jeyaratnam 1990) (see Chapters 4 and 11).

A solution to the unsustainable and excessive use of insecticides is conservation biological control where natural enemies have the potential to achieve up to 90% of the biological control occurring in agricultural settings (Pimentel 2005). Estimates based on the economic costs of insect damage to crops and that of insecticides themselves have indicated that reduced insect pest populations by natural enemies can save annually up to US$13.6 billion in the USA (Losey and Vaughan 2006). The economic value specifically attributed to arthropod natural enemies has been assessed at US$4.5 billion per year, since insect parasitoids and predators achieve about 33% of biological control of crop pests (Hawkins *et al.* 1999). However, as insecticide use is still widespread in insect pest management, it has a confounding effect on the insect pest management achieved by using natural enemies and thus it is difficult to quantify directly the rate of biological control due to natural enemies and the rate of insecticide used (Kremen and Chaplin-Kramer 2007).

7.3 Importance of Multitrophic Interactions to Biological Pest Control

7.3.1 Importance of Multitrophic Chemical Interactions

A new frontier in biological control using unmanaged vegetation is the potential to manipulate information exchanged between organisms in agricultural landscapes to better enhance insect pest management. Insect pests are constantly monitoring their environment to detect information on the occurrence of natural enemies and altering their behaviour, for example detecting visual (Jones and Dornhaus 2011) and/or chemical cues (Huryn and Chivers 1999; Ninkovic *et al.* 2013) to identify the presence of predators and therefore avoid being consumed. They adjust their behaviour in response to these cues, altering patterns of movement (Lee *et al.* 2011), feeding (Reigada and Godoy 2012) and reproduction (Vonesh and Blaustein 2010). In the presence of predators, insect herbivores frequently drop from plants (Nelson and Rosenheim 2006), consume less or lower quality food (Schmitz *et al.* 1997) and have elevated stress responses (Janssens and Stoks 2013) that combine to limit reproduction (McCauley *et al.* 2011). Furthermore, these effects can reduce herbivore population growth to an equal or greater extent than direct attack by natural enemies (Preisser *et al.* 2005).

Unmanaged vegetation can attract, fail to attract or even repel natural enemies and chemical properties of the associated plants are crucial to the response of parasitoids and predators (Wäckers 2004). Previous researchers have mostly focused on flowers in terms of effects on parasitoid fitness and to a lesser extent flower attractiveness for foraging parasitoids (Belz *et al.* 2013; Foti *et al.* 2017). Also, experimental testing of the chemical attractiveness of unmanaged vegetation has occurred. For example, hedgerows of *Eulaliopsis binata* (Retz.) C. E. Hubbard volatiles had a repellent effect on the aphid pests *Sitobion avenae* (F.) and *Rhopalosiphum padi* (L.), but attracted generalist predator species of oxyopid spider (Shi *et al.* 2011).

Plants repelling pests and attracting natural enemies have also been tested in field studies (Schader *et al.* 2005), suggesting that some plants may have dual effects of repelling insect pests and attracting natural enemies (Beizhou *et al.* 2011). This dual effect is exploited in push-pull manipulation of non-crop habitat, insect pests and their natural enemies to maximize biological control (Cook *et al.* 2007). Push-pull habitat manipulation exploits the chemicals released by plants when insect herbivores attack them. These chemicals are called herbivore-induced plant volatiles (HIPVs). These molecules can directly deter insect pest attack, attract natural enemies to help defend the plant and inform other plants of impending damage (James *et al.* 2012; Khan *et al.* 2008). Synthetic HIPVs can also be combined with unmanaged vegetation such as floral resources to 'attract and reward' natural enemies (Khan *et al.* 2008; Simpson *et al.* 2011a, 2011b) to enhance their survival and diversity, leading to increased efficacy of biological control (Gordon *et al.* 2013; Simpson *et al.* 2011b). In *Zea mays* L., *Brassica oleracea* L. and *Vitis vinifera* L. crops, *F. esculentum* was used as a 'reward' plant to provide resources, combined with synthetic HIPVs of methyl jasmonate, methyl salicylate or methyl anthranilate to attract natural enemies, resulting in increased numbers of parasitoids and predators (Simpson *et al.* 2011a, 2011b).

Exploiting these chemical cues to attract natural enemies and reduce insect pest damage using unmanaged vegetation remains a largely unutilized but important goal for biological control. Future research needs to develop such user-friendly 'recipes' involving unmanaged vegetation that maximize the efficiency of repellent plants, trap plants and attracting natural enemies to reduce insect pest damage at a landscape scale. To reduce EDS, the web of chemical cues between the multiple plants, natural enemies and insect pests needs also to be considered when developing these protocols for biological control.

7.3.2 Importance of Intraguild Interactions to Biological Control

Researchers have often reviewed the effects of predators and parasitoids separately due to the diversity and complexity of natural enemy assemblages (Murdoch and Briggs 1996; Sunderland and Samu 2000; Symondson *et al.* 2002). However, a more community ecology approach is required for future research, as parasitoids and predators may interact (intra- or interguild). This phenomenon could be synergistic or lead to interference competition and mortality of natural enemies by other natural enemies. The diversity of arthropod natural enemy communities, which can be closely linked to the community composition of the surrounding unmanaged vegetation, is a characteristic that has often been shown to be beneficial for insect pest management (Letourneau *et al.* 2009; Tscharntke *et al.* 2005; Vance-Chalcraft *et al.* 2007).

There are several ways by which species-rich natural enemy communities can positively contribute to the ES of reduced insect pest populations. For example, a diverse enemy community could reduce the populations of a greater number of pest species on diverse crops (Kremen and Chaplin-Kramer 2007). In addition, natural enemies can display species complementarity when they exhibit additive or synergistic effects, leading to insect pest suppression being greater than the sum of the individual natural enemy contributions (Losey and Denno 1998). Furthermore, having a high diversity of natural enemies can buffer biological control effectiveness against disturbance where some natural enemies may be more susceptible to specific disturbance events than others and therefore take longer to recover. Hence, a diverse natural enemy community can maintain sufficient biological control during and after disturbances when particular natural enemies have reduced populations (Tscharntke *et al.* 2005). However, a diverse community of arthropod natural enemies does not always lead to positive interactions as negative effects due to natural enemy intraguild competition and predation can result in EDS (see Chapter 8).

In complex natural enemy communities, both positive and negative species interactions can be displayed and the resulting impact on insect pest populations will depend on the overall effects. For example, Snyder and Ives (2003) found that additive effects of insect pest suppression by parasitoids and predators were stronger than intraguild predation between the two groups of natural enemies leading to enhanced insect pest suppression in complex systems. This study is of particular relevance as parasitoids caused little immediate reduction in insect pest population growth but their effect became significant after a delay, corresponding to their generation time. In contrast, predators caused an immediate decline in the aphid population growth rate but they did not show density-dependent pest suppression (Snyder and Ives 2003). Therefore, interactions between natural enemies need to be considered when determining their efficacy and how to improve it (Peri *et al.* 2014), because increased net reduction in insect pest populations is the desired outcome.

7.4 Importance of Unmanaged Vegetation for Biological Control

7.4.1 Importance of Unmanaged Vegetation for Arthropod Natural Enemies

Unmanaged vegetation can provide multiple benefits to natural enemies involved in biological control. These benefits can involve woody and herbaceous plants inhabiting non-crop areas which may provide pollen and nectar as food sources which can increase the longevity (Pfannenstiel 2012; Zhu *et al.* 2015), fecundity (Amorós-Jiménez *et al.* 2014; Pfannenstiel 2012), parasitism rate (Mathews *et al.* 2007) or predation (Belz *et al.* 2013; Zhu *et al.* 2014) as well as the F1 sex ratio in parasitoids which have fed on nectar (Berndt and Wratten 2005). Important fitness proxies such as fecundity and longevity strongly depend on the amount and quality of sugar resources (Costamagna and Landis 2004; Wäckers 2001). Natural enemies can obtain sugars from floral nectar (Casas *et al.* 2003; Lavandero *et al.* 2005; Lee *et al.* 2006; Steppuhn and Wäckers 2004; Winkler *et al.* 2009), extrafloral nectar (Géneau *et al.* 2013; Heil 2015) and honeydew (Hogervorst *et al.* 2007; Tena *et al.* 2013). Peach trees with extra-floral nectars can support more parasitoids in spring, increasing their parasitism of *Grapholita molesta* (Busck) later in the season compared to those without nectar. This reduced fruit injury from *G. molesta* by 90% (Mathews *et al.* 2007). This is one of many examples of such resources impacting the effectiveness of natural enemies in reducing pest populations (Landis *et al.* 2000; Tena *et al.* 2015; Tylianakis *et al.* 2004).

Floral and extra-floral resources can attract natural enemies from the surrounding areas (Foti *et al.* 2017; Turlings and Wäckers 2004), and these can subsequently migrate into nearby crops (Hickman and Wratten 1996; Long *et al.* 1998; Nicholls *et al.* 2001) where they may suppress insect pest populations (Landis *et al.* 2000; Tylianakis *et al.* 2004). For example, Tschumi *et al.* (2015) sowed a seed mixture of seven flowering plant species in 3m wide strips in winter wheat fields in Europe to provide resources for natural enemies. This allowed ladybirds, lacewings, hoverflies and other natural enemy groups to reduce the cereal leaf beetle *Oulema* sp. densities by over 40% (to below the economic threshold) and reduced crop damage by 61%. Using non-crop vegetation outside the planted area (rice in this case) to provide floral and extra-floral resources to enhance natural enemy efficiency can be so effective that insecticides are no longer required (Gurr *et al.* 2016). However, this requires careful study to prevent any EDS and determine the optimum ratio of plant species, spacing and sowing rate for the targeted insect pests in a given landscape.

The non-crop vegetation of unmanaged habitats can also provide alternative hosts/prey, which can sustain parasitoids and generalist predators when insect pests are at low density or not present (Frank 2010; Huang *et al.* 2011). This approach using non-crop habitat systems such as beetle banks and banker plants has been adopted in Europe, Japan, USA and Canada (Huang *et al.* 2011; Thomas *et al.* 1991). Banker plants are non-crop habitat that is deliberately inoculated with non-pest herbivores that sustain natural enemy populations (Frank 2010; Huang *et al.* 2011). For example, *Carica papaya* (L.) is used as a banker plant with the alternative whitefly host *Trialeurodes variabilis* (Quaintance) for the parasitoid *Encarsia sophia* (Girault and Dodd) to enhance the control of the whitefly pest *Bemisia tabaci* (Gennadius) in greenhouse tomato production (Xiao *et al.* 2011). The possibility of natural enemy population build-up is important for the ES in order to achieve high parasitism or prey consumption rates in the crop at the beginning of pest invasion (Bianchi and van der Werf 2004; Östman 2004). There is an increasing body of research using non-crop habitat such as banker plant systems to provide alternative hosts and prey (Huang *et al.* 2011). However, similar to other ES of non-crop habitat, in conservation biological control there is a lack of consensus on the optimal non-crop habitat systems to use even for frequently targeted insect pests and crops. Therefore, a research priority is to generate an understanding of how non-crop habitat, crop species and alternative hosts interact to affect natural enemy preference, dispersal and abundance (Frank 2010). These aspects need to be incorporated into the development 'recipes' of non-crop habitat that farmers can implement to sustainable control insect pests using conservation biological control. However, buckwheat is usually the most effective and Foti *et al.* (2017) give possible mechanisms for this.

Unmanaged vegetation can provide arthropod natural enemies with favourable environments in which they mate, reproduce, find shelter and overwinter (Bianchi *et al.* 2006; Zhang *et al.* 2007). Non-crop areas that mainly consist of woody habitats can provide a more favourable microclimate than field centres, particularly in annual crops (Forman and Baudry 1984, but see Gavish-Regev *et al.* 2008 for arid areas). This may enhance the activity of parasitoids, which are particularly sensitive to stressful temperatures (Hance *et al.* 2007) and allow natural enemies to persist and recolonize crops after disturbances such as insecticide spraying (Landis *et al.* 2000; Pollard and Holland 2006). Parasitoids and predators often overwinter in non-crop habitats (Corbett and Rosenheim 1996; Jmhasly and Nentwig 1995) such as beetle banks (Collins *et al.* 2002). These natural enemies can subsequently migrate into surrounding fields in spring (Corbett and Rosenheim 1996; Dennis and Fry 1992; Thomas *et al.* 1991), where they may reduce insect pest population densities (Collins *et al.* 2002;

Landis and van der Werf 1997; Menalled *et al.* 1999). Beetle banks are raised strips of semi-permanent tussock grasses that are refuges for a range of natural enemies in the middle of crop fields where the former can access insect pests not only around the edge of the crop but also in the centre (Collins *et al.* 2002; Thomas *et al.* 1991). Using beetle banks can reduce aphid pests by 87% after 2 years (Holland *et al.* 2012). This allows natural enemy populations to persist on farmland in unfavourable conditions so they can migrate into the field early in the crop season to ensure high recruitment rates (Bianchi *et al.* 2006; Cook *et al.* 2007) and prevent insect pest outbreaks (Landis *et al.* 2000; Thomas *et al.* 1991). However, acceptance of non-crop habitat having a key role in facilitating the ES of insect pest management at the landscape level (Bianchi *et al.* 2006; Zhang *et al.* 2007) must not ignore the fact that EDS can also be associated with unmanaged vegetation because pests are also capable of exploiting such habitats (see Chapter 8).

7.4.2 Importance of the Landscape Scale

Manipulation of non-crop unmanaged vegetation at the landscape level has been suggested as a strategy to affect the dynamics of both insect pests and their associated arthropod natural enemies (Jonsson *et al.* 2010; Landis *et al.* 2000). Both insect pest invasion and pest suppression by natural enemies have strong links to landscape structure, connectivity and complexity, which do not depend only on in-field situations (Veres *et al.* 2013; With *et al.* 2002). Up-scaling from a field to a landscape spatial scale is important in order to take into account the fact that several insect pests and natural enemies can move at the landscape scale to search for resources. For example, aphids that change hosts use different resources during their life-cycle. Similarly, parasitoids and other sugar-feeding natural enemies use short-lived floral resources that are scattered in the landscape matrix (Olson and Wäckers 2007).

Species can respond differently to the proportion of unmanaged vegetation in the landscape according to their mobility. The abundance of some predators such as ballooning spiders reacts at a scale of several kilometres (Schmidt and Tscharntke 2005) and aphid parasitoids at a scale ranging 0.5–2 km (Thies *et al.* 2005). The differential dispersal ability and habitat exploitation displayed by arthropod natural enemies are likely to impact species interactions, community structures and conservation biological control at the landscape scale. Furthermore, some arthropod natural enemies are likely to suffer more from habitat fragmentation than insect pests. This is because natural enemies are more susceptible to disturbance as they have slower reproductive strategies, are more sensitive to insecticides and can be less mobile than their associated herbivores (Cronin 2004; Kruess and Tscharntke 2000; Zabel and Tscharntke 1998). For example, abundance and species richness of parasitoid assemblages can decrease when the distance from unmanaged vegetation increases, and this in turn can lead to reduced parasitism rates. This is particularly true for specialist natural enemies, which respond more strongly to unmanaged vegetation at smaller scales than generalist species (Chaplin-Kramer *et al.* 2011). As habitat connectivity is likely to be lower in simplified landscapes than in complex ones, large-scale landscapes characterized by low connectivity may incur an increased risk of insect pest outbreaks due to insufficient conservation biological control (Bianchi *et al.* 2006).

Agricultural landscapes are considered to be very dynamic environments in which only semi-natural elements such as woodlots or hedgerows are temporally stable habitats (Burel and Baudry 1990; Petit *et al.* 2002). Most annual crops are subject to a high frequency of disturbance due to soil and pest management, harvest or crop rotation (Bianchi *et al.* 2006; Menalled *et al.* 1999). Due to the lack of stability typical of agro-ecosystems, it is thus expected that the impact of the landscape matrix on insect pest abundance and pest suppression by natural enemies is likely to fluctuate within and between years (Menalled *et al.* 2003). However, these temporal dynamics are frequently overlooked and researchers often characterize agricultural landscapes by their spatial features, such as the proportions of different land covers or the ratio between farmed areas and non-crop cover (Forman and Godron 1986). The need for up-scaling from the field to the landscape level has triggered an increasing number of studies that report relationships between landscape complexity (often expressed as the proportion of semi-natural area or unmanaged vegetation), insect pest abundance or pest suppression by natural enemies.

7.4.3 Relationship between Landscape Complexity and Natural Enemy Efficacy

The relationship between landscape complexity and the ES of reduced pest insect populations is a topic that has received increased interest in recent years. This is clear from an analysis of the number of papers retrieved from the ISI Web of Science database as of December 2015 using as keywords 'landscape' AND 'crop' AND ['predator' OR 'parasitoid' OR 'pest' OR 'biological control']. This retrieved 895 scientific articles with 301 papers being published in the last 3 years (Figure 7.2). The majority of the studies assessed the response of arthropod natural enemies when landscape complexity increases (Frank and Reichhart 2004; French and Elliot 2001; Kruess 2003; Langer and Hance 2004; Martin *et al.* 2013; Thies *et al.* 2003, 2005). In several cases, such

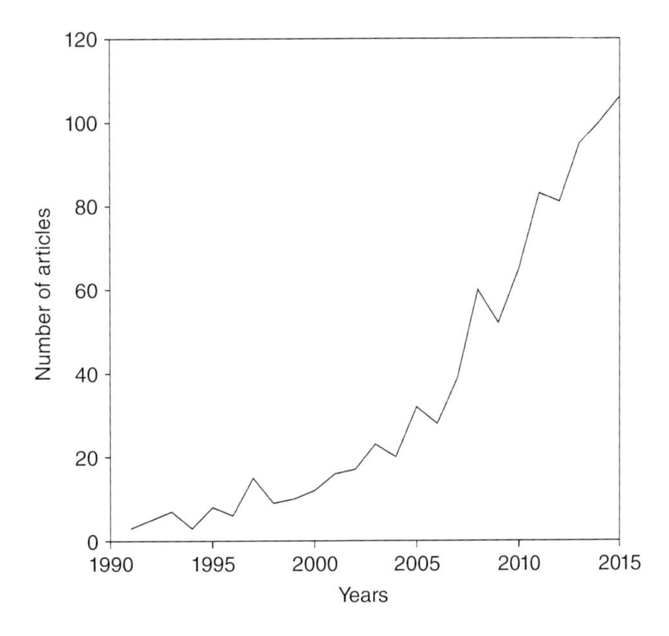

Figure 7.2 Number of scientific articles that report on investigations of the relationship between landscape features and crop protection found in the ISI Web of Science database in December 2015.

complexity can positively affect the impact of natural enemies through an effect on their abundance and/or richness (Bianchi *et al.* 2006; Chaplin-Kramer *et al.* 2011; Kremen and Chaplin-Kramer 2007; Shackelford *et al.* 2013; Veres *et al.* 2013). In a literature review of 28 studies, it was found that natural enemy populations in complex landscapes were higher in 74% of the studies compared to mono-crops (Bianchi *et al.* 2006). Interestingly, beneficial effects were found regardless of the structure of the landscape, with both herbaceous and woody habitats being able to provide similar positive effects to arthropod natural enemies. In a meta-analysis of 46 studies, it was found that natural enemies had a strong positive response when the complexity of the landscape increased, but this effect occurred at smaller scales for specialist species compared with generalist ones (Chaplin-Kramer *et al.* 2011). This indicates that landscape complexity implications need to be assessed at different scales depending on the natural enemies and associated insect pests of interest.

A number of studies have also investigated the response of insect pests to landscape structure (Bennett and Gratton 2012; Morandin *et al.* 2014; Östman *et al.* 2001; Roschewitz *et al.* 2005; Thies and Tscharntke 1999; Thies *et al.* 2003, 2005; Traugott *et al.* 2012). There is some evidence suggesting that increasing the landscape complexity leads to a decrease in pest populations. For example, the review by Bianchi *et al.* (2006) showed that pest density was reduced in 45% of the studies. In contrast, the meta-analysis done by Chaplin-Kramer *et al.* (2011) mentioned that there is no significant link between pest abundance and landscape complexity. However, despite the results of that study, some experimental works have shown that a positive relationship between habitat complexity and natural enemies can cascade into reduced pest density (Thies and Tscharntke 1999; Thies *et al.* 2003). Therefore, it seems that the above conflicting conclusions are context dependent. In particular, the herbivore species under investigation seems to play an important role. For example, rape pollen beetle abundance has been shown to decrease with an increase in landscape complexity (Thies and Tscharntke 1999; Thies *et al.* 2003) whereas aphids do not consistently show the same pattern (Östman *et al.* 2001; Thies *et al.* 2005). Östman *et al.* (2001) found that aphid populations were negatively affected by landscape complexity. Conversely, Thies *et al.* (2005) found that, in complex landscapes, there were increased parasitism rates but aphid establishment rates were also higher in crops, leading to similar aphid abundance regardless of landscape complexity.

Thus, it is possible that the lack of insect pest management in many instances where complex landscapes involve non-crop habitat is a result of a positive response by both natural enemies and their associated pests so that the overall net effect on pest abundance remains neutral. Additionally, the lack of response of pest density to landscape complexity could be due to the temporal aspects of pest invasion. Indeed, herbivore insects are not always pests during the crop growing season as outbreaks generally occur at particular times in relationship to crop phenology. However, such information is often unavailable as the majority of the studies focus exclusively on the spatial relation between pest management and landscape complexity. Therefore, future studies are encouraged to measure the abundance of pests and their associated natural enemies over time in order to gain a more complete understanding of relationships between landscape complexity and insect pest management at the landscape scale (Jonsson *et al.* 2015).

7.5 Landscape Use to Maximize Biological Control

Natural enemies can benefit in multiple ways from unmanaged vegetation in the agricultural landscape (see earlier), but an increase in natural enemy abundance does not always translate to reduced insect pest populations (Bianchi *et al.* 2006). Floral resources of unmanaged vegetation can support both natural enemies and insect pests (Baggen *et al.* 1999; Wäckers 2001). Therefore, the composition of areas with flowering plants in unmanaged vegetation has important management implications as it can selectively favour natural enemies, which generally require different flowering species from those required by insect pests (Patt *et al.* 1997; Wäckers 2004).

Having only a few patches or strips of unmanaged vegetation may not be enough. A network of interconnected 'corridors' of unmanaged vegetation across one or more farms may be more effective (Bianchi *et al.* 2006; Gamez-Virués *et al.* 2015; Holland *et al.* 2012). This would allow increased dispersal of natural enemies between food sources and increase the persistence of natural enemy populations during unfavourable conditions and disturbances, improving overall insect pest suppression in the landscape (Bianchi *et al.* 2006; Holland *et al.* 2012).

As discussed earlier, the spatial scale of response of insect pests and natural enemies is an important factor to take into consideration for management. This is because insects and other organisms providing both ES and EDS to agriculture are not restricted to the boundaries of crop fields but also may move within the landscape between natural habitats, hedgerows and fields (Zhang *et al.* 2007). In the case of natural enemies, the spatial scale of response is linked to their degree of specialization to their hosts or prey (Chaplin-Kramer *et al.* 2011). When generalist species contribute the most to reducing insect pest populations below threshold levels, insect pest management actions should require co-operative approaches or be landscape based since generalist species respond at larger spatial scales than specialist species (Chaplin-Kramer *et al.* 2011). In contrast, if specialist natural enemies are primarily responsible for reducing insect pest populations, then individual actions taken at the farm scale can be effective since specialist species tend to be more effective when there is high landscape complexity near the field margins (Chaplin-Kramer *et al.* 2011; Tylianakis *et al.* 2007).

Therefore, it is important to characterize the community of natural enemies associated with insect pests to understand how to manage unmanaged vegetation in the landscape to maximize the ES provided by natural enemies.

Farmer decisions to conserve or restore unmanaged vegetation to maximize the ES and thus reduce insect pest populations are usually based on private economic interests that affect the farm itself, and such decisions may be in contrast with the provision of alternative ES such as food production. After all, farmers will not take valuable land out of production to provide habitats for beneficial insects if it is not economically viable. Pywell *et al.* (2015) illustrated, in central England, that creating non-crop habitat on field margins (up to 8% of each field) maintains and, for some crops, increases overall yield over a 5-year crop rotation.

However, on-farm unmanaged vegetation needs to provide multiple ES to balance any trade-offs and ultimately lead to increased food production. For instance, in Thailand, Vietnam and China, Gurr *et al.* (2016) grew strips of nectar-producing plants around rice fields over 4 years and monitored levels of pest infestation, insecticide use and yields. During this time, two key pest populations were reduced, parasitoids and predators of the main rice pests, together with detritivores, were more abundant, insecticide applications were reduced by 70%, grain yields increased by 5% and the programme delivered an overall economic advantage of 7.5%. Another example is shelterbelt of *Miscanthus × giganteus*, a sterile hybrid bioenergy grass that grows 4 m tall, on dairy farms where it can provide at least 15 ES including an 18% increase in grass production and shelter for beneficial insects (Littlejohn *et al.* 2015).

Furthermore, when farmers decide to reserve land for restoring/conserving unmanaged vegetation, other farmers may benefit by improved insect pest management as such ES are provided at the landscape scale. In this case, neighbours who do not take out land for pest control may enjoy the benefits from higher landscape complexity without facing a reduction in crop production. Such so-called economic externalities (see Chapters 3 and 16) imply that the first farmer, acting alone, will not set aside the amount of habitat required for both the farmer and the neighbours. Trade-offs between private and social interests clearly occur and when the value of unmanaged vegetation for the group is greater than for those farmers surrounding the unmanaged vegetation, subsidies should be considered. However, it is difficult to assess economically the value of ES associated with such habitats at different spatial scales.

Without financial incentives, farmers have no immediate interest in contributing to landscape complexity. Public policies aimed at creating incentives for farmers to act on behalf of the collective interest should foster a co-operative approach as the provision of several ES to the general public goes beyond the farm scale (Bianchi *et al.* 2013; Zhang *et al.* 2007). Furthermore, to achieve farmer acceptance and uptake of agro-ecology-based habitat management that provides reduced insect pest populations and other ES, scientists and policymakers need to discuss with farmers the challenges that they face and their current management methods, then develop habitat management solutions that enhance multiple ES simultaneously while increasing food production and/or reducing costs. Once protocols are developed based on sound scientific research, these need to be taught by 'teacher' farmers to other local farmers where terminology that the local farmers understand is used and there is a level of trust and respect between individuals (Warner 2007). However, local, regional and national subsidies are often involved, such as in European agri-environment schemes (Batáry *et al.* 2015).

7.6 Conclusions

The goal of restoring, conserving or creating beneficial unmanaged vegetation in order to minimize insecticide application in the farming landscape could be achieved by maximizing ES to control insect pest populations by natural enemies. This is especially true in those situations characterized by low landscape complexity because the benefit of increasing vegetation complexity of unmanaged habitats in complex landscapes would be negligible (Tscharntke *et al.* 2002, 2011). Most of the studies investigating the impact of landscape complexity in crop protection have analysed the response of arthropod natural enemies, whereas studies investigating the impact of landscape on crop performance are still scarce. However, the key aspect of reduced pest populations is the prevention or reduction in crop damage, and this aspect should be incorporated in studies investigating how landscape complexity can affect insect pest responses (Gurr *et al.* 2016; Thies and Tscharntke 1999). Such economic assessments are crucial to make economic comparisons between different ES relevant to agriculture and to make decisions about which ES would be relatively more important than others to be maximized in the perspective of multiple ES (Bianchi *et al.* 2013; Zhang *et al.* 2007). In order to restore, conserve or create unmanaged vegetation targeting multiple ES, it is crucial to know whether enhancing one service impacts other services in a positive, negative or neutral manner (Shackelford *et al.* 2013).

Beneficial insects are responsible for delivering multiple ES, among which reduction of insect pest populations, enhancement of crop pollination and improvement of soil properties by ground-dwelling arthropod decomposers are of fundamental importance as regulating ES. It seems that landscape complexity can impact pollinators and natural enemies in similar positive ways, suggesting that it may be possible to enhance both simultaneously (Kremen and Chaplin-Kramer 2007; Shackelford *et al.* 2013). However, further comparative studies should be conducted to determine how to mitigate any negative interactions between multiple ES and potential EDS (see Chapter 8).

To date, the majority of studies had been conducted in annual crops with managed non-crop vegetation in Western Europe and North America. Future research is needed for the incorporation of non-crop vegetation in agro-ecosystems in other regions of the globe. This is desirable because unmanaged vegetation does not bear ongoing costs. The vast potential ES provided by unmanaged vegetation is underutilized, and many ES and the ecological functions that regulate them are known to be context dependent (Kremen 2005).

Finally, the mechanistic aspects of the ES provision are still poorly understood and scarce knowledge is available about how species in natural ecosystems interact to generate services important for agriculture.

References

Amorós-Jiménez, R., Pineda, A., Fereres, A. and Marcos-García, M.Á. (2014) Feeding preferences of the aphidophagous hoverfly *Sphaerophoria rueppellii* affect the performance of its offspring. *BioControl* **59**: 427–435.

Baggen, L.R., Gurr, G.M. and Meats, A. (1999) Flowers in tritrophic systems: mechanisms allowing selective exploitation by insect natural enemies for conservation biological control. *Entomologia Experimentalis et Applicata* **91**: 155–161.

Barbosa, P.A. (1998) *Conservation Biological Control*. Academic Press, San Diego, CA, USA.

Batáry, P., Dicks, L.V., Kleijn, D. and Sutherland, W.J. (2015) The role of agri-environment schemes in conservation and environmental management. *Conservation Biology* **29**: 1006–1016.

Beizhou, S., Jie, Z., Jinghui, H., Hongying, W., Yun, K. and Yuncong, Y. (2011) Temporal dynamics of the arthropod community in pear orchards intercropped with aromatic plants. *Pest Management Science* **67**: 1107–1114.

Belz, E., Kölliker, M. and Balmer, O. (2013) Olfactory attractiveness of flowering plants to the parasitoid *Microplitis mediator*: potential implications for biological control. *BioControl* **58**: 163–173.

Bennett, A.B. and Gratton, C. (2012) Measuring natural pest suppression at different spatial scales affects the importance of local variables. *Environmental Entomology* **41**: 1077–1085.

Berndt, L.A. and Wratten, S.D. (2005) Effects of alyssum flowers on the longevity, fecundity, and sex ratio of the leafroller parasitoid *Dolichogenidea tasmanica*. *Biological Control* **32**: 65–69.

Bianchi, F.J.J.A. and van der Werf, W. (2004) Model evaluation of the function of prey in non-crop habitats for biological control by ladybeetles in agricultural landscapes. *Ecological Modelling* **171**: 177–193.

Bianchi, F.J.J.A., Booij, C.J.H. and Tscharntke, T. (2006) Sustainable pest regulation in agricultural landscapes: a review on landscape composition, biodiversity and natural pest control. *Proceedings of the Royal Society B: Biological Sciences* **273**: 1715–1727.

Bianchi, F., Mikos, V., Brussaard, L., Delbaere, B. and Pulleman, M.M. (2013) Opportunities and limitations for functional agrobiodiversity in the European context. *Environmental Science and Policy* **27**: 223–231.

Blouin, M., Hodson, M.E., Delgado, E.A., *et al.* (2013) A review of earthworm impact on soil function and ecosystem services. *European Journal of Soil Science* **64**: 161–182.

Burel, F. and Baudry, J. (1990) Structural dynamic of a hedgerow network landscape in Brittany, France. *Landscape Ecology* **4**: 197–210.

Cardinale, B.J., Duffy, J.E., Gonzalez, A., *et al.* (2012) Biodiversity loss and its impact on humanity. *Nature* **486**: 59–67.

Casas, J., Driessen, G., Mandon, N., *et al.* (2003) Energy dynamics in a parasitoid foraging in the wild. *Journal of Animal Ecology* **72**: 691–697.

Chaplin-Kramer, R., O'Rourke, M.E., Blitzer, E.J. and Kremen, C. (2011) A meta-analysis of crop pest and natural enemy response to landscape complexity. *Ecology Letters* **14**: 922–932.

Collins, K.L., Boatman, N.D., Wilcox, A., Holland, J.M. and Chaney, K. (2002) Influence of beetle banks on cereal aphid predation in winter wheat. *Agriculture, Ecosystems and Environment* **93**: 337–350.

Cook, S.M., Khan, Z.R. and Pickett, J.A. (2007) The use of push-pull strategies in integrated pest management. *Annual Review of Entomology* **52**: 375–400.

Corbett, A. and Rosenheim, J.A. (1996) Impact of a natural enemy overwintering refuge and its interaction with the surrounding landscape. *Ecological Entomology* **21**: 155–164.

Costamagna, A.C. and Landis, D.A. (2004) Effect of food resources on adult *Glyptapanteles militaris* and *Meteorus communis* (Hymenoptera: Braconidae), parasitoids of *Pseudaletia unipuncta* (Lepidoptera: Noctuidae). *Environmental Entomology* **33**: 128–137.

Costanza, R., D'Arge, R., de Groot, R., *et al.* (1997) The value of the world's ecosystem services and natural capital. *Nature* **387**: 253–260.

Costanza, R., de Groot, R., Sutton, P., *et al.* (2014) Changes in the global value of ecosystem services. *Global Environmental Change* **26**: 152–158.

Cronin, J.T. (2004) Host–parasitoid extinction and colonization in a fragmented prairie landscape. *Oecologia* **139**: 503–514.

Cullen, R., Warner, K.D., Jonsson, M. and Wratten, S.D. (2008) Economics and adoption of conservation biological control. *Biological Control* **45**: 272–280.

Culliney, T.W. (2014) Crop losses to arthropods. In: Pimentel, D. and Peshin, R. (eds) *Integrated Pest Management*. Springer, Dordrecht, The Netherlands, pp. 201–225.

Daily, G.C. (1997) *Nature's Services: Societal Dependence on Natural Ecosystems*. Island Press, Washington, DC, USA.

Dennis, P. and Fry, G.L.A. (1992) Field margins: can they enhance natural enemy population densities and general arthropod diversity on farmland? *Agriculture, Ecosystems and Environment* **40**: 95–115.

De Schutter, O.D. (2010) Report submitted by the Special Rapporteur on the right to food. United Nations General Assembly. Available at: http://www2.ohchr.org/english/issues/food/docs/A-HRC-16-49.pdf (accessed 3 March 2017).

Földesi, R., Kovács-Hostyánszki, A., Kőrösi, Á., *et al.* (2016) Relationships between wild bees, hoverflies and pollination success in apple orchards with different landscape contexts. *Agricultural and Forest Entomology* **18**: 68–75.

Forman, R.T.T. and Baudry, J. (1984) Hedgerows and hedgerow networks in landscape ecology. *Environmental Management* **8**: 495–510.

Forman, R.T.T. and Godron, M. (1986). *Landscape Ecology*. John Wiley and Sons, New York, USA.

Foti, M.C., Rostás, M., Peri, E., *et al.* (2017) Chemical ecology meets conservation biological control: Identifying plant volatiles as predictors of floral resource suitability for an egg parasitoid of stink bugs. *Journal of Pest Science* **90**: 299–310.

Frank, S.D. (2010) Biological control of arthropod pests using banker plant systems: past progress and future directions. *Biological Control* **52**: 8–16.

Frank, S.D., Wratten, S.D., Sandhu, H.S. and Shrewsbury, P.M. (2007) Video analysis to determine how habitat strata affect predator diversity and predation of *Epiphyas postvittana* (Lepidoptera: Tortricidae) in a vineyard. *Biological Control* **41**: 230–236.

Frank, T. and Reichhart, B. (2004) Staphylinidae and Carabidae overwintering in wheat and sown wildflower areas of different age. *Bulletin of Entomological Research* **94**: 209–217.

French, B.W. and Elliot, N.C. (2001) Species diversity, richness, and evenness of ground beetles (Coleoptera: Carabidae) in wheat fields and adjacent grasslands and riparian zones. *Southwestern Entomologist* **26**: 315–324.

Fuentes-Contreras, E., Basoalto, E., Franck, P., Lavandero, B., Knight, A.L. and Ramírez, C.C. (2014) Measuring local genetic variability in populations of codling moth (Lepidoptera: Tortricidae) across an unmanaged and commercial orchard interface. *Environmental Entomology* **43**: 520–527.

Gamez-Virués, S., Perović, D.J., Gossner, M.M., *et al.* (2015) Landscape simplification filters species traits and drives biotic homogenization. *Nature Communications* **6**: 8568.

Gavish-Regev, E., Lubin, Y. and Coll, M. (2008) Migration patterns and functional groups of spiders in a desert agroecosystem. *Ecological Entomology* **33**: 202–212.

Géneau, C.E., Wäckers, F.L., Luka, H. and Balmer, O. (2013) Effects of extrafloral and floral nectar of *Centaurea cyanus* on the parasitoid wasp *Microplitis mediator*: olfactory attractiveness and parasitization rates. *Biological Control* **66**: 16–20.

Godfray, H.C.J. and Garnett, T. (2014) Food security and sustainable intensification. *Philosophical Transactions of the Royal Society B: Biological Sciences* **369**: 20120273.

Gordon, G.O., Wratten, S.D., Jonsson, M., Simpson, M. and Hale, R. (2013) 'Attract and reward': combining a herbivore-induced plant volatile with floral resource supplementation – multi-trophic level effects. *Biological Control* **64**: 106–115.

Gurr, G.M., Wratten, S.D., Snyder, W.E. and Read, D.M.Y. (2012) *Biodiversity and Insect Pests: Key Issues for Sustainable Management*. Wiley-Blackwell, Chichester, UK.

Gurr, G.M., Lu, Z., Zheng, X., *et al.* (2016) Multi-country evidence that crop diversification promotes ecological intensification of agriculture. *Nature Plants* **2**: 16014.

Hance, T., van Baaren, J., Vernon, P. and Boivin, G. (2007) Impact of extreme temperatures on parasitoids in a climate change perspective. *Annual Review of Entomology* **52**: 107–126.

Hawkins, B.A., Mills, N.J., Jervis, M.A. and Price, P.W. (1999) Is the biological control of insects a natural phenomenon? *Oikos* **86**: 493–506.

Heil, M. (2015) Extrafloral nectar at the plant-insect interface: a spotlight on chemical ecology, phenotypic plasticity, and food webs. *Annual Review of Entomology* **60**: 213–232.

Heimpel, G.E., Yang, Y., Hill, J.D. and Ragsdale, D.W. (2013) Environmental consequences of invasive species: greenhouse gas emissions of insecticide use and the role of biological control in reducing emissions. *PLoS ONE* **8(8)**: e72293.

Hickman, J.M. and Wratten, SD. (1996) Use of *Phacelia tanacetifolia* strips to enhance biological control of aphids by hoverfly larvae in cereal fields. *Journal of Economic Entomology* **89**: 832–840.

Hogervorst, P.A.M., Wäckers, F.L. and Romeis, J. (2007) Detecting nutritional state and food source use in field-collected insects that synthesize honeydew oligosaccharides. *Functional Ecology* **21**: 936–946.

Holland, J.M., Oaten, H., Moreby, S., *et al.* (2012) Agri-environment scheme enhancing ecosystem services: a demonstration of improved biological control in cereal crops. *Agriculture, Ecosystems and Environment* **155**: 147–152.

Huang, N., Enkegaard, A., Osborne, L.S., *et al.* (2011) The banker plant method in biological control. *Critical Reviews in Plant Sciences* **30**: 259–278.

Huryn, A.D. and Chivers, D.P. (1999) Contrasting behavioral responses by detritivorous and predatory mayflies to chemicals released by injured conspecifics and their predators. *Journal of Chemical Ecology* **25**: 2729–2740.

James, D.G., Orre-Gordon, S., Reynolds, O.L. and Simpson, M. (2012) Employing chemical ecology to understand and exploit biodiversity for pest management. In: Gurr, G., Wratten. S.D., Snyder, W.E. and Read, D.M.Y. (eds) *Biodiversity and Insect Pests: Key Issues For Sustainable Management.* John Wiley and Sons, Chichester, UK, pp. 185–195.

Janssens, L. and Stoks, R. (2013) Predation risk causes oxidative damage in prey. *Biology Letters* **9**: 20130350.

Jeyaratnam, J. (1990) Acute pesticide poisoning: a major global health problem. *World Health Statistics Quarterly* **43**:139–144.

Jmhasly, P. and Nentwig, W. (1995) Habitat management in winter wheat and evaluation of subsequent spider predation on insect pests. *Acta Oecologica* **16**: 389–403.

Jones, E. and Dornhaus, A. (2011) Predation risk makes bees reject rewarding flowers and reduce foraging activity. *Behavioural Ecology and Sociobiology* **65**: 1505–1511.

Jonsson, M., Wratten, S.D., Landis, D.A. and Gurr, G.M. (2008) Recent advances in conservation biological control of arthropods by arthropods. *Biological Control* **45**: 172–175.

Jonsson, M., Wratten, S.D., Landis, D.A., Tompkins, J.M.L. and Cullen, R. (2010) Habitat manipulation to mitigate the impacts of invasive arthropod pests. *Biological Invasions* **12**: 2933–2945.

Jonsson, M., Straub, C.S., Didham, R.K., *et al.* (2015) Experimental evidence that the effectiveness of conservation biological control depends on landscape complexity. *Journal of Applied Ecology* **52**: 1274–1282.

Khan, Z.R., James, D.G., Midega, C.A.O. and Pickett, J.A. (2008) Chemical ecology and conservation biological control. *Biological Control* **45**: 210–224.

Kremen, C. (2005) Managing ecosystem services: what do we need to know about their ecology? *Ecology Letters* **8**: 468–479.

Kremen, C. and Chaplin-Kramer, R. (2007) Insects as providers of ecosystem services: crop pollination and pest control. In: Stewart, A.J.A., New, T.R. and Lewis, O.T. (eds) *Insect Conservation Biology*. CABI International, Wallingford, UK, pp. 349–382.

Kruess, A. (2003) Effects of landscape structure and habitat type on a plant-herbivore-parasitoid community. *Ecography* **26**: 283–290.

Kruess, A. and Tscharntke, T. (2000) Species richness and parasitism in a fragmented landscape: experiments and field studies with insects on *Vicia sepium*. *Oecologia* **122**: 129–137.

Landis, D.A. and van der Werf, W. (1997) Early-season predation impacts the establishment of aphids and spread of beet yellows virus in sugar beet. *Entomophaga* **42**: 499–516.

Landis, D.A., Wratten, S.D. and Gurr, G.M. (2000) Habitat management to conserve natural enemies of arthropod pests in agriculture. *Annual Review of Entomology* **45**: 175–201.

Langer, A. and Hance, T. (2004) Enhancing parasitism of wheat aphids through apparent competition: a tool for biological control. *Agriculture Ecosystems and Environment* **102**: 205–212.

Lavandero, B., Wratten, S.D., Shishehbor, P. and Worne, S. (2005) Enhancing the effectiveness of the parasitoid *Diadegma semiclausum* (Helen): movement after use of nectar in the field. *Biological Control* **34**: 152–158.

Lee, D.H., Nyrop, J.P. and Sanderson, J.P. (2011) Avoidance of natural enemies by adult whiteflies, *Bemisia argentifolii*, and effects on host plant choice. *Biological Control* **58**: 302–309.

Lee, J.C., Andow, D.A. and Heimpel, G.E. (2006) Influence of floral resources on sugar feeding and nutrient dynamics of a parasitoid in the field. *Ecological Entomology* **31**: 470–480.

Letourneau, D.K., Jedlicka, J.A., Bothwell. S.G. and Moreno, C.R. (2009) Effects of natural enemy biodiversity on the suppression of arthropod herbivores in terrestrial ecosystems. *Annual Review of Ecology Evolution and Systematics* **40**: 573–592.

Littlejohn, C.P., Curran, T.J., Hofmann, R.W. and Wratten, S.D. (2015) Farmland, food, and bioenergy crops need not compete for land. *Solutions* **6**: 34–48.

Long, R.F., Corbett, A., Lamb, C., Reberg-Horton, C., Chandler, J. and Stimmann, M. (1998) Beneficial insects move from flowering plants to nearby crops. *California Agriculture* **52**: 23–26.

Losey, J.E. and Denno, R.F. (1998) Positive predator–predator interactions: enhanced predation rates and synergistic suppression of aphid populations. *Ecology* **79**: 2143–2152.

Losey, J.E. and Vaughan, M. (2006) The economic value of ecological services provided by insects. *BioScience* **56**: 311–323.

Luck, G.W., Daily, G.C. and Ehrlich, P.R. (2003) Population diversity and ecosystem services. *Trends in Ecology and Evolution* **18**: 331–336.

Martin, E.A., Reineking, B., Seo, B. and Steffan-Dewenter, I. (2013) Natural enemy interactions constrain pest control in complex agricultural landscapes. *Proceedings of the National Academy of Sciences of the USA* **110**: 5534–5539.

Mathews, C.R., Brown, M.W. and Bottrell, D.G. (2007) Leaf extrafloral nectaries enhance biological control of a key economic pest, *Grapholita molesta* (Lepidoptera: Tortricidae), in peach (Rosales: Rosaceae). *Environmental Entomology* **36**: 383–389.

McCauley, S.J., Rowe, L. and Fortin, M.J. (2011) The deadly effects of 'nonlethal' predators. *Ecology* **92**: 2043–2048.

Menalled, F.D., Marino, P.C., Gage, S.H. and Landis, D.A. (1999) Does agricultural landscape structure affect parasitism and parasitoid diversity? *Ecological Applications* **9**: 634–641.

Menalled, F.D., Costamagna, A.C., Marino, P.C. and Landis, D.A. (2003) Temporal variation in the response of parasitoids to agricultural landscape structure. *Agriculture, Ecosystems and Environment* **96**: 29–35.

Morandin, L.A., Long, R.F. and Kremen, C. (2014) Hedgerows enhance beneficial insects on adjacent tomato fields in an intensive agricultural landscape. *Agriculture Ecosystems and Environment* **189**: 164–170.

Murdoch, W.W. and Briggs, C.J. (1996) Theory for biological control: recent developments. *Ecology* 77: 2001–2013.

Nelson, E.H. and Rosenheim, J.A. (2006) Encounters between aphids and their predators: the relative frequencies of disturbance and consumption. *Entomologia Experimentalis et Applicata* **118**: 211–219.

Nicholls, C.I., Parrella, M. and Altieri, M.A. (2001) The effects of a vegetational corridor on the abundance and dispersal of insect biodiversity within a northern California organic vineyard. *Landscape Ecology* **16**: 133–146.

Ninkovic, V., Feng, Y.R., Olsson, U. and Pettersson, J. (2013) Ladybird footprints induce aphid avoidance behaviour. *Biological Control* **65**: 63–71.

Olson, D.M. and Wäckers, F.L. (2007) Management of field margins to maximize multiple ecological services. *Journal of Applied Ecology* **44**: 13–21.

Östman, Ö. (2004) The relative effects of natural enemy abundance and alternative prey abundance on aphid predation rates. *Biological Control* **30**: 281–287.

Östman, Ö., Ekbom, B., Bengtsson, J. and Weibull, A.C. (2001) Landscape complexity and farming practice influence the condition of polyphagous carabid beetles. *Ecological Applications* **11**: 480–488.

Patt, J.M., Hamilton, G.C. and Lashomb, J.H. (1997) Foraging success of parasitoid wasps on flowers: interplay of insect morphology, floral architecture and searching behavior. *Entomologia Experimentalis et Applicata* **83**: 21–30.

Peri, E., Cusumano, A., Amodeo, V., Wajnberg, E. and Colazza, S. (2014) Intraguild interactions between two egg parasitoids of a true bug in semi-field and field conditions. *PLoS ONE* **9**(6): e99876.

Petit, S., Howard, D.C., Smart, S.M. and Firbank, L.G. (2002) Biodiversity in British agroecosystems: the changing regional landscape context. In: British Crop Protection Council (ed.) *The BCPC Conference: Pests and Diseases, Volumes 1 and 2*. Proceedings of an International Conference held at the Brighton Hilton Metropole Hotel, Brighton, UK, 18–21 November 2002, pp. 957–964.

Pfannenstiel, R.S. (2012) Direct consumption of cotton pollen improves survival and development of *Cheiracanthium inclusum* (Araneae: Miturgidae) spiderlings. *Annals of the Entomological Society of America* **105**: 275–279.

Pimentel, D. (2005) Environmental and economic costs of the application of pesticides primarily in the United States. *Environment, Development and Sustainability* 7: 229–252.

Pimentel, D., Stachow, U., Takacs, D.A., *et al.* (1992) Conserving biological diversity in agricultural/forestry systems: most biological diversity exists in human-managed ecosystems. *BioScience* **42**: 354–362.

Platt, J.O., Caldwell, J.S. and Kok, L.T. (1999) Effect of buckwheat as a flowering border on populations of cucumber beetles and their natural enemies in cucumber and squash. *Crop Protection* **18**: 305–313.

Pollard, K.A. and Holland, J.M. (2006) Arthropods within the woody element of hedgerows and their distribution pattern. *Agricultural and Forest Entomology* **8**: 203–211.

Power, A.G. (2010) Ecosystem services and agriculture: tradeoffs and synergies. *Philosophical Transactions of the Royal Society B: Biological Sciences* **365**: 2959–2971.

Preisser, E.L., Bolnick, D.I. and Benard, M.F. (2005) Scared to death? The effects of intimidation and consumption in predator–prey interactions. *Ecology* **86**: 501–509.

Pywell, R.F., Heard, M.S., Woodcock, B.A., *et al.* (2015) Wildlife friendly farming increases crop yield: evidence for ecological intensification. *Proceedings of the Royal Society B* **282**: 20151740.

Reid, W., Mooney, H.A., Cropper, A., *et al.* (2005) *Millennium Ecosystem Assessment Synthesis: Ecosystems and Human Well-being: Synthesis.* United Nations, Island Press, Washington, DC, USA.

Reigada, C. and Godoy, W.A.C. (2012) Direct and indirect top-down effects of previous contact with an enemy on the feeding behaviour of blowfly larvae. *Entomologia Experimentalis et Applicata* **142**: 71–77.

Reyes, M., Franck, P., Olivares, J., Margaritopoulos, J., Knight, A., and Sauphanor, B. (2009) Worldwide variability of insecticide resistance mechanisms in the codling moth, *Cydia pomonella* L.(Lepidoptera: Tortricidae). *Bulletin of Entomological Research* **99**: 359–369.

Roberts, L., Brower, A., Kerr, G., *et al.* (2015) *The Nature of Wellbeing: How Nature's Ecosystem Services Contribute to the Wellbeing of New Zealand and New Zealanders.* Department of Conservation, Wellington, New Zealand.

Roschewitz, I., Hucker, M., Tscharntke, T. and Thies, C. (2005) The influence of landscape context and farming practices on parasitism of cereal aphids. *Agriculture Ecosystems and Environment* **108**: 218–227.

Sandhu, H., Wratten, S.D., Costanza, R., Pretty, J., Porter, J.R. and Reganold, J. (2015) Significance and value of non-traded ecosystem services on farmland. *Peer Journal* **3**: e762.

Scarratt, S.L., Wratten, S.D. and Shishehbor, P. (2008) Measuring parasitoid movement from floral resources in a vineyard. *Biological Control* **46**: 107–113.

Schader, C., Zaller, J.G. and Köpke, U. (2005) Cotton–basil intercropping: effects on pests, yields and economical parameters in an organic field in Fayoum, Egypt. *Biological Agriculture and Horticulture* **23**: 59–72.

Schmidt, M.H. and Tscharntke, T. (2005) Landscape context of sheetweb spider (Araneae: Linyphiidae) abundance in cereal fields. *Journal of Biogeography* **32**: 467–473.

Schmitz, O.J., Beckerman, A.P. and O'Brien, K.M. (1997) Behaviorally mediated trophic cascades: effects of predation risk on food web interactions. *Ecology* **78**: 1388–1399.

Shackelford, G., Steward, P.R., Benton, T.G., *et al.* (2013) Comparison of pollinators and natural enemies: a meta-analysis of landscape and local effects on abundance and richness in crops. *Biological Reviews* **88**: 1002–1021.

Shi, G.Q., Lin, C.W., Liu, Z.Y., *et al.* (2011) Effects of plant hedgerow on population dynamics of wheat aphid and its natural enemies. *Journal of Applied Ecology* **22**: 3265–3271.

Simpson, M., Gurr, G.M., Simmons, A.T., *et al.* (2011a) Attract and reward: combining chemical ecology and habitat manipulation to enhance biological control in field crops. *Journal of Applied Ecology* **48**: 580–590.

Simpson, M., Gurr, G.M., Simmons, A.T., *et al.* (2011b) Field evaluation of the 'attract and reward' biological control approach in vineyards. *Annals of Applied Biology* **159**: 69–78.
Snyder, W.E. and Ives, A.R. (2003) Interactions between specialist and generalist natural enemies: parasitoids, predators, and pea aphid biocontrol. *Ecology* **84**: 91–107.
Sparks, T.C. and Nauen, R. (2015) IRAC: Mode of action classification and insecticide resistance management. *Pesticide Biochemistry and Physiology* **121**: 122–128.
Steppuhn, A. and Wäckers, F.L. (2004) HPLC sugar analysis reveals the nutritional state and the feeding history of parasitoids. *Functional Ecology* **18**: 812–819.
Sunderland, K. and Samu, F. (2000) Effects of agricultural diversification on the abundance, distribution, and pest control potential of spiders: a review. *Entomologia Experimentalis et Applicata* **95**: 1–13.
Symondson, W.O.C., Sunderland, K.D. and Greenstone, M.H. (2002) Can generalist predators be effective biocontrol agents? *Annual Review of Entomology* **47**: 561–594.
Tena, A., Pekas, A., Wäckers, F. and Urbaneja, A. (2013) Energy reserves of parasitoids depend on honeydew from non-hosts. *Ecological Entomology* **38**: 278–289.
Tena, A., Pekas, A., Cano, D., Wäckers, F.L. and Urbaneja, A. (2015) Sugar provisioning maximizes the biocontrol service of parasitoids. *Journal of Applied Ecology* **52**: 795–804.
Thies, C. and Tscharntke, T. (1999) Landscape structure and biological control in agroecosystems. *Science* **285**: 893–895.
Thies, C., Steffan-Dewenter, I. and Tscharntke, T. (2003) Effects of landscape context on herbivory and parasitism at different spatial scales. *Oikos* **101**: 18–25.
Thies, C., Roschewitz, I. and Tscharntke, T. (2005) The landscape context of cereal aphid-parasitoid interactions. *Proceedings of the Royal Society B: Biological Sciences* **272**: 203–210.
Thomas, M.B., Wratten, S.D. and Sotherton, N.W. (1991) Creation of island habitats in farmland to manipulate populations of beneficial arthropods: predator densities and emigration. *Journal of Applied Ecology* **28**: 906–917.
Tilman, D. (1999) Global environmental impacts of agricultural expansion: the need for sustainable and efficient practices. *Proceedings of the National Academy of Sciences of the USA* **96**: 5995–6000.
Traugott, M., Bell, J.R., Raso, L., Sint, D. and Symondson, W.O.C. (2012) Generalist predators disrupt parasitoid aphid control by direct and coincidental intraguild predation. *Bulletin of Entomological Research* **102**: 239–247.
Tscharntke, T., Steffan-Dewenter, I., Kruess, A. and Thies, C. (2002) Contribution of small habitat fragments to conservation of insect communities of grassland-cropland landscapes. *Ecological Applications* **12**: 354–363.
Tscharntke, T., Klein, A.M., Kruess, A., Steffan-Dewenter, I. and Thies, C. (2005) Landscape perspectives on agricultural intensification and biodiversity – ecosystem service management. *Ecology Letters* **8**: 857–874.
Tscharntke, T., Batary, P. and Dormann, C.F. (2011) Set-aside management: how do succession, sowing patterns and landscape context affect biodiversity? *Agriculture Ecosystems and Environment* **143**: 37–44.
Tschumi, M., Albrecht, M., Entling, M.H. and Jacot, K. (2015) High effectiveness of tailored flower strips in reducing pests and crop plant damage. *Proceedings of the Royal Society B: Biological Science* **282**: 20151369.
Turlings, T.C. and Wäckers, F. (2004) Recruitment of predators and parasitoids by herbivore-injured plants. *Advances in Insect Chemical Ecology* **2**: 21–75.

Tylianakis, J.M., Didham, R.K. and Wratten, S.D. (2004) Improved fitness of aphid parasitoids receiving resource subsidies. *Ecology* **85**: 658–666.
Tylianakis, J.M., Tscharntke, T. and Lewis, O.T. (2007) Habitat modification alters the structure of tropical host-parasitoid food webs. *Nature* **445**: 202–205.
Vance-Chalcraft, H.D., Rosenheim, J.A., Vonesh, J.R., Osenberg, C.W. and Sih, A. (2007) The influence of intraguild predation on prey suppression and prey release: a meta-analysis. *Ecology* **88**: 2689–2696.
Veres, A., Petit, S., Conord, C. and Lavigne, C. (2013) Does landscape composition affect pest abundance and their control by natural enemies? A review. *Agriculture Ecosystems and Environment* **166**: 110–117.
Vonesh, J.R. and Blaustein, L. (2010) Predator-induced shifts in mosquito oviposition site selection: a meta-analysis and implications for vector control. *Israel Journal of Ecology and Evolution* **56**: 263–279.
Wäckers, F.L. (2001) A comparison of nectar- and honeydew sugars with respect to their utilization by the hymenopteran parasitoid *Cotesia glomerata*. *Journal of Insect Physiology* **47**: 1077–1084.
Wäckers, F.L. (2004). Assessing the suitability of flowering herbs as parasitoid food sources: flower attractiveness and nectar accessibility. *Biological Control* **29**: 307–314.
Wang, X.G., Stewart, T.J., Biondi, A., *et al.* (2016) Population dynamics and ecology of *Drosophila suzukii* in Central California. *Journal of Pest Science* **89**: 701–712.
Warner, K.D. (2007) *Agroecology in Action: extending alternative agriculture through social networks*. MIT Press, Cambridge, MA, USA.
Winkler, K., Wäckers, F.L. and Pinto, D.M. (2009) Nectar-providing plants enhance the energetic state of herbivores as well as their parasitoids under field conditions. *Ecological Entomology* **34**: 221–227.
With, K.A., Pavuk, D.M., Worchuck, J.L., Oates, R.K. and Fisher, J.L. (2002) Threshold effects of landscape structure on biological control in agroecosystems. *Ecological Applications* **12**: 52–65.
Wratten, S.D., Tomasetto, F. and Dinnis, D.M. (2014). Land use: management for biodiversity and conservation. In: van Alfen, N.K. (ed.) *Encyclopedia of Agriculture and Food Systems*. Elsevier, London, pp. 134–138.
Xiao, Y., Chen, J., Cantliffe, D., Mckenzie, C., Houben, K. and Osborne, L.S. (2011) Establishment of papaya banker plant system for parasitoid, *Encarsia sophia* (Hymenoptera: Aphilidae) against *Bemisia tabaci* (Hemiptera: Aleyrodidae) in greenhouse tomato production. *Biological Control* **58**: 239–247.
Zabel, J. and Tscharntke, T. (1998) Does fragmentation of *Urtica* habitats affect phytophagous and predatory insects differentially? *Oecologia* **116**: 419–425.
Zhang, W., Ricketts, T.H., Kremen, C., Carney, K. and Swinton, S.M. (2007) Ecosystem services and dis-services to agriculture. *Ecological Economics* **64**: 253–260.
Zhu, P., Lu, Z., Heong, K., *et al.* (2014) Selection of nectar plants for use in ecological engineering to promote biological control of rice pests by the predatory bug, *Cyrtorhinus lividipennis*, (Heteroptera: Miridae). *PLoS ONE* **9(9)**: e108669.
Zhu, P., Wang, G., Zheng, X., Tian, J. and Lu, Z. (2015) Selective enhancement of parasitoids of rice Lepidoptera pests by sesame (*Sesamum indicum*) flowers. *BioControl* **60**: 157–167.

8

The Role of Ecosystem Disservices in Pest Management

Mark A.K. Gillespie and Steve D. Wratten

8.1 Introduction

The ecosystem services (ES) provided by unmanaged habitats outlined in the previous chapter largely have a positive impact on human well-being. Whether it is through pest population reduction or other services such as carbon sequestration, enhanced pollination or biodiversity conservation, these can be the intended or secondary benefits of conserving or restoring natural and semi-natural habitats in agricultural landscapes. However, when attempting to harness these 'benefits of nature', unintended negative impacts that are harmful to human well-being can often occur. Such unwanted effects can be termed 'negative ecosystem services' or 'ecosystem disservices' (EDS). This chapter explores how these EDS may occur in relation to conserving, restoring or neglecting unmanaged habitats in agricultural settings, and what may be done to try to prevent them or mitigate their effects.

According to von Dohren and Haase (2015), the concept of EDS is not yet well defined, and has not yet been studied with the same intensity as ES. However, as this chapter will demonstrate, the two are inextricably linked, with the same ecosystems and processes benefiting and costing society at the same time. Nevertheless, it is possible to adapt the well-known definition of ecosystem services (ES) of Daily (1997), and define EDS as 'the conditions and processes through which natural ecosystems, and the species that make them up, are harmful to human life'. They may be harmful directly such as infectious diseases arising from pathogens and their vectors, or in tropical countries as extreme as human mortality or injury caused by wild animals associated with natural ecosystems (Dunn 2010). Alternatively, a more prosaic and indirect pathway of 'harm' is when the services arising from an ecosystem or its species lead to economic costs to society or individuals. As a simple example, while urban forests provide aesthetic, recreational and pollution-reducing ES, they can also require a high level of maintenance, cause damage to infrastructure and pose a social nuisance to hayfever sufferers (Escobedo *et al.* 2011).

In agriculture, EDS are often expressed as reducing productivity (and therefore profit) or leading to increases in operational costs (Zhang *et al.* 2007). Pests and weeds can themselves generate EDS, being species that reduce productivity and often lead to an increase in the costs of chemical inputs, mechanical removal or integrated programmes

Environmental Pest Management: Challenges for Agronomists, Ecologists, Economists and Policymakers, First Edition. Edited by Moshe Coll and Eric Wajnberg.

Table 8.1 The components of unmanaged habitat and the possible ecosystem services and disservices that they generate. The lists here are not exhaustive and the extent of their occurrence will depend on a range of factors including the plant and animal species composition of the unmanaged habitat, the scale under consideration and the background landscape context.

Component of unmanaged habitat	Ecosystem services generated via	Ecosystem disservices generated via
Flowering plants	• Food for pollinators • Food for adult natural enemies • Conservation of threatened species	• Spread of weed species • Food for adults of pests • Food for fourth trophic level (intraguild predators, hyperparasitoids)
Structural complexity	• Refuge for natural enemies • Reproductive habitat for natural enemies • Overwintering habitat for natural enemies • Prevention of pest movement into crop • Beneficial microclimate conditions	• Refuge for pest species • Alternative food sources for pest species capable of "switching" • Overwintering habitat for pest species • Prevention of movement of natural enemies into crop
Late successional vegetation	• Stable habitat for natural enemies • Increased genetic diversity for natural enemy populations	• Stable habitat for pest species • Increased genetic diversity for pest species
Connectivity	• Aids movement of natural enemies	• Aids movement of pests
Tall vegetation	• Provides perches for beneficial birds • Barrier to movement of pests	• Provides perches for pest birds • Competition for light and water with crop • Barrier to movement of natural enemies
Plant species diversity	• Often results in diverse natural enemy community • Possible facilitation effects (one natural enemy enhances the effect of another)	• Diverse pest community • May increase the potential of antagonistic interactions (competition, predation, avoidance behaviour) between natural enemies • Spread of weed species

that prevent their reaching outbreak status. These control methods also themselves generate EDS, whether it is the pollution associated with pesticides or the increase in pest diversity that is sometimes associated with some natural biocontrol methods. This illustrates a key point about ES and EDS: both will always be present to some degree, so there will always be a need to balance the costs of managing the EDS against the profit from the ES (Table 8.1). To make matters more complicated, an ES at one scale may be an EDS at another, or one person's ES can be another person's EDS (for example, a 'weed' species can be costly to remove for a farmer (EDS) but if that species is nationally rare, there are clear conservation benefits of the plant's habitat (ES)). Multiple trade-offs between the ES to enhance and the EDS to avoid exist in an agricultural landscape. Part of the major challenge in managing landscapes characterized by a mosaic of agricultural and natural or semi-natural ecosystems is therefore in balancing the relative impacts of multiple ES and EDS to maximize productivity, profit and overall benefit and minimize cost, loss and overall harm (Zhang *et al.* 2007).

In many cases, EDS may arise inadvertently when ES have been targeted. The broadest example is agriculture itself: the basic ES flowing from agriculture is the food, fuel and fibre it provides, but unintentional EDS such as water pollution, soil erosion and reduced soil fertility often result from intensive farming practices (Daily 1997). The local conditions created by arable agriculture are also ideal for certain animal species to become pests or certain plant species to become weeds and the damage caused by such organisms makes them an important global EDS. More specifically, the range of plant resources associated with non-crop habitats such as nectar, pollen, alternative food, refuges and mating habitat benefits the insects and birds that provide the ES of pest regulation (Landis *et al.* 2000). However, pest species may also benefit from the same resources (Thies *et al.* 2005), or the predators and parasitoids of natural enemies of pests may also gain an advantage (Paredes *et al.* 2013). Thus, the conservation of non-crop habitats to provide ES may also result in EDS flowing from the same ecosystem.

Despite this, natural and semi-natural habitats are largely considered to be important for agricultural production. Apart from the ES contributing to pest control as outlined in the previous chapter, unmanaged habitats provide ES such as pollination (Kremen *et al.* 2002), cover and food for game and other mammal and bird species (Thomas *et al.* 2001) and a host of supporting services such as water and climate regulation, erosion prevention, and soil structure and fertility (Garcia-Feced *et al.* 2015; Power 2010; Zhang *et al.* 2007). On balance, crops surrounded by a greater amount of natural or semi-natural habitat receive higher levels of ES such as pest suppression (Veres *et al.* 2013) and pollination (Carvalheiro *et al.* 2011). Nevertheless, EDS flowing from these habitats can pose significant economic problems (Triplett *et al.* 2012) which are rarely accounted for when the economics of ES to and from agriculture are reviewed (Power 2010). Furthermore, a lack of knowledge of the extent and real costs of EDS may be a barrier to farmer adoption of agri-environment schemes that encourage the protection of unmanaged habitats (Bianchi *et al.* 2013). It is important, therefore, that a better understanding of the extent and cost of EDS is gained, so that an integrated framework of ES and EDS can be developed (von Dohren and Haase 2015), and management strategies can be employed to support and enhance populations of beneficial organisms without increasing pest problems.

This chapter aims to highlight the main pathways of pest-related EDS flowing from unmanaged habitats (summarized in Table 8.1) before considering the ways in which they can be prevented, minimized and/or accounted for in management decision making. Then, particular consideration is given to the importance of scale factors associated with the flow of EDS from unmanaged habitat. We then outline the current knowledge of EDS management and conclude with emphasis on the need to account for these types of EDS in cost–benefit analysis and to incorporate a broader understanding of EDS into the ES framework.

8.2 EDS and Unmanaged Habitats

For the purposes of this chapter, unmanaged habitats are defined as any non-crop habitat that is not under regular farm management. This includes anything from small patches to large areas of natural or semi-natural vegetation such as semi-natural grasslands and forests, as well as marginal unfarmed features such as hedgerows, woodlots and riparian habitat, but excludes managed non-crop habitat in fields and field margins such as grass banks, flower strips and buffer strips. Often, similar principles will apply to these latter areas, but they are excluded here for the purposes of brevity. Throughout the chapter, the terms 'unmanaged habitat' and 'semi-natural habitat' will be used interchangeably. In addition, the terms 'habitat diversity' and 'landscape complexity' will be used to illustrate the negative impacts of a high proportion of unmanaged habitat in the landscape. Many studies use these terms as measures to describe the spectrum of landscape types between very simple, monocultural arable areas (low habitat diversity or a low proportion of semi-natural habitat cover) and very complex, polycultural areas with multiple patches of semi-natural habitat (high habitat diversity or a high proportion of habitat cover). However, it should be noted that a high proportion of natural habitat does not necessarily equate to high habitat diversity. For example, a large area of native coniferous forest may also be very simple and species poor, or a large reserve of semi-natural grassland may have low landscape complexity at a large scale but accommodate high floral and animal diversity at finer scales. Care should therefore be taken when considering the impacts of unmanaged habitats *per se*, and those arising from the composition, diversity and configuration of unmanaged habitats in the landscape (Fahrig *et al.* 2011). Nevertheless, when referring to high diversity or complex landscapes, the examples given here associate these terms with a high proportion of diverse non-crop habitat types.

As discussed in the previous chapter, there are numerous positive links between unmanaged habitats and ES, but successful pest management is not just a case of providing unmanaged habitat and biodiversity enhancement *per se*, as this is no guarantee of successful biological control (Bianchi *et al.* 2006). Some pests, such as some aphid species, also spend part of their life cycle in non-crop areas, particularly if they make use of alternative hosts outside the crop over winter (Thies *et al.* 2005). While a meta-analysis of studies found that semi-natural habitat usually has a positive effect on pest suppression, there are exceptions and the analysis by no means covered all possible pest–crop interactions (Veres *et al.* 2013). The scales at which pests and natural enemies confer their EDS or ES in the farmland environment are likely to range from the individual field to the landscape as a whole, and will differ from species to species depending on their dispersal abilities (Bianchi *et al.* 2006). Furthermore, the impact of unmanaged habitats will vary across these scales and will also depend on the composition of the habitats, the connectivity between them and their cover and distribution in the landscape (Mitchell *et al.* 2014).

A high level of semi-natural habitat cover is thought to buffer the damaging effects of the intensive agricultural disturbance regime to a certain extent, and conservation biological control in particular aims to create conditions in agro-ecosystems that are less harmful to beneficial insects. However, any refuge for denizens of one trophic level may also be beneficial to others, from predators and parasitoids of natural enemies

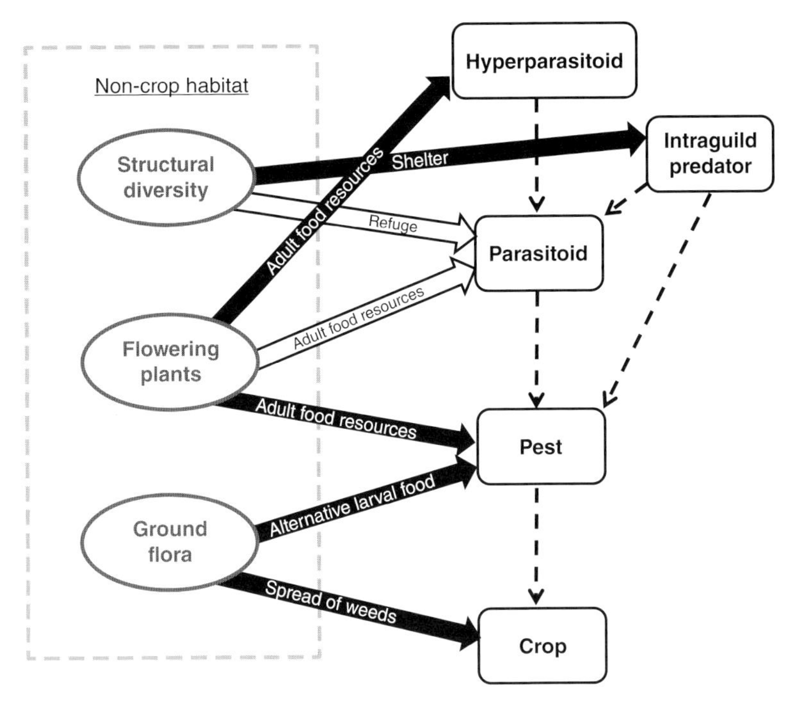

Figure 8.1 A schematic example of the possible pathways of ecosystem services (ES) and ecosystem disservices (EDS) flowing from an area of unmanaged habitat to a simple agricultural food web. Dotted arrows indicate links between components of the food web (*black boxes*), the white solid arrows represent ES and the black solid arrows represent EDS flowing from components of the non-crop habitat.

(termed 'secondary' or 'hyperparasitoids'; species that lay eggs in or on hosts already parasitized by members of the third (beneficial) trophic level), to the main pest and alternative pests, and even to the primary producer level in the form of weed species (Langridge 2011; Roschewitz *et al.* 2005). The EDS associated with unmanaged habitats can take many forms (Figure 8.1). In relation to pests, these may arise from direct benefits to them through the provision of alternative food, shelter or water, or a series of interactions may prevent the natural enemies of pests from successfully suppressing pest populations in the crop. Alternatively, interactions between natural enemies and other species such as pollinators can also affect their ability to deliver ES. Depending on their structure and distribution in the agricultural matrix, natural habitats may also prevent movement of certain important species while assisting the movement of others. All these aspects of EDS are considered in detail in the sections that follow.

8.2.1 Benefits of Unmanaged Habitats to Pests

One of the most common ways in which unmanaged habitat can benefit pest species is through food resources. Adults of pest insects such as moths and butterflies can use a range of flower species to obtain the carbohydrate and protein needed to live longer and lay more eggs (Kehrli and Bacher 2008). Often the hymenopteran and dipteran natural enemies of these pests use a different range of plant species, preferring a set of morphologically distinct flowers or species with a differing nutrient composition. Knowledge of these preferences forms the basis of flower selection when designing seed mixes for adding non-crop plant species to field margins to attract natural enemies and encourage pest suppression services. This selectivity is now a well-developed science following the recognition that pest species can sometimes gain more benefits than the natural enemies (Baggen and Gurr 1998). The same can be true of flowers found in natural or semi-natural habitats, the flowers of weed species or even the flowering crop itself (Rusch *et al.* 2011). If the suite of flowering plants associated with unmanaged habitats suits local pests more than natural enemies, the net effect is likely to be an EDS.

Unmanaged habitat is perhaps especially beneficial for pest species that are not restricted to unique hosts or a small range of hosts. Particularly successful species of this type of insect can make use of the crop when it is optimally available, for example when it is at peak growth stage, and when it reaches the harvest stage can switch to alternative host plants located outside the crop. Although most aphid species are specialists, some are particularly adept at this life strategy, being able to develop populations on both the crop and related weed species. For example, the aphid *Sitobion avenae* Fabricius (Hemiptera: Aphididae) can colonize cereals and other uncultivated grass species (Hand 1989) and genotypes capable of switching between crop and non-crop species pose a formidable challenge to pest management programmes. Some studies have attributed this kind of host switching to the positive relationship found between landscape complexity and pest abundance (Thies *et al.* 2005). However, Vialatte *et al.* (2005) studied the genetic exchange between populations of *S. avenae* on crop and non-crop plants in France and found that the two population groups remained largely independent and did not mix, suggesting that the populations of aphids developing on uncultivated grass species are not a great threat or a source of colonizing pest aphids. Despite this, they could not completely exclude the possibility of uncultivated areas acting as a reservoir of genetic diversity for this aphid species, which may subsequently help the pest to adapt to changing agricultural practices and crop variants.

Semi-natural habitat can also provide a number of other resources to insects. When the conditions in the arable field become inhospitable to insects due to the phenology of the crop or the timing of agricultural practices, unmanaged habitats may act as refuges for pests and natural enemies alike. Perennial vegetation such as woodland or grassland surrounding oil seed rape fields, for example, can provide overwintering habitat for stem weevils (*Ceutorhynchus* sp., Coleoptera: Curculionidae) that pupate in the soil, the brassica pod midge (*Dasineura brassicae* Winn, Diptera: Cecidomyiidae) that overwinters in brassica plants and pollen beetles (*Meligethes* sp., Coleoptera: Nitidulidae) that hibernate in woods and other sheltered places. Zaller *et al.* (2008) studied these pests in Austrian agro-ecosystems in relation to field and landscape characteristics and found that pest abundance was negatively related to the area of oil seed rape in the landscape. Abundances of all three pests were also positively linked to the amount of woodland, which the authors attribute to the availability and diversity of alternative host plants and overwintering sites. Woodlands could also act to modify the local microclimate by reducing exposure to strong winds (Foggo *et al.* 2001).

Natural habitat is also useful to pest species of birds and other animals. For example, Schackermann *et al.* (2015) studied almond and sunflower crops in Israel to test the impact of natural shrublands and semi-natural planted forest habitats on the seed predation by birds and reptiles. They found that the cover of semi-natural forests within a 1 km radius was strongly linked to bird species richness, probably due to the height of the vegetation in forests providing protection and perches for birds. Both bird abundance and species richness were also strongly correlated with seed predation. Natural shrublands did not confer the same benefit because the height of the vegetation was much lower. This suggests that natural shrubland habitats can be conserved around almond and sunflower crops, but that planted pine forests should be managed carefully or that almond and sunflower crops should not be grown near such habitats. However, as with the other examples highlighted here, it is not this simple. Some bird species associated with these habitats are also known to deliver ES in the form of invertebrate pest control (Dix *et al.* 1995). Triplett *et al.* (2012) summarized the overall costs and benefits of birds to farmers, and noted that many pest bird species make use of natural vegetation for perching, roosting, nesting or for vantage points over the crop to avoid predation. Birds may also aid weed dispersal from uncultivated areas into the crop, but some species provide pest control, pollination (in tropical regions), seed dispersal (of plants important for erosion control, for example) and waste disposal services. Clearly, detailed information specific to a particular situation about the relative costs and benefits of both beneficial and harmful bird species is needed to design effective management strategies.

8.2.2 EDS Derived from Unmanaged Habitat Affecting the Enemies of Beneficial Species: Hyperparasitoids

In addition to pests and their natural enemies, semi-natural habitats can benefit the fourth trophic level, for example species that feed on beneficial insects. Zhao *et al.* (2013) sampled parasitoid species in structurally simple or complex landscapes in the wheat-growing region near Yinchuan, north-west China, and found that the assemblages were different for the two landscape types. Those with high habitat diversity characterized by more hedgerows, woodlands and grasslands and a higher diversity of plant species had higher diversity of both parasitoids and hyperparasitoids. This is perhaps unsurprising because hyperparasitoids rely on the same non-host food sources as the parasitoids they attack (Araj *et al.* 2008), and their density and diversity are likely to be closely linked to their hosts spatially and temporally (Banks *et al.* 2008). Higher trophic levels are also thought to be more sensitive to environmental disturbance than lower ones (Holt *et al.* 1999), suggesting that less intensive agricultural areas will be more hospitable to hyperparasitoids, increasing the likelihood of this form of EDS. Furthermore, there is evidence that the sensitivity of higher trophic levels to disturbance will be less pronounced if they are generalists that feed opportunistically (Holt *et al.* 1999). This is supported by work by Rand *et al.* (2012) who demonstrated that generalist hyperparasitoids are likely to benefit more from additional resources in unmanaged habitats surrounding crops, and to respond more rapidly to increases in landscape complexity than specialist hyperparasitoids.

Another example of this ecosystem function is the study by Gagic *et al.* (2012) of winter wheat fields in contrasting landscapes in Germany, involving the temporal and spatial analysis of 64 aphid-parasitoid-hyperparasitoid food webs. Similar to the work by Zhao *et al.* (2013), the differences in community assemblages were distinct between low- and high-complexity landscapes. In particular, food webs in simple landscapes were dominated by aphid species that favour high concentrations of leaf nitrogen

(*Metopolophium dirhodum* Walker, *Rhopalosiphum padi* L., Hemiptera: Aphididae), whereas the food webs of complex landscapes were dominated by *Sitobion avenae* which tends to benefit from a high cover of grasslands and the overwintering sites they provide (Thies *et al.* 2005). These differences seemed to drive the identity of the dominant species at the third and fourth trophic levels. The structure of food webs also changed as the crop matured, highlighting the need to understand the organisms involved in the cropping system and the importance of the landscape context. Importantly for biological control, both parasitism and hyperparasitism rates were higher in landscapes with high habitat diversity (complex landscapes) which the authors suggest are linked to the availability of alternative resources in these areas which help these species persist and actively search for hosts. Overall, the findings from the study supported two key points: (1) primary parasitism rates were negatively related to community complexity (i.e. the biodiversity of the food web), suggesting that the parasitoids are better biocontrol agents in simple food webs (Tylianakis *et al.* 2007), and (2) hyperparasitoids were positively linked to community complexity, supporting the idea that ecosystem functioning increases with greater biodiversity (Hooper *et al.* 2005). The implications of these findings for ES and EDS in relation to the unmanaged aspects of the agro-ecosystem are yet to be studied, but in theory, the effect of pest suppression ES may have the potential to be cancelled out in complex situations.

The evidence provided by the studies above does not yet result in conclusions that hyperparasitoids in complex landscapes confer a net EDS, possibly because the late seasonal peaks of hyperparasitoid attacks mean that the impacts on primary parasitoids are likely to occur in subsequent years (Rand *et al.* 2012). However, the possibility of EDS occurring in this way has been explored by a small number of previous studies that have shown no effect (Nofemela 2013) or a negative effect of hyperparasitoids on primary parasitoid abundance or parasitism of aphids (Holler *et al.* 1993), or a positive effect on aphid densities (Brodeur and Rosenheim 2000; Rosenheim 1998). Holler *et al.* (1993) concluded from a study of food webs in cereal crops in Germany that primary parasitoid females left the host patch when hyperparasitoids were present, and that other mortality factors contributed to low parasitoid abundance and parasitism rate. Conversely, Gomez-Marco *et al.* (2015) found that a single primary parasitoid attacked the spirea citrus aphid *Aphis spiraecola* Patch (Hemiptera: Aphididae) on clementine in Spain, but at least six hyperparasitoid species attacked the primary parasitoid throughout the growing season. The disruption to biological control by the fourth trophic level is clearly feasible in some cases, but these negative effects have not yet been directly related to the extent of unmanaged habitat in the landscape.

8.2.3 EDS Derived from Unmanaged Habitat Affecting the Enemies of Beneficial Species: Intraguild Predation

Pest herbivores may be shared by many natural enemies and this is more likely in diverse communities such as those in or around large complex patches of vegetation (Martin *et al.* 2013; Muller and Brodeur 2002; Traugott *et al.* 2012; Vance-Chalcraft *et al.* 2007). A diverse community of natural enemies can lead to a range of interactions between carnivorous species. Ideally, for the pest manager, natural enemies will have either additive or synergistic effects: the overall pest suppression service will be equal to or greater than the sum of the individual natural enemy pressures. In many cases, this is true in practice: species-rich natural enemy communities provide better pest regulation services than species-poor ones (Letourneau *et al.* 2009; Vance-Chalcraft *et al.* 2007). However, antagonistic interactions between shared natural enemies of herbivore pests can occur, such as hyperparasitism described in the previous section, intraguild predation and other behavioural disruptions. Intraguild predation occurs when two or more natural enemies attack a herbivore, but one of them (i.e. the intraguild predator) also feeds on or parasitizes its competitor (i.e. the intraguild prey). The strength of the intraguild predation impact and the effects on pest suppression depend on the competitive strength of the natural enemies involved and on the abundance of the herbivore (Muller and Brodeur 2002).

Janssen *et al.* (2007) conducted a meta-analysis of the literature concerning intraguild predation and 'habitat structure' (studies where the structure of farmland habitats was manipulated). They found that the negative effect of intraguild predators on other natural enemies was reduced by high levels of manipulated habitat. Furthermore, they found that intraguild predation had no effect on the shared prey (the pest) when the intraguild prey was a better competitor than the intraguild predator, but a negative effect when the intraguild predator was the best competitor. Overall, the conclusions were positive for pest managers because the primary natural enemies benefited from habitat structure but the shared prey did not, although it should be noted that the review included studies from aquatic food webs and other non-agricultural habitats. Conversely, in another meta-analysis of intraguild effects, Letourneau *et al.* (2009) found that a species-rich community of natural enemies had a negative effect on pest suppression in 30% of studies (80 published cases) only. Although this review also included studies in natural habitats and non-agricultural settings, as well as manipulative field cage or laboratory studies, the potential for pest control disruption clearly exists even if explicit empirical evidence does not. The inconsistency between these two meta-analyses is likely to be due to the different focus of the two studies and the variation in habitats considered.

The potential for disruption to pest control ES was shown in a study by Martin *et al.* (2013), a rare example of the study of multiple natural enemies in relation to habitat complexity. They examined the impact of different functional guilds of natural enemies of herbivores infesting cabbage plants in South Korea by erecting over plots exclosures that restricted access completely, or restricted access to one of three guilds of natural enemies: flying insects, birds and ground-dwelling arthropods, or combinations of these three. In terms of pest pressure, landscapes with a high percentage cover of semi-natural habitats had greater pest densities than simple landscapes. This point highlights the first EDS detailed above and the authors note that the availability of overwintering habitat, refuges and alternative resources is the probable reason for the trend. All natural enemies reduced pest densities significantly, and the degree of pest suppression was higher with increasing landscape complexity. However, some aspects of pest control were affected by antagonistic interactions between the guilds. For example, in landscapes with a high level of semi-natural habitat cover (>25%), crop damage reduction by flying insects was negatively impacted by birds, probably through intraguild predation. By contrast, birds contributed positively to the pest control service in more simple landscapes. This additional EDS could also help to explain the higher initial pest densities recorded in complex landscapes. The higher diversity of both birds and flying insects (parasitoids) in complex landscapes is likely to partly explain the prevalence of intraguild predation in this system. The lack of a similar effect in simple landscapes may be related to the limited coincidence of birds and flying insect enemies in space and time, forcing birds to forage for herbivores. Furthermore, the significant differences in crop yield between treatments provide a clear measure to value the costs associated with intraguild predation in this system. Similar experimental designs in other landscape contexts and cropping systems should be employed to provide further evidence of the impact of multiple enemies on pest control.

Other examples of intraguild predation impacts include studies by Bennett and Gratton (2012). They studied the effects of landscape-scale variables on pest suppression along a rural to urban landscape gradient in central Wisconsin, USA, and found that higher levels of flower species diversity reduced pest suppression by natural enemies, suggesting that intraguild predation and/or natural enemy distraction (e.g. natural enemies spend more time foraging for nectar and pollen than attacking pests) could explain the trend. However, while there is evidence for diverse natural enemy assemblages, such as those associated with complex landscapes, disrupting pest suppression (Traugott *et al.* 2012) or enhancing it (Morandin *et al.* 2014), few studies have directly tested the impact of unmanaged habitats on pest suppression via intraguild predation. Nevertheless, a wide range of interactions between enemies clearly exists, and these are likely to become more complex as the landscape complexity increases. A good understanding of the key components of biodiversity is therefore required when making strategic decisions.

8.2.4 Unmanaged Habitats as Corridors or Barriers to Movement

An interesting series of interactions was demonstrated in a study of multiple ecosystem services provided by forest fragments surrounding soya bean fields in Quebec, Canada, by Mitchell *et al.* (2014). They examined the importance of both distance to forest and the level of isolation of the forest fragment to ES including crop yield, aphid population regulation and herbivory regulation in general. Forest fragments can confer combined ES and EDS: they can intercept the long-distance dispersal of aphids, for example (Irwin *et al.* 2007), and compete with crops for water and light (Mitchell *et al.* 2014), but can also accommodate a higher density and diversity of natural enemies and pollinators (Tscharntke *et al.* 2005). Generally, Mitchell *et al.* (2014) found that crop yield was lowest close to forests, probably because the tall forest vegetation competed with the crop for resources such as light and water resources. Yield was also lower when forest fragments were isolated, which is likely to be related to lower pollination and pest regulation services associated with non-crop habitat isolation. Aphid population regulation depended on the density of aphids. In a year with high aphid density, aphid regulation was higher in well-connected landscapes, suggesting that aphid predators (including parasitoids) were benefiting from these areas (Tscharntke *et al.* 2005). However, in a year with low aphid numbers, aphid regulation increased with fragment isolation and the same pattern was found for herbivory regulation in general. The authors suggest that this is because well-connected forest fragments assist with herbivore dispersal, as also discussed by Bianchi *et al.* (2006). Finally, they found that different ES were optimized at different distances from forest fragments, indicating that there is no perfect distance or level of connectivity. In the system that the authors studied, there are therefore likely to be important trade-offs depending on the value of the different ES flowing from unmanaged habitats. However, such trade-offs will be different for each agro-ecosystem setting. The differences in patterns between years also highlight the difficulty in making generalizations.

Shackelford *et al.* (2013) also identified a need to consider 'cultural species' when managing for ES. These species are defined as those that benefit from areas of crop, rather than of non-crop habitat. For example, some pest species can move through the landscape more easily with a higher proportion of crop cover in the landscape (Rand *et al.* 2014), and such species would be classed as cultural species. One study in Switzerland found that 24% of spider species, 42% of beetle species and 17% of bees, ants and wasps did not rely on semi-natural habitats and so could fall into the category of cultural species (Duelli and Obrist 2003). For any natural enemy species fitting this category, the proportion of natural habitat in the landscape may be detrimental to their abundance and ability to deliver pest regulation services by removing areas of potential prey or creating barriers to movement. Thus, if the most important natural enemies are cultural species, the presence of unmanaged habitat may lead to an indirect EDS. Knowing the most effective ES providers in the agro-ecosystem and how they respond to natural habitats is clearly important in this context.

8.2.5 Conservation of Unmanaged Habitats versus Restoration

The decision to conserve or restore semi-natural habitats or to continue to farm an area of land is a key economic decision. For example, floodplains are agricultural areas with high economic potential because of high soil fertility, but the restoration of riparian ecosystems can be vital for the conservation of threatened species and the provision of important agricultural ES. Restoration or creation of semi-natural habitat often necessarily involves the removal of agricultural land from production. The transitional stage that follows will be characterized by early successional vegetation, which can be a source of pests and weeds and can contribute to the negative perception of restored habitat (Langridge 2011).

Weeds can themselves be a source of insect pests by providing food sources for larval stages or nectar and pollen sources to the adult stages (Weber *et al.* 1990), or for herbivores such as slugs which can be vectors of plant diseases (Kollmann and Bassin 2001). These molluscs can even facilitate weed seed dispersal (Fischer *et al.* 2011). Of course, the same plant resources can be important for natural enemies of insect pests in terms of alternative hosts or prey and adult food sources (Altieri *et al.* 2015). The spread of unwanted plant species from marginal habitats such as hedgerows is minimal and restricted to the outer edges of the field (Wilson and Aebischer 1995), but weed infestation from larger natural areas is perceived to be much greater, although this has not been studied in great detail.

Langridge (2011) studied the spread of weed seeds from restored riparian vegetation around the Sacramento River in California, USA, into walnut orchards, and compared this to the seed spillover from remnant riparian forest and other agricultural land. Weed seed abundance was greater in orchards adjacent to the restored habitat than the other two vegetation types, but only in the area immediately adjacent to the habitat. However, the study also showed that, with increasing age of the restoration site, weed seed dispersal into the agricultural fields increased. This is contrary to the expectation that more mature restored sites are dominated by late-successional, native and non-weedy species as shown by other studies (Blumenthal *et al.* 2003). While the overall impact of seed spillover from restored habitat was limited to 1% of the agricultural land along the river, this example highlights the importance of scale in considering EDS. To the farmers with land immediately adjacent to the restored habitat, the weed management costs associated with this location are higher than those associated with a site further away from the habitat. However, this must be weighed against (1) the savings associated with ES flowing to the adjacent farms from the habitat, and (2) the wider-scale ES flowing to the agricultural community and society at large.

8.3 Landscape Context and the EDS from Unmanaged Habitats

It is clear from a number of studies of ES and EDS that simultaneous consideration of local management practices and the context of the landscape is needed when designing management strategies (Kleijn *et al.* 2004; Tscharntke *et al.* 2005). For example, Fischer *et al.* (2011) examined the impact of local and landscape factors on weed seed removal by mammal and invertebrate seed predators. They found that, in organic fields, seed predation increased with increasing landscape complexity, because the surrounding semi-natural habitats provided a source of seed predators. However, in conventionally managed fields, seed predation decreased with increasing landscape complexity, suggesting that, in simple landscapes, seed predators had larger home ranges and spent more time foraging due to low background food levels. In complex landscapes, the movement of seed predators into conventional fields was lower because more food was available in unmanaged habitats. While this does not typically represent an EDS of unmanaged habitats, it does point to a reduction in the ES due to the presence of unmanaged habitats in certain settings.

The landscape scale is important because a patch of semi-natural habitat may deliver EDS if the landscape context is inappropriate for the delivery of ES. For example, if an area of unmanaged habitat is maintained in an agricultural landscape but there is no connection to other sources of natural enemies, pest species may benefit more because they may make better use of the crop as a connecting habitat (Rand *et al.* 2014). Alternatively, a highly complex landscape will allow natural enemies to move from one patch to another as they seek resources and refuges from adverse environmental conditions. Recommendations for sustainable management of agro-ecosystems often include the notion that restoring, conserving or creating semi-natural habitats will be more efficient in simple than in complex landscapes, because in the latter the difference made by adding additional complexity will be negligible (Tscharntke *et al.* 2011) (Figure 8.2).

Another important point about landscape complexity is that the scale of habitat diversity will vary from species to species, may be different for the pest and the natural enemies and will be dependent on whether they are specialist or generalist natural enemies (Chaplin-Kramer *et al.* 2011). Specialist natural enemies tend to have poorer dispersal abilities, can be more sensitive to environmental disturbance and are more affected by complexity factors at the field or farm scale such as local plant diversity, compared to generalist natural enemies (Chaplin-Kramer *et al.* 2011; Letourneau *et al.* 2011). Furthermore, the impact of the semi-natural habitats on both pests and their natural enemies will depend on the level of isolation of the habitat, the particular

(a)

(b)

Figure 8.2 Examples of the differing scales of unmanaged habitat. The ES and EDS flowing from a number of small patches of unmanaged habitat in a relatively complex landscape (a) are likely to differ widely from those flowing from a large patch of unmanaged habitat in a simple landscape (b). Consideration of landscape context is therefore essential when managing for ES and EDS.

vegetation composition and age and the distance of a crop from a patch of habitat (Martin *et al.* 2013; Mitchell *et al.* 2014). In short, information on a range of scales is required on a host of variables before sustainable management strategies can be designed to balance ES and EDS from unmanaged habitats.

8.4 Managing for EDS from Unmanaged Habitats

The total elimination of EDS from agriculture generally, and in relation to unmanaged habitats in particular, is unlikely and probably undesirable (e.g. local extinction of a pest species may also mean local extinction of the specialist natural enemies). Instead, researchers and farm managers are faced with minimizing EDS and managing trade-offs between important ES (Power 2010). There are unlikely to be universal rules governing these trade-offs because many ES and EDS are dependent on context (Kremen 2005; Zhang *et al.* 2007). There are also trade-offs between private financial costs and benefits and the common good. For example, a decision to control pests by converting local natural habitat to arable and exclusively using pesticides may benefit the individual farmer's net profit, but subsequently may damage human health and reduce the pollination and pest control services enjoyed by landowners surrounding that farm. Public policy therefore needs to provide the right amount of incentive to maintain the right amount of natural habitat to balance ES and EDS, and this is a matter for further targeted research (Zhang *et al.* 2007).

Unfortunately, not enough is known about the net effects of unmanaged habitats to truly incorporate their conservation or conversion to agriculture into economic-based decision making (Langridge 2011). Furthermore, the true importance of the unmanaged habitat is often entangled in a web of direct and indirect ES and EDS flowing at various scales. For example, the Sacramento River in western USA was historically characterized by a wide riparian habitat, but agricultural conversion has led to a loss of all but 4% of the original habitat. This degradation has had important negative impacts on fisheries and water quality, leading to state legislature requiring restoration of the habitat. While this restoration may contribute to pest and weed problems, the local ES to farmers and the widespread restored ES to fisheries along with the watershed communities also need to be taken into account when determining the economic success or failure of the project (Langridge 2011). Landscape-scale studies are now beginning to get to grips with complex situations like this, attempting to incorporate the varying scales of ES and EDS to provide an ecosystem-, landscape- or watershed-based view of net ES.

A key challenge in this line of study is 'who pays?'. If the value of ES flowing to society and the agricultural community from an unmanaged habitat is greater than the net benefits flowing to a farmer immediately adjacent to a natural habitat, should society subsidize those farmers for the net costs they incur? For example, the costs to the individual farmer of establishing semi-natural habitat are initially greater than the savings from using less pesticide, but the provision of additional services to this and other farms is likely to justify public compensation, particularly if the surrounding farms do not have to incur similar costs (Bianchi *et al.* 2013; Zhang *et al.* 2007). Without such payments, the farmer restoring the semi-natural habitat has no financial incentive to do so. However, the amount of compensation is difficult to judge until valuation is available for all ES and EDS occurring at multiple scales.

Positive ecosystem outputs have been defined as ES largely so that they can be comprehensively mapped and accounted for economically and therefore conceptualized as a market good to be traded or enhanced (Boyd and Banzhaf 2007). It follows, therefore, that EDS as a concept should be considered in the same way, as a negative externality of ecosystems (see Chapter 16). However, this is an extremely difficult task given the complexity described throughout this chapter.

Triplett *et al.* (2012) attempted to quantify the net cost or benefit of birds to farmers only. They assessed the cost of crop or infrastructure damage by the loss of economic yield or the increase in costs of production such as control techniques and their associated impacts. However, direct quantification is extremely difficult, relying on the subjective valuation of farmers themselves and on their ability to differentiate between the damage caused by birds and damage caused by other agents (wind, frost, insects, etc.), as well as lack of yield associated with poor pollination, partial crop damage or suboptimal growing conditions. Further, the effectiveness of control techniques is rarely tested and quantified through well-designed and replicated experiments. This kind of accounting may also suffer from an error known as 'double-counting' in ES valuation (Fu *et al.* 2011). For example, the costs to one farmer may include the need to manage uncultivated land on the farm to restrict the alternative resources of a particularly damaging bird species. However, the actions of this farmer may benefit other neighbouring farmers, so technically this cost should be shared among the farmers who take advantage of it. This problem is particularly relevant to such a mobile group of species as birds because, like insects, they rarely restrict their impacts to farm boundaries.

Despite the difficulties with valuation, there are few other methods available to balance the negative and positive aspects of a divisive issue such as unmanaged habitat conservation or restoration, and importantly to decide on the appropriate trade-offs between conservation and active management. As with the broader scale aspects of pest management, several authors have concluded that the future of cost–benefit analysis and trade-off management will require a co-ordinated multiple-actor and multi-scale approach (Triplett *et al.* 2012). However, a detailed framework is not usually offered because the number of variables involved are so context dependent.

Current agri-environment schemes are criticized for failing to consider the wider landscape, and calls have been made for research into co-ordinated action across a collective of land managers. Such programmes can incentivize habitat conservation by farmers across a broad landscape and improve the delivery of a broad range of ES and minimize a number of EDS (Bianchi *et al.* 2013). However, all these schemes rely on the translation of scientific research into practices that are meaningful and specific to individual landscapes, crops, climates and pest–natural enemy complexes. This could perhaps be assisted by gearing agri-environment and compensation schemes more towards ecological ends and developing a more collaborative environment between farmers, agronomists, scientists and policy makers.

8.5 Conclusions and Future Research

Shortly after the concept of ES was introduced, research began to question whether multiple ES or 'stacked ES' could be enhanced simultaneously through simple habitat management practices (Campbell *et al.* 2012; Gurr *et al.* 2003; Olson and Wackers 2007;

Shackelford *et al.* 2013; Wratten *et al.* 2012). Demonstrating this potential and the associated savings in operational costs could be a key to increasing farmer adoption of such practices. However, at around the same time, research also began to acknowledge the presence of trade-offs between ecosystem services. For example, there is clearly a trade-off between food production and biodiversity conservation and balance of the two will depend on local needs and cultural values. There will also be important trade-offs between different ES, and a key challenge in making accurate judgements about the balance between them is in understanding the effects ES have on each other and how ES and EDS flow from different practices across multiple scales. In short, if agroecosystems are to be managed sustainably for multiple ES such as food production, biodiversity conservation, pest regulation and pollination, knowledge of whether each service has positive, negative or neutral effects on the others is needed (Shackelford *et al.* 2013). If unmanaged habitats are to be conserved or restored for one service (e.g. pollination, biodiversity conservation), how will they impact other services (e.g. pest regulation) or EDS (e.g. pest outbreaks, weed dispersal)?

Improving all the ES in one field, farm or landscape is likely to be difficult (Mitchell *et al.* 2014) and, by the same token, minimizing all EDS simultaneously will be equally challenging. Co-ordination and communication between landowners within a landscape and a thorough understanding of local and regional patterns of multi-scale ES and EDS provision are likely to be key to effective and sustainable agricultural management (Bommarco *et al.* 2013; Mitchell *et al.* 2014). Landscape-wide or regional approaches to ES and EDS management will rely on co-ordinated relevant information on the services flowing from different habitat types and their associated organisms, the impact of various management practices and the importance of habitat composition, configuration and distribution.

References

Altieri, M.A., Nicholls, C.I., Gillespie, M.A.K., *et al.* (2015) *Crops, Weeds and Pollinators: Understanding Ecological Interaction for Better Management.* Food and Agriculture Organization of the United Nations, Rome, Italy.

Araj, S.E., Wratten, S., Lister, A. and Buckley, H. (2008) Floral diversity, parasitoids and hyperparasitoids – a laboratory approach. *Basic and Applied Ecology* **9**: 588–597.

Baggen, L.R. and Gurr, G.M. (1998) The influence of food on *Copidosoma koehleri* (Hymenoptera: Encyrtidae), and the use of flowering plants as a habitat management tool to enhance biological control of potato moth, *Phthorimaea operculella* (Lepidoptera: Gelechiidae). *Biological Control* **11**: 9–17.

Banks, J.E., Bommarco, R. and Ekbom, B. (2008) Population response to resource separation in conservation biological control. *Biological Control* **47**: 141–146.

Bennett, A.B. and Gratton, C. (2012) Measuring natural pest suppression at different spatial scales affects the importance of local variables. *Environmental Entomology* **41**: 1077–1085.

Bianchi, F.J.J.A., Booij, C.J.H. and Tscharntke, T. (2006) Sustainable pest regulation in agricultural landscapes: a review on landscape composition, biodiversity and natural pest control. *Proceedings of the Royal Society B: Biological Sciences* **273**: 1715–1727.

Bianchi, F., Mikos, V., Brussaard, L., Delbaere, B. and Pulleman, M.M. (2013) Opportunities and limitations for functional agrobiodiversity in the European context. *Environmental Science and Policy* **27**: 223–231.

Blumenthal, D.M., Jordan, N.R. and Svenson, E.L. (2003) Weed control as a rationale for restoration: the example of tallgrass prairie. *Conservation Ecology* **7**: article no 6.

Bommarco, R., Kleijn, D. and Potts, S.G. (2013) Ecological intensification: harnessing ecosystem services for food security. *Trends in Ecology and Evolution* **28**: 230–238.

Boyd, J. and Banzhaf, S. (2007) What are ecosystem services? The need for standardized environmental accounting units. *Ecological Economics* **63**: 616–626.

Brodeur, J. and Rosenheim, J.A. (2000) Intraguild interactions in aphid parasitoids. *Entomologia Experimentalis et Applicata* **97**: 93–108.

Campbell, A.J., Biesmeijer, J.C., Varma, V. and Wackers, F.L. (2012) Realising multiple ecosystem services based on the response of three beneficial insect groups to floral traits and trait diversity. *Basic and Applied Ecology* **13**: 363–370.

Carvalheiro, L.G., Veldtman, R., Shenkute, A.G., *et al.* (2011) Natural and within-farmland biodiversity enhances crop productivity. *Ecology Letters* **14**: 251–259.

Chaplin-Kramer, R., O'Rourke, M.E., Blitzer, E.J. and Kremen, C. (2011) A meta-analysis of crop pest and natural enemy response to landscape complexity. *Ecology Letters* **14**: 922–932.

Daily, G.C. (1997) *Nature's Services: societal dependence on natural ecosystems.* Island Press, Washington, DC, USA.

Dix, M.E., Johnson, R.J., Harrell, M.O., *et al.* (1995) Influences of trees on abundance of natural enemies of insect pests: a review. *Agroforestry Systems* **29**: 303–311.

Duelli, P. and Obrist, M.K. (2003) Regional biodiversity in an agricultural landscape: the contribution of seminatural habitat islands. *Basic and Applied Ecology* **4**: 129–138.

Dunn, R.R. (2010) Global mapping of ecosystem disservices: the unspoken reality that nature sometimes kills us. *Biotropica* **42**: 555–557.

Escobedo, F.J., Kroeger, T. and Wagner, J.E. (2011) Urban forests and pollution mitigation: analyzing ecosystem services and disservices. *Environmental Pollution* **159**: 2078–2087.

Fahrig, L., Baudry, J., Brotons, L., *et al.* (2011) Functional landscape heterogeneity and animal biodiversity in agricultural landscapes. *Ecology Letters* **14**: 101–112.

Fischer, C., Thies, C. and Tscharntke, T. (2011) Mixed effects of landscape complexity and farming practice on weed seed removal. *Perspectives in Plant Ecology Evolution and Systematics* **13**: 297–303.

Foggo, A., Ozanne, C.M.P., Speight, M.R. and Hambler, C. (2001) Edge effects and tropical forest canopy invertebrates. *Plant Ecology* **153**: 347–359.

Fu, B.J., Su, C.H., Wei, Y.P., Willett, I.R., Lu, Y.H. and Liu, G.H. (2011) Double counting in ecosystem services valuation: causes and countermeasures. *Ecological Research* **26**: 1–14.

Gagic, V., Haenke, S., Thies, C., Scherber, C., Tomanovic, Z. and Tscharntke, T. (2012) Agricultural intensification and cereal aphid-parasitoid-hyperparasitoid food webs: network complexity, temporal variability and parasitism rates. *Oecologia* **170**: 1099–1109.

Garcia-Feced, C., Weissteiner, C.J., Baraldi, A., *et al.* (2015) Semi-natural vegetation in agricultural land: European map and links to ecosystem service supply. *Agronomy for Sustainable Development* **35**: 273–283.

Gomez-Marco, F., Urbaneja, A., Jaques, J.A., Rugman-Jones, P.F., Stouthamer, R. and Tena, A. (2015) Untangling the aphid-parasitoid food web in citrus: can hyperparasitoids disrupt biological control? *Biological Control* **81**: 111–121.

Gurr, G.M., Wratten, S.D. and Luna, J.M. (2003) Multi-function agricultural biodiversity: pest management and other benefits. *Basic and Applied Ecology* **4**: 107–116.

Hand, S.C. (1989) The overwintering of cereal aphids on gramineae in southern England, 1977–1980. *Annals of Applied Biology* **115**: 17–29.

Holler, C., Borgemeister, C., Haardt, H. and Powell, W. (1993) The relationship between primary parasitoids and hyperparasitoids of cereal aphids – an analysis of field data. *Journal of Animal Ecology* **62**: 12–21.

Holt, R.D., Lawton, J.H., Polis, G.A. and Martinez, N.D. (1999) Trophic rank and the species-area relationship. *Ecology* **80**: 1495–1504.

Hooper, D.U., Chapin, F.S., Ewel, J.J., *et al.* (2005) Effects of biodiversity on ecosystem functioning: a consensus of current knowledge. *Ecological Monographs* **75**: 3–35.

Irwin, M.E., Kampmeier, G.E. and Weisser, W.W. (2007) Aphid movement: process and consequences. In: van Emden, H.F. and Harrington, R. (eds) *Aphids as Crop Pests*. CABI International, Wallingford, UK, pp. 153–186.

Janssen, A., Sabelis, M.W., Magalhaes, S., Montserrat, M. and van der Hammen, T. (2007) Habitat structure affects intraguild predation. *Ecology* **88**: 2713–2719.

Kehrli, P. and Bacher, S. (2008) Differential effects of flower feeding in an insect host-parasitoid system. *Basic and Applied Ecology* **9**: 709–717.

Kleijn, D., Berendse, F., Smit, R., *et al.* (2004) Ecological effectiveness of agri-environment schemes in different agricultural landscapes in the Netherlands. *Conservation Biology* **18**: 775–786.

Kollmann, J. and Bassin, S. (2001) Effects of management on seed predation in wildflower strips in northern Switzerland. *Agriculture Ecosystems and Environment* **83**: 285–296.

Kremen, C. (2005) Managing ecosystem services: what do we need to know about their ecology? *Ecology Letters* **8**: 468–479.

Kremen, C., Williams, N.M. and Thorp, R.W. (2002) Bee diversity and pollination services in an agro-natural landscape. *Ecological Society of America Annual Meeting Abstracts* **87**: 183–184.

Landis, D.A., Wratten, S.D. and Gurr, G.M. (2000) Habitat management to conserve natural enemies of arthropod pests in agriculture. *Annual Review of Entomology* **45**: 175–201.

Langridge, S.M. (2011) Limited effects of large-scale riparian restoration on seed banks in agriculture. *Restoration Ecology* **19**: 607–616.

Letourneau, D.K., Jedlicka, J.A., Bothwell, S.G. and Moreno, C.R. (2009) Effects of natural enemy biodiversity on the suppression of arthropod herbivores in terrestrial ecosystems, *Annual Review of Ecology Evolution and Systematics* **40**: 573–592.

Letourneau, D.K., Armbrecht, I., Salguero Rivera, B., *et al.* (2011) Does plant diversity benefit agroecosystems? A synthetic review. *Ecological Applications* **21**: 9–21.

Martin, E.A., Reineking, B., Seo, B. and Steffan-Dewenter, I. (2013) Natural enemy interactions constrain pest control in complex agricultural landscapes. *Proceedings of the National Academy of Sciences of the USA* **110**: 5534–5539.

Mitchell, M.G.E., Bennett, E.M. and Gonzalez, A. (2014) Forest fragments modulate the provision of multiple ecosystem services. *Journal of Applied Ecology* **51**: 909–918.

Morandin, L.A., Long, R.F. and Kremen, C. (2014) Hedgerows enhance beneficial insects on adjacent tomato fields in an intensive agricultural landscape. *Agriculture Ecosystems and Environment* **189**: 164–170.
Muller, C.B. and Brodeur, J. (2002) Intraguild predation in biological control and conservation biology. *Biological Control* **25**: 216–223.
Nofemela, R.S. (2013) The effect of obligate hyperparasitoids on biological control: differential vulnerability of primary parasitoids to hyperparasitism can mitigate trophic cascades. *Biological Control* **65**: 218–224.
Olson, D.M. and Wackers, F.L. (2007) Management of field margins to maximize multiple ecological services. *Journal of Applied Ecology* **44**: 13–21.
Paredes, D., Cayuela, L., Gurr, G.M. and Campos, M. (2013) Effect of non-crop vegetation types on conservation biological control of pests in olive groves. *Peer Journal* **1**: 15.
Power, A.G. (2010) Ecosystem services and agriculture: tradeoffs and synergies. *Philosophical Transactions of the Royal Society B: Biological Sciences* **365**: 2959–2971.
Rand, T.A., van Veen, F.J.F. and Tscharntke, T. (2012) Landscape complexity differentially benefits generalized fourth, over specialized third, trophic level natural enemies. *Ecography* **35**: 97–104.
Rand, T.A., Waters, D.K., Blodgett, S.L., Knodel, J.J. and Harris, M.O. (2014) Increased area of a highly suitable host crop increases herbivore pressure in intensified agricultural landscapes. *Agriculture Ecosystems and Environment* **186**: 135–143.
Roschewitz, I., Gabriel, D., Tscharntke, T. and Thies, C. (2005) The effects of landscape complexity on arable weed species diversity in organic and conventional farming. *Journal of Applied Ecology* **42**: 873–882.
Rosenheim, J.A. (1998) Higher-order predators and the regulation of insect herbivore populations. *Annual Review of Entomology* **43**: 421–447.
Rusch, A., Valantin-Morison, M., Sarthou, J.P. and Roger-Estrade, J. (2011) Multi-scale effects of landscape complexity and crop management on pollen beetle parasitism rate. *Landscape Ecology* **26**: 473–486.
Schackermann, J., Mandelik, Y., Weiss, N., von Wehrden, H. and Klein, A.M. (2015) Natural habitat does not mediate vertebrate seed predation as an ecosystem dis-service to agriculture. *Journal of Applied Ecology* **52**: 291–299.
Shackelford, G., Steward, P.R., Benton, T.G., *et al.* (2013) Comparison of pollinators and natural enemies: a meta-analysis of landscape and local effects on abundance and richness in crops. *Biological Reviews* **88**: 1002–1021.
Thies, C., Roschewitz, I. and Tscharntke, T. (2005) The landscape context of cereal aphid-parasitoid interactions. *Proceedings of the Royal Society B: Biological Sciences* **272**: 203–210.
Thomas, S.R., Goulson, D. and Holland, J.M. (2001) Resource provision for farmland gamebirds: the value of beetle banks. *Annals of Applied Biology* **139**: 111–118.
Traugott, M., Bell, J.R., Raso, L., Sint, D. and Symondson, W.O.C. (2012) Generalist predators disrupt parasitoid aphid control by direct and coincidental intraguild predation. *Bulletin of Entomological Research* **102**: 239–247.
Triplett, S., Luck, G.W. and Spooner, P.G. (2012) The importance of managing the costs and benefits of bird activity for agricultural sustainability. *International Journal of Agricultural Sustainability* **10**: 268–288.

Tscharntke, T., Klein, A.M., Kruess, A., Steffan-Dewenter, I. and Thies, C. (2005) Landscape perspectives on agricultural intensification and biodiversity – ecosystem service management. *Ecology Letters* **8**: 857–874.
Tscharntke, T., Batary, P. and Dormann, C.F. (2011) Set-aside management: how do succession, sowing patterns and landscape context affect biodiversity? *Agriculture Ecosystems and Environment* **143**: 37–44.
Tylianakis, J.M., Tscharntke, T. and Lewis, O.T. (2007) Habitat modification alters the structure of tropical host-parasitoid food webs. *Nature* **445**: 202–205.
Vance-Chalcraft, H.D., Rosenheim, J.A., Vonesh, J.R., Osenberg, C.W. and Sih, A. (2007) The influence of intraguild predation on prey suppression and prey release: a meta-analysis. *Ecology* **88**: 2689–2696.
Veres, A., Petit, S., Conord, C. and Lavigne, C. (2013) Does landscape composition affect pest abundance and their control by natural enemies? A review. *Agriculture Ecosystems and Environment* **166**: 110–117.
Vialatte, A., Dedryver, C.A., Simon, J.C., Galman, M. and Plantegenest, M. (2005) Limited genetic exchanges between populations of an insect pest living on uncultivated and related cultivated host plants. *Proceedings of the Royal Society B: Biological Sciences* **272**: 1075–1082.
von Dohren, P. and Haase, D. (2015) Ecosystem disservices research: a review of the state of the art with a focus on cities. *Ecological Indicators* **52**: 490–497.
Weber, D.C., Mangan, F.X., Ferro, D.N. and Marsh, H.V. (1990) Effect of weed abundance on European corn borer (Lepidoptera, Pyralidae) infestation of sweet corn. *Environmental Entomology* **19**: 1858–1865.
Wilson, P.J. and Aebischer, N.J. (1995) The distribution of dicotyledonous arable weeds in relation to distance from the field edge. *Journal of Applied Ecology* **32**: 295–310.
Wratten, S.D., Gillespie, M., Decourtye, A., Mader, E. and Desneux, N. (2012) Pollinator habitat enhancement: benefits to other ecosystem services. *Agriculture Ecosystems and Environment* **159**: 112–122.
Zaller, J.G., Moser, D., Drapela, T., Schmoeger, C. and Frank, T. (2008) Insect pests in winter oilseed rape affected by field and landscape characteristics. *Basic and Applied Ecology* **9**: 682–690.
Zhang, W., Ricketts, T.H., Kremen, C., Carney, K. and Swinton, S.M. (2007) Ecosystem services and dis-services to agriculture. *Ecological Economics* **64**: 253–260.
Zhao, Z.H., Liu, J.H., He, D.H., Guan, X.Q. and Liu, W.H. (2013) Species composition and diversity of parasitoids and hyper-parasitoids in different wheat agro-farming systems. *Journal of Insect Science* **13**: 162.

Part IV

Effects of Global Changes on Pest Management

9

Effect of Climate Change on Insect Pest Management

Nigel R. Andrew and Sarah J. Hill

9.1 Introduction

Insect responses to environmental change are crucial for understanding how agro-ecosystems will respond to climate change. Many insect species are pests of crops, but they also play crucial roles as parasitoids and predators of key pest species. Changes in an insect population's physiology, biochemistry, biogeography and population dynamics may occur among populations across their distribution, among the growing seasons, and among crop types. An insect population's response to a rapidly changing climate may also be variable when insects interact with different competitors, predators and parasitoids and impose costs at different life stages. This also can influence the overall food production systems that can be at critical risk from the impacts of climate change (IPCC 2014).

Here we will focus on key herbivore pests from major cropping regions worldwide and key natural enemies (parasitoids and predators) of these pests. We assess the current knowledge of the impact that climate change is having on pest management, particularly assessing biologically based methods (such as Integrated Pest Management, IPM). A key focus of the chapter is on ecological, physiological and behavioural responses of organisms. Changes in the physiological tolerances and population depletion of beneficial parasitoids and predators, or non-pest competitive species, could cause major population restructure of currently common species, leading to the collapse of trophic interactions and depletion of ecosystem services. These issues are critically assessed throughout the chapter, identifying major gaps in our current knowledge.

The chapter is divided into six sections. First, the different types of climate changes influencing agro-ecosystems; second, the possible types of insect responses to these climate changes: adaptation, change of geographic distribution, extinction; third, the current state of climate change research being carried out on insects in agro-ecosystems; fourth, how these climate changes and insect responses may affect various ecological processes important for plant protection (e.g. pest population dynamics, pest–crop plant interactions, intraguild predation, tritrophic interactions, pest–pathogen and pest–symbiont interactions); fifth, we assess IPM approaches and how they may be affected by climate change; and finally, an assessment of key areas in which knowledge needs to be rapidly built on in the future.

Environmental Pest Management: Challenges for Agronomists, Ecologists, Economists and Policymakers, First Edition. Edited by Moshe Coll and Eric Wajnberg.

9.2 Observed Climate Changes Influencing Agro-Ecosystems

Human impacts on our climate systems are becoming better understood, and these 'trend effects' (Jentsch *et al.* 2007), such as general changes in temperature, precipitation, wind and solar radiation, are being seen across the globe. As indicated by the fifth IPCC report (IPCC 2013), it is clear that greenhouse gas emission (primarily carbon dioxide) is causing ocean and air temperatures to rise, glaciers to melt and sea levels to rise. These changes are having stronger influences on our weather systems, including a higher incidence of severe and extreme weather events, changing rainfall regimes and changes in seasonal averages, all of which are modifying the climates of agro-ecosystems.

Desertification occurs when trees and plant cover are removed, particularly in drylands due to both climatic events and human activities (UNCCD 2008). Dryland systems make up more than 40% of the world's land mass and 44% of the world's food production systems. Desertification and drought have reduced arable land available for food production at 30–35 times the historical rate, which is equivalent to 12 million ha per year (UNCCD 2008). Such rapid changes over the last two centuries will have serious consequences for human agricultural systems.

Agriculture has developed under fairly predictable and stable climatic regimes from approximately 10 000 years ago (Feynman and Ruzmaikin 2007). Before this period, the climate was highly variable and restricted agricultural development (Feynman and Ruzmaikin 2007). Climate variability also had a significant effect on agricultural civilizations around 4000 years ago in the Neolithic culture of central China (Wang *et al.* 2005), the Egyptian Old Kingdom (Butzer 1976) and Akkadia in Mesopotamia (Cullen *et al.* 2000). Changes in the current agro-ecosystem climate have the potential to cause substantial destabilization of current farming practices.

9.3 Insect Responses to Climate Change

Many species may not be able to extend or move out of their current distribution due to a number of factors such as dispersal constraints and restrictions due to parasitoid/predator/symbiont relationships, and they are essentially stuck within the realized niche and not able to expand into their broader fundamental niche. Microclimate may buffer or amplify macroclimates and this can occur along multiple axes. These microclimatic variation axes can be defined as abiotic/biotic axes, amplification versus buffering axes, and long versus short temporal and spatial scale axes (Woods *et al.* 2015). Biotic environments are influenced by nearby organisms, such as social insect nests, insect herbivores influenced by leaf surface temperature and humidity via stomatal opening, and leaf miners positioned under the leaf lamina. Abiotic environments are influenced by different structures such as rocks, soils, topography and plant canopies. Both abiotic and biotic environments can be manipulated to some extent by organisms to find their most favourable microclimate, making responses to macroclimate warming more difficult to assess and predict.

Spatial and temporal extents of microclimates are critical, especially in relation to the organisms being assessed. For example, within crops, wingless aphids will stay on their maternal plant (Gia and Andrew 2015), whereas gregarious desert locusts may swarm and travel hundreds of kilometres and have been responsible for complete crop losses affecting over eight million people (Latchininsky *et al.* 2011). Invertebrates crossing between different crops and areas of natural vegetation, or living within a complex topography, are more likely to expose themselves to a wider range of microclimates. Heterogeneity and spatial structure within an organism's microenvironment are critical when assessing thermoregulation, movement and energetics of invertebrates (Sears and Angilletta 2015).

Changing behaviour to thermally regulate within varying microclimates is also key for invertebrates. Moving to an environment where individuals are not exposed to extreme temperatures is critical, and this would include localities in their current distribution in which they are not exposed to temperatures above their thermal safety margin for long periods of time. An example of this is the western horse lubber grasshopper (*Taeniopoda eques*, Romaleidae: Orthoptera) which moves between vegetation and soil during the day to attain its optimum temperature of 35.2 °C: roosting on the plants overnight, moving to the ground in the mornings (to warm up), then returning to the vegetation during the middle of the day (to stay cool), back to open ground in the afternoon (for warmth), and then returning to vegetation at dusk for protection (Whitman 1987). If insects were unable to access their preferred thermal environments, they may have reduced reproductive output, reduced physiological efficiencies, or be less competitive against congeners or less responsive to natural enemies.

Behavioural adaptation to climate change by insects is understudied (Andrew *et al.* 2013b) but is a critical aspect of an insect's response to climate change. Insect pests may be able to move to the underside of a leaf and deal with a cooler microclimate, or be able to reduce their exposure to extreme temperatures throughout the hottest parts of the day or season, thus enabling them to find the most optimum thermal environment available within a small spatial scale.

Behavioural changes of insects to find optimum thermal environments will also change interactions among species. When the cereal crop aphid *Sitobion avenae* (Aphididae: Hemiptera) and its primary natural enemy, the parasitoid *Aphidius rhopalosiphi* (Aphidiidae: Hymenoptera), were exposed to 5 °C variation in temperature, interactions between the two species changed (Le Lann *et al.* 2014). Parasitoids had the highest oviposition rate at the 'resting' temperature of 20 °C. When temperatures were increased to 25 °C, aphid metabolic rates exhibited a stronger increase compared to the parasitoid, as did aphid defence against parasitoid attacks. This can then lead to a reduction in the parasitoid efficiency of aphid control. In a predator–prey system, including predatory ground beetles (Carabidae), mobile adult prey (*Drosophila*; Drosophilidae: Diptera) and resident prey (a larval *Alphitobius*; Tenebrionidae: Coleoptera) species, attacks on mobile species increased along a temperature gradient (from 5 °C to 30 °C) whereas attacks on the resident prey species remained consistent along the gradient (Vucic-Pestic *et al.* 2011). For the predators, there was a reduction in energetic efficiency with warming, as the ratio of feeding to metabolic rate could not stay constant, which over longer time periods could lead to starvation. In aquatic systems with different water temperatures (15–30 °C), reductions in attack speeds of predators (sea bass) and increases in escape speeds of prey (mosquitofish) at high temperatures result in a lower predation pressure (Grigaltchik *et al.* 2012).

Population abundances of pests, beneficial insects, competitors and symbionts may go through substantive changes with a changing climate. These changes can cause interactions (positive, negative and neutral) to become more or less intense (Lankau and Strauss 2011). For example, if a pest species is released from competitive interactions with a congeneric, its abundance may increase with a changing climate and it may become more invasive and impact on a wider number of species within its realized niche (Bolnick *et al.* 2010).

Interactions between populations, species and communities as temperature increases may also change. Such interactions, including competition, predation and symbiotic relationship, as well as dispersal differences, may be put under or released from pressures, ultimately changing interaction dynamics (Gilman *et al.* 2010; Urban *et al.* 2012). If crops are planted in a new environment to keep within their climatic envelope, then natural enemy and pest species may show different responses, as outlined by Gilman *et al.* (2010) for native species along a climatic gradient. The simple models provided in this study are performed when the natural enemy and pest species have different dispersal and ecological restrictions. In a simple pest-specialized enemy relationship, the enemy's range is limited as it is restricted by the distance moved by the pest. Hence, the pest can disperse further than its specialized enemy and thus increase its abundance and damage. Under apparent competition with two pest species and a natural enemy, one pest species can have a positive effect on the natural enemy to reduce the abundance of the other pest species. Once the pest species expand their range beyond the range of the natural enemy, both pest species increase in their abundance. For keystone attack between a natural enemy and two asymmetric competitive pest species, once both move beyond their keystone natural enemy, the superior competitive species will exclude the subordinate.

Climate change will also impact on the phenology of organisms. Across trophic levels these differences may be weaker or stronger, again causing changes in behavioural interactions. In a 17-year study within a Netherlands oak forest ecosystem, Both *et al.* (2009) found that budburst advanced by 0.17 days per year, as did herbivorous caterpillars (0.75 days per year) and passerine birds (four species between 0.36 and 0.50 days per year). However, the keystone raptor predators did not show any advance in hatching dates.

Insect crop pests and their natural enemies may also have their physiological responses modified to cope with warmer temperatures. Over short time periods with exposure to extreme temperatures (both minimum and maximum temperatures), insects may acclimate or produce heat shock proteins, cryoprotectants and osmolyte compounds, among others, within their bodies to survive short-term exposure to high and low temperatures (Colinet *et al.* 2015; Ghaedi and Andrew 2016).

The life-history response of a pest species to climate change has been well studied in the sap-sucking green vegetable bug *Nezara viridula* (Pentatomidae: Hemiptera). In Japan, reproductive diapause and body colour changes are controlled by photoperiod (Musolin 2012). Critical to the overwintering success of *N. viridula* is the timing of adult emergence, diapause induction and adult size. As climate conditions poleward have become more favourable, *N. viridula* has shifted its northern range limit and become more successful. In Australia, the lethal temperatures for *N. viridula* did not differ between a coastal and an inland site (Chanthy *et al.* 2012), and the bugs responded to warmer temperatures by increasing their upper lethal limits. The interaction between temperature and humidity was also found to change reproductive performance and longevity as well as nymphal duration and survival, making it critically important to assess insect responses to climate change at different life stages (Chanthy *et al.* 2015) in different geographical locations.

Animals can respond to climate change via phenotypic plasticity (Bradshaw and Holzapfel 2008; Cleland *et al.* 2007) or evolve a genetic adaptation (Chevin *et al.* 2010; Hoffmann and Sgro 2011). Genetic adaptation to climate change is critical for insects to survive in novel environments (Gienapp *et al.* 2008; Hoffmann and Sgro 2011; van Asch *et al.* 2013). Such changes are evident with rapidly changing allele frequencies in insects exploiting new conditions they are exposed to with climate change (Bradshaw and Holzapfel 2008; Hoffmann and Willi 2008; Merilä 2012).

Species adaptation rate is critical (Visser 2008) and this will change according to the type of environment that both a pest and natural enemy are found within, the regional forces being exerted on both species, photoperiod length (Bradshaw and Holzapfel 2001, 2008) and the pace at which both the host species and herbivore adapt over generations (Bridle *et al.* 2014; Franks *et al.* 2007; van Asch *et al.* 2013). Adaptive trait variation has been shown in *Drosophila* (Umina *et al.* 2005). However, observing evolutionary responses to climate change is more difficult; the right traits need to be followed which assist the population and species to adapt to either persist locally or shift their distribution (Davis *et al.* 2005; Thompson *et al.* 2013a).

It has been argued that hard physiological boundaries exist that constrain the evolution of high temperature tolerances of terrestrial organisms, but not so for cold tolerance (Araújo *et al.* 2013). This study suggests that climate change risk exposure is inflated as thermal tolerance calculation, based on the niches used by cold-adapted species, is underestimated. However, species living in niches close to their thermal tolerance thresholds will be more affected by global warming (Araújo *et al.* 2013). This is because they are exposed to more extreme temperature events that reduce both their warming tolerance (an organism's critical thermal maxima minus the habitat temperature) and thermal safety margins (an organism's optimal temperature minus habitat temperature) (Andrew *et al.* 2013a; Deutsch *et al.* 2008).

Species geographical ranges may also be modified with evolutionary responses. Species have been shown to evolve a photoperiod response to climate change enabling them to invade new areas and expand their geographical range, for example Japanese and US populations of the Asian tiger mosquito *Aedes albopictus* (Culcidae: Diptera) developing photoperiodic control of seasonal development (Urbanski *et al.* 2012). In contrast, as climate restricts arthropods invading particular environments (Arndt and Perner 2008), evolving species will be limited in what regions they can invade (Hoffmann 2010).

As many insect pest species are generalist feeders, particularly R-strategist species that go through boom–bust cycles, they are expected to increase their ranges much more readily than more specialist secondary pest species (Warren *et al.* 2001). Secondary pests would need to develop more generalist phenotypes to expand their host plant and climate range or overcome their ecological barriers to dispersal (Lavergne *et al.* 2010). As environmental conditions change and ranges expand, so does the ability of species to disperse further, enabling disperser species to use a wider range of resources (Hill *et al.* 2011; Kuussaari *et al.* 2000). As the climate rapidly changes, generalist species should find it relatively easier to spread to these new habitats and also cause a shift in the population ecology of the species from locally adapted to regionally adapted (Bridle *et al.* 2014). This could make pest outbreaks and their control more predictable, changing regional opportunities to implement area-wide management strategies. Transplant experiments have been shown to be a powerful tool for assessing the future capability of insects to adapt to novel conditions and novel host plants (Nooten and Andrew 2017; Nooten *et al.* 2014).

Endosymbionts can also enable insect species to extend their geographical and altitudinal distributions. A symbiotic micro-organism (endosymbiont) is defined as a persistent relationship that a micro-organism has with an insect. This relationship can be mutualistic, commensal or parthenogenic (Douglas 2007). Many insect pests depend on micro-organisms to upregulate essential nutrients that are not readily available in the insect's diet (Douglas and van Emden 2007). Bacterial symbionts can confer plasticity to insects that enhances their thermal adaptation, or if they are long-term bacterial mutualists this may constrain adaptation, enhance genome deterioration and increase vulnerability to thermal stresses (Wernegreen 2012). For example, in aphids the endosymbiont *Buchnera* (Enterobacteriaceae: Enterobacteriales) is an obligate bacterial symbiont that is required to synthesize nutrients (Douglas 1998). Dunbar *et al.* (2007) identified a mutation in the transcriptional promoter of the heat shock protein ipbA in aphids. Individuals with this mutation who were exposed to low-level heat lost most or all symbionts but, at lower temperatures, those aphids exhibiting the mutation had a reproductive advantage. This indicated that vertically transmitted microbial symbionts play a key role in the thermal tolerance of insects, particularly when environmental conditions change across a species distribution (Dunbar *et al.* 2007).

If populations cannot adapt to the changing conditions within their current environment, or if they cannot move to stay within their climatic envelope, then they will go locally extinct. Although insect pest species are unlikely to go extinct, their natural enemies in localized areas may be at higher risk. This is particularly so for specialist parasitoids that are reliant on a single pest species for their life cycle development, and when the adult is susceptible to changes in microclimatic fluctuations.

9.4 Overview of Insect Pests in Agro-Ecosystems and Climate Change

We assessed 638 publications that identified climate change and insect pests in their title, abstracts or key words from the Web of Science database accessed in June 2015. Of these 638 publications, 260 were found to have attributes that enabled an assessment of their content directly related to climate change and insect pests in agricultural systems (hereafter called CCIPs). The database of papers can be found on Figshare (Andrew and Hill 2016). We excluded forestry papers associated with wood production, but included agroforestry crops such as bananas, coffee and palm oil plantations. For this assessment of research, we followed the protocols outlined in Andrew *et al.* (2013). Apart from key bibliographic information, we also extracted information relating to the continent in which the study was carried out (referred to as region), publication type (lab experiment, desktop survey, field experiment/survey, modelling, review), habitat and host plant (crop, grassland, laboratory artificial), insect order, family and species assessed, climate change mechanisms studied (temperature, precipitation, carbon dioxide, UV-B, agricultural intensification, genetic changes, El Niño southern oscillation, season) and the insect traits measured (abundance, phenology, distribution, herbivory, physiology, behaviour, interactions, genetics/genomics, assemblage/community changes, use in IPM, development time, survival, body weight, morphology, reproductive output and pesticide exposure). From this, we identified current trends in the literature and areas that are comparatively well studied using Cytoscape network analysis software version 3.2.1 (Shannon *et al.* 2003) and generated the network analysis shown Figures 9.2, 9.3

and 9.4. Cytoscape visualizes interaction networks and pathways and integrates a range of information (data on insect pest climate change publications in this case).

In terms of publication type, reviews were the most common (86 publications), followed by model papers (78); there is a large gap between these and lab experiments (44) and those assessing field surveys (25) and field experiments (18) (Figure 9.1a). European studies dominate those in the Web of Science literature (62 publications) with a large gap between this group and studies worldwide (38), from North America (37) and Asia (34) (Figure 9.1b). Of the taxa assessed, multiple orders per paper were most common (64), followed by Lepidoptera (53) and Hemiptera (51). In 32 papers taxa were

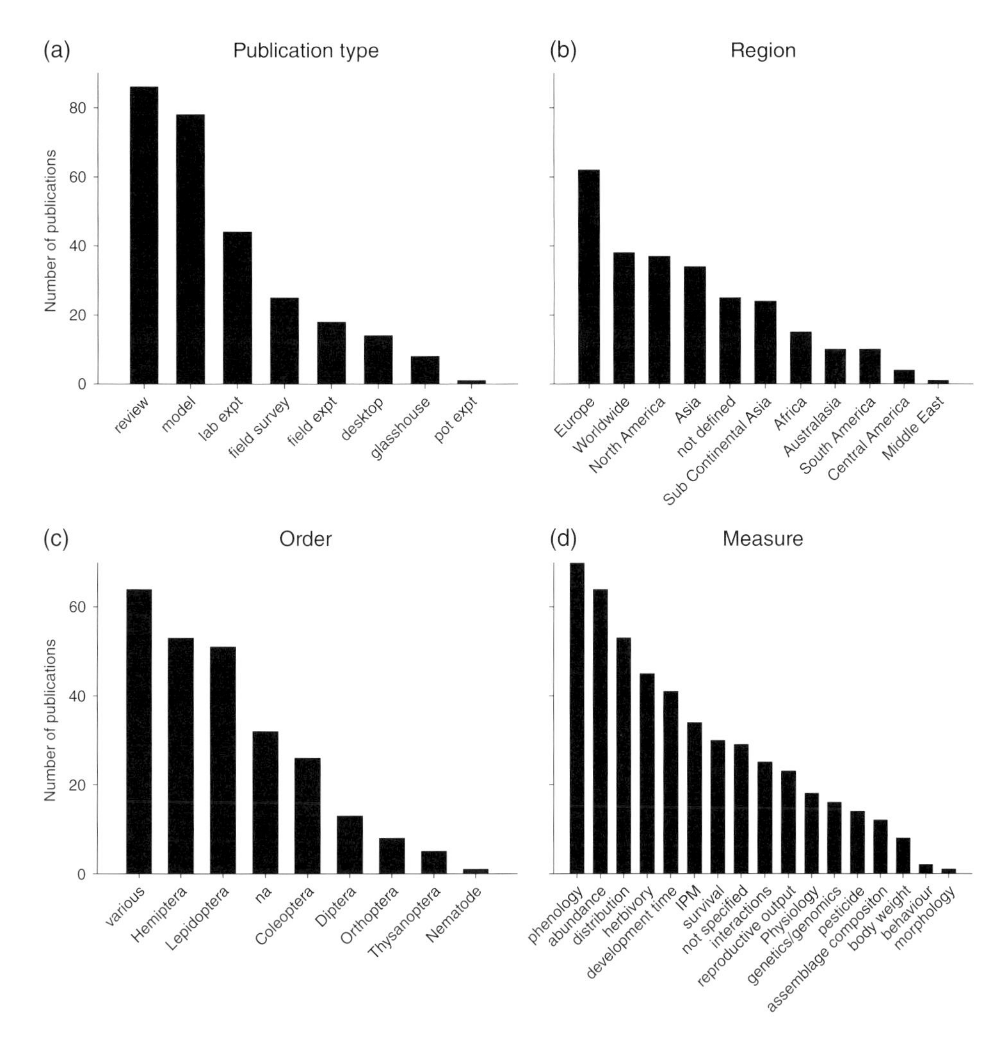

Figure 9.1 Summary of publications from the literature in different groupings ranked in decreasing order per class. (a) Publication type: lab expt = experiments conducted indoors; glasshouse = experiments conducted in closed spaces exposed to solar radiation; pot expt = experiments outside but in a constrained substrate. (b) Region of world where data were collected from. (c) Insect order assessed (including nematodes; 'na' indicates order not identified). (d) Trait measured in the manuscript to assess climate change impacts. Papers having multiple groupings (e.g. two orders were assessed or four traits were measured) were counted in each of the appropriate groups.

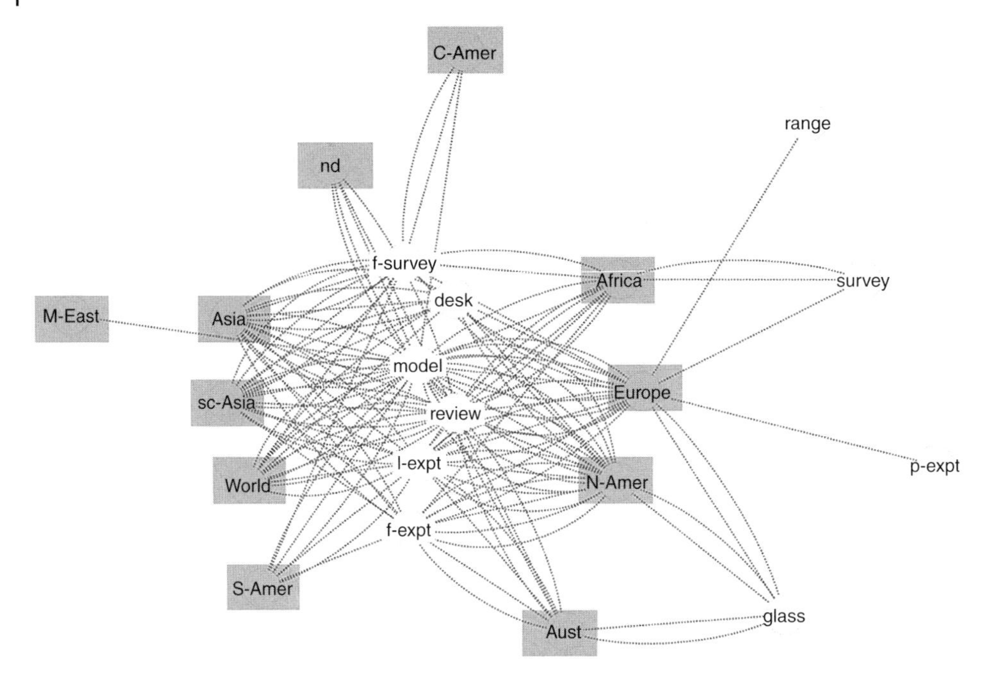

Figure 9.2 Network analysis exhibiting regions (source interaction node = shaded box), order (interaction type = dotted line) and publication type (target interaction node). Region name abbreviations: Aust = Australasia/Pacific; C Amer = Central America; M-East = Middle East; sc-Asia = subcontinental Asia; N-Amer = North America; S-Amer = South America; nd = region not defined. Publication type abbreviations: f-survey = field survey; desk = desktop analysis; l-expt = lab experiment; f-expt = field experiment; glass = glasshouse experiment; p-expt = pot experiment. To increase clarity of network, insect orders (*dotted line*) and number of publications per order are not labelled; see text for more detail. Distances between regions and publication type are to maximize spatial clarity of interactions. More centralized nodes have a more diverse range of interactions.

not explicitly identified, with Coleoptera (26) the next most abundant (Figure 9.1c). Insect pest phenology was most commonly assessed in studies (70 publications), followed by abundance (64), distribution (53), herbivory (45), development (41), IPM (34) and survival (30) (Figure 9.1d).

With respect to study region, publication type and order studied, the most common publications were reviews collating data from worldwide sources assessing a range of taxa (18 publications); second highest publications (14) were modelling of Lepidoptera in Europe; and third highest (with eight publications) were four different groupings: lab experiments of Hemiptera in Asia; range expansion of various taxa in Europe; and either reviews with no specific region in mind and mention of no specific insect taxa, or with a variety of taxa identified. Of the top 20 types in these categories, eight were based in Europe, nine were models and seven encompassed a range of taxa (Figure 9.2). When we considered the region, publication type and climate change mechanisms tested, 26 publications modelled temperature in Europe, 17 modelled temperature in North America and 17 reviewed temperature but with no defined region. Of the top 20 types in this category 14 assessed temperatures (Figure 9.3). When we considered publication type, what was measured and climate change

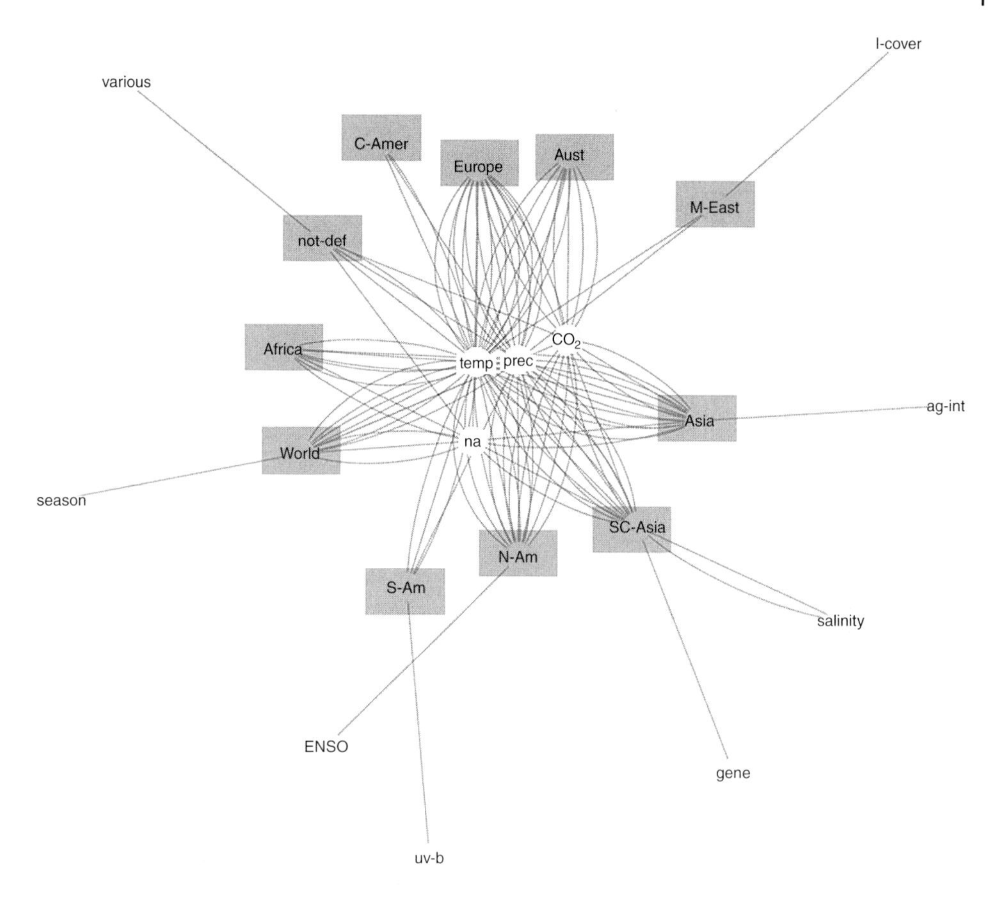

Figure 9.3 Network analysis exhibiting regions (source interaction node = shaded box), publication type (interaction type = dotted line) and climate change measure (target interaction node). Region name abbreviations: Aust = Australasia/Pacific; C Amer = Central America; M-East = Middle East; sc-Asia = subcontinental Asia; N-Am = North America; S-Am = South America; not-def = region not defined. Climate change measure abbreviations: gene = genetics; uv-b = ultraviolet radiation b; prec = precipitation; temp = temperature; CO_2 = carbon dioxide; ENSO = El Niño southern oscillation; ag-int = agricultural intensity; l-cover = landcover; na = climate change mechanisms not identified. To increase clarity of network, publication type (*dotted line*) and associated numbers are not labelled; see text for more detail. Distances between regions and measures are to maximize spatial clarity of interactions. More centralized nodes have a more diverse range of interactions.

mechanism tested, 36 publications modelled phenological changes in relation to temperature, 26 modelled distribution changes in relation to temperatures and 17 were both reviews of temperature impacts on abundance and on distribution, and 15 publications assessed survival related to temperature using lab experiments (Figure 9.4).

From these network analyses, a few clear trends become evident. First, field data are key to better understanding the impacts of climate change, and to better identify if there are predictable responses or if crops, pests and natural enemies will respond idiosyncratically. An increase in the relative pool of field-based data collection and publications (both surveys and experiments; see Figures 9.1a and 9.2) is also required

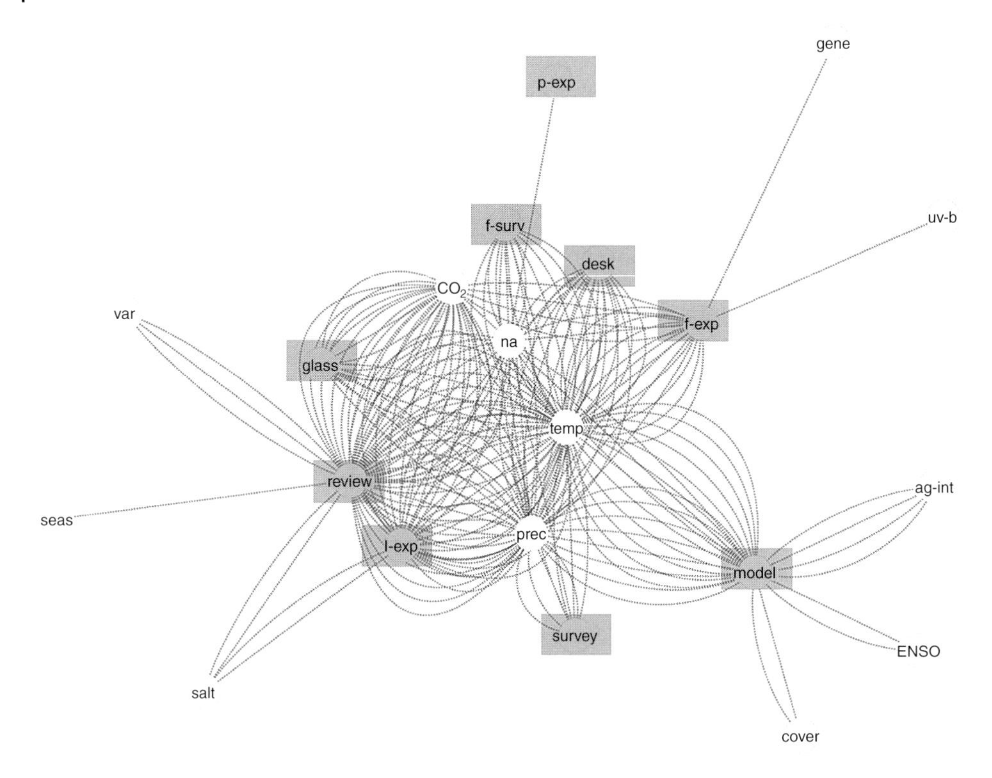

Figure 9.4 Network analysis exhibiting publication type (source interaction node = shaded box), measure (interaction type = dotted line) and climate change mechanism (target interaction node). Publication type abbreviations: f-surv = field survey; desk = desktop analysis; l-exp = lab experiment; f-exp = field experiment; glass = glasshouse experiment; p-exp = pot experiment. Climate change mechanism abbreviations: gene = genetics; uv-b = ultraviolet radiation b; prec = precipitation; temp = temperature; CO_2 = carbon dioxide; ENSO = El Niño southern oscillation; ag-int = agricultural intensity; seas = season; salt = salinity; cover = landcover; na = climate change mechanisms not identified. To increase clarity of network, measure (*dotted line*) and associated numbers are not labelled; see text for more detail. Distances between regions and measures are to maximize spatial clarity of interactions. More centralized nodes have a more diverse range of interactions.

to enhance modelling and review articles and identify predictable responses. Second, a relatively larger research effort needs to take place in regions which are at the bottom third of the region publication rankings (see Figure 9.1b). Third, although much information could be found in 'grey' literature, there is a critical need for peer-reviewed and easily accessible data and information for researchers and policy makers worldwide. This will ideally lead to a more efficient use of funding resources to answer critical questions, and reduce overlap of projects essentially answering the same key questions, reviewing/modelling the same datasets and 'reinventing the wheel' (see Figure 9.1b). Finally, research also needs to address the impacts of climate change particularly in relation to land cover, land use change and agricultural intensity which have been relatively neglected to this point (e.g. outliers in Figures 9.3 and 9.4). Such studies may play important roles in promoting insect diversity conservation (Oliver *et al.* 2016).

9.5 How Climate Change and Insect Responses May Affect Various Ecological Processes Important for Plant Protection

9.5.1 Pest Population Dynamics

The integration of thermal biology into understanding insect pest population dynamics, fluctuations and demography has been a significant positive addition to plant protection regimes. This research area will continue to provide insights into forecasting pest outbreaks under a warming and more fluctuating climate in future decades (Bale and Hayward 2010; Colinet *et al.* 2015; Denlinger and Lee 2010; Kingsolver 1989).

The occurrence of high temperature days is increasing in frequency in agricultural cropping areas (Zhang *et al.* 2015). Exposure of different developmental stages to extreme hot temperature events may influence insect pest ontogeny, preventing survival to maturity and affecting feeding and growth rates and adult reproduction (Kingsolver *et al.* 2011), which can then lead to changes in interactions between insect pests and their associated natural enemies. Following this, the impact of cumulative stresses on insect pests at multiple life stages is poorly known. However, for the few assessments conducted, high temperature exposure at different life stages can have serious consequences on population dynamics (Ma *et al.* 2004a; Zani *et al.* 2005; Zhang *et al.* 2015). For the aphid *Metopolophium dirhodum* (Aphididae: Hemiptera), high temperature exposure in individuals older than the third instar reduces their longevity and number of total offspring produced, as did a temperature pulse (over 29 °C) lasting longer than a week compared to a 1- or 2-day exposure to this temperature (Ma *et al.* 2004b). For the diamondback moth *Plutella xylostella* (Plutelliae: Lepidoptera), first instar caterpillars were most susceptible to heat stress, but if they survived, the exposure did not influence their development into adults. Third instar caterpillars were most resistant to heat stress but it reduced their capacity to pupate and develop into full adults if heat exposure persisted for longer than 16 hours, and heat exposure reduced reproductive capacity in older caterpillars (Zhang *et al.* 2015).

Overwintering responses of insects is key in assessing pest population dynamics (Bale 2010), with seasonal variation in physiological responses, including diapause and cold tolerance, and acclimation key strategies for survival (Bale and Hayward 2010; Hoffmann *et al.* 2003). Winter temperatures, variability and snow cover changes can have profound impacts on insects, including energy and water balance, cold injury, phenology and interaction changes (Williams *et al.* 2014). For example, freeze-tolerant larvae of the goldenrod gall fly (*Eurosta solidaginis*; Tephritidae: Diptera) exposed to warmer temperatures when overwintering have higher winter metabolic rates, which in turn reduced their survival and fecundity compared to conspecifics overwintering in colder sites (Irwin and Lee 2003).

9.5.2 Pest–Crop Plant Interactions

The nutritional quality of agricultural products is expected to change with modified CO_2 concentrations, higher temperatures and more variable moisture regimes, and these changes will be complex, especially in terms of palatability for insect pest species. For example, the effects of elevated CO_2 and reduced soil water modified the major hormone signalling pathways of jasmonic acid (JA), salicylic acid (SA) and ethylene before and after herbivore damage by the Japanese beetle *Popillia japonica* (Scarabaeidae: Coleoptera). After exposure to elevated CO_2, leaf sugars, SA and herbivory increased, whereas JA signalling transcripts decreased. Induction of JA and ethylene was increased after soil water was reduced and after herbivory. JA and ethylene transcript expression was suppressed with the interaction of elevated CO_2 and reduced soil water but there was no increase in herbivore susceptibility (Casteel *et al.* 2012). Such a variety of different responses between independent factors suggests that interactions between the host plant and environment are critical to assessing the impacts of climate change on pest dynamics.

9.5.3 Intraguild Predation

Intraguild predation occurs when a parasitoid or predator attacks both the host/prey species in which it is thought to be the primary target of biocontrol (e.g. the pest), as well as feeding on other parasitoid/predatory species within the system (Polis *et al.* 1989). Increased temperature (+2 °C) increased predator abundance (carabid beetles) in a spring-sown wheat crop, but a 10% precipitation increase caused no change in the predator's abundance (Berthe *et al.* 2015). However, predatory staphylinid beetles were negatively affected by increasing temperatures. Even though an increase in major predators is seen as a major benefit to agricultural systems, there may also be an increase in intraguild predation, which may reduce the predation benefit in total (Berthe *et al.* 2015; Raso *et al.* 2014) and change the interaction networks among pests and their natural enemies (Bohan *et al.* 2013).

Introduced biological control agents can have serious consequences for intraguild predation and have negative impacts on native species found in agro-ecosystems (Rosenheim *et al.* 1995). The recent introduction and establishment of a primary aphid-feeder *Harmonia axyridis* (Pallas) (Coleoptera: Coccinellidae) in North America, South America and Europe have reduced populations of native coccinellids. Based on laboratory colonies, it is known to feed on other predatory species (Pell *et al.* 2008). With a changing climate and further movements of introduced biological control agents into new habitats, their impacts may exacerbate competition and enhance intraguild predation of native species.

9.5.4 Tritrophic Interactions

As part of a complexity of interactions, parasitoids are influenced by host plant and herbivore dynamics (Harvey *et al.* 2003), and climate change will have substantive impacts on these interactions (Facey *et al.* 2014). Stireman *et al.* (2005) assessed caterpillar–parasitoid interactions across 15 databases from central Brazil to Canada. They found that, as climatic variability increases, the ability of parasitoids to track host populations decreases. They predict that as precipitation and temperature become more variable with climate change, there will be more herbivore outbreaks in managed systems, as natural enemy–herbivore dynamics are disrupted. However, responses across all landscapes are still highly unpredictable; individual species and their changes in interactions among trophic levels can be highly idiosyncratic as different responses have been elicited when temperature and carbon dioxide are modified independently (which has been undertaken in the majority of studies) (Facey *et al.* 2014; Jamieson *et al.* 2012).

There is a paucity of studies assessing changes in tropic interactions with the effects of interrelated climate factors (Facey *et al.* 2014; Jamieson *et al.* 2012), particularly in terms of the effect on tritrophic interactions after exposure to changes in climatic variation and increased carbon dioxide. In turn, this makes it extremely difficult to predict the effects of climate change on tritrophic interactions across systems.

9.5.5 Pest–Pathogen Interactions

Agronomic plant diseases transferred by insect pests have been relatively underassessed as part of climate change impacts, in particular their role in reducing global food security and their negative interactions with CO_2 fertilization (Chakraborty and Newton 2011; Juroszek and von Tiedemann 2013). It is expected that, as the geographical ranges of pests start to shift, the pathogens associated with these pests will also move into new regions, exposing different crop varieties to a range of new viruses and phytopathogens that are transferred by insect pests (Coakley *et al.* 1999). Since 1960, hundreds of pests and pathogens have moved their distributions on average $2.7 \pm 0.8\,\text{km}\,\text{yr}^{-1}$, which is similar to movements of wild species, independently of land use changes, variation in crop varieties or agricultural technologies (Bebber *et al.* 2013). In addition, during the period 2001–2003, at a global scale, 37.4% of the rice harvest and 40.3% of the potato harvest were lost to damage: 15.1% (rice) and 10.9% (potatoes) to pests, 10.8% (rice) and 14.5% (potatoes) to pathogens, 1.4% (rice) and 6.6% (potatoes) to viruses, and 10.2% (rice) and 8.3% (potatoes) to weeds (Oerke 2006). It is estimated that 10–16% of the global harvest is lost to plant diseases (including those transferred by insects), which equated to a US$220 billion loss, and then a further 6–12% postharvest crop loss (Oerke 2006; Strange and Scott 2005). With a changing climate, and increased insect pest activity and movement, these economic losses are expected to increase.

9.5.6 Pest–Symbiont Interactions

When reared at high constant temperatures (30 °C), two stink bug species *Acrosternum hilare* and *Murgantia histionica* (both Pentatomidae: Hemiptera) lost their gut-associate symbionts, had lower survivorship and reproductive rates than at 25 °C (Prado *et al.* 2010). The intracellular symbiont *Blochmannia* (Enterobacteriaceae: Enterobacteriales) in ants contributes to nitrogen recycling and nutrient biosynthesis, but is nearly completely depleted in minor workers and unmated queens when the ants are exposed to a higher temperature of 37.7 °C for 4 weeks (Fan and Wernegreen 2013).

The symbiont assemblage within a pest can also have complex interactions. In the whitefly *Bemisia tabaci* complex (Aleyrodidae: Hemiptera), Shan *et al.* (2014) assessed the responses of three symbionts to high and low temperature exposures: two intracellular *Candidatus* species (Rhizobiaceae: Rhizobiales) and a body cavity resident *Rickettsia* species (Rickettsiaceae: Rickettsiales). As duration of a temperature exposure to 40 °C continued, the infection rates of *C. Hamiltonella defensa* (Enterobacteriaceae: Enterobacteriales) reduced, the infection of *C. Portiera aleyrodidarum* (Enterobacteriaceae: Enterobacteriales) was less affected initially but exhibited a more drastic reduction after 3–5 days, while *Rickettsia* infection rates were not significantly impacted (Shan *et al.* 2014). As microbial 'mycetocyte symbionts' are essential for the insect's full development, and they are obligately transmitted via insect ovaries in a range of taxa including planthoppers and aphids, any reduction in symbiont numbers reduces reproductive output and lifespan of the insects (Douglas 2007).

In summary, there is a critical need to better understand fundamental crop, pest and natural enemy biology (i.e. ecology, physiology, behaviour), as well as how interactions between and among species at a range of trophic levels may be modified with a changing climate. Of most interest will be identifying critical life stages where both pest species and natural enemies are most vulnerable to the abiotic and biotic changes associated with a rapidly changing climate. This will enable more focused IPM approaches to be used to control pests and to extend and increase the efficacy of natural enemies.

9.6 Climate Change and IPM Approaches

With CO_2 levels and temperatures increasing, precipitation becoming more variable and non-native insect species moving into new ranges, changes in insect–plant interactions and IPM regimes will be substantive and less predictable (Trumble and Butler 2009). It is generally expected that insect chewing herbivores will consume more leaf tissue as plant nutrition is reduced (more carbon per unit of nitrogen), many insect pest species will develop quicker as they are ectotherms (or regional heterotherms) and as their internal temperature varies considerably and they respond quickly to increased temperatures.

It is generally anticipated that a changing climate and more variable weather patterns will make pests (and pathogen) attacks more unpredictable and their amplitude larger. Combined with the uncertainty of how climate change will directly impact on crop yields, the insect–plant interactions in this system remain unclear along with what effect this will have on crop productivity (Gregory *et al.* 2009). One key ecosystem change may be that glasshouse pests could become more problematic in open pastures and fields (Laštůvka 2009). It is also thought that population growth and longevity of short-lived species, including insect pests, may be enhanced (Morris *et al.* 2008). Relaxed cold limitation could be one of the key drivers for exacerbating the expansion of insect pests into new regions, and a longer growing season in current regions (Diffenbaugh *et al.* 2008).

9.6.1 Cultural

Changing farming and adaptive management strategies will be required to reduce the impact that agricultural pests have on crops (Thomson *et al.* 2010). This may include: (1) planting different plant varieties, (2) planting at different times of the year to minimize exposure to pest outbreaks, and (3) increasing the diversity of habitat on edges to promote natural enemy numbers. All of these strategies are used to minimize pest impact at the farm scale. Other relatively simple strategies include mulching, raised beds and shelters to conserve soil moisture, protecting crops from heavy rains, high temperatures and flooding, and preventing soil degradation. At the farm level and the microclimate level, changing farming strategies is most critical.

9.6.2 Crop Rotation and Diversification

Crop rotation and diversification can build a higher level of resilience into agricultural production by reducing pest outbreaks and pathogen transmission, and buffering crop production from more frequent extreme climatic events as well as higher levels of climatic variability (Lin 2011). Increased diversity within agro-ecosystems will increase the functional ecosystem diversity of the landscape as well as increasing redundancy if species do become locally extinct. This is critical under a rapidly changing climate as biotic (e.g. pest, pathogens) and abiotic (e.g. solar radiation, temperature and precipitation) pressures are likely to change (Lin 2011; Vandermeer *et al.* 1998).

Crop rotation can assist in suppressing diseases, which are predicted to increase in prevalence under a changing climate. For example, planting oilseed, pulse and forage crops within a cereal cropping system disrupts disease cycles (Krupinsky *et al.* 2002). Increasing genetic diversity can also suppress diseases, such as fungal blast occurrence among different rice varieties. Disease-susceptible rice varieties exhibited a 89% yield increase in the Yunnan Province of China when planted in mixtures with resistant varieties, and rice blast (the major disease of rice) was reduced by 94% (Zhu *et al.* 2000).

Structural diversity can also suppress pests. Unharvested lucerne refuge strips provide habitat for natural enemies of *Helicoverpa* (Noctuidae: Lepidoptera), and the unharvested refuge strips are ideally placed 30 m apart to allow natural enemies to work as effective biological controls in the harvested strips (Hossain *et al.* 2002). Non-crop vegetation can be used to develop 'beetle banks' at field margins that can be used as overwintering habitat for natural enemies (Collins *et al.* 2002; Thomas *et al.* 1991). With a warmer and drier climate, these refuges can also increase the microclimate diversity of a farm, providing opportunities for climatic respite associated with extreme temperatures in a relatively homogenous production landscape.

Having a polyculture can assist with climate change buffering. In dealing with local variability and disturbance, small holder farmers in varying cropping regions (e.g. north-east Tanzania and east-central Sweden) use wild varieties and a diversity of crops, spatially and temporally, to enhance their capacity to deal with agro-ecosystem changes (Tengö and Belfrage 2004).

9.6.3 Biological Factors

Methodologically, most assessments of parasitoids and predators have been done at constant temperatures. Bahar *et al.* (2012) found that fluctuating temperatures in laboratory conditions (particularly lower temperatures) can substantially change the developmental period of pest herbivores (in their case the diamondback moth) and its parasitoid. Short-term temperature fluctuations can cause substantial stress on both pest species and their natural enemies, which can then have substantive influences on their interactions (Chidawanyika *et al.* 2012). Insect biology of both pests and natural enemies in agro-ecosystems, including generation times, sex ratio, lifespan, fecundity, activity, distribution and survival, are all affected by temperature extremes and fluctuations (Duale 2005; Hance *et al.* 2007; Kalyebi *et al.* 2005; Liu *et al.* 1995; Sorribas *et al.* 2012). There may also be spatial and temporal mismatches between pests and their natural enemies which will reduce the efficacy of biocontrol agents, and predicting these impacts will be difficult without a thorough understanding of the tritrophic interactions among species (Thomson *et al.* 2010).

9.6.4 Pesticides

With a doubling of maize, wheat and rice production worldwide since the 1960s, there has been a 15–20-fold increase in pesticide use (Oerke 2006) (see also Chapter 4). Additionally, as crop yield has increased, due to the use of high-yielding varieties, soil and water management, fertilization and cultivation methods, there has been an increase in crop loss due to pests. Many new varieties of crops are more reliant on pesticides as they have lower tolerance to competitors and herbivory, as much of the inbuilt resilience is bred out (Oerke 2006). With the expectation of more insect pest outbreaks and that global food production needs to increase by 50% to meet the 2050 global population needs, it is assumed that food security using a range of pesticides will be one of the more sought-after tools of management (Chakraborty and Newton 2011).

Pesticide applications are the primary method of managing pests in the industrialized world (Ziska 2014) (see also Chapter 4). The application of pesticides is correlated with temperature at sites and site minimum temperature can serve as a proxy for pesticide application. For example, Ziska (2014) assessed pesticide applications on soybean along a 2100 km latitudinal gradient in the USA and found that soybean yields did not vary over the gradient, while total pesticide application increased from 4.3 kg ha^{-1} active ingredient in Minnesota (having a minimum daily temperature of –28.6 °C) to 6.5 kg ha^{-1} active ingredient in Louisiana (–5.1 °C minimum daily temperature). The authors of this study suggest that, with a changing climate, herbicide use will increase in the more temperate regions, whereas there will be a greater increase in insecticide and fungicide use closer to the tropics (Ziska 2014). This is due to the fact that, in temperate regions, warming enhances growth and insect reproductive output, as well as survival (Patterson *et al.* 1999).

In some cases, exposure to sublethal concentrations of pesticide could lead to cross-tolerance of temperature and the insecticide. An example of this is the brown planthopper (*Nilaparvata lugens*) which attacks rice crops in Asia (Ge *et al.* 2013). When brown planthoppers were exposed to sublethal concentrations of the commonly used insecticide triazophos (40 ppm) at 40 °C, mortality was reduced from 94% to 50% and lethal mean time (LT_{50} based on a Gompertz model) was increased by over 17 hours, compared to a control (tap water and the non-active substances dimethyl sulfoxide and emulsifier). The authors found that, when insecticide usage increased, Hsp70 and arginine kinase were upregulated, both being critical for the brown planthopper's survival and thermotolerance. This indicates that a sublethal stress induced by an insecticide can initiate cross-tolerance to temperature. From the agricultural perspective, this indicates that the brown planthopper population that is exposed to sublethal concentrations of triazophos will increase cross-tolerance and reproductive potential. If pesticides become a trigger for induced thermotolerance, then pests may be able to survive hotter temperatures and cause more damage to sensitive crops.

9.6.5 Semiochemicals

The signalling chemicals (semiochemicals) which cause changes in the behaviours of other living organisms (Dicke and Sabelis 1988) play a critical role in IPM. The use of pheromones (which act between individuals of the same species) and allelochemicals (acting between species, including kairomones which benefit the receiver, allomones which benefit the emitter and synomones which benefit both) is a key method that insects use to sense their environment. Their use in monitoring, trapping, mating disruption, push-pull strategies and biological controls makes them ideal for a range of IPM techniques (Heuskin *et al.* 2011; Wajnberg and Colazza 2013). Temperature, humidity and air speed can have critical impacts on the effectiveness of semiochemicals (Heuskin *et al.* 2011). For example, Cork *et al.* (2008) used PVC-resin controlled-release formulations to deliver sex pheromones to the yellow rice stem borer at a range of temperatures (from 22 °C to 34 °C). The temperature used highly influenced pheromone rates, with half-lives of the sex pheromone decreasing with an increase in temperature.

Temperature has also been shown as the critical environmental variable influencing volatile release rates in moth sex pheromones (van der Kraan and Ebbers 1990), light brown apple moth pheromones (Bradley *et al.* 1995), tsetse fly kairomones (Torr *et al.* 1997) and waterbuck odours to control tsetse fly (Shem *et al.* 2009), oriental fruit moth pheromone (Atterholt *et al.* 1999) and sawfly sex pheromones (Johansson *et al.* 2001). As the annual climate warms across agricultural landscapes, and as microclimates become more variable, it would be anticipated that the use of these volatiles in their current forms may become less effective and may require a synergist or other compounds to reduce their volatility under high temperature regimes.

9.6.6 Reproductive Control

The sterile insect technique (SIT) is a critical method used to control insects (Knipling 1959) which releases radiation-induced sterile males into wild populations to reduce the number of offspring after mating with wild females. It is a key method used to control *Ceratitis capitata* (Tephritidae: Diptera) worldwide (Robinson 2002). One of the strains of *C. capitata* has a temperature sensitivity gene, *tsl*, which makes the homozygous female embryos sensitive to high temperature mortality (compared to males) after 24 hours of development (Fisher 1998; Robinson 2002). Females remain sensitive to temperature throughout their lifetime, but the impact of the *tsl* gene mutation or the effect of irradiation on released males in the field are currently unknown (Nyamukondiwa *et al.* 2013). In South Africa, populations increase once sufficient degree days have accumulated, and they decrease as temperatures fall below minimum critical temperatures. Individuals expressing the *tsl* mutation exhibit a higher critical thermal maximum and greater longevity in the field compared to wild-type individuals, indicating that the sterile insect technique may be more effective in a warming climate (Nyamukondiwa *et al.* 2013). This advantage of lab-reared sterile males could enhance their usefulness as a pest management tool under a warming climate.

9.6.7 Long-term Monitoring

One of the key requirements to determine if climate change is changing the population dynamics of pest species is having access to long-term data (Yamamura *et al.* 2006). Without this key baseline data, it is extremely difficult to fully assess changes in pest and beneficial populations with changing climate regimes and predict future population dynamics. However, data covering population dynamics of populations over 50 years are very sparse, with only a few examples, such as annual light trap catches in Japanese rice paddy fields for 50 years (Yamamura *et al.* 2006), aphid suction trap catches at Rothamsted, UK, also for 50 years (Bell *et al.* 2015), and a 1910 year record of locust outbreaks in China based on a reconstructed time and abundance series (Tian *et al.* 2011). A lack of long-term data makes predicting pest outbreaks extremely difficult across most agro-ecological regions, and makes modelling population dynamics tenuous when attempting to align with changing climate regimes in different regions. In addition, any long-term assessment of parasitoid/predator–host/prey interactions and changes in trophic level interactions is not available, making predictions of community assemblage changes in agro-ecosystems with climate change even more difficult.

9.7 Directions for Future Research

Agricultural impact assessment based on changing yields due to increased pressures from pests due to climate change is still in its infancy (Gregory *et al.* 2009; Scherm 2004). However, it is clear that human-induced climate change will have impacts on all aspects of IPM systems, pest outbreaks, pollinator synchrony with flowers, efficiency of crop protection technologies, and parasitoid and predator effectiveness (Sharma 2014). Biological responses to climate change, particularly changes in temperature, can be based on threshold-level responses rather than linear responses, and when interactions occur with other climatic changes and biological adaptation, responses at all levels will be complex (Benedetti-Cecchi *et al.* 2006; Gutschick and BassiriRad 2003; Thompson *et al.* 2013b). There is a critical need for continued assessment of biological responses to climate change within and among species (Andrew 2013; Andrew and Terblanche 2013), particularly in the field and at critical life history stages which are vulnerable to the abiotic and biotic impacts of climate change.

For many crops, pesticides are still the main form of pest control (Nash and Hoffmann 2012) (see also Chapter 4). Under a changing climate, insect pests are likely to become more damaging, especially if the current worldwide broad-spectrum spraying regimes continue. For IPM to be adopted more fully within cropping systems, regimes that increase management strategy flexibility, such as those outlined by Nash and Hoffmann (2012), need to be implemented. This requires a greater understanding of pest population dynamics, thermal physiology, ecology, behaviour and core IPM priorities of host plant resistance, area-wide management, emergency chemical control when required, and predictive modelling tools when controlling pests in a more variable climate (Nguyen *et al.* 2014; Sutherst *et al.* 2011). A more holistic inclusion of different management regimes including resistant cultivars, preservation of natural enemy activity, utilizing thresholds, use of pheromones, use of selective insecticides in preference to broad-spectrum usage, landscape manipulation, tillage management, crop rotation, biological control (naturally occurring and safely introduced, classic, mass-reared natural enemies) within an adaptive management context will be critical for managing insect pests in agro-ecosystems within a rapidly changing climate.

Acknowledgements

We thank the editors, reviewers and Bianca Bishop for improving the clarity of this review.

References

Andrew, N.R. (2013) Population dynamics of insects: impacts of a changing climate. In: Rohde, K. (ed.) *The Balance of Nature and Human Impact*. Cambridge University Press, Cambridge, UK, pp. 311–323.

Andrew, N.R. and Hill, S.J. (2016) Effect of climate change on insect pest management: dataset. Figshare. Available at: https://dx.doi.org/10.6084/m9.figshare.3206599.v1 (accessed 7 March 2017).

Andrew, N.R. and Terblanche, J.S. (2013) The response of insects to climate change. In: Salinger, J. (ed.) *Living in a Warmer World: how a changing climate will affect our lives*. David Bateman Ltd, Auckland, New Zealand, pp. 38–50.

Andrew, N.R., Hart, R.A., Jung, M.P., Hemmings, Z., and Terblanche, J.S. (2013a) Can temperate insects take the heat? A case study of the physiological and behavioural responses in a common ant, *Iridomyrmex purpureus* (Formicidae), with potential climate change. *Journal of Insect Physiology* **59**: 870–880.

Andrew, N.R., Hill, S.J., Binns, M., *et al.* (2013b) Assessing insect responses to climate change: what are we testing for? Where should we be heading? *Peer Journal* **1**: e11.

Araújo, M.B., Ferri-Yáñez, F., Bozinovic, F., Marquet, P.A., Valladares, F. and Chown, S.L. (2013) Heat freezes niche evolution. *Ecology Letters* **16**: 1206–1219.

Arndt, E. and Perner, J. (2008) Invasion patterns of ground-dwelling arthropods in Canarian laurel forests. *Acta Oecologica* **34**: 202–213.

Atterholt, C.A., Delwiche, M.J., Rice, R.E. and Krochta, J.M. (1999) Controlled release of insect sex pheromones from paraffin wax and emulsions. *Journal of Controlled Release* **57**: 233–247.

Bahar, M.H., Soroka, J.J. and Dosdall, L.M. (2012) Constant versus fluctuating temperatures in the interactions between *Plutella xylostella* (Lepidoptera: Plutellidae) and its larval parasitoid *Diadegma insulare* (Hymenoptera: Ichneumonidae). *Environmental Entomology* **41**: 1653–1661.

Bale, J.S. (2010) Implications of cold-tolerance for pest management. In: Deninger, D.L. and Lee, R.E. (eds) *Low Temperature Biology of Insects*. Cambridge University Press, Cambridge, pp. 342–372.

Bale, J.S. and Hayward, S.A.L. (2010) Insect overwintering in a changing climate. *Journal of Experimental Biology* **213**: 980–994.

Bebber, D.P., Ramotowski, M.A.T. and Gurr, S.J. (2013) Crop pests and pathogens move polewards in a warming world. *Nature Climate Change* **3**: 985–988.

Bell, J.R., Alderson, L., Izera, D., *et al.* (2015) Long-term phenological trends, species accumulation rates, aphid traits and climate: Five decades of change in migrating aphids. *Journal of Animal Ecology* **84**: 21–34.

Benedetti-Cecchi, L., Bertocci, I., Vaselli, S. and Maggi, E. (2006) Temporal variance reverses the impact of high mean intensity of stress in climate change experiments. *Ecology* **87**: 2489–2499.

Berthe, S.C.F., Derocles, S.A.P., Lunt, D.H., Kimball, B.A. and Evans, D.M. (2015) Simulated climate-warming increases Coleoptera activity-densities and reduces community diversity in a cereal crop. *Agriculture, Ecosystems and Environment* **210**: 11–14.

Bohan, D.A., Raybould, A., Mulder, C., *et al.* (2013) Networking agroecology: integrating the diversity of agroecosystem interactions. In: Guy, W. and David, A.B. (eds) *Advances in Ecological Research Vol. 49: Ecological Networks in an Agricultural World*. Academic Press, London, UK, pp. 1–67.

Bolnick, D.I., Ingram, T., Stutz, W.E., Snowberg, L.K., Lau, O.L. and Paull, J.S. (2010) Ecological release from interspecific competition leads to decoupled changes in population and individual niche width. *Proceedings of the Royal Society of London B: Biological Sciences* **277**: 1789–1797.

Both, C., van Asch, M., Bijlsma, R.G., van den Burg, A.B. and Visser, M.E. (2009) Climate change and unequal phenological changes across four trophic levels: constraints or adaptations? *Journal of Animal Ecology* **78**: 73–83.

Bradley, S.J., Suckling, D.M., McNaughton, K.G., Wearing, C.H. and Karg, G. (1995) A temperature-dependent model for predicting release rates of pheromone from a polyethylene tubing dispenser. *Journal of Chemical Ecology* **21**: 745–760.

Bradshaw, W.E. and Holzapfel, C.M. (2001) Genetic shift in photoperiodic response correlated with global warming. *Proceedings of the National Academy of Sciences* **98**: 14509–14511.

Bradshaw, W.E. and Holzapfel, C.M. (2008) Genetic response to rapid climate change: it's seasonal timing that matters. *Molecular Ecology* **17**: 157–166.

Bridle, J.R., Buckley, J., Bodsworth, E.J. and Thomas, C.D. (2014) Evolution on the move: specialization on widespread resources associated with rapid range expansion in response to climate change. *Proceedings of the Royal Society B: Biological Sciences* **281**: e20131800.

Butzer, K.W. (1976) *Early Hydraulic Civilization In Egypt*. University of Chicago Press, Chicago, IL, USA.

Casteel, C.L., Niziolek, O.K., Leakey, A.D.B., Berenbaum, M.R. and DeLucia, E.H. (2012) Effects of elevated CO_2 and soil water content on phytohormone transcript induction in *Glycine max* after *Popillia japonica* feeding. *Arthropod-Plant Interactions* **6**: 439–447.

Chakraborty, S. and Newton, A.C. (2011) Climate change, plant diseases and food security: an overview. *Plant Pathology* **60**: 2–14.

Chanthy, P., Martin, B., Gunning, R. and Andrew, N.R. (2012) The effects of thermal acclimation on lethal temperatures and critical thermal limits in the green vegetable bug, *Nezara viridula* (L.) (Hemiptera: Pentatomidae). *Frontiers in Physiology* **3**: 465.

Chanthy, P., Martin, B., Gunning, R. and Andrew, N.R. (2015) Influence of temperature and humidity regimes on the developmental stages of Green Vegetable Bug, *Nezara viridula* (L.) (Hemiptera: Pentatomidae) from inland and coastal populations in Australia. *General and Applied Entomology* **43**: 37–55.

Chevin, L.M., Lande, R. and Mace, G.M. (2010) Adaptation, plasticity, and extinction in a changing environment: towards a predictive theory. *PLoS Biology* **8(4)**: e1000357.

Chidawanyika, F., Mudavanhu, P. and Nyamukondiwa, C. (2012) Biologically based methods for pest management in agriculture under changing climates: challenges and future directions. *Insects* **3**: 1171–1189.

Cleland, E.E., Chuine, I., Menzel, A., Mooney, H.A. and Schwartz, M.D. (2007) Shifting plant phenology in response to global change. *Trends in Ecology and Evolution* **22**: 357–365.

Coakley, S.M., Scherm, H. and Chakraborty, S. (1999) Climate change and plant disease management. *Annual Review of Phytopathology* **37**: 399–426.

Colinet, H., Sinclair, B.J., Vernon, P. and Renault, D. (2015) Insects in fluctuating thermal environments. *Annual Review of Entomology* **60**: 123–140.

Collins, K.L., Boatman, N.D., Wilcox, A., Holland, J.M. and Chaney, K. (2002) Influence of beetle banks on cereal aphid predation in winter wheat. *Agriculture, Ecosystems and Environment* **93**: 337–350.

Cork, A., de Souza, K., Hall, D.R., Jones, O.T., Casagrande, E., Krishnaiah, K. and Syed, Z. (2008) Development of PVC-resin-controlled release formulation for pheromones and use in mating disruption of yellow rice stem borer, *Scirpophaga incertulas*. *Crop Protection*, **27**: 248–255.

Cullen, H.M., deMenocal, P.B., Hemming, S., *et al.* (2000) Climate change and the collapse of the Akkadian empire: evidence from the deep sea. *Geology* **28**: 379–382.

Davis, M.B., Shaw, R.G. and Etterson, J.R. (2005) Evolutionary responses to changing climate. *Ecology* **86**: 1704–1714.

Denlinger, D.L. and Lee, R.E. (2010) *Low Temperature Biology of Insects*. Cambridge University Press, Cambridge, UK.

Deutsch, C.A., Tewksbury, J.J., Huey, R.B., *et al.* (2008) Impacts of climate warming on terrestrial ectotherms across latitude. *Proceedings of the National Academy of Sciences* **105**: 6668–6672.

Dicke, M. and Sabelis, M.W. (1988) Infochemical terminology: based on cost-benefit analysis rather than origin of compounds? *Functional Ecology* **2**: 131–139.

Diffenbaugh, N.S., Krupke, C.H., White, M.A. and Alexander, C.E. (2008) Global warming presents new challenges for maize pest management. *Environmental Research Letters* **3**: 044007.

Douglas, A.E. (1998) Nutritional interactions in insect-microbial symbioses: aphids and their symbiotic bacteria *Buchnera*. *Annual Review of Entomology* **43**: 17–37.

Douglas, A.E. (2007) Symbiotic microorganisms: untapped resources for insect pest control. *Trends in Biotechnology* **25**: 338–342.

Douglas, A.E. and van Emden, H.F. (2007) Nutrition and symbiosis. In: van Emden H.F. and Harrington R. (eds) *Aphids as Crop Pests*. CABI, Wallingford, UK, pp. 115–134.

Duale, A.H. (2005) Effect of temperature and relative humidity on the biology of the stem borer parasitoid *Pediobius furvus* (Gahan) (Hymenoptera: Eulophidae) for the management of stem borers. *Environmental Entomology* **34**: 1–5.

Dunbar, H.E., Wilson, A.C.C., Ferguson, N.R. and Moran, N.A. (2007) Aphid thermal tolerance is governed by a point mutation in bacterial symbionts. *PLoS Biology* **5(5)**: e96.

Facey, S.L., Ellsworth, D.S., Staley, J.T., Wright, D.J. and Johnson, S.N. (2014) Upsetting the order: how climate and atmospheric change affects herbivore-enemy interactions. *Current Opinion in Insect Science* **5**: 66–74.

Fan, Y. and Wernegreen, J. (2013) Can't take the heat: high temperature depletes bacterial endosymbionts of ants. *Microbial Ecology* **66**: 727–733.

Feynman, J. and Ruzmaikin, A. (2007) Climate stability and the development of agricultural societies. *Climate Change* **84**: 295–311.

Fisher, K. (1998) Genetic sexing strains of mediterranean fruit fly (Diptera: Tephritidae): optimizing high temperature treatment of mass-reared temperature-sensitive lethal strains. *Journal of Economic Entomology* **91**: 1406–1414.

Franks, S.J., Sim, S. and Weis, A.E. (2007) Rapid evolution of flowering time by an annual plant in response to a climate fluctuation. *Proceedings of the National Academy of Sciences* **104**: 1278–1282.

Ge, L.Q., Huang, L.J., Yang, G.Q., *et al.* (2013) Molecular basis for insecticide-enhanced thermotolerance in the brown planthopper *Nilaparvata lugens* Stål (Hemiptera:Delphacidae). *Molecular Ecology* **22**: 5624–5634.

Ghaedi, B. and Andrew, N.R. (2016) The physiological consequences of varied heat exposure events in adult *Myzus persicae*: a single prolonged exposure compared to repeated shorter exposures. *Peer Journal* **4e**: 2290.

Gia, M.H. and Andrew, N.R. (2015) Performance of the cabbage aphid *Brevicoryne brassicae* (Hemiptera: Aphididae) on canola varieties. *General and Applied Entomology* **43**: 1–10.

Gienapp, P., Teplitsky, C., Alho, J.S., Mills, J.A. and Merilä, J. (2008) Climate change and evolution: disentangling environmental and genetic responses. *Molecular Ecology* **17**: 167–178.

Gilman, S.E., Urban, M.C., Tewksbury, J., Gilchrist, G.W. and Holt, R.D. (2010) A framework for community interactions under climate change. *Trends in Ecology and Evolution* **25**: 325–331.

Gregory, P.J., Johnson, S.N., Newton, A.C. and Ingram, J.S.I. (2009) Integrating pests and pathogens into the climate change/food security debate. *Journal of Experimental Biology* **60**: 2827–2838.

Grigaltchik, V.S., Ward, A.J.W. and Seebacher, F. (2012) Thermal acclimation of interactions: differential responses to temperature change alter predator–prey relationship. *Proceedings of the Royal Society B: Biological Sciences* **279**: 4058–4064.

Gutschick, V.P. and BassiriRad, H. (2003) Extreme events as shaping physiology, ecology, and evolution of plants: toward a unified definition and evaluation of their consequences. *New Phytologist* **160**: 21–42.

Hance, T., van Baaren, J., Vernon, P. and Boivin, G. (2007) Impact of extreme temperatures on parasitoids in a climate change perspective. *Annual Review of Entomology* **52**: 107–126.

Harvey, J.A., van Dam, N.M. and Gols, R. (2003) Interactions over four trophic levels: foodplant quality affects development of a hyperparasitoid as mediated through a herbivore and its primary parasitoid. *Joural of Animal Ecology* **72**: 520–531.

Heuskin, S., Verheggen, F.J., Haubruge, E., Wathelet, J.P. and Lognay, G. (2011) The use of semiochemical slow-release devices in integrated pest management strategies. *Biotechnologie, Agronomie, Société et Environnement* **15**: 459–470.

Hill, J.K., Griffiths, H.M. and Thomas, C.D. (2011) Climate change and evolutionary adaptations at species' range margins. *Annual Review of Entomology* **56**: 143–159.

Hoffmann, A.A. (2010) A genetic perspective on insect climate specialists. *Australian Journal of Entomology* **49**: 93–103.

Hoffmann, A.A. and Sgro, C.M. (2011) Climate change and evolutionary adaptation. *Nature* **470**: 479–485.

Hoffmann, A.A. and Willi, Y. (2008) Detecting genetic responses to environmental change. *Nature Reviews Genetics* **9**: 421–432.

Hoffmann, A.A., Sørensen, J.G. and Loeschcke, V. (2003) Adaptation of *Drosophila* to temperature extremes: bringing together quantitative and molecular approaches. *Journal of Thermal Biology* **28**: 175–216.

Hossain, Z., Gurr, G.M., Wratten, S.D. and Raman, A. (2002) Habitat manipulation in lucerne *Medicago sativa*: arthropod population dynamics in harvested and 'refuge' crop strips. *Journal of Applied Ecology* **39**: 445–454.

IPCC (2013) Summary for policymakers. In: Stocker, T.F., Qin, D., Plattner, G.K., *et al.* (eds) *Climate Change 2013: The Physical Science Basis. Contribution of Working Group I to the Fifth Assessment Report of the Intergovernmental Panel on Climate Change*. Cambridge University Press, Cambridge, UK, pp. 1–28.

IPCC (2014) *Climate Change 2014: Synthesis Report. Contribution of Working Groups I, II and III to the Fifth Assessment Report of the Intergovernmental Panel on Climate Change.* IPCC, Geneva, Switzerland.

Irwin, J.T. and Lee, J.R.E. (2003) Cold winter microenvironments conserve energy and improve overwintering survival and potential fecundity of the goldenrod gall fly, *Eurosta solidaginis. Oikos* **100**: 71–78.

Jamieson, M.A., Trowbridge, A.M., Raffa, K.F. and Lindroth, R.L. (2012) Consequences of climate warming and altered precipitation patterns for plant-insect and multitrophic interactions. *Plant Physiology* **160**: 1719–1727.

Jentsch, A., Kreyling, J. and Beierkuhnlein, C. (2007) A new generation of climate-change experiments: events, not trends. *Frontiers in Ecology and the Environment* **5**: 365–374.

Johansson, B.G., Anderbrant, O., Simandl, J., *et al.* (2001) Release rates for pine sawfly pheromones from two types of dispensers and phenology of *Neodiprion sertifer. Journal of Chemical Ecology* **27**: 733–745.

Juroszek, P. and von Tiedemann, A. (2013) Plant pathogens, insect pests and weeds in a changing global climate: a review of approaches, challenges, research gaps, key studies and concepts. *Journal of Agricultural Science* **151**: 163–188.

Kalyebi, A., Sithanantham, S., Overholt, W.A., Hassan, S.A. and Mueke, J.M. (2005) Parasitism, longevity and progeny production of six indigenous Kenyan trichogrammatid egg parasitoids (Hymenoptera: Trichogrammatidae) at different temperature and relative humidity regimes. *Biocontrol Science and Technology* **15**: 255–270.

Kingsolver, J.G. (1989) Weather and the population dynamics of insects: integrating physiological and population ecology. *Physiological Zoology* **62**: 314–334.

Kingsolver, J.G., Arthur Woods, H., Buckley, L.B., Potter, K.A., MacLean, H.J. and Higgins, J. K. (2011) Complex life cycles and the responses of insects to climate change. *Integrative and Comparative Biology* **51**: 719–732.

Knipling, E.F. (1959) Sterile-male method of population control: successful with some insects, the method may also be effective when applied to other noxious animals. *Science* **130**: 902–904.

Krupinsky, J.M., Bailey, K.L., McMullen, M.P., Gossen, B.D. and Turkington, T.K. (2002) Managing plant disease risk in diversified cropping systems. *Agronomy Journal* **94**: 198–209.

Kuussaari, M., Singer, M. and Hanski, I. (2000) Local specialization and landscape-level influence on host use in an herbivorous insect. *Ecology* **81**: 2177–2187.

Lankau, R.A. and Strauss, S.Y. (2011) Newly rare or newly common: evolutionary feedbacks through changes in population density and relative species abundance, and their management implications. *Evolutionary Applications* **4**: 338–353.

Laštůvka, Z. (2009) Climate change and its possible influence on the occurrence and importance of insect pests. *Plant Protection Science* **45**: S53–S62.

Latchininsky, A., Sword, G., Sergeev, M., Cigliano, M.M. and Lecoq, M. (2011) Locusts and grasshoppers: behavior, ecology, and biogeography. *Psyche: a Journal of Entomology* 578327.

Lavergne, S., Mouquet, N., Thuiller, W. and Ronce, O. (2010) Biodiversity and climate change: integrating evolutionary and ecological responses of species and communities. *Annual Review of Ecology, Evolution, and Systematics* **41**: 321–350.

Le Lann, C., Lodi, M. and Ellers, J. (2014) Thermal change alters the outcome of behavioural interactions between antagonistic partners. *Ecological Entomology* **39**: 578–588.

Lin, B.B. (2011) Resilience in agriculture through crop diversification: adaptive management for environmental change. *BioScience* **61**: 183–193.

Liu, S.S., Zhang, G.M. and Zhu, J. (1995) Influence of temperature variations on rate of development in insects: analysis of case studies from entomological literature. *Annals of the Entomological Society of America* **88**: 107–119.

Ma, C.S., Hau, B. and Poehling, H.M. (2004a) Effects of pattern and timing of high temperature exposure on reproduction of the rose grain aphid, *Metopolophium dirhodum*. *Entomologia Experimentalis et Applicata* **110**: 65–71.

Ma, C.S., Hau, B. and Poehling, H.M. (2004b) The effect of heat stress on the survival of the rose grain aphid, *Metopolophium dirhodum* (Hemiptera : Aphididae). *European Journal of Entomology* **101**: 327–331.

Merilä, J. (2012) Evolution in response to climate change: in pursuit of the missing evidence. *BioEssays* **34**: 811–818.

Morris, W.F., Pfister, C.A., Tuljapurkar, S., *et al.* (2008) Longevity can buffer plant and animal populations against changing climatic variability. *Ecology* **89**: 19–25.

Musolin, D.L. (2012) Surviving winter: diapause syndrome in the southern green stink bug *Nezara viridula* in the laboratory, in the field, and under climate change conditions. *Physiological Entomology* **37**: 309–322.

Nash, M.A. and Hoffmann, A.A. (2012) Effective invertebrate pest management in dryland cropping in southern Australia: the challenge of marginality. *Crop Protection* **42**: 289–304.

Nguyen, C., Bahar, M.H., Baker, G. and Andrew, N.R. (2014) Thermal tolerance limits of diamondback moth in ramping and plunging assays. *PLoS ONE* **9(1)**: e87535.

Nooten, S.S. and Andrew, N.R. (2017) Transplant experiments – a powerful method to study climate change impacts. In: Johnson, S. and Jones, H. (eds) *Invertebrates and Global Climate Change*. Wiley-Blackwell, Oxford, UK, pp. 46–67.

Nooten, S.S., Andrew, N.R. and Hughes, L. (2014) Potential impacts of climate change on insect communities: a transplant experiment. *PLoS ONE* **9(1)**: e85987.

Nyamukondiwa, C., Weldon, C.W., Chown, S.L., le Roux, P.C. and Terblanche, J.S. (2013) Thermal biology, population fluctuations and implications of temperature extremes for the management of two globally significant insect pests. *Journal of Insect Physiology* **59**: 1199–1211.

Oerke, E.C. (2006) Crop losses to pests. *Journal of Agricultural Science* **144**: 31–43.

Oliver, I., Dorrough, J., Doherty, H. and Andrew, N.R. (2016) Landscape adaptation for biodiversity conservation in a changing climate: additive and synergistic effects of land cover, land use and climate on insect biodiversity. *Landscape Ecology* **31**: 2415.

Patterson, D.T., Westbrook, J.K., Joyce, R.J.V., Lingren, P.D. and Rogasik, J. (1999) Weeds, insects, and diseases. *Climate Change* **43**: 711–727.

Pell, J., Baverstock, J., Roy, H., Ware, R. and Majerus, M.N. (2008) Intraguild predation involving *Harmonia axyridis*: a review of current knowledge and future perspectives. *BioControl* **53**: 147–168.

Polis, G.A., Myers, C.A. and Holt, R.D. (1989) The ecology and evolution of intraguild predation: potential competitors that eat each other. *Annual Review of Ecology, Evolution, and Systematics* **20**: 297–330.

Prado, S.S., Hung, K.Y., Daugherty, M.P. and Almeida, R.P.P. (2010) Indirect effects of temperature on stink bug fitness, via maintenance of gut-associated symbionts. *Applied Environmental Microbiology* **76**: 1261–1266.

Raso, L., Sint, D., Mayer, R., Plangg, S., *et al.* (2014) Intraguild predation in pioneer predator communities of alpine glacier forelands. *Molecular Ecology* **23**: 3744–3754.
Robinson, A.S. (2002) Genetic sexing strains in medfly, *Ceratitis capitata*, sterile insect technique programmes. *Genetica* **116**: 5–13.
Rosenheim, J.A., Kaya, H.K., Ehler, L.E., Marois, J.J. and Jaffee, B.A. (1995) Intraguild predation among biological-control agents: theory and evidence. *Biological Control* **5**: 303–335.
Scherm, H. (2004) Climate change: can we predict the impacts on plant pathology and pest management? *Canadian Journal of Plant Pathology* **26**: 267–273.
Sears, M.W. and Angilletta, M.J. (2015) Costs and benefits of thermoregulation revisited: both the heterogeneity and spatial structure of temperature drive energetic costs. *American Naturalist* **185**: E94–E102.
Shan, H.W., Lu, Y.H., Bing, X.L., Liu, S.S. and Liu, Y.Q. (2014) Differential responses of the whitefly *Bemisia tabaci* symbionts to unfavorable low and high temperatures. *Microbial Ecology* **68**: 472–482.
Shannon, P., Markiel, A., Ozier, O., *et al.* (2003) Cytoscape: a software environment for integrated models of biomolecular interaction networks. *Genome Research* **13**: 2498–2504.
Sharma, H.C. (2014) Climate change effects on insects: implications for crop protection and food security. *Journal of Crop Improvement* **28**: 229–259.
Shem, P.M., Shiundu, P.M., Gikonyo, N.K., Ali, A.H. and Saini, R.K. (2009) Release kinetics of a synthetic tsetse allomone derived from waterbuck odour from a tygon silicon dispenser under laboratory and semi field conditions. *American-Eurasian Journal of Agricultural and Environmental Science* **6**: 625–636.
Sorribas, J., van Baaren, J. and Garcia-Marí, F. (2012) Effects of climate on the introduction, distribution and biotic potential of parasitoids: applications to biological control of California red scale. *Biological Control* **62**: 103–112.
Stireman, J.O. III, Dyer, L.A., Janzen, D.H., *et al.* (2005) Climatic unpredictability and parasitism of caterpillars: implications of global warming. *Proceedings of the National Academy of Sciences* **102**: 17384–17387.
Strange, R.N. and Scott, P.R. (2005) Plant disease: a threat to global food security. *Annual Review of Phytopathology* **43**: 83–116.
Sutherst, R.W., Constable, F., Finlay, K.J., Harrington, R., Luck, J. and Zalucki, M.P. (2011) Adapting to crop pest and pathogen risks under a changing climate. *Wiley Interdisciplinary Reviews: Climate Change* **2**: 220–237.
Tengö, M. and Belfrage, K. (2004) Local management practices for dealing with change and uncertainty: a cross-scale comparison of cases in Sweden and Tanzania. *Ecology and Society* **9**: 4.
Thomas, M.B., Wratten, S.D. and Sotherton, N.W. (1991) Creation of 'island' habitats in farmland to manipulate populations of beneficial arthropods: Predator densities and emigration. *Journal of Applied Ecology* **28**: 906–917.
Thompson, J., Charpentier, A., Bouguet, G., *et al.* (2013a) Evolution of a genetic polymorphism with climate change in a Mediterranean landscape. *Proceedings of the National Academy of Sciences* **110**: 2893–2897.
Thompson, R.M., Beardall, J., Beringer, J., Grace, M. and Sardina, P. (2013b) Means and extremes: building variability into community-level climate change experiments. *Ecology Letters* **16**: 799–806.

Thomson, L.J., Macfadyen, S. and Hoffmann, A.A. (2010) Predicting the effects of climate change on natural enemies of agricultural pests. *Biological Control* **52**: 296–306.
Tian, H., Stige, L.C., Cazelles, B., *et al.* (2011) Reconstruction of a 1,910-y-long locust series reveals consistent associations with climate fluctuations in China. *Proceedings of the National Academy of Sciences* **108**: 14521–14526.
Torr, S.J., Hall, D.R., Phelps, R.J. and Vale, G.A. (1997) Methods for dispensing odour attractants for tsetse flies (Diptera: Glossinidae). *Bulletin of Entomological Research* **87**: 299–311.
Trumble, J.T. and Butler, C.D. (2009) Climate change will exacerbate California's insect pest problems. *California Agriculture* **63**: 73–78.
Umina, P.A., Weeks, A.R., Kearney, M.R., McKechnie, S.W. and Hoffmann, A.A. (2005) A rapid shift in a classic clinal pattern in *Drosophila* reflecting climate change. *Science* **308**: 691–693.
UNCCD (2008) The 10-year strategic plan and framework to enhance the implementation of the Convention (2008–2018). United Nations Convention to Combat Desertification. Available at: www.unccd.int/Lists/SiteDocumentLibrary/10YearStrategy/Strategy-leaflet-eng.pdf (accessed 7 March 2017).
Urban, M.C., Tewksbury, J.J. and Sheldon, K.S. (2012) On a collision course: competition and dispersal differences create no-analogue communities and cause extinctions during climate change. *Proceedings of the Royal Society B: Biological Sciences* **282**: 1–9.
Urbanski, J., Mogi, M., O'Donnell, D., DeCotiis, M., Toma, T. and Armbruster, P. (2012) Rapid adaptive evolution of photoperiodic response during invasion and range expansion across a climatic gradient. *American Naturalist* **179**: 490–500.
Van Asch, M., Salis, L., Holleman, L.J.M., van Lith, B. and Visser, M.E. (2013) Evolutionary response of the egg hatching date of a herbivorous insect under climate change. *Nature Climate Change* **3**: 244–248.
Van der Kraan, C. and Ebbers, A. (1990) Release rates of tetradecen-1-ol acetates from polymeric formulations in relation to temperature and air velocity. *Journal of Chemical Ecology* **16**: 1041–1058.
Vandermeer, J., van Noordwijk, M., Anderson, J., Ong, C. and Perfecto, I. (1998) Global change and multi-species agroecosystems: concepts and issues. *Agriculture, Ecosystems and Environment* **67**: 1–22.
Visser, M.E. (2008) Keeping up with a warming world; assessing the rate of adaptation to climate change. *Proceedings of the Royal Society B: Biological Sciences*, **275**: 649–659.
Vucic-Pestic, O., Ehnes, R.B., Rall, B.C. and Brose, U. (2011) Warming up the system: higher predator feeding rates but lower energetic efficiencies. *Global Change Biology* **17**: 1301–1310.
Wajnberg, E. and Colazza, S. (2013) *Chemical Ecology of Insect Parasitoids*. Wiley-Blackwell, Oxford, UK.
Wang, Y., Cheng, H., Edwards, R.L., *et al.* (2005) The holocene Asian monsoon: links to solar changes and North Atlantic climate. *Science* **308**: 854–857.
Warren, M.S., Hill, J.K., Thomas, J.A., *et al.* (2001) Rapid responses of British butterflies to opposing forces of climate and habitat change. *Nature* **414**: 65–69.
Wernegreen, J.J. (2012) Mutualism meltdown in insects: bacteria constrain thermal adaptation. *Current Opinion in Microbiology* **15**: 255–262.
Whitman, D.W. (1987) Thermoregulation and daily activity patterns in a black desert grasshopper, *Taeniopoda eques*. *Animal Behaviour* **35**: 1814–1826.

Williams, C.M., Henry, H.A.L. and Sinclair, B.J. (2014) Cold truths: how winter drives responses of terrestrial organisms to climate change. *Biology Review* **90**: 214–235.

Woods, H.A., Dillon, M.E. and Pincebourde, S. (2015) The roles of microclimatic diversity and of behavior in mediating the responses of ectotherms to climate change. *Journal of Thermal Biology* **54**: 86–97.

Yamamura, K., Yokozawa, M., Nishimori, M., Ueda, Y. and Yokosuka, T. (2006) How to analyze long-term insect population dynamics under climate change: 50-year data of three insect pests in paddy fields. *Population Ecology* **48**: 31–48.

Zani, P.A., Cohnstaedt, L.W., Corbin, D., Bradshaw, W.E. and Holzapfel, C.M. (2005) Reproductive value in a complex life cycle: heat tolerance of the pitcher-plant mosquito, *Wyeomyia smithii*. *Journal of Evolutionary Biology* **18**: 101–105.

Zhang, W., Chang, X.Q., Hoffmann, A.A., Zhang, S. and Ma, C.S. (2015) Impact of hot events at different developmental stages of a moth: the closer to adult stage, the less reproductive output. *Scientific Reports* **5**: 10436.

Zhu, Y., Chen, H., Fan, J., *et al.* (2000) Genetic diversity and disease control in rice. *Nature* **406**: 718–722.

Ziska, L.H. (2014) Increasing minimum daily temperatures are associated with enhanced pesticide use in cultivated soybean along a latitudinal gradient in the mid-western United States. *PLoS ONE* **9(6)**: e98516.

10

Effects of Biological Invasions on Pest Management

George K. Roderick and Maria Navajas

10.1 Invasion Science

Invasive alien species are typically defined as non-indigenous species 'whose introduction and/or spread threatens biological diversity' (Convention on Biological Diversity 2016) or causes other environmental (Wilcove *et al.* 1998) or economic impact (Pimentel *et al.* 2000, 2005). Researchers debate as to whether harmful impact is important for the definition of invasive alien species. For example, should the term include all non-indigenous widespread species, or only those non-indigenous species exhibiting continued range expansion (Sax *et al.* 2007; Valéry *et al.* 2008)? Researchers do agree on two features that characterize invasive species: (1) their geographical range is expanding and (2) they now play an important ecological role in both managed and natural ecological communities (Colautti and Lau 2015; Davis *et al.* 2011). Because of the unique processes associated with the spread and impact of invasive species and their widespread ecological significance, the study of invasive species has been considered as its own subdiscipline of ecology and pest management (Elton 1958; Hill *et al.* 2016; Simberloff *et al.* 2013).

Biological invasion is a process with several identifiable key stages (Figure 10.1) (Blackburn *et al.* 2011; Facon *et al.* 2006; Kolar and Lodge 2001; Lambrinos 2004; Sakai *et al.* 2001). In the invasion process, organisms are introduced through human activity (anthropogenic) or arrive unaided, and a fraction of those species become established. Of the established species, a smaller proportion spreads, and of those species, only some cause some type of environmental or economic impact.

Often, but not always, invasive populations may take some time to reach a population size large enough so that it is noticed or causes a significant environmental or economic problem. This time period between introduction and crisis is called the 'lag phase'. At each stage in this process, ecological and evolutionary forces operate as filters, resulting in only a small proportion of species surviving to the next stage. This phenomenon has been termed the '10's rule', as an arbitrarily small proportion (i.e. 10%) of species make it to the next stage (Williamson 1996). Community ecological interactions, such as competition, predation and mutualism, are all important in the success of invading species and, depending on the interaction, these effects can have a positive or negative effect on invasion success. In addition, evolutionary forces that change the frequencies of genes within populations, such as migration, genetic drift, mutation and selection,

Environmental Pest Management: Challenges for Agronomists, Ecologists, Economists and Policymakers, First Edition. Edited by Moshe Coll and Eric Wajnberg.

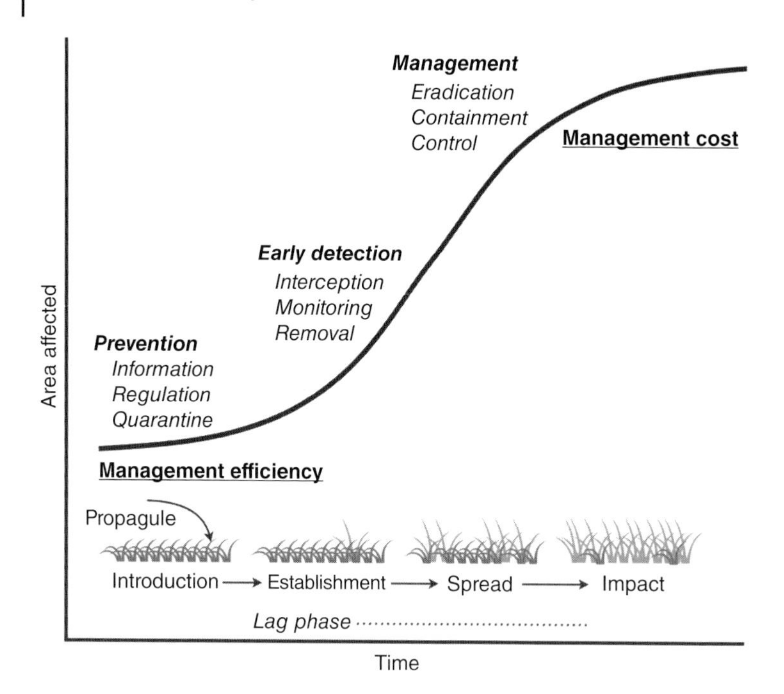

Figure 10.1 The invasion process and associated management strategies at each stage (after Agriculture Victoria 2015; Harvey and Mazzotti 2014; Kolar and Lodge 2001; Sakai *et al.* 2001; Simberloff *et al.* 2013). Management efficiency transitions to greater cost with time and area affected as management options become limited.

also play a significant role in the invasion process, but are less well understood than ecological effects (Hill *et al.* 2016; Roderick *et al.* 2012).

Knowledge of how invasive species evolve in new environments and what limits their capacity to adapt to new environments can provide important insights into real-world challenges, such as predicting whether organisms introduced for biological control will switch to non-target hosts, how invasive species will respond to management strategies, such as pesticide use leading to resistance, and in what way species will respond to global changes in climate and land/water (Hill *et al.* 2016) (see Chapter 9).

10.1.1 Introduction Stage

Efforts to categorize biological invasions have enabled a better understanding of the biology of invasive species as well as a means to better predict their spread and impact. Spread of species to a new geographic location requires dispersal. Some organisms disperse in association with human activities, some rely on other organisms as vectors, and some disperse unaided on their own (Falk-Petersen *et al.* 2006). Hulme *et al.* (2008) considered two mechanisms of dispersal associated with human activities: (1) purposeful importation of a commodity and (2) arrival of a vector involved in transportation, and a third mechanism of (3) natural dispersal from another region (Table 10.1). Each of these mechanisms is associated with one or more pathways through which dispersal

Table 10.1 Mechanisms and pathways of dispersal proposed by Hulme *et al.* (2008). Mechanisms involving commodities and transportation are associated with human activities.

Mechanisms	Associated Pathways
Purposeful introduction of a commodity	Release, escape or as contaminants
Vector involved in transportation	Stowaways
Natural dispersal	Dispersal corridors or unguided

occurs. Knowledge of both dispersal mechanisms and pathways is necessary for understanding interacting biological and socioeconomic factors responsible for colonization, spread and impact, as well as for management (Liebhold *et al.* 2016).

10.1.1.1 Commodities: Release, Escape, Contaminants

Many researchers (Howarth 1996; Hulme *et al.* 2008; Kiritani and Yamamura 2003; Sax *et al.* 2005; Yano *et al.* 1999) have documented the importance of movement of commodities (i.e. articles of trade) in facilitating biological invasions. At least three invasion pathways are associated with commodities: release, escape, or as contaminants. Many beneficial species are released purposefully as commodities, and these species may become invasive. Organisms released for biological control are one example, and while current programmes of biological control are increasingly tightly regulated, including pre-release quarantine and host range testing, some, particularly early, introductions of biological control organisms have resulted in severe negative impacts (Roderick and Howarth 1999; Simberloff and Stiling 1996; Snyder *et al.* 2004) (see also Chapter 5).

For such species, adaptation to novel conditions or unrecognized phenotypic plasticity, particularly associated with novel hosts/prey or new physical environments, is often difficult to predict through pre-release testing (Hajek *et al.* 2016; Roderick *et al.* 2012). For example, following use in glasshouses for biological control, the Asian harlequin ladybird, *Harmonia axyridis*, became invasive in Britain and elsewhere (Lombaert *et al.* 2010; Majerus *et al.* 2006; Roy and Wajnberg 2008). Another example of an invasion associated with introduction of a commodity is the release and spread of the excitable Africanized honeybee, *Apis mellifera scutellata*, in the Americas (Hall and Muralidharan 1989). Other live commodities may escape their intended range or confines, such as purposefully introduced horticultural plants that escape as weeds outside garden boundaries. Because of the prevalence and impact of escaped ornamental plants, accessible 'green lists' of non-invasive ornamentals have been constructed to aid horticulturalists (Dehnen-Schmutz 2011).

Finally, many invasive alien species are spread as contaminants of commodities, and move through global trade or other human transport (Hulme *et al.* 2008). Noted examples include *Bemisia* whiteflies that have been transported worldwide on ornamental poinsettia and other plant species (Hadjistylli *et al.* 2010, 2016; Perring *et al.* 1993) and the glassy-winged sharpshooter, *Homalodisca vitripennis* (see case studies, below), that spread internationally through movement of citrus and vine hosts (Petit *et al.* 2008). Contaminants of food products include the recent invasions of Asian citrus psyllid,

Diaphorina citri (Hall *et al.* 2013), numerous species of fruit flies, *Bactrocera* spp. (Clarke *et al.*, 2005), the red tomato spider mite, *Tetranychus evansi* (Boubou *et al.* 2012), and widow spiders that move around the world associated with produce, such as bananas. There are also many examples of microbes and fungi associated with movement of plants (Desprez-Loustau *et al.* 2007). Movement of wood products, including wooden shipping pallets and lumber, is thought to have spread many wood-associated species, including the Asian gypsy moth, *Lymantria dispar asiatica*, pine beetles, *Dendroctonus* spp., the Formosan subterranean termite, *Coptotermes formosanus*, Asian long-horned beetle, *Anoplophora glabripennis*, and the hemlock woolly adelgid, *Adelges tsugae*, to name only a few. Insect, plant and microbe contaminants are also common in grain supplies, seeds, animal feed, stored food products and soil (Hulme *et al.* 2008). With an increase in global trade, controlling the spread of contaminants of commodities will continue to be a worldwide challenge.

10.1.1.2 Vectors: Stowaways

Many invasive alien species of plants and animals have been transported to new geographical areas as stowaways associated with some vehicle or animal vector (Liebhold *et al.* 2016). For example, the Polynesian tiger mosquito, *Aedes polynesiensis*, is thought to have stowed away in water containers transported by ancient Polynesians voyaging across the Pacific, a pathway also exploited more recently by the Asian tiger mosquito, *Aedes albopictus*, that spread in water that collects inside discarded automobile tyres (Benedict *et al.* 2007). The glassy-winged sharpshooter, *Homalodisca vitripennis*, and many other invasive alien species have been observed in cargo holds of airplanes (Liebhold *et al.* 2006). Bedbugs, *Cimex lectularius*, move with human belongings (Saenz *et al.* 2012), presumably in luggage. Flightless gypsy moths, *Lymantria dispar dispar*, have been spread through transport on cars and trucks in North America (Johnson *et al.* 2006), a pathway that has also been proposed for spread of the horse-chestnut leafminer, *Cameraria ohridella*, in Europe (Gilbert *et al.* 2005).

In addition to vehicles as vectors, many organisms are transported to new habitats in association with other organisms. For example, the distribution of ticks carrying Lyme disease is associated with vertebrate hosts (Ostfeld *et al.* 2006; Swei *et al.* 2011). Other species are vectored by birds, sometimes over great distances, either as external stowaways or in bird guts inside seeds (Gillespie *et al.* 2012). Finally, humans themselves transport their own domesticated parasites and symbionts, especially lice and mites.

10.1.1.3 Dispersal: Corridors or Unguided

Many organisms disperse naturally on their own, either along dispersal corridors or unguided (Hulme *et al.* 2008). Rivers, streams and other waterways provide dispersal corridors that channel invasions of freshwater aquatic species. In terrestrial systems, dispersal corridors include disturbed roadsides, railways and walking trails. For example, the invasive Argentine ant, *Linepithema humile*, moves along roads and trails into native forest in Hawaii where it is a serious ecological pest (Krushelnycky and Gillespie 2008). Other invasive alien species spread easily in the absence of corridors from one area to another. As one might expect, this pathway is common for species with great aerial dispersal ability, such as the rice brown planthopper, *Nilaparvata lugens*, which is known to move seasonally between tropical and temperate regions in South-East Asia (Denno and Roderick 1990; Mun *et al.* 1999; Zhu *et al.* 2000). Spiders ballooning on threads of silk are another example of natural dispersal, accounting for the fact that spiders are predictably among the first to arrive in newly opened habitats, such as rice fields (Gillespie *et al.* 2012; Matteson *et al.* 1994; Way and Heong 1994) and on volcanic islands (Gillespie *et al.* 2012). Less naturally dispersive species can also move surprisingly efficiently, locally and regionally. Examples include the Colorado potato beetle (Grapputo *et al.* 2005), various ladybird coccinellid beetles (Lombaert *et al.* 2010; Majerus *et al.* 2006; Roy and Wajnberg 2008; Snyder *et al.* 2004) and many species of ants (Holway *et al.* 2002).

10.1.2 Establishment Stage

10.1.2.1 Propagule Pressure and Founding Population Size

Of critical importance for the success of colonizers is how many individuals colonize, but also when and where (Simberloff 2009). A measure of the magnitude of dispersal from a source area to a new habitat over time is termed propagule pressure and is defined as 'the number of individuals released into a region to which they are not native' (Lockwood *et al.* 2005). As noted above, propagules can disperse in many ways, often reflecting the nature of available habitat but also the mechanisms and pathways of introduction. Propagules can include one or more individuals that colonize at one time, or several or many colonization events in one place over time or at many places. Indeed, some pathways of invasion are conducive to large numbers of propagules, such as introductions associated with commodities or cultivation (Liebhold *et al.* 2016).

Of particular interest for management (both prevention and control) is whether the probability of establishment continues to increase with increasing propagule numbers or rather, whether likelihood of establishment levels off, such that continued numbers of propagules do not contribute to a higher likelihood of establishment (Figure 10.2). Understanding this relationship is also important in

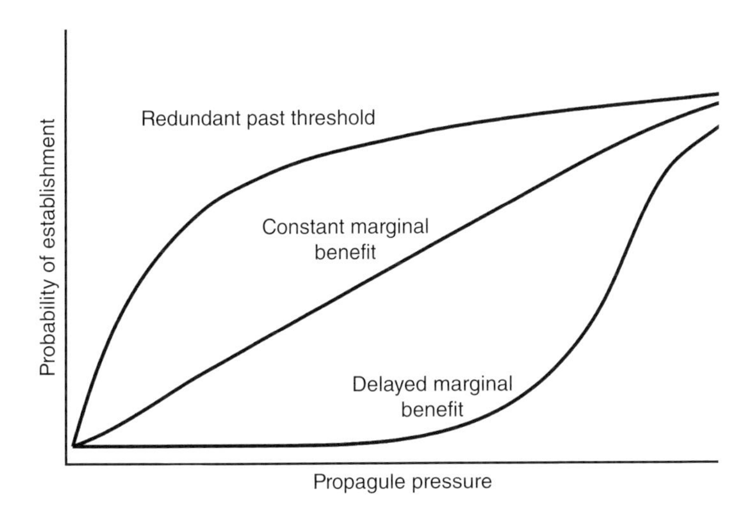

Figure 10.2 Effect of propagule pressure on probability of establishment. A constant marginal benefit response (*middle curve*) suggests that the probability of establishment will continue to increase with additional propagules. A delayed marginal benefit (*lower curve*) might result from Allee effects or other factors, in which the probability of establishment increases with propagule numbers only greater than a particular population size (see text). The redundant past threshold model shows that after some level of propagule pressure, the probability of establishment no longer increases at the same rate. Source: Adapted from Lockwood *et al.* (2005).

predicting the success of purposeful introductions, such as for biological control. For example, how many individuals are necessary to ensure a successful introduction and will additional releases continue to increase the probability of successful establishment?

The likelihood that a colonizing population will survive and establish depends in part on the size of the population, for both ecological and genetic reasons (Simberloff 2009). Small populations will be more likely to go extinct as a result of purely demographic stochasticity or random fluctuations in population size (Lande 1988). In addition, small populations may not have the numbers to achieve maximum growth rates, for example if the likelihood of finding mates is low, a phenomenon known as the Allee effect (Fauvergue *et al.* 2012). Environmental conditions also vary (i.e. environmental stochasticity) and even larger populations can be affected by severe weather events and natural disturbances. As a rule of thumb, researchers consider effective sizes of populations affected by demographic stochasticity to be in the order of 10^1–10^2, with the risk decreasing with increasing population size while sizes affected by environmental stochasticity can be orders of magnitude larger.

Establishment is also affected by genetic variation. Carlquist (1966) noted that 'difficulties of establishment seem much greater than those of transport. To establish, the number of founding individuals and their genetic makeup becomes important'. It is well known that small numbers of founders carry only a subset of the genetic variation in the source population (examples include tephritid fruit flies, mites, mosquitoes, and ants) (Davies *et al.* 1999a; Fonseca *et al.* 2000; Holway *et al.* 2002; Navajas and Boursot 2003; Navajas *et al.* 2009; Roderick and Navajas 2003), and such founder events may be associated with inbreeding depression or limited ability for adaptation. That small invasive populations can be successful despite low genetic diversity is a paradox, and not predicted by theory (Bock *et al.* 2015; Sax *et al.* 2007; Simberloff 2009). For example, the invasive Japanese knotweed, *Fallopia japonica*, in Britain appears to be a single genetic clone. Another example of success with reduced genetic variation is seen in the parasitic *Varroa* mite, which jumped to feed on the European honeybee, *Apis mellifera*, from the eastern honeybee, *A. cerana*, when the former was introduced into Asia. The parasite has spread worldwide in 30 years. Surprisingly, only two of the 18 original Asian haplotypes switched to *A. mellifera*, and only one of these two has been successful in colonizing new geographical regions (Navajas *et al.* 2009; Solignac *et al.* 2005).

There are several solutions to explain this paradox. First, in many invasive species, invading populations originate from populations that were invasive elsewhere, in a pattern of serial invasions, also known as a bridgehead effect. An example of this phenomenon is the medfly, *Ceratitis capitata*, which originated in sub-Saharan Africa and spread to the Mediterranean, and then worldwide (Davies *et al.* 1999a). Another example is the spread of the glassy-winged sharpshooter from one island to the next within and between island archipelagoes in French Polynesia (Petit *et al.* 2008). Such invasive species may have overcome limitations of small populations and low genetic diversity through selection in previous colonization episodes. Indeed, one predictor of which plants are likely to be invasive weeds is whether the species has been observed to be invasive elsewhere (Bock *et al.* 2015; Lonsdale 1999).

What studies of such species are not able to observe are all the introductions that failed to establish, which is no doubt a considerably high number (Carlquist 1966). Thus, similar to a 'non-reporting bias' in statistical sampling, we record as invasive only those species that are able to establish for whatever reasons. Second, many invading populations have been shown be the result of not one but multiple cryptic invasions (Bock *et al.* 2015). For example, molecular genetic studies of the medfly, *Ceratitis capitata*, the olive fly, *Bactrocera oleae*, the oriental fruit fly, *Bactrocera dorsalis*, and other tephritid fruitflies show that multiple, often cryptic invasions are common (Clarke *et al.* 2005; Davies *et al.* 1999a; Malacrida *et al.* 2007; Nardi *et al.* 2005, 2010). Similar results have been found in *Bemesia* whiteflies (Hadjistylli *et al.* 2010, 2016), the glassy-winged sharpshooter, *Homalodisca vitripennis* (Petit *et al.* 2008) and the Colorado potato beetle, *Leptinotarsa decemlineata* (Grapputo *et al.* 2005), as well as other terrestrial arthropods, particularly mites (Boubou *et al.* 2012). Mixing or hybridization associated with multiple colonization events may increase the genetic variation in colonizing populations, which, in theory, should contribute to the ability of invasive populations to adapt to novel conditions.

Finally, when colonists come from different geographic locations, the genetic pool in the invasive population can be sizeable. A now classic example is that of multiple colonizations of a Caribbean lizard, *Anolis sagrei*, from different locations in the Caribbean into urban south Florida (Kolbe *et al.* 2004). When *A. sagrei* individuals from these different sources mixed, the genetic diversity in the resulting invasive population was highly significantly greater than any of the sources individually. Such admixture of individuals from multiple populations may be far more important than realized in contributing to genetic variation in invasive species, and may allow invasive populations to have high levels of genetic variation very early in the invasion process. In a similar way, individuals of invasive species can mix genetically, or hybridize, with individuals of indigenous species or other invaders. Examples of invasive species exhibiting hybridization include *Bemisia* whiteflies, *Rhagoletis* flies, *Anas* dabbling ducks and *Spartina* plants, among many others (Vellend *et al.* 2007). Taken together, both theory and observations suggest that the factors limiting success of small invasive populations will be ecological and not genetic (Lande 1988; Simberloff 2009).

Another key to establishment is the extent to which invasive alien species are already adapted to the novel habitat. Such preadaptation can be a result of the new environment closely matching the invasion source, for example in climate variables or similarities in the biotic composition, including symbionts, competitors or predators. Examples include Mediterranean species that invade areas of similar climates worldwide (Barrett 2015). In managed systems, environmental similarities of native and invaded areas may include aspects of pest management, including similar pesticide use. Alternatively, conditions in the source environment may have preadapted the organism for high fitness in a new different set of environmental conditions found in the novel habitat, often in unpredictable ways. Another form of preadaptation is phenotypic plasticity, in which species can naturally cope with a diverse or changing set of natural conditions, as seen for at least some traits of the soapberry bug, when presented with an introduced plant (Carroll and Boyd 1992). Phenotypic plasticity also provides the opportunity for an expression of diverse phenotypes upon which selection can act (Lande 1988; Migeon *et al.* 2015).

10.1.2.2 Species Interactions

Species interactions, or lack thereof, are critical in invasions, especially escape from competitors and enemies, at least initially (Torchin and Mitchell 2004; Torchin *et al.* 2003) or in species-poor communities, such as following a disturbance, with few competitors and predators (Williamson 1996). Alternatively, the presence of some interacting species can facilitate range expansion, while their absence can limit spread, as for example the presence of plants that fix nitrogen (Vitousek *et al.* 1996). Likewise, the presence of a particular plant may allow the invasion of specialized herbivores (Hurley *et al.* 2016); the Colorado potato beetle is thought to have moved from native solanaceous species to potatoes with the arrival of European settlers in the American West (Grapputo *et al.* 2005). A pollinator may facilitate invasion of its plant symbiont, and vice versa (Leong *et al.* 2014; Olesen *et al.* 2002). Mutualistic interactions, particularly involving microbes and fungi, are little appreciated in the success of invasive species (Simberloff 2006), but are becoming increasingly recognized as important. In sap-feeding insects, for example, microbe symbionts provide essential amino acids necessary to adapt to new host species (McFall-Ngai *et al.* 2013).

10.1.3 Spread Stage

Once established, invasive species typically expand their range geographically. It has been recognized since Darwin that certain ecological features are critical in species range expansion, including ecological attributes of either the invasive species themselves or the invaded habitats, and/or their interaction (Elton 1958; Sax *et al.* 2007). Once established, individuals may spread to new areas of the same type of environment or to different types of habitats, where there may be selection as a result of the new environmental conditions. In the absence of connectivity through regular dispersal, populations in different areas may become genetically differentiated, either through genetic drift or through local adaptation (Vellend *et al.* 2007). Genetic variation may again be increased when individuals from such populations hybridize upon later contact. By contrast, if founding populations are small and deleterious mutations accumulate in expanding populations, expansion success could be limited. This little-studied phenomenon is termed 'expansion load' (Peischl *et al.* 2015). Several other related evolutionary questions concerning geographical spread remain unanswered, including the relative importance of adaptation at this phase compared to preadaptation, the importance of local adaptation in farther range expansion, and the extent to which the population structure of invasive populations resembles that of species with a longer history in the environment (Barrett 2015; Gillespie *et al.* 2008).

When populations are expanding geographically, they are necessarily also expanding in numbers. Often, the time from introduction to the time of a significant population size can be long, for example tens or even hundreds of generations, resulting in the lag-phase noted above (see Figure 10.1). Examples include invasions of the gypsy moth, *Lymantria dispar dispar* (Johnson *et al.* 2006) and the light brown apple moth, *Epiphyas postvittana* (Suckling and Brockerhoff 2010), among others. There are both ecological and evolutionary explanations for this lag phase. Increased population growth of invasive species can result from invasive populations responding sufficiently in numbers to new biotic pressures including competitors and predators, or in overcoming Allee effects at small population size as noted above. Large-scale monitoring programmes measuring the same characteristics across a wide invasive range, such as implemented for the garlic mustard, *Alliaria petiola* (Colautti *et al.* 2014), offer a comparative framework for understanding the role of novel environments. It may also be that additive genetic variation has increased since establishment through a variety of possible mechanisms, including mutation, genome rearrangements and horizontal transfer of novel genetic elements. With available genetic variation, selection can be critically important at the phase of geographical spread; selection can act to favour those individuals that are able to make it to new environments and/or reach such environments before other individuals. For example, extensive studies of the cane toad, *Rhinella marina*, show selection for morphology and life history traits that favour dispersal, especially at the invasion front (Phillips and Shine 2006; Shine 2010). The process of geographic spread itself creates conditions that may accelerate evolution, through genetic drift, selection or genetic rearrangements (Kirkpatrick and Barrett 2015). New genomic tools offer unprecedented opportunities to understand the relative importance of each genetic mechanism in the ability of invasive alien species to expand geographically.

10.1.4 Impact Stage

Eventually, invasive populations build in numbers and geographic area to have a noticeable impact on ecological communities or on human activities, with possible economic consequences. Most management activities are implemented during this phase, assuming action is still possible (Simberloff *et al.* 2013). During this phase, predators and other natural enemies may respond numerically in response to abundant food sources provided by the invasive alien species, often with profound consequences for ecological communities (Davis *et al.* 2011). Also, selection can continue to act on life history variation, including dispersal, in response either to biological or environmental conditions in the novel habitat, or anthropogenic conditions, such as pest control. For example, resistance to pesticides, including antibiotics, is well known throughout the diversity of life, including microbes, plants and animals, allowing many organisms to spread uncontrolled.

10.2 Invasions – A Natural Process?

Dispersal is an important part of the life cycle of organisms and many species are adapted to colonizing new habitats. Some species are extremely well adapted for colonization, such as those that dominate newly disturbed habitats or communities that characterize early ecological succession. Given time, even remote habitats will be colonized. For example, the remote archipelago of Hawaii has been colonized naturally numerous times. For terrestrial arthropods alone, at least 400 or so colonization events account for the native arthropod diversity of 10 000 or so species (Howarth and Mull 1992). For these reasons, if dispersal and colonization is a natural process and will continue, if not accelerate, why should governments and management agencies bother to control invasive species?

It is true that large-scale biotic exchange has occurred frequently in the past, such as when continents came together for the first time. However, while colonizations and large-scale biotic exchange have been ongoing in geological time, recent human activities have accelerated the rate of exchange. Not only is species turnover more rapid (Burns 2015), but also both biological communities and species richness have become more homogenized among regions (Sax and Gaines 2006; van Kleunen *et al.* 2015). Second, studies of species diversity in invaded habitats show that the number of indigenous species and non-indigenous species seems to be correlated, suggesting that biological communities are not saturated (Sax and Gaines 2006).

All environments are not experiencing the same influx of invasive species. For example, through both natural and anthropogenic means, more species are invading low temperate latitudes, although with smaller geographical ranges. By contrast, fewer species are invading at higher latitudes, though with larger geographic ranges. Islands appear to be disproportionally affected (Sax and Gaines 2008).

Finally, the spread and impact of biological invasions depend on the species involved and their interactions. For example, climate and dispersal limitation explain only a portion of the distribution of invasive species, suggesting a key role of species interactions (Capinha *et al.* 2015). From these observations, one can conclude that, although dispersal is a natural process, globalization has now accelerated biological homogenization of ecological communities. Importantly, these effects will impact different regions differently, suggesting that targeting invasive species in particular regions or establishing quarantines – for example, focusing attention on managed habitats at low temperate latitudes and islands – will be most productive.

10.3 Perception and Value of Introduced and Invasive Alien Species

Invasive alien species have a bad name in conservation biology and as pests of managed systems, but does excluding non-indigenous species make for effective policy? In a thought-provoking article, Davis *et al.* (2011) argued that unreasonable attention has been given to whether a species is native or not, rather than to understanding its role in the ecological community. These authors argue that ecosystems are quickly changing as a result of important ecological 'drivers, such as climate change, nitrogen eutrophication, increased urbanization, and other land use changes.' Thus, management of species needs to focus on the problems and not on whether a species is indigenous or not. Many examples exist of large investments in time and funding devoted to control of invasive alien species, only to have invasive species persist. One example is the devil's claw plant, *Martynia annua*, which invaded Australia from Mexico and was removed from the Gregory National Park, only to persist in nearby cattle stations. Davis *et al.* (2011) also point to numerous invasive alien species that now provide ecosystem services. They conclude, 'We urge conservationists and land managers to organize priorities around whether species are producing benefits or harm to biodiversity, human health, ecological services and economies.'

Of course, the modern perspective outlined by Davis *et al.* (2011) is a reflection of the fact that human culture has always been associated with the manipulation of the distribution of other species and their traits. The purposes of such introductions have included sources of food and fiber, but also for value of biological control, ecological restoration and ecosystem services, such as pollination (see Chapter 7). In modern times, we have the advantage of assessing both the benefits for society as well as potential risks, including potential negative impacts on the environment and human health.

10.4 When to Act, and Why?

Recognizing that biological invasions are a process and that the number of invasive alien species is likely to continue to grow, one can ask at what point in the process is the best time to control invasive alien species. Economic studies of the impact of invasive species and cost of management are not only accumulating but are now influencing decisional policy. For example, the Convention on Biological Diversity (2016) emphasizes the need to monitor pathways of invasion, leading to prevention. A recent analysis by Simberloff *et al.* (2013) notes that the optimal strategy for dealing with invasive species 'evolves with time since introduction, with management efficiency decreasing and management costs increasing with time since introduction'. Thus, management efforts will be rewarded early in the invasion process, and particularly if introductions are prevented (Convention on Biological Diversity 2016). Economic studies also suggest that eradication, though expensive, can be much less expensive than long-term management. Thus, the initial panic that often accompanies the outbreak of an invasive species may be entirely appropriate, as a call for eradication rather than facing a long-term prospect of continued management. Numerous eradication programmes have been organized against insect pests from several orders and their frequency is steadily increasing, mainly due to increased movement of pests favoured by trade and travel (on containment of invasive species, see Bloem *et al.* 2014).

10.5 How Best to Control Invasive Species?

The use of highly destructive means, including pesticides, herbicides and vegetation cutting, has been advocated to suppress populations of invasive alien species (Bloem *et al.* 2014). In doing so, those advocating such approaches are scaling the cost of action and its impact with the eventual costs of not acting. In some cases, the relative impacts of control versus not acting are obvious. In others, it is less obvious. In biological control, for example, economic analyses suggest that classic biological control is a highly efficient strategy of pest control (Lodge 1993; Simberloff *et al.* 2013). However, many ecological consequences remain difficult to predict (Jennings *et al.*) (see Chapter 5).

Several new genetic approaches have become available that potentially can lead to the complete elimination of a pest species, including manipulation of genetic material coupled with molecular mechanisms of gene drive that will push the modification through the entire population (Sheppard *et al.* 2004). Such efforts have focused on the possible elimination of mosquito species (Sinkins and Gould 2006). One question that is quick to arise is, if an entire species is removed, what will be the ecological impact? Already, researchers are manipulating strains of symbiotic *Wolbachia* in mosquitoes in the laboratory and field (Fang 2010), both to reduce the likelihood that they will transmit diseases like Zika, dengue and malaria, but also to reduce numbers of mosquitoes in sterile insect technique-type programmes. Such efforts do provide an opportunity to study effects on food webs under controlled conditions. Whether the public will allow such research to proceed is not certain. Currently, introductions for biological control purposes are allowed in many countries, although with sharp regulatory differences. Genetically modified organisms (GMOs) are some of the world's most controversial technologies (McMeniman *et al.* 2009; Walker *et al.* 2011; Xi *et al.* 2005) (see Chapter 12), with their use banned over large regions (e.g. most of the EU countries).

The best course of action will also depend on the type of habitat, not only weighing benefits and risks associated with environmental concerns and human health but also recognizing the nature of the ecological communities. For example, natural habitats potentially have a wealth of natural enemies that can and do attack invasive alien species (Torchin and Mitchell 2004). That natural enemies may respond numerically and functionally to the presence of new resources in the form of invasive alien species may explain why some invasive alien species eventually decline in numbers without aggressive management. One example is the two-spotted leafhopper, *Sophonia rufofascia*, described from China but first noticed in the Hawaiian Islands in 1987. After several years of concern for both agricultural crops and rare plant in natural environments, the leafhopper declined in numbers before a biological control programme was started. The insect is now difficult to find in Hawaii. The option of doing nothing may make more sense environmentally and economically when the ecological food webs are more complex, as found in natural compared to highly managed systems.

10.6 Case Studies

10.6.1 The Glassy-winged Sharpshooter, *Homalodisca vitripennis*

The glassy-winged sharpshooter, *Homalodisca vitripennis* (Germar) (= *H. coagulata* [Say]) (Hemiptera: Cicadellidae), is a xylem-feeding leafhopper and the vector for the *Xylella fastidiosa* bacterium. In California, USA, this bacterium causes Pierce's disease in grapes and citrus and a different strain of this bacterium has recently spread in olives in Europe. The glassy-winged sharpshooter is native to south-eastern North America, from north-east Mexico into the south-eastern USA. In the late 1980s, it spread to California, where it vectors the *Xylella* bacterium (National Academies of Sciences 2016). Though *Xylella* was in California prior to the glass-winged sharpshooter invasion, the existing leafhoppers were not as effective as a vector for many reasons, including the relatively large size of the glassy-winged sharpshooter, which in the process of feeding inserts the bacterium farther into the host plants. As the sharpshooter extended its range, the bacterium caused extensive damage to both the citrus and grape industries, including the well-known wine area of Napa Valley, California.

Control of the glassy-winged sharpshooter in California was achieved through cultural means but also insecticides, though the bacterium persists. A biocontrol programme was also initiated, though this was not successful because the seasonality of the sharpshooter in California disrupted the tight associations between populations of hosts and parasitoids necessary for effective biological control.

In 1999, the glassy-winged sharpshooter invaded Tahiti, French Polynesia, probably associated with introduction of plant material from California, perhaps citrus, and rapidly spread on Tahiti and neighbouring islands (Purcell and Saunders 1999). Numbers increased rapidly, and although the bacterium associated with Pierce's disease was not introduced, the species became a public nuisance and the possibility of Pierce's disease meant a constant threat for crops and native plants, but also foreign trade. Following an extensive ecological study to determine potential non-target effects (Petit *et al.* 2008), a biological control programme was initiated involving the host-specific egg parasitoid, *Gonatocerus ashmeadi* Girault (Hymenoptera: Mymaridae). Following arrival of the parasitoid, the glassy-winged sharpshooter populations were reduced by over 95%, and remain low, though seasonal fluctuations persist (Grandgirard *et al.* 2007). Importantly, although the parasitoid was not introduced to islands other than Tahiti and neighbouring Moorea, it spread to other islands in the Society archipelago and other French Polynesian archipelagos, probably associated with unmonitored movement of plants (Grandgirard *et al.* 2008, 2009).

Recently, a different strain of *Xylella* has caused damage to olives in Europe, especially Italy. It appears that this strain was introduced from Central America and is not associated with the glassy-winged sharpshooter.

The invasion of the glassy-winged sharpshooter illustrates the potential importance of symbiotic relationships. In California, the invasion of this pest facilitated the spread of a damaging bacterium. In Tahiti, the absence of the bacterium meant that the invasion of the glassy-winged sharpshooter resulted in social and recreational problems, rather than agricultural loss. That the introduced parasitoid was able to spread to widely dispersed archipelagos demonstrated the lack of biosecurity in the region.

10.6.2 The Red Tomato Spider Mite, *Tetranychus evansi*

The red tomato spider mite, *Tetranychus evansi*, is a pest of solanaceous crops, particularly tomatoes and eggplant. The mite probably originated in South America where it was not considered a pest, but in the past 15 years has emerged as a new destructive pest present in many countries, i.e. in sub-Saharan Africa and around the Mediterranean basin, as well as in several parts of South-East Asia (Petit *et al.* 2009). Its distribution area is expected to expand under climate change scenarios according to modelling studies (Meynard *et al.* 2013). It is probably transported with vegetables or other plant materials. While biocontrol by predatory mites is one option for management, predators in the newly introduced areas are unable to control the invasive mite (Navajas *et al.* 2013).

Recent molecular genetic work coupled with tests of alternative invasion scenarios show that the invasion history of this mite is complex. For example, populations in Europe and Africa resulted from at least three independent introductions from South America and involved mites from two distinct sources in Brazil (Boubou *et al.* 2012). The study also provided evidence for the 'bridgehead' effect in which one invasive population gave rise to others. Furthermore, invasive populations were demonstrated to be the result of multiple colonization events, with evidence of subsequent admixture.

The invasion of the red tomato spider mite illustrates many characteristics of invasive species, including serial colonization history, where one invasive population is the source of others, each with limited genetic diversity, but that multiple cryptic colonization events can result in invasive populations of increased genetic diversity, often associated with admixture. Studies of this species show that genetic tools and newly developed computational methods can resolve complex invasion pathways that historical collections failed to detect.

10.6.3 The Yellow Fever Mosquito, *Aedes aegypti*

The yellow fever mosquito, *Aedes aegypti* (L.), is a vector for arboviruses including yellow fever virus, dengue virus, chikungunya virus and Zika virus (ECDC 2016).

A. aegypti has its origins in Africa (ECDC 2016). Two forms of the species exist: *A. aegypti aegypti*, which is noted for breeding in man-made habitats and is associated with dengue epidemics worldwide, and *A. a. formosus* which is less associated with human activities and found in natural habitats, though it transmits dengue locally. *A a. aegypti* is thought to have spread repeatedly to the New World from West Africa with the African slave trade starting in the 15th century. The species spread to Asia multiple times in the 18th and 19th centuries associated with trade and then worldwide following World War II. In South America, many control programmes were initiated in the early 1900s in response to continued colonization. Though largely controlled in many areas, including Brazil, by the 1970s, *A. aegypti* reinvaded these countries in the late 1970s.

Aedes aegypti has an interesting negative association with the congener species *A. albopictus*, which is native to South-East Asia (Mousson *et al.* 2005). The rise in numbers of *A. aegypti* in South-East Asia in the first half of the 20th century was associated with the decline of *A. albopictus*, while the later introduction of *A. albopictus* into the Americas in the 1980s was associated with the decline of *A. aegypti*. Both *A. aegypti* and *A. albopictus* can transmit the Zika virus (Braks *et al.* 2004). *Aedes aegypti* is of great concern because of the human arboviruses it transmits. Though a relatively poor disperser on its own (Chouin-Carneiro *et al.* 2016; Diagne *et al.* 2015), it has been distributed worldwide through human activities. It shows a pattern of multiple colonizations and recurrent spreading, though distinct populations retain important biological differences, such as the propensity to transmit human viruses (Takahashi *et al.* 2005). The rapid spread of diseases, like that caused by the Zika virus, shows the importance of a mosquito species already in the area that can transmit the virus.

10.7 Conclusions

With ever increasing globalization of economies and biodiversity, invasive species will continue to be important both economically and ecologically. Knowledge of the biology of invasive species is also growing rapidly, but gaps remain. Active areas of research include the use of molecular population genetics facilitated by high-throughput DNA sequencing to infer the origins of colonization events and other features of demographic history (Chouin-Carneiro *et al.* 2016; Diagne *et al.* 2015). The increased accessibility to 'omics' technologies is providing new opportunities for novel methods of control (Davies *et al.* 1999b; Estoup and Guillemaud 2010; Grbić *et al.* 2011; van Leeuwen *et al.* 2013), including emerging approaches based on RNA interference (RNAi) with high potential to be used to control insect pests in crops (Fang 2010; National Academies of Sciences 2016; Sinkins and Gould 2006). Biodiversity collections are proving invaluable as sources of DNA for studies of origins in addition to providing documentation of historical ranges (Carey 1991; Malacrida *et al.* 2007; Marsico *et al.* 2010; Suarez and Tsutsui 2004). Collections can also provide information on food webs and other ecological interactions, such as through examination of pollen or stable isotopes (Hobson *et al.* 2012).

Recent advances in making predictions of range expansion associated with global change, especially with changes in climate and land use, are possible through using collection data, online databases, niche modelling and integral projection models (DAISIE 2015; Merow *et al.* 2014; Meynard *et al.* 2013; Rapacciuolo *et al.* 2012; Suarez and Tsutsui 2004; UC Berkeley 2014; Vilà *et al.* 2010) (see also Chapter 9). Coupled with availability of information, international collaboration and networks have proven to be an important element in control of invasive alien species, as for example in the development of biological traits of the ladybird beetle *Harmonia axyridis* (Roy *et al.* 2016). Likewise, collaborations among researchers within scientific disciplines, such as the biological control community (IOBC 2016), help to integrate and transfer information more effectively.

Finally, citizen science is allowing the public to participate in large scientific endeavours and at the same time benefit from new knowledge. Noted examples of citizen science include identification tools, such as Discover Life (Pickering 2009) and iNaturalist (Ueda and Loarie 2013), as well as targeted research focusing on changing geographic distributions, such as the Lost Ladybug Project (Cornell University 2014) or the AGIIR app for alerting on invasive insect occurrences in France (AGIIR 2016). Data from such efforts are now available for use in scientific research (Sullivan *et al.* 2009).

Understanding the mechanisms of invasion and associated pathways remains critical for management, including monitoring, interception and policies to restrict trade (Hulme 2006; Petit *et al.* 2009). According to the International Plant Protection Convention (IPPC 2016), any measure aimed at preventing the introduction and spread of new pests must be justified by a science-based pest risk analysis (Jeger *et al.* 2012). Knowledge of pathways necessary to predict future spread and impact is also part of the process to provide scientific and economic evidence on whether an organism qualifies as a quarantine pest and how it should be managed. Where invasions involve mechanisms associated with commodities or human activity, such information can also aid in understanding the process of invasion. For example, when invasive species are contaminants of commodities, the occurrence and traits of contaminants can be at least partially understood by the commodity itself (Hulme *et al.* 2008). Likewise, understanding vectors of transportation provides testable hypotheses for the spread of species associated with those vectors (Carey 1991). Airplane and shipping routes, coupled with climate matching, predict some aspects of biological invasions (Liebhold *et al.* 2006; Tatum and Hay 2007).

Risk assessment associated with predicting the spread and impact of invasive species continues to be difficult, but necessary (Shogren 2000), and guidance to countries on the application of biosecurity measures to protect plants from pests that can be transported by commodity trade is well established (Hulme and Weser 2011; Jeger *et al.* 2012). Ecological systems are inherently complex but, in addition, a changing environment and novel sets of species interactions create new uncertainties. Moreover, invasive propagules are typically rare (Drake and Lodge 2006; Gillespie *et al.* 2012; Simberloff 2009). While it is clear that managed, urban and natural systems are necessarily connected, there are few incentives for studying risk of invasions for non-economic species. For example, national and international initiatives such as the European Food Safety Authority (EFSA 2015) and USDA (USDA APHIS 2015) focus on species of commercial interest but are less concerned about risks to natural environments (but see Gilioli *et al.* 2014). Risk assessment will also be critical in making informed decisions regarding the use of new genetic technologies, including GMOs, new sterile release strategies and the potential elimination of pest species (Roy *et al.* 2013).

As invasive species continue to drive changes in ecosystems worldwide (Barnosky *et al.* 2012), managing invasive species will necessarily require global policies and co-operation for more biosecurity in trade and travel.

Acknowledgements

We thank the editors Moshe Coll and Eric Wajnberg, and Helen Roy and an anonymous reviewer for encouragement and suggestions. This work was supported by the INRA-ACCAF programme, the French Agence Nationale de la Recherche (ID grant ANR-14-JFAC-0006-01, action FACCE-ERA-NET+ funded by the European Commission; EC contract 618105) and the US Department of Agriculture, UC Agricultural Experiment Station, and the France-Berkeley Fund.

References

AGIIR (2016) Alerter – Gérer les insectes invasifs et/ou ravageurs. Available at : http://ephytia.inra.fr/fr/Products/view/128/Agiir (accessed 7 March 2017).

Agriculture Victoria (2015) Invasive Plants and Animals Policy Framework. Available at: http://agriculture.vic.gov.au/agriculture/pests-diseases-and-weeds/protecting-victoria-from-pest-animals-and-weeds/invasive-plants-and-animals/invasive-plants-and-animals-policy-framework (accessed 7 March 2017).

Barnosky, A.D., Hadly, E.A., Bascompte, J., *et al.* (2012) Approaching a state shift in Earth's biosphere. *Nature* **486**: 52–58.

Barrett, S.C.H. (2015) Foundations of invasion genetics: the Baker and Stebbins legacy. *Molecular Ecology* **24**: 1927–1941.

Benedict, M.Q., Levine, R.S., Hawley, W.A. and Lounibos, L.P. (2007) Spread of the tiger: global risk of invasion by the mosquito *Aedes albopictus. Vector-Borne and Zoonotic Diseases* **7**: 76–85.

Blackburn, T.M., Pysek, P., Bacher, S., *et al.* (2011) A proposed unified framework for biological invasions. *Trends in Ecology and Evolution* **26**: 333–339.

Bloem, K., Brockerhoff, E.G., Mastro, V., Simmons, G.S., Sivinski, J. and Suckling, D.M. (2014) Insect eradication and containment of invasive alien species. In: Gordh, G. and McKirdy, S. (eds) *The Handbook of Plant Biosecurity*. Springer, Dordrecht, The Netherlands, pp. 417–446.

Bock, D.G., Caseys, C., Cousens, R.D., *et al.* (2015) What we still don't know about invasion genetics. *Molecular Ecology* **24**: 2277–2297.

Boubou, A., Migeon, A., Roderick, G.K., *et al.* (2012) Test of colonisation scenarios reveals complex invasion history of the red tomato spider mite *Tetranychus evansi. PLoS ONE* **7(4)**: e35601.

Braks, M.A.H., Honório, N., Lounibos, L., Lourenço-de-Oliveira, R. and Juliano, S. (2004) Interspecific competition between two invasive species of container mosquitoes, *Aedes aegypti* and *Aedes albopictus* (Diptera: Culicidae), in Brazil. *Annals of the Entomological Society of America* **97**: 130–139.

Burns, K.C. (2015) A theory of island biogeography for exotic species. *American Naturalist* **186**: 441–451.

Capinha, C., Essl, F., Seebens, H., Moser, D. and Pereira, H.M. (2015) The dispersal of alien species redefines biogeography in the Anthropocene. *Science* **348**: 1248–1251.

Carey, J.R. (1991) Establishment of the Mediterranean fruit fly in California. *Science* **253**: 1369–1373.

Carlquist, S. (1966) The biota of long-distance dispersal. I. Principles of dispersal and evolution. *Quarterly Review of Biology* **41**: 247–270.

Carroll, S.P. and Boyd, C. (1992) Host race radiation in the soapberry bug: natural history with the history. *Evolution* **46**: 1052–1069.

Chouin-Carneiro, T., Vega-Rua, A., Vazeille, M., *et al.* (2016) Differential susceptibilities of *Aedes aegypti* and *Aedes albopictus* from the Americas to Zika virus. *PLoS Neglected Tropical Diseases* **10(3)**: e0004543.

Clarke, A. R., Armstrong, K.F., Carmichael, A.E., *et al.* (2005) Invasive phytophagous pests arising through a recent tropical evolutionary radiation: the *Bactrocera dorsalis* complex of tropical fruit flies. *Annual Review of Entomology* **50**: 293–319.

Colautti, R.I. and Lau, J.A. (2015) Contemporary evolution during invasion: evidence for differentiation, natural selection, and local adaptation. *Molecular Ecology* **24**: 1999–2017.

Colautti, R.I., Franks, S.J., Hufbauer, R.A., *et al.* (2014) *The Global Garlic Mustard Field Survey (GGMFS): Challenges and Opportunities of a Unique, Large-Scale Collaboration for Invasion Biology*. Paper presented at the 7th NEOBIOTA Conference, Pontevedra, Spain. NeoBiota.

Convention on Biological Diversity (2016) Invasive Alien Species. Glossary of Terms. Available at: www.cbd.int/invasive/terms.shtml (accessed 7 March 2017).

Cornell University (2014) Lost Ladybug Project. Available at: www.lostladybug.org/ (accessed 7 March 2017).

DAISIE (2015) Delivering Alien Invasive Species Inventories for Europe. Available at: www.europe-aliens.org (accessed 7 March 2017).

Davies, N., Villablanca, F.X. and Roderick, G.K. (1999a) Bioinvasions of the medfly, *Ceratitis capitata*: source estimation using DNA sequences at multiple intron loci. *Genetics* **153**: 351–360.

Davies, N., Villablanca, F.X. and Roderick, G.K. (1999b) Determining the sources of individuals in recently founded populations: multilocus genotyping in non-equilibrium genetics. *Trends in Ecology and Evolution* **14**: 17–21.

Davis, M.A., Chew, M.K., Hobbs, R.J., *et al.* (2011) Don't judge species on their origins. *Nature* **474**: 153–154.

Dehnen-Schmutz, K. (2011) Determining non-invasiveness in ornamental plants to build green lists. *Journal of Applied Ecology* **48**: 1374–1380.

Denno, R.F. and Roderick, G.K. (1990) Population biology of planthoppers. *Annual Review of Entomology* **35**: 489–520.

Desprez-Loustau, M.L., Robin, C., Buee, M., *et al.* (2007) The fungal dimension of biological invasions. *Trends in Ecology and Evolution* **22**: 472–480.

Diagne, C.T., Diallo, D., Faye, O., *et al.* (2015) Potential of selected Senegalese *Aedes* spp. mosquitoes (Diptera: Culicidae) to transmit Zika virus. *BMC Infectious Diseases* **15**: 492.

Drake, J.M. and Lodge, D.M. (2006) Allee effects, propagule pressure and the probability of establishment: risk analysis for biological invasions. *Biological Invasions* **8**: 365–375.

ECDC (2016) *Aedes aegypti*. Available at: http://ecdc.europa.eu/en/healthtopics/vectors/mosquitoes/Pages/aedes-aegypti.aspx (accessed 7 March 2017).

EFSA (2015) Available at: www.efsa.europa.eu (accessed 7 March 2017).

Elton, C. (1958) *The Ecology of Invasions by Animals and Plants*. Chapman and Hall, London, UK.

Estoup, A. and Guillemaud, T. (2010) Reconstructing routes of invasion using genetic data: why, how and so what? *Molecular Ecology* **19**: 4113–4130.

Facon, B., Genton, B.J., Shykoff, J., Jarne, P., Estoup, A. and David, P. (2006) A general eco-evolutionary framework for understanding invasions. *Trends in Ecology and Evolution* **21**: 130–135.

Falk-Petersen, J., Bøhn, T. and Sandlund, O.T. (2006) On the numerous concepts in invasion biology. *Biological Invasions* **8**: 1409–1424.

Fang, J. (2010) Ecology: a world without mosquitoes. *Nature* **466**: 432–434.

Fauvergue, X., Vercken, E., Malausa, T. and Hufbauer, R.A. (2012) The biology of small, introduced populations, with special reference to biological control. *Evolutionary Applications* **5**: 424–443.

Fonseca, D.M., LaPointe, D.A. and Fleischer, R.C. (2000) Bottlenecks and multiple introductions: population genetics of the vector of avian malaria in Hawaii. *Molecular Ecology* **9**: 1803–1814.

Gilbert, M., Guichard, S., Freise, J., *et al.* (2005) Forecasting *Cameraria ohridella* invasion dynamics in recently invaded countries: from validation to prediction. *Journal of Applied Ecology* **42**: 805–813.

Gilioli, G., Schrader, G., Baker, R., *et al.* (2014) Environmental risk assessment for plant pests: a procedure to evaluate their impacts on ecosystem services. *Science of the Total Environment* **468**: 475–486.

Gillespie, R.G., Claridge, E.M. and Roderick, G.K. (2008) Biodiversity dynamics in isolated island communities: interaction between natural and human-mediated processes. *Molecular Ecology* **17**: 45–57.

Gillespie, R.G., Baldwin, B.G., Waters, J.M., Fraser, C.I., Nikula, R. and Roderick, G.K. (2012) Long-distance dispersal: a framework for hypothesis testing. *Trends in Ecology and Evolution* **27**: 52–61.

Grandgirard, J., Hoddle, M.S., Petit, J.N., Percy, D.M., Roderick, G.K. and Davies, N. (2007) Pre-introductory risk assessment of *Gonatocerus ashmeadi* (Hymenoptera: Mymaridae) for use as a classical biological control agent against *Homalodisca vitripennis* (Hemiptera: Cicadellidae) in the Society Islands of French Polynesia. *Biocontrol Science and Technology* **17**: 809–822.

Grandgirard, J., Hoddle, M.S., Petit, J.N., Roderick, G.K. and Davies, N. (2008) Engineering an invasion: classical biological control of the glassy-winged sharpshooter, *Homalodisca vitripennis*, by the egg parasitoid *Gonatocerus ashmeadi* in Tahiti and Moorea, French Polynesia. *Biological Invasions* **10**: 135–148.

Grandgirard, J., Hoddle, M.S., Petit, J.N., Roderick, G.K. and Davies, N. (2009) Classical biological control of the glassy-winged sharpshooter, *Homalodisca vitripennis*, by the egg parasitoid *Gonatocerus ashmeadi* in the Society, Marquesas and Australs archipelagos of French Polynesia. *Biological Control* **48**: 155–163.

Grapputo, A., Boman, S., Lindstroem, L., Lyytinen, A. and Mappes, J. (2005) The voyage of an invasive species across continents: genetic diversity of North American and European Colorado potato beetle populations. *Molecular Ecology* **14**: 4207–4219.

Grbić, M., van Leeuwen, T., Clark, R.M., *et al.* (2011) The genome of *Tetranychus urticae* reveals herbivorous pest adaptations. *Nature* **479**: 487–492.

Hadjistylli, M., Brown, J.K. and Roderick, G.K. (2010) Tools and recent progress in studying gene flow and population genetics of the *Bemisia tabaci* sibling species group. In: Stansly, P.A. and Naranjo, S.E. (eds) *Bemisia: Bionomics and Management of a Global Pest*. Springer, Dordrecht, The Netherlands, pp. 69–103.
Hadjistylli, M., Roderick, G.K. and Brown, J.K. (2016) Global population structure of a worldwide pest and virus vector: genetic diversity and population history of the *Bemisia tabaci* sibling species group. *PLoS ONE* **11(11)**: e0165105.
Hajek, A.E., Hurley, B.P., Kenis, M., *et al.* (2016) Exotic biological control agents: a solution or contribution to arthropod invasions? *Biological Invasions* **18**: 953–969.
Hall, D.G., Richardson, M.L., Ammar, E.D. and Halbert, S.E. (2013) Asian citrus psyllid, *Diaphorina citri*, vector of citrus huanglongbing disease. *Entomologia Experimentalis et Applicata* **146**: 207–223.
Hall, H.G. and Muralidharan, K. (1989) Evidence from mitochondrial DNA that African honey bees spread as continuous maternal lineages. *Nature* **339**: 211–213.
Harvey, R. and Mazzotti, F. (2014) The Invasion Curve: A Tool for Understanding Invasive Species Management in South Florida. University of Florida, IFAS Extension. Available at: http://edis.ifas.ufl.edu/uw392 (accessed 7 March 2017).
Hill, M.P., Susana Clusella-Trullas, S., Terblanche, J.S. and Richardson, D.M. (2016) Drivers, impacts, mechanisms and adaptation in insect invasions. *Biological Invasions* **18**: 883–891.
Hobson, K.A., Soto, D.X., Paulson, D.R., Wassenaar, L.I. and Matthews, J.H. (2012) A dragonfly (δ2H) isoscape for North America: a new tool for determining natal origins of migratory aquatic emergent insects. *Methods in Ecology and Evolution* **3**: 766–772.
Holway, D.A., Lach, L., Suarez, A.V., Tsutsui, N.D. and Case, T.J. (2002) The causes and consequences of ant invasions. *Annual Review of Ecology and Systematics* **33**: 181–233.
Howarth, F.G. (1996) *The Major Taxonomic Groups that Become Invasive Alien Pests in Hawaii and the Characteristics that Make them Pestiferous*. Technical Report Number 1996.022. Hawaii Biological Survey, Honolulu, Hawaii, USA.
Howarth, F.G. and Mull, W.P. (1992) *Hawaiian Insects and Their Kin*. University of Hawaii Press, Honolulu, Hawaii, USA.
Hulme, P.E. (2006) Beyond control: wider implications for the management of biological invasions. *Journal of Applied Ecology* **43**: 835–847.
Hulme, P.E. and Weser, C. (2011) Mixed messages from multiple information sources on invasive species: a case of too much of a good thing? *Diversity and Distributions* **17**: 1152–1160.
Hulme, P.E., Bacher, S., Kenis, M., *et al.* (2008) Grasping at the routes of biological invasions: a framework for integrating pathways into policy. *Journal of Applied Ecology* **45**: 403–414.
Hurley, B., Garnas, J., Wingfield, M., Branco, M., Richardson, D. and Slippers, B. (2016) Increasing numbers and intercontinental spread of invasive insects on eucalypts. *Biological Invasions* **18**: 921–933.
IOBC (2016) Available at: www.iobc-global.org (accessed 7 March 2017).
IPPC (2016) International Plant Protection Convention. Available at: www.wto.org/english/thewto_e/coher_e/wto_ippc_e.htm (accessed 7 March 2017).
Jeger, M., Schans, J., Lövei, G.L., *et al.* (2012) Risk assessment in support of plant health. *Efsa Journal* **10(10)**: s1012.
Johnson, D.M., Liebhold, A.M., Tobin, P.C. and Bjørnstad, O.N. (2006) Allee effects and pulsed invasion by the gypsy moth. *Nature* **444**, 361–363.

Kiritani, K. and Yamamura, K. (2003) Exotic insects and their pathways for invasion. In: Ruiz, G.M. and Carlton, J.T. (eds) *Invasive Species: Vectors and Management Strategies*. Island Press, Washington, DC, USA, pp. 44–67.
Kirkpatrick, M. and Barrett, B. (2015) Chromosome inversions, adaptive cassettes and the evolution of species' ranges. *Molecular Ecology* **24**: 2046–2055.
Kolar, C.S. and Lodge, D.M. (2001) Progress in invasion biology: predicting invaders. *Trends in Ecology and Evolution* **16**: 199–204.
Kolbe, J.J., Glor, R.E., Schettino, L.R., Lara, A.C., Larson, A. and Losos, J.B. (2004) Genetic variation increases during biological invasion by a Cuban lizard. *Nature* **431**: 177–181.
Krushelnycky, P.D. and Gillespie, R.G. (2008) Compositional and functional stability of arthropod communities in the face of ant invasions. *Ecological Applications* **18**: 1547–1562.
Lambrinos, J.G. (2004) How interactions between ecology and evolution influence contemporary invasion dynamics. *Ecology* **85**: 2061–2070.
Lande, R. (1988) Genetics and demography in biological conservation. *Science* **24**: 1455–1460.
Leong, M., Kremen, C. and Roderick, G.K. (2014) Pollinator interactions with yellow starthistle (*Centaurea solstitialis*) across urban, agricultural, and natural landscapes. *PLoS ONE* **9**(1): e86357.
Liebhold, A.M., Work, T.T., McCullough, D.G. and Cavey, J.F. (2006) Airline baggage as a pathway for alien insect species invading the United States. *American Entomologist* **52**: 48–54.
Liebhold, A., Yamanaka, T., Roques, A., *et al.* (2016) Global compositional variation among native and non-native regional insect assemblages emphasizes the importance of pathways. *Biological Invasions* **18**:893–905.
Lockwood, J.A., Cassey, P. and Blackburn, T. (2005) The role of propagule pressure in explaining species invasions. *Trends in Ecology and Evolution* **20**: 223–228.
Lodge, D.M. (1993) Biological invasions: lessons for ecology. *Trends in Ecology and Evolution* **8**: 133–137.
Lombaert, E., Guillemaud, T., Cornuet, J.M., Malausa, T., Facon, B. and Estoup, A. (2010) Bridgehead effect in the worldwide invasion of the biocontrol harlequin ladybird. *PLoS ONE* **5**(3): e9743.
Lonsdale, W.M. (1999) Concepts and synthesis: global patterns of plant invasions, and the concept of invasibility. *Ecology* **80**: 1522–1536.
Majerus, M., Strawson, V. and Roy, H. (2006) The potential impacts of the arrival of the harlequin ladybird, *Harmonia axyridis* (Pallas) (Coleoptera: Coccinellidae), in Britain. *Ecological Entomology* **31**: 207–215.
Malacrida, A.R., Gomulski, L.M., Bonizzoni, M., Bertin, S., Gasperi, G. and Guglielmino, C. R. (2007) Globalization and fruitfly invasion and expansion: the medfly paradigm. *Genetica* **131**: 1–9.
Marsico, T.D., Burt, J.W., Espeland, E.K., *et al.* (2010) Underutilized resources for studying the evolution of invasive species during their introduction, establishment, and lag phases. *Evolutionary Applications* **3**: 203–219.
Matteson, P.C., Gallagher, K.D. and Kenmore, P.E. (1994) Extension of integrated pest management for planthoppers in Asian irrigated rice: empowering the user. In: Denno, R. and Perfect, T. (eds) *Planthoppers – Their Ecology and Management*. Springer, Dordrecht, The Netherlands, pp. 656–685.

McFall-Ngai, M., Hadfield, M.G., Bosch, T.C.G., *et al.* (2013) Animals in a bacterial world, a new imperative for the life sciences. *Proceedings of the National Academy of Sciences* **110**: 3229–3236.

McMeniman, C.J., Lane, R.V., Cass, B.N., *et al.* (2009) Stable introduction of a life-shortening *Wolbachia* infection into the mosquito *Aedes aegypti*. *Science* **323**: 141–144.

Merow, C., Dahlgren, J.P., Metcalf, C.J.E., *et al.* (2014) Advancing population ecology with integral projection models: a practical guide. *Methods in Ecology and Evolution* **5**: 99–110.

Meynard, C.N., Migeon, A. and Navajas, M. (2013) Uncertainties in predicting species distributions under climate change: a case study using *Tetranychus evansi* (Acari: Tetranychidae), a widespread agricultural pest. *PLoS ONE* **8(6)**: e66445.

Migeon, A., Auger, P., Hufbauer, R. and Navajas, M. (2015) Genetic traits leading to invasion: plasticity in cold hardiness explains current distribution of an invasive agricultural pest, *Tetranychus evansi* (Acari: Tetranychidae). *Biological Invasions* **17**: 2275–2285.

Mousson, L., Dauga, C., Garrigues, T., Schaffner, F., Vazeille, M. and Failloux, A.B. (2005) Phylogeography of *Aedes* (*Stegomyia*) *aegypti* (L.) and *Aedes* (*Stegomyia*) *albopictus* (Skuse) (Diptera: Culicidae) based on mitochondrial DNA variations. *Genetical Research* **86**: 1–11.

Mun, J.H., Song, Y.H., Heong, K.L. and Roderick, G.K. (1999) Genetic variation among Asian populations of rice planthoppers, *Nilaparvata lugens* and *Sogatella furcifera* (Hemiptera: Delphacidae): mitochondrial DNA sequences. *Bulletin of Entomological Research* **89**: 245–253.

Nardi, F., Crapelli, A., Dallai, R., Roderick, G.K. and Frati, F. (2005) Population structure and colonization history of the olive fly, *Bactrocera oleae* (Diptera, Tephritidae). *Molecular Ecology* **14**: 2729–2738.

Nardi, F., Carapelli, A., Boore, J.L., Roderick, G.K., Dallai, R. and Frati, F. (2010) Domestication of olive fly through a multi-regional host shift to cultivated olives: comparative dating using complete mitochondrial genomes. *Molecular Phylogenetics and Evolution* **57**: 678–686.

National Academies of Sciences (2016) *Genetically Engineered Crops: Experiences and Prospects*. National Academies Press, Washington, DC, USA.

Navajas, M. and Boursot, P. (2003) Nuclear ribosomal differentiation between two closely mite species polyphyletic for mitochondrial DNA. *Proceedings of the Royal Society of London B: Biological Sciences* **270**: S124–S127.

Navajas, M., Anderson, D., de Guzman, L., *et al.* (2009) New Asian types of *Varroa destructor*: a potential new threat for world apiculture. *Apidologie* **41**: 181–193.

Navajas, M., de Moraes, G.J., Auger, P. and Migeon, A. (2013) Review of the invasion of *Tetranychus evansi*: biology, colonization pathways, potential expansion and prospects for biological control. *Experimental and Applied Acarology* **59**: 43–65.

Olesen, J.M., Eskildsen, L.I. and Venkatasamy, S. (2002) Invasion of pollination networks on oceanic islands: importance of invader complexes and endemic super generalists. *Diversity and Distributions* **8**: 181–192.

Ostfeld, R.S., Canham, C.D., Oggenfuss, K., Winchcombe, R.J. and Keesing, F. (2006) Climate, deer, rodents, and acorns as determinants of variation in Lyme-disease risk. *PLoS Biology* **4(6)**: e145.

Peischl, S., Kirkpatrick, M. and Excoffier, L. (2015) Expansion load and the evolutionary dynamics of a species range. *American Naturalist* **185**: E81–E93.

Perring, T.M., Cooper, A.D., Rodriquez, R.J., Farrar, C.A. and Bellows, T.S. Jr. (1993) Identification of a whitefly species by genomic and behavioral studies. *Science* **259**: 74–77.

Petit, J.N., Hoddle, M.S., Grandgirard, J., Roderick, G.K. and Davies, N. (2008) Invasion dynamics of the glassy-winged sharpshooter *Homalodisca vitripennis* (Germar) (Hemiptera: Cicadellidae) in French Polynesia. *Biological Invasions* **10**: 955–967.

Petit, J.N., Hoddle, M.S., Grandgirard, J., Roderick, G.K. and Davies, N. (2009) Successful spread of a biocontrol agent reveals a biosecurity failure: elucidating long distance invasion pathways for *Gonatocerus ashmeadi* in French Polynesia. *BioControl* **54**: 485–495.

Phillips, B.L. and Shine, R. (2006) An invasive species induces rapid adaptive change in a native predator: cane toads and black snakes in Australia. *Proceedings of the Royal Society of London B: Biological Sciences* **273**: 1545–1550.

Pickering, J. (2009) Discover life. Available at: www.discoverlife.org (accessed 7 March 2017).

Pimentel, D., Lack, L., Suniga, R. and Morrison, D. (2000) Environmental and economic costs of nonindigenous species in the United States. *BioScience* **50**: 53–65.

Pimentel, D., Zuniga, R. and Morrison, D. (2005) Update on the environmental and economic costs associated with alien-invasive species in the United States. *Ecological Economics* **52**: 273–288.

Purcell, A. and Saunders, S. (1999) Glassy-winged sharpshooters expected to increase plant disease. *California Agriculture* **53**: 26–27.

Rapacciuolo, G., Roy, D.B., Gillings, S., Fox, R., Walker, K. and Purvis, A. (2012) Climatic associations of British species distributions show good transferability in time but low predictive accuracy for range change. *PLoS ONE* **7**(7): e40212.

Roderick, G.K. and Howarth, F.G. (1999) Invasion genetics: natural colonizations, non-indigenous species, and classical biological control. In: Yano, E., Matsuo, K., Shiyomi, M. and Andow, D. (eds) *Biological Invasions of Pests and Beneficial Organisms (Volume NIAES Series 3)*. National Institute of Agro-Environmental Sciences, Tsukuba, Japan, pp. 98–108.

Roderick, G.K. and Navajas, M. (2003) Genes in novel environments: genetics and evolution in biological control. *Nature Reviews Genetics* **4**: 889–899.

Roderick, G.K., Hufbauer, R.A. and Navajas, M. (2012) Evolution and biological control. *Evolutionary Applications* **5**: 419–423.

Roy, H.E. and Wajnberg, E. (2008) *Biological Control to Invasion: The Laydbird Harmonia axyridis as a Model Species.*Springer, Dordrecht, The Netherlands.

Roy, H., Schonrogge, K., Dean, H., *et al.* (2013) Invasive alien species – framework for the identification of invasive alien species of EU concern. Available at: http://ec.europa.eu/environment/nature/invasivealien/docs/Final%20report_12092014.pdf (accessed 7 March 2017).

Roy, H.E., Brown, P.M., Adriaens, T., *et al.* (2016) The harlequin ladybird, *Harmonia axyridis*: global perspectives on invasion history and ecology. *Biological Invasions* **18**: 997–1044.

Saenz, V.L., Booth, W., Schal, C. and Vargo, E.L. (2012) Genetic analysis of bed bug populations reveals small propagule size within individual infestations but high genetic diversity across infestations from the eastern United States. *Journal of Medical Entomology* **49**: 865–875.

Sakai, A.K., Allendorf, F.W., Holt, J.S., *et al.* (2001) The population biology of invasive species. *Annual Review of Ecology and Systematics* **32**: 305–332.

Sax, D.F. and Gaines, S.D. (2006) The biogeography of naturalized species and the species-area relationship: reciprocal insights to biogeography ans invasion biology.

In: McMahon, S.M., Fukami, T. and Cadotte, M.W. (eds) *Conceptual Ecology and Invasion Biology: Reciprocal Approaches to Nature*. Springer, Dordrecht, The Netherlands, pp. 449–480.

Sax, D.F. and Gaines, S.D. (2008) Species invasions and extinction: the future of native biodiversity on islands. *Proceedings of the National Academy of Sciences* **105**: 11490–11497.

Sax, D.F., Stchowicz, J.J. and Gaines, S.D. (2005) *Species Invasions: Insights into Ecology, Evolution, and Biogeography*. Sinauer Associates, Sunderland, MA, USA.

Sax, D.F., Stachowicz, J.J., Brown, J.H., *et al.* (2007) Ecological and evolutionary insights from species invasions. *Trends in Ecology and Evolution* **22**: 465–471.

Sheppard, A., Hill, R., DeClerck-Floate, R., *et al.* (2004) *A Global Review of Risk–Cost–Benefit Assessments for Introductions of Biological Control Agents Against Weeds: A Crisis in The Making?* Paper presented at the XI International Symposium on Biological Control of Weeds, Canberra, Austrialia.

Shine, R. (2010) The ecological impact of invasive cane toads (*Bufo marinus*) in Australia. *Quarterly Review of Biology* **85**: 253–291.

Shogren, J. F. (2000) Risk reduction strategies against the 'explosive invader'. In: Perrings, C., Williamson, M. and Dalmazzone, D. (eds) *The Economics of Biological Invasions*. Edward Elgar, Cheltenham, UK, pp. 56–67.

Simberloff, D. (2006) Invasional meltdown 6 years later: important phenomenon, unfortunate metaphor, or both? *Ecology Letters* **9**: 912–919.

Simberloff, D. (2009) The role of propagule pressure in biological invasions. *Annual Review of Ecology, Evolution, and Systematics* **40**: 81–102.

Simberloff, D. and Stiling, P. (1996) Risks of species introduced for biological control. *Biological Conservation* **78**: 185–192.

Simberloff, D., Martin, J.L., Genovesi, P., *et al.* (2013) Impacts of biological invasions: what's what and the way forward. *Trends in Ecology and Evolution* **28**: 58–66.

Sinkins, S.P. and Gould, F. (2006) Gene drive systems for insect disease vectors. *Nature Reviews Genetics* **7**: 427–435.

Snyder, W.E., Clevenger, G.M. and Eigenbrode, S.D. (2004) Intraguild predation and successful invasion by introduced ladybird beetles. *Oecologia* **140**: 559–565.

Solignac, M., Cornuet, J.M., Vautrin, D., *et al.* (2005) The invasive Korea and Japan types of *Varroa destructor*, ectoparasitic mites of the Western honeybee (*Apis mellifera*), are two partly isolated clones. *Proceedings of the Royal Society of London B: Biological Sciences* **272**: 411–419.

Suarez, A.V. and Tsutsui, N.D. (2004) The value of museum collections for research and society. *BioScience* **54**: 66–74.

Suckling, D. and Brockerhoff, E. (2010) Invasion biology, ecology, and management of the light brown apple moth (Tortricidae). *Annual Review of Entomology* **55**: 285–306.

Sullivan, B.L., Wood, C.L., Iliff, M.J., Bonney, R.E., Fink, D. and Kelling, S. (2009) eBird: a citizen-based bird observation network in the biological sciences. *Biological Conservation* **142**: 2282–2292.

Swei, A., Ostfeld, R.S., Lane, R.S. and Briggs, C.J. (2011) Impact of the experimental removal of lizards on Lyme disease risk. *Proceedings of the Royal Society B: Biological Sciences* **278**: 2970–2978.

Takahashi, L.T., Maidana, N.A., Ferreira, W.C. Jr, Pulino, P. and Yang, H.M. (2005) Mathematical models for the *Aedes aegypti* dispersal dynamics: travelling waves by wing and wind. *Bulletin of Mathematical Biology* **67**: 509–528.

Tatum, A.J. and Hay, S.I. (2007) Climatic similarity and biological exchange in the worldwide airline transportation network. *Proceedings of the Royal Society of London, B: Biological Sciences* **274**: 1489–1496.

Torchin, M.E. and Mitchell, C.E. (2004) Parasites, pathogens, and invasions by plants and animals. *Frontiers in Ecology and the Environment* **2**: 183–190.

Torchin, M.E., Lafferty, K.D., Dobson, A.P., McKenzie, V.J. and Kuris, A.M. (2003) Introduced species and their missing parasites. *Nature* **421**: 628–630.

UC Berkeley (2014) Holos, Berkeley Ecoinformatics Engine. Available at: https://ecoengine.berkeley.edu (accessed 7 March 2017).

Ueda, K. and Loarie, S. (2013) iNaturalist. Available at: www.inaturalist.org (accessed 7 March 2017).

USDA APHIS (2015) United States Department of Agriculture, Animal and Plant Health Inspection Service. Available at: www.aphis.usda.gov/wps/portal/aphis/home/ (accessed 7 March 2017).

Valéry, L., Fritz, H., Lefeuvre, J.C. and Simberloff, D. (2008) In search of a real definition of the biological invasion phenomenon itself. *Biological Invasions* **10**: 1345–1351.

Van Kleunen, M., Dawson, W., Essl, F., *et al.* (2015) Global exchange and accumulation of non-native plants. *Nature* **525**: 100–103.

Van Leeuwen, T., Dermauw, W., Grbic, M., Tirry, L. and Feyereisen, R. (2013) Spider mite control and resistance management: does a genome help? *Pest Management Science* **69**: 156–159.

Vellend, M., Harmon, L.J., Lockwood, J.L., *et al.* (2007) Effects of exotic species on evolutionary diversification. *Trends in Ecology and Evolution* **22**: 481–488.

Vilà, M., Basnou, C., Pyšek, P., *et al.* (2010) How well do we understand the impacts of alien species on ecosystem services? A pan European, cross-taxa assessment. *Frontiers in Ecology and the Environment* **8**: 135–144.

Vitousek, P.M., d'Antonio, C.M., Loope, L.L. and Westbrooks, R. (1996) Biological invasions as global environmental change. *American Scientist* **84**: 468–478.

Walker, T., Johnson, P., Moreira, L., *et al.* (2011) The wMel *Wolbachia* strain blocks dengue and invades caged *Aedes aegypti* populations. *Nature* **476**: 450–453.

Way, M. and Heong, K. (1994) The role of biodiversity in the dynamics and management of insect pests of tropical irrigated rice – a review. *Bulletin of Entomological Research* **84**: 567–588.

Wilcove, D.S., Rothstein, D., Dubow, J., Phillips, A. and Losos, E. (1998) Quantifying threats to imperiled species in the United States. *BioScience* **48**: 607–615.

Williamson, M. (1996) *Biological Invasions*. Chapman and Hall, London, UK.

Xi, Z., Khoo, C.C. and Dobson, S.L. (2005) *Wolbachia* establishment and invasion in an *Aedes aegypti* laboratory population. *Science* **310**: 326–328.

Yano, E., Matsuo, K., Shiyomi, M. and Andow, D. (1999) *Biological Invasions of Pests and Beneficial Organisms (Volume NIAES Series 3)*. National Institute of Agro-Environmental Sciences, Tsukuba, Japan.

Zhu, M., Song, Y.H., Uhm, K.B., Turner, R.W., Lee, J.H. and Roderick, G.K. (2000) Simulation of the long range migration of brown planthopper, *Nilaparvata lugens* (Stål), by using a boundary layer atmospheric model and geographic information system. *Journal of Asia-Pacific Entomology* **3**: 25–32.

Part V

Pest Control and Public Health

11

Pesticides and Human Health

Jane A. Hoppin and Catherine E. LePrevost

11.1 Introduction

Pesticides are a broad class of compounds designed to kill or control pests ranging from insects and weeds to micro-organisms and rodents (Alavanja 2009). The umbrella term 'pesticide' (see also Chapters 4 and 15) encompasses functional groups which include herbicides, insecticides, fungicides and fumigants. Plant regulators, defoliants and desiccants, in addition to nitrogen stabilizers, may also be classified as pesticides. Traditionally, pesticides have been evaluated as a whole, by functional group based on the pest that they are designed to kill (e.g. herbicide, fungicide, insecticide) or by chemical class (e.g. organochlorine, organophosphate and pyrethroid for insecticides and phenoxy and triazine for herbicides). Figure 11.1 displays some common agricultural pesticides with their functional groups and classes, demonstrating the breadth of chemicals identified as pesticides. This figure does not include all pesticide classes on the market but illustrates the diversity in pesticide active ingredients and chemical properties. Some of the newer classes of pesticides such as the neonicotinoids are not included. Pesticide products include not only the active ingredient but also so-called inert ingredients – such as fragrances, dyes, emulsifiers, solvents and carriers that are added to the pesticide – that do not have pesticidal properties. Active ingredients appear on pesticide product labels, but inert ingredients are regarded as 'confidential business information' and as such are not displayed or otherwise publicly available. 'Inert' ingredients are not toxicologically inert and may have adverse human health effects.

11.2 Human Exposure to Pesticides

Individuals may encounter pesticides through various scenarios, including intentional and unintentional exposures. Intentional exposure to pesticides is common: self-poisoning using pesticides accounts for one-third of suicides worldwide (Bertolote *et al.* 2006; Gunnell *et al.* 2007). Unintentional exposures to pesticides may occur through occupational and take-home pathways, home and garden use, public health applications such as use for vector control, and residues in food and water.

Occupational exposures are the most significant in terms of concentration, frequency and duration. Pesticide formulations used in occupational settings are also those most

Environmental Pest Management: Challenges for Agronomists, Ecologists, Economists and Policymakers, First Edition. Edited by Moshe Coll and Eric Wajnberg.

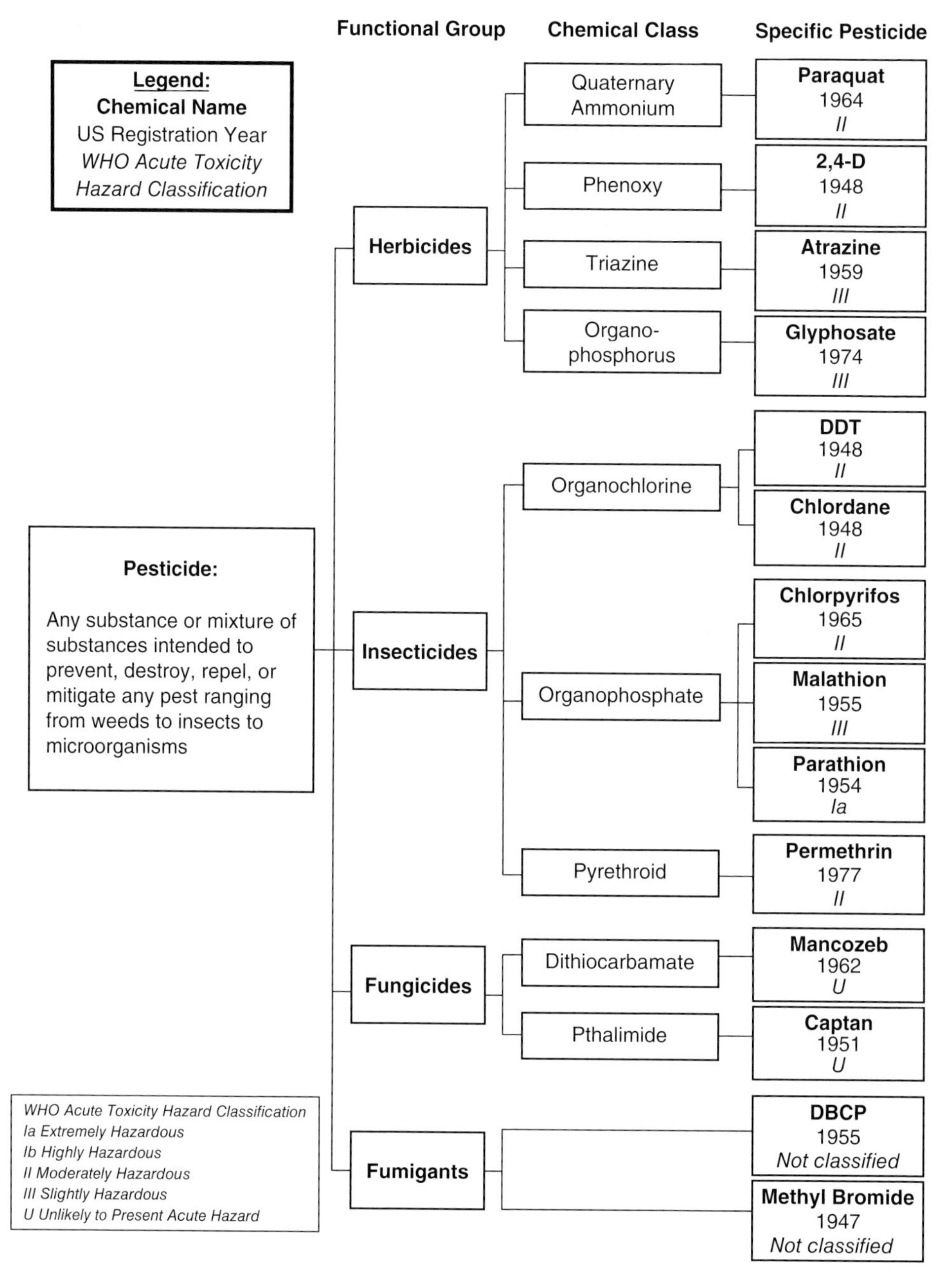

Figure 11.1 Pesticide classification hierarchy. This figure illustrates the relationship between functional group and chemical class for some commonly used and historical pesticides. Information is provided with the pesticide registration year along with WHO Acute Toxicity Hazard Classification (WHO - www.who.int/ipcs/publications/pesticides_hazard/en/). The figure does not include all pesticides or all classes of chemicals, but rather a subset of pesticides. DBCP, dibromochloropropane; DDT, dichlorodiphenyltrichloroethane.

toxic to humans. Similar chemicals may be used in both agricultural and residential settings. Agricultural workers, maintenance and groundskeepers, exterminators and animal handlers are among those at risk for work-related pesticide exposure. Furthermore, pesticide-exposed workers may bring pesticides into the home and unintentionally expose family members by transporting residues on their clothing, skin and shoes, as well as through family vehicles (i.e. 'take-home pathway'; Curl *et al.* 2002).

Pesticides can enter the body via three routes: ingestion (oral), inhalation and dermal (skin and eyes) (Krieger 2010). For agricultural workers, oral exposure may occur when there is hand-to-mouth activity (e.g. eating, drinking, smoking or other tobacco use) with unwashed hands after the worker has handled pesticides or pesticide-treated plants or animals, as well as through accidental consumption due to improper storage of pesticides in food and drink containers. Inhalation exposure is possible from breathing dusts and vapours when handling pesticides that are in a granular form, mixing or loading volatile pesticides into application equipment or applying a pesticide without use of adequate respiratory personal protective equipment. Dermal exposure may result from spills and splashes, contaminated and unwashed clothing, and contact with pesticide application equipment and pesticide-treated plants and animals. Off-site pesticide drift can result in both inhalation exposure and dermal exposure for agricultural workers, their families and bystanders. Overall, dermal exposure is the most common route among agricultural workers (Krieger 2010).

The general population may be exposed to pesticides when they or others apply pesticide products to homes and gardens, or as a result of spray drift from nearby agricultural fields. Residential pesticide exposure is a particular concern for children because they exhibit greater susceptibility and higher rates of exposure (due to hand-to-mouth behaviours and increased floor time) and spend more time at home (Garry 2004). In the USA, 42% of calls to poison control centres regarding pesticide exposure involved children 5 years of age and younger (Langley and Mort 2012). The general population may also be exposed to pesticides or their metabolites via food. Recent studies have shown that shifting children to organic diets results in lower levels of pesticide metabolites in their urine, but the health implications of this have not been evaluated (Bradman *et al.* 2015; Lu *et al.* 2006).

Agricultural workers and their family members have a high probability of pesticide exposure due to their involvement in pesticide application, crop production and animal handling activities, as well as through living in areas where pesticides are applied on a regular basis. In agricultural cohort studies in the USA, France and Norway, pesticide use estimates ranged from 63% to 99%, with pesticide use ranging from 1 to 200 days per year (Alavanja *et al.* 1996; Brouwer *et al.* 2016). In addition to exposure as a result of normal work activities, agricultural workers may have higher than intended pesticide exposures which may or may not result in poisonings.

Non-occupational, environmental pesticide exposures can occur through diet, as described above, as well as through the environmental media of air, water and soil (Hodgson 2010; Krieger 2010). In residential settings, outdoor and indoor air, drinking water, and soils and dusts may be sources of exposure to active pesticide compounds. For the general public, the most significant source of pesticide exposure is through pesticide use at home, with typically higher indoor than outdoor exposures (Krieger 2010). Contributors to non-occupational, residential pesticide exposures include pet pesticide products (e.g. flea and tick prevention medications) and treated pets, home gardens in which pesticides are applied, treatments to the home to prevent or control infestations (e.g. structural pest control for termites, cockroaches), and pesticide-treated turf around homes, parks and athletics fields (Krieger 2010).

11.3 Acute Toxicity

Pesticides are designed to kill or limit the growth of a variety of pests. As such, exposure to these chemicals may also pose a risk to human health. Regulatory agencies classify pesticide products, composed of both active and inert ingredients, according to their acute toxicity. For example, the United States Environmental Protection Agency (USEPA) assigns toxicity signal words to approved pesticides to be displayed on product labels (USEPA 2015). Acute toxicity is characterized by high-level, short-term exposure resulting in immediate adverse effects, usually within 24 hours. Acute toxicity is chemical specific, varying across and within functional groups and classes of pesticides. The acute toxicities of the pesticides shown in Figure 11.1 are based on the World Health Organization (WHO) hazard categories for listed active ingredients. Insecticides are of particular concern for acute toxicity to humans because they commonly target the insect nervous system, which is similar structurally and functionally to that of humans (Reigart and Roberts 2013).

Acute toxicity and the corresponding classifications are evaluated according to laboratory experiments with test animals, typically mice and rats, with death being the toxic effect of interest. In these experiments, acute toxicity in laboratory animals is assessed according to the various routes of exposure (USEPA 2002). It is important to note that the WHO classification scheme is based on laboratory experiments in rats that evaluate death by oral and dermal routes of exposure only. Because the inhalation route of exposure is not evaluated in the WHO pesticide hazard classification, fumigants, which are highly volatile and thus are likely to be inhaled through concentrations in the air, are not classified, despite their generally having high levels of acute toxicity (WHO 2010).

Acute pesticide poisoning is regarded as a serious public health concern for agricultural workers globally, particularly in the developing world (Jeyaratnam 1990). There are no worldwide monitoring efforts, and it is widely assumed that pesticide poisoning is underdiagnosed and underreported. Therefore, quantifying acute pesticide poisoning is challenging. Published estimates of annual poisonings worldwide are highly variable, ranging from 3 million cases of overall pesticide poisonings to 25 million cases of occupational poisonings among agricultural workers alone (Jeyaratnam 1985, 1990). Rates of pesticide poisoning are substantially higher among agricultural workers than other occupations (Calvert *et al.* 2013). Agricultural workers may not seek medical care for pesticide poisonings because of acceptance of pesticide exposure and illness as 'part of the job', failure to recognize symptoms as being related to pesticides, concern about missed work time and limited access to healthcare services. The Agricultural Health Study, a large prospective cohort study of US farmers and their spouses, has provided detailed pesticide-specific information for over 89 000 individuals (Figure 11.2). In this study, approximately 2% of participants reported a physician-diagnosed pesticide poisoning (Starks *et al.* 2012a). Additionally, more than 20% of US Agricultural Health Study pesticide applicators reported at least one 'high pesticide exposure event' in their

Agricultural Health Study

Started in 1993

Population:
- 52 394 private pesticide applicators (farmers) from Iowa and North Carolina, USA (>80% of all private applicators in these states)
- 32 345 spouses of private applicators (75% of married applicators had their spouse enroll)
- 4916 commercial pesticide applicators from Iowa

Health Outcomes:
- Continuous follow-up for cancer and mortality
- Information for non-cancer outcomes updated regularly

Pesticide Information:
- Pesticide use information for more than 50 pesticides at enrollment
- Detailed information on over 300 current use pesticides from follow-up interviews in 1999–2003 and 2005–2010

Further Reading:
- Over 100 peer-reviewed papers on specific pesticides and human health outcomes
- http://aghealth.nih.gov/
- Alavanja *et al.* 1996; Hoppin *et al.* 2014; Koutros *et al.* 2010; Waggoner *et al.* 2011

Figure 11.2 Background on the United States Agricultural Health Study, the largest and longest prospective study of farmers and their spouses focusing on the human health effects of pesticides.

lifetime. High pesticide exposure events were self-reported events defined as an incident or experience resulting in an unusually high exposure while handling pesticides (Starks *et al.* 2012a). These high pesticide exposure events have been linked to serious long-term adverse health effects, including asthma, chronic bronchitis and impairment of the central nervous system (Hoppin *et al.* 2007, 2009; Starks *et al.* 2012a).

Acute pesticide poisoning symptoms range from mild to severe, including death. Most pesticides share common poisoning symptoms. Yet these common symptoms are also observed in other occupational illnesses among agricultural workers, including nicotine poisoning and heat stress. Overall, the symptoms of pesticide poisoning are non-specific and include headache, dizziness, abdominal pain, nausea, vomiting and diarrhea. Pesticides may also cause contact-related irritation symptoms, such as eye, throat and skin irritation not related to allergies. The non-specific nature of many pesticide poisoning symptoms makes the relatively few poisoning cases that result in a visit to a healthcare facility difficult to diagnose.

In contrast, exposure to certain classes of pesticides, for example organophosphate insecticides, is associated with a constellation of poisoning symptoms, termed 'toxidrome'. Organophosphate insecticides are used in agriculture, gardening, homes and veterinary medicine, and this class of pesticides has historically been associated with a significant percentage of reported pesticide poisoning cases in the USA (Reigart and Roberts 2013). Because organophosphate insecticides inhibit the activity of the neurological enzyme acetylcholinesterase, the body produces excess fluids in the case of poisoning, including eye tearing, nasal drainage, drooling and profuse sweating.

Table 11.1 Online pesticide information resources.

Resource	Description	Web address
CDMS, Inc.	Database of pesticide labels, searchable by manufacturer, brand name and common name	www.cdms.net/
Dictionary of Agromedicine	Online dictionary of more than 2200 terms related to the intersection of human health and agriculture, including pesticides	http://agromedicinedictionary.ces.ncsu.edu/
Pesticide Action Network (PAN) pesticide database	Database of toxicity and regulatory information for pesticides	www.pesticideinfo.org/
National Pesticide Information Centre at Oregon State University (USA)	Clearing house for the public and professionals of pesticide information resources, including pesticide fact sheets	http://npic.orst.edu/
'Recognition and Management of Pesticide Poisonings' (Reigart and Roberts 2013)	Manual for clinicians and others containing pesticide poisoning and treatment information, organized by chemical class	www.epa.gov/pesticide-worker-safety/recognition-and-management-pesticide-poisonings
WHO 'Recommended Classification of Pesticides by Hazard'	Resource that identifies acute toxicity for pesticides by active ingredient	www.who.int/ipcs/publications/pesticides_hazard/en/

The poisoned individual may experience muscle tremors or convulsions, and the pupils become constricted. Severe organophosphate insecticide poisoning can result in death due to respiratory failure. It is important to note that poisoned individuals may experience the non-specific symptoms listed above in addition to the classic toxidrome for organophosphate insecticides (Reigart and Roberts 2013). Readers are advised to refer to the USEPA's 'Recognition and Management of Pesticide Poisonings' (Reigart and Roberts 2013) as well as other online resources given in Table 11.1.

Traditionally, information on the health effects of pesticides has been limited to information on poisonings and assessments of pesticides or functional groups as a whole. However, recent research has shifted in focus to chronic pesticide toxicity, characterized by prolonged, lower-level exposure resulting in delayed effects, consistent with long-term low-level use. As part of the pesticide registration process, data on chronic toxicity in rodents, including effects on reproduction and development, as well as genotoxicity and carcinogenicity are considered. Potential human exposure is based on prediction models of environmental fate and transport for pesticide registration. These models include information on the physical and chemical properties of the pesticide and the proposed application method, as well as persistence in the environment. Until pesticides are released onto the market, there is no way to assess the actual extent of human exposure. Low-level exposure may contribute to chronic health effects that may take years to develop. Therefore, assessing the potential human health effects involves both detailed exposure characterization as well as long-term follow-up to assess for the wide range of adult chronic diseases that contribute greatly to morbidity but are not evaluated in animal studies, such as neurodegenerative and respiratory diseases and diabetes. Below, we present a summary of the human data for cancer, neurological outcomes, respiratory health, diabetes, birth outcomes and cardiovascular diseases, as these areas represent the bulk of the chronic human health data for pesticides.

11.4 Chronic Human Health Effects

Understanding of the chronic human health effects of pesticides has improved greatly in the past decade due to better characterization of specific chemical exposures, rather than pesticides or insecticides as a functional group. Chronic health effects are those health effects that develop over time and may have long-term health consequences. Characterizing exposure to pesticides can be challenging as few individuals know what chemicals they are exposed to and few of the pesticides currently on the market are long-lived enough to be detectable later in biological tissues. Pesticides like DDT (dichlorodiphenyltrichloroethane) and other organochlorines persist in both the environment and biological tissues, so researchers have been able to assess historical exposures using biological markers. More recent pesticides like the organophosphate insecticides are short-lived in the environment and biological tissues, so investigators have to rely on surrogate measures of exposure, such as questionnaire reports of use or linkage to pesticide use registries to obtain measures of exposure. For individuals who know what pesticides they use, questionnaires have advantages because data on use of specific chemicals can be collected on a large number of individuals inexpensively. For populations who do not know the chemicals they are exposed to, such as farm workers or the general population, questionnaires can be of limited value in the identification of specific chemical exposure.

The vast majority of research has focused on individuals exposed to the parent compounds, either through direct application or bystander or neighbouring exposure. Large prospective studies of farmers, such as the US Agricultural Health Study, have provided chemical-specific information on potential human health effects of pesticides to farmers and their spouses.

Over the past decade (2006–2015), research on the human health effects of pesticides has expanded greatly, with over 1000 peer-reviewed articles published. With better information on specific pesticides and exposure characterization, these articles have focused not only on cancer and neurological consequences but also on a wide array of common health outcomes, including respiratory diseases, diabetes and cardiovascular outcomes. Below, we summarize the current state of the science on the chronic human health effects of pesticides, for a subset of health outcomes.

11.4.1 Cancer

Cancer is a common health outcome, with approximately one in three people receiving a diagnosis of cancer in their lifetime (ACS 2016). Prior to registration, all pesticides are evaluated in animals for potential human carcinogenicity. However, people continue to be concerned about the potentially carcinogenic effects of pesticides. In 1990, the International Agency for Research on Cancer (IARC) classified occupational spraying and application of non-arsenical insecticides as a Group 2A carcinogen: probable human carcinogen (IARC 1991). Specific pesticides were not identified. In 2015, IARC hosted two meetings to evaluate the potential for human carcinogenicity of organophosphorus compounds, organochlorine insecticides and phenoxy herbicides based on results from both human and animal evaluations. As a result of these evaluations, lindane was classified as a known human carcinogen and glyphosate, diazinon and DDT as probable human carcinogens. Table 11.2 lists the carcinogenicity classifications for pesticides by the IARC and the US National Toxicology Program (Guyton *et al.* 2015; Loomis *et al.* 2015; NTP 2014). These evaluations focus on hazard identification using both human and animal data, but do not provide information on the nature of the exposure–response relationship.

11.4.1.1 Childhood Cancers

Pesticides have been investigated as a potential contributor to childhood cancers, though the evidence for specific chemicals is limited. Childhood cancers are rare cancers occurring in approximately 17 out of 100 000 children in the USA annually. Given the rarity of these cancers, it has been difficult to evaluate the role of specific pesticides. A comprehensive review of the literature on pesticides and childhood cancer in 2007 indicated evidence of an association between pesticide exposure both prenatally and during childhood and the development of cancer, but lacked detail on specific pesticide active ingredients (Infante-Rivard and Weichenthal 2007).

Three recent meta-analyses have evaluated parental occupational pesticide exposure as well as residential pesticide exposure and the risk of childhood leukaemia. Childhood leukaemia was associated with prenatal maternal, but not paternal, pesticide exposure with a more than two-fold risk associated with insecticides and a more than three-fold risk associated with herbicides, in a meta-analysis that included data from 31 studies (Wigle *et al.* 2009). Residential pesticide exposure during pregnancy was associated with a two-fold risk of childhood leukaemia, with the risk highest for indoor use of insecticides in two meta-analyses that incorporated 26 different studies (Chen *et al.* 2015; van Maele-Fabry *et al.* 2011). Outdoor exposure to herbicides was not associated with leukaemia in one analysis (van Maele-Fabry *et al.* 2011), but was in a subsequent analysis (Chen *et al.* 2015). Residential pesticide use was also associated with childhood lymphomas (Chen *et al.* 2015). In a similar analysis for childhood brain tumours with data from 20 studies, occupational exposure to pesticides during the prenatal period was associated with a 30–50% increased risk of childhood brain tumours (van Maele-Fabry *et al.* 2013). At this point, the data are insufficient to identify specific agents. However, the data are sufficient to warrant caution regarding the use of pesticides by individuals during pregnancy.

11.4.1.2 Cancer in Adults

The evidence for pesticides and cancer is stronger for adults than children, although it is inconclusive for many pesticides, with data coming both from studies of occupationally exposed individuals and from population-based studies using biological markers to assess historical exposure to organochlorine compounds.

Lymphohaematopoietic cancers, such as non-Hodgkin lymphoma and leukaemia, have been associated with pesticide use over time. In a review of 13 different studies, non-Hodgkin lymphoma was significantly associated with occupational exposure to pesticides, with the strongest association seen among individuals who had worked with

Table 11.2 Pesticides evaluated for carcinogenicity by the International Agency for Research on Cancer (IARC) and the US National Toxicology Program (NTP) as of December 2015.

Pesticide	IARC group*	NTP**	Current use on food crops?
Insecticides			
Lead arsenate	1	–	No
Organochlorine compounds			
DDT	2A	Reasonably anticipated	No
Hexachlorobenzene	–	Reasonably anticipated	No
Kepone	–	Reasonably anticipated	No
Lindane	1	Reasonably anticipated	No
Mirex	–	Reasonably anticipated	No
Toxaphene	–	Reasonably anticipated	No
TCDD (dioxin, contaminant of 2,4,5-T and 2,4,5-TP)	1	–	No
Organophosphate insecticides			
Diazinon	2A	–	Yes
Malathion	2A	–	Yes
Parathion	2B	–	Yes
Tetrachlorvinphos	2B	–	Yes
Herbicides			
2,4-D	2B	–	Yes
Amitrole	–	Reasonably anticipated	No
Glyphosate	2A	–	Yes
Nitrofen	–	Reasonably anticipated	No
Fungicides			
Captafol	2A	Reasonably anticipated	No
Fumigants			
Dibromochloropropane	–	Reasonably anticipated	No
Ethylene Dibromide	2A	Reasonably anticipated	No

* IARC classifications: Group 1 = known human carcinogen; Group 2A = probable human carcinogen; Group 2B = possible human carcinogen.
** NTP classifications: known or reasonably anticipated to cause cancer in humans.
– Not evaluated.
References: IARC (1991), NTP (2014), Guyton *et al.* (2015), Loomis *et al.* (2015).

pesticides for more than 10 years (Merhi *et al.* 2007). Pesticide manufacturing workers from 37 cohorts had a two-fold increased risk of dying of lymphoma, with a similar level of risk for phenoxy herbicide manufacturers (Jones *et al.* 2009). The herbicides 2,4-D and glyphosate and the insecticides carbaryl, carbofuran, diazinon, malathion, DDT and lindane were significantly associated with non-Hodgkin lymphoma in a meta-analysis that included the results from 44 different studies (Schinasi and Leon 2014). In the US Agricultural Health Study, use of the organochlorine insecticide lindane for more than 22 days was associated with a doubling of non-Hodgkin lymphoma risk (Weichenthal *et al.* 2010). Also in this cohort, the herbicide alachlor and the insecticides chlorpyrifos, diazinon and permethrin were associated with lymphohaematopoietic cancers as a group. Permethrin was also associated with multiple myeloma, and the organochlorine insecticides chlordane/heptachlor, chlorpyrifos, diazinon, EPTC (s-ethyl dipropylthiocarbamate) and fonofos were associated with leukaemia (Weichenthal *et al.* 2010). In all cases, there was evidence of an exposure–response relationship, such that individuals who used pesticides for longer were at higher risk.

A number of other cancers have been evaluated for an association with pesticide exposure. Over 100 articles evaluated the link between DDT or its metabolite DDE (dichlorodiphenyldichloroethylene) and cancer, with most of the results inconclusive. Of particular interest has been the association of organochlorine pesticides, in particular DDT, and breast cancer, and yet, more than 40 articles published since 1993 have not found a clear association (Loomis *et al.* 2015). There is some evidence of an association of DDT with liver and testicular cancers (Loomis *et al.* 2015). Pesticide manufacturing workers had a higher risk of testicular cancer, but this was not statistically significant (Jones *et al.* 2009). Prostate cancer is elevated in farmers, so researchers have explored potential associations with pesticides. To date, the organophosphate insecticide fonofos has the strongest and most consistent association with prostate cancer in the US Agricultural Health Study (Weichenthal *et al.* 2010). Other cancers that have been associated with specific pesticides in the Agricultural Health Study include lung, pancreatic, colon, rectum, bladder and brain cancers, as well as melanoma (Weichenthal *et al.* 2010). Growing evidence suggests that use of specific pesticides may increase the risk of specific cancers in adults. Laboratory animal studies suggest possible mechanisms (Alavanja *et al.* 2013). Data from non-occupationally exposed individuals are limited.

11.4.2 Neurological Consequences

Long-term neurological consequences in both children and adults have been a concern due to both the acute human health effects of exposure and the neurotoxic mechanisms of action for most insecticides (Kamel and Hoppin 2004). Pesticides have been evaluated for impacts on growth and development of children as well as neurological diseases in children such as autism and attention deficit hyperactivity disorder (ADHD). In adults, researchers have investigated the association of pesticides with non-specific neurological symptoms, neurophysiological function, depression and suicide as well as the neurodegenerative diseases of Parkinson disease and amyotrophic lateral sclerosis (ALS). The pesticides of interest have included the organophosphate and organochlorine insecticides, as well as the herbicide paraquat. Other chemicals have been evaluated but the data are more limited and are not reviewed here.

11.4.2.1 Childhood Development

Among children, extensive studies have evaluated the impact of organochlorine and organophosphate insecticides on child neurological and behavioural development. Outcomes of interest have included cognitive function, motor development and behavioural deficits using standardized neuropsychological scales. For organochlorine insecticides, prenatal exposure has been evaluated using the concentration of DDE in blood. Other studies have also considered breast milk as a potential exposure source. Although researchers have investigated the impact of organochlorine insecticides on childhood development in prospective studies over multiple continents and time periods, the evidence for an association of DDT with these outcomes is inconsistent, with the majority of the effects being attributed to polychlorinated biphenyls (PCBs), another environmentally and biologically persistent organochlorine compound (Korrick and Sagiv 2008).

For organophosphate insecticides, the evidence of developmental effects is much stronger than for organochlorine insecticides, with data from animal studies providing potential biological mechanisms to support the observational results for human populations exposed both agriculturally and residentially (Gonzalez-Alzaga *et al.* 2014; London *et al.* 2012; Munoz-Quezada *et al.* 2013; Perera *et al.* 2005). Most studies assessed exposure using urinary metabolites of organophosphate insecticides, the dialkylphosphates, which are not chemically specific. Some have used a urinary metabolite of chlorpyrifos called TCPy (3,5,6-trichloro-2-pyridinol). A 2014 review evaluated 20 articles assessing organophosphate insecticide exposure and neurodevelopmental effects in children (Gonzalez-Alzaga *et al.* 2014). Another review assessed 27 articles on this topic (Munoz-Quezada *et al.* 2013), and all but one paper showed some adverse effects with organophosphate exposure and neurobehavioural development. Overall, prenatal exposure was associated with motor deficits in neonates, attention deficits in toddlers and cognitive deficits at age 7 (Munoz-Quezada *et al.* 2013). Occupational exposure to organophosphate insecticides during pregnancy was associated with decreased visual memory and motor co-ordination at age 6–8 years (Gonzalez-Alzaga *et al.* 2014). ADHD has been associated with prenatal organophosphate insecticide exposure and in a cross-sectional study of children in the USA (Gonzalez-Alzaga *et al.* 2014; London *et al.* 2012). For other common pesticides, only the pyrethroid chemicals have been evaluated for neurodevelopmental outcomes, and while it is possible that they have some impact on neurodevelopment, at this point the majority of the research is still focused on organophosphate insecticides.

11.4.2.2 Autism

Autism, a common neurodevelopmental disorder, has been associated with agricultural pesticide exposure during pregnancy (Shelton *et al.* 2014). Autism was more common in children whose mothers lived within 1.5 km of agricultural fields treated with organophosphate insecticides. The association was greater for exposure to chlorpyrifos during the second trimester. Pyrethroid exposure in the third trimester also increased autism risk. Future studies are needed to better characterize these risks given that these are bystander exposures and not through personal occupational exposure.

11.4.2.3 Adult Neurological Outcomes

Given the neurotoxic effects of acute pesticide exposures, extensive research has been done to characterize the impact of low and moderate pesticide exposure, consistent with regular, routine use of pesticides (Kamel and Hoppin 2004). Neurological outcomes assessed in adults include measures of central and peripheral nervous system function as well as potential neurological symptoms. While studies of objective measures such as performance on cognitive function tests and nerve conduction velocity have not demonstrated strong associations with pesticide use (Kamel and Hoppin 2004; Starks *et al.* 2012b, c), evaluation of non-specific symptoms, such as headache and nausea, has been associated with pesticide use (Kamel and Hoppin 2004).

11.4.2.4 Depression and Suicide

Pesticides have been associated with depressive symptoms in adults (Kamel and Hoppin 2004; London *et al.* 2012), and there is concern that pesticide exposure may contribute to suicide risk. Among women in the US Agricultural Health Study, pesticide poisoning was predictive of incident depression, but use of specific chemicals was not (Beard *et al.* 2013). Among applicators in the US Agricultural Health Study, fungicides, fumigants and the organochlorine insecticides were associated with depression. Some but not all of the organophosphate insecticides (diazinon, malathion and parathion) were associated with incident depression as well (Beard *et al.* 2014). In South African farm workers, cumulative organophosphate exposure was associated with impulsivity, depression or suicide, with a history of past pesticide poisoning a risk factor as well (London *et al.* 2012).

The association of pesticides with suicide is poorly understood. In some communities, pesticides may be the most available agent for suicide. In the US Agricultural Health Study, there was no evidence of an association with agricultural pesticide use and suicide (Beard *et al.* 2011). It is possible that the demographic characteristics of this cohort (e.g. well-educated, land-owning farmers) make them more resilient with regard to suicide. However, this analysis was the largest study to date to evaluate the evidence for specific pesticide exposure and suicide.

11.4.2.5 Neurodegenerative Diseases

The strongest evidence for pesticides and any chronic human health outcomes is for Parkinson disease. Multiple observational studies in humans as well as laboratory studies in animals have identified specific pesticides associated with Parkinson disease including paraquat, rotenone and maneb (Costello *et al.* 2009; Tanner *et al.* 2011; Wang *et al.* 2011). The evidence for paraquat comes from laboratory studies, studies of occupationally exposed farmers, and from rural residents exposed to agricultural pesticides at their homes and work places (Baltazar *et al.* 2014; Berry *et al.* 2010; Costello *et al.* 2009; Dinis-Oliveira *et al.* 2006; Tanner *et al.* 2011). Other pesticides have been investigated for Parkinson disease risk, including the insecticide rotenone, the organophosphate insecticides (Manthripragada *et al.* 2010; Wang *et al.* 2014) and the fungicide maneb (Brown *et al.* 2006; Elbaz and Tranchant 2007; Freire and Koifman 2012; Hatcher *et al.* 2008; Tanner *et al.* 2011). Given that pesticide use patterns differ around the world, the consistency of the evidence for these pesticides and Parkinson disease is quite strong.

Amyotrophic lateral sclerosis (ALS), a highly fatal neurodegenerative disease, has been associated with pesticide exposure but the evidence is not strong. Due to the rarity of this illness and the high case fatality rate, it is difficult to assemble sufficient sample sizes to explore the role of specific chemicals. Most studies to date have evaluated pesticides as a whole and have observed an increased risk associated with pesticide exposure in men (Trojsi *et al.* 2013). In the US Agricultural Health Study, organochlorine insecticides (including DDT), pyrethroid insecticides, herbicides and fumigants were associated with ALS, but not other pesticide groups including the organophosphate insecticides (Kamel *et al.* 2012).

11.4.3 Respiratory Outcomes

Respiratory symptoms and disease are a common cause of morbidity in both developing and developed countries. Growing evidence suggests that pesticide exposure in both residential and agricultural settings may influence respiratory health of children and adults, although data on specific chemicals are limited (Fieten *et al.* 2009; Hoppin *et al.* 2002, 2006, 2007, 2008, 2009, 2014). Some pesticides may contribute to the development of chronic respiratory diseases, such as asthma or chronic obstructive pulmonary disease, while other exposures may act as irritants to contribute to respiratory symptoms or asthma exacerbation. Animal models describe potential mechanisms by which specific pesticides, including organophosphate insecticides (Proskocil *et al.* 2008, 2013), phenoxy herbicides (Colosio *et al.* 1999; Fukuyama *et al.* 2009) and paraquat (Cho *et al.* 2008), may contribute to respiratory and allergic outcomes in humans.

Of all the respiratory studies to date, evidence is strongest for specific pesticides and adult-onset asthma. Specific chemicals involved include paraquat, pyrethroid and organophosphate insecticides. Paraquat poisoning can result in fibrotic lung disease regardless of route of exposure (dermal, oral). Paraquat is also associated with respiratory symptoms in farm workers in South Africa and Nicaragua (Dalvie *et al.* 1999; Schenker *et al.* 2004). Pyrethroid insecticides (Lessenger 1992; Newton and Breslin 1983; Wagner 2000) and fungicides including mancozeb (Boers *et al.* 2008; Chatzi *et al.* 2007; Draper *et al.* 2003; Honda *et al.* 1992; Shelton *et al.* 1992) are associated with case reports of occupational asthma. Permethrin use by farm women is associated with elevated non-allergic asthma risk (Hoppin *et al.* 2008). Use of chlorpyrifos has been associated with wheeze and asthma in US farmers (Hoppin *et al.* 2002, 2006, 2009) and with wheeze among non-smoking indigenous women in Costa Rica (Fieten *et al.* 2009). Specific organophosphate insecticides, in particular chlorpyrifos and parathion, have been associated with wheeze in farmers (Hoppin *et al.* 2002) and commercial pesticide applicators (Hoppin *et al.* 2006) in an exposure-dependent fashion. Additionally, parathion and coumaphos were associated with a two-fold increased prevalence of allergic asthma in both farmers (Hoppin *et al.* 2009) and their wives (Hoppin *et al.* 2008). Additionally, chlorpyrifos was associated with allergic asthma in an exposure-dependent fashion (Hoppin *et al.* 2009). Organophosphates as measured by acetylcholinesterase inhibition have been associated with increased respiratory symptoms in Kenya (Ohayo-Mitoko *et al.* 2000), South Africa (Ndlovu *et al.* 2014), India (Chakraborty *et al.* 2009; Chitra *et al.* 2006) and Spain (Hernandez *et al.* 2008).

The data for children are limited but there is some evidence that prenatal and early life exposure may contribute to respiratory symptoms, asthma and poorer lung growth as measured by spirometry. Early life exposure to herbicides and insecticides was associated with early-onset persistent asthma in southern California, USA (Salam *et al.* 2004). Prenatal exposure to pyrethroid insecticides has been associated with respiratory symptoms in young children in New York City (Liu *et al.* 2012; Reardon *et al.* 2009).

Prenatal and early life (<5 years) exposure to non-specific organophosphate insecticide has been associated with respiratory symptoms and decreased lung function among children in California (Raanan *et al.* 2016, 2015). Prenatal exposure to DDE has been associated with increased wheeze and asthma, but not with atopy (Sunyer *et al.* 2005). While the evidence is still building, data from both human and animal studies suggest that pesticides have respiratory consequences regardless of exposure route.

11.4.4 Diabetes

Diabetes is a growing health burden worldwide, with rates increasing in both developed and developing nations. In 2009–2012 data from the USA, approximately 13.6% of adults had diabetes (CDC 2014). While obesity and sedentary lifestyle are known risk factors for adult-onset diabetes, exposure to environmental chemicals, including pesticides, may also play a role.

The majority of evidence of an association between pesticides and diabetes is for the insecticides, both the organochlorine and organophosphate classes. Organochlorine pesticides and their metabolites persist in biological tissues, so researchers have used stored blood samples to assess the association with diabetes in both occupationally exposed and general population groups. The first evidence of an association with organochlorine compounds and diabetes risk came from studies of Vietnam veterans who were exposed to the herbicide mix Agent Orange with the contaminant of 2,3,7,8-tetrachlordibenzodioxin (TCDD) (Starling and Hoppin 2015). Among participants in the US Agricultural Health Study, chlordane and heptachlor were associated with adult-onset diabetes in men, and dieldrin and the herbicides 2,4,5-T and 2,4,5-TP, which may have been contaminated with TCDD, were associated with diabetes in women (Starling and Hoppin 2015). The herbicides 2,4,5-T and 2,4,5-TP were also associated with gestational diabetes in this population (Starling and Hoppin 2015). DDT has also been associated with diabetes (Magliano *et al.* 2014).

The data for organophosphate and other current use insecticides are more limited, but evidence is increasing. In the US Agricultural Health Study, use of the organophosphate insecticides coumaphos, phorate, terbufos and trichlorfon were associated with adult-onset diabetes in men (Montgomery *et al.* 2008), and fonofos, parathion and phorate were associated with diabetes in women (Starling *et al.* 2014). Phorate and diazinon were also associated with gestational diabetes (Starling and Hoppin 2015). Studies in other countries have started to evaluate potential diabetes risk with organophosphate insecticides and pyrethroid pesticides, but to date the studies have been small and have used relatively crude exposure measures (e.g. working at a pyrethroid factory) (Jaacks and Staimez 2015). Currently, while there is evidence that some pesticides may contribute to obesity, it appears that the association between pesticides and diabetes is independent of obesity, meaning that there may be a direct impact of pesticides on developing diabetes.

11.4.5 Birth Outcomes

For many years, people have been concerned about the potential impact of pesticides on birth outcomes, including birth defects, low birth weight, stillbirth, preterm birth and miscarriage. Exposures of interest have included personal use of pesticides and proximity to treated fields. Pesticides evaluated include organochlorine, organophosphate and carbamate insecticides as well as a number of herbicides.

Despite this extensive evaluation, there is no strong evidence of an association between specific pesticides and these outcomes. An extensive review evaluating all peer-reviewed articles from 1966 to 2005 found that there was limited or inadequate evidence to support causality for all associations examined (Weselak *et al.* 2007). These authors critically examined the articles and applied a weight of evidence approach for specific chemicals and specific outcomes. Another review focusing on populations living near agricultural pesticide applications had similar findings for the 25 studies that were evaluated (Shirangi *et al.* 2011). The authors felt that while 'residential proximity to agricultural pesticide applications may be an important source of ambient, environmental exposure', the strength of the evidence was generally weak due primarily to challenges in exposure characterization (Shirangi *et al.* 2011). Because specific birth defects are rare and timing of exposure is critical for development of a specific type of birth defect, there is often limited statistical power to assess this relationship. While both reviews indicated that the current evidence is inadequate to suggest a causal relationship, both felt that better exposure characterization would facilitate better science on this topic.

11.4.6 Cardiovascular Outcomes

While specific pesticide exposure has been associated with a number of common adult chronic diseases, not all pesticides have been associated with all diseases and, for some diseases, there is no evidence that pesticides play a role. For example, in two studies on heart attack occurrence and heart attack mortality from the US Agricultural Health Study, there was no evidence that any of the pesticides were associated with these outcomes (Dayton *et al.* 2010; Mills *et al.* 2009). Similarly, there was no evidence for pesticide exposure and stroke mortality in this cohort (Rinsky *et al.* 2013). While pesticide poisonings do have cardiac symptoms, at this point the data on chronic low-level pesticide exposure do not support an association with cardiovascular outcomes.

11.5 Conclusions

The totality of potential human health effects of pesticides is unknown, though current epidemiological studies are greatly enhancing our knowledge. Lack of animal models for the full range of human diseases makes premarket testing impossible. Additionally, until pesticides are released into the environment, understanding the potential human exposure is difficult. The information on human health effects of pesticides comes from epidemiological studies of individuals exposed to pesticides at work or at home. These are not experimental studies, so exact characterization of the exposure–response relationship or complete control for variation in the population is not possible. Additionally, it is impossible in these studies to tease out whether it is the pesticide active ingredient or some other part of the pesticide product that contributes to the observed health effects. However, the studies to date are of high quality, have been conducted in multiple populations and suggest areas of potential concern that should be considered by pesticide users and their physicians. Because epidemiological studies can only characterize health effects of chemicals after years on the market, the data for newer chemicals like the neonicotinoid insecticides are more limited. While newer chemicals are often believed to be safer, many of the older chemicals, such as DDT and paraquat, continue to be used around the world under conditions that may contribute to human exposure.

In order to prevent pesticide-related health effects, current understanding of the impacts of pesticides on human health supports the following recommendations. Globally, efforts should be made to enhance monitoring and improve healthcare providers' recognition and management of acute pesticide poisonings. To reduce the number of poisonings, improvements need to be made in pesticide storage, particularly in rural areas where pesticides may be more likely to be used in suicides (Gunnell *et al.* 2007). Technological advances should focus on both reducing the toxicity of new pesticide products and decreasing human exposure, particularly to the most toxic pesticides, through the development of user-friendly personal protective equipment, improved pesticide packaging and innovative engineering controls (e.g. closed pesticide application systems).

Training and hazard communication programmes for agricultural workers are needed worldwide and should include information about acute and chronic health outcomes, safe pesticide handling techniques, strategies for minimizing take-home and bystander exposures, and integrated pest management, in which pesticides are one of a combination of approaches implemented to mitigate pests. Agricultural workers can reduce personal exposure by utilizing good hygiene practices of washing their hands, bodies and clothing after working around pesticides. Pesticide applicators should follow pesticide label instructions for diluting and applying pesticides, re-entering treated areas after pesticide application, and using personal protective equipment and clothing, which minimally include long trousers, long-sleeved shirts and boots.

Language and literacy are important considerations in the communication of pesticide risks to agricultural workers in both less and more economically developed countries, particularly among immigrant workers. Worldwide, agricultural workers include at-risk individuals, such as immigrants, ethnic minorities, women, children and individuals with low incomes and limited formal education (Donham and Thelin 2006). In the areas of training and hazard communication, healthcare access and provision, and health and safety protection in the work environment, the unique sociocultural needs of vulnerable pesticide-exposed populations must be recognized and addressed.

Pesticides are important agricultural chemicals. In this chapter, we have highlighted some of the key human health outcomes associated with pesticides, but this should not be regarded as comprehensive. While pesticides are associated with common chronic health outcomes, greater concern and focus should be on the prevention of pesticide poisonings. Ultimately, the overarching issue should be proper and safe use of pesticides in all populations.

References

Alavanja, M.C. (2009) Introduction: pesticides use and exposure extensive worldwide. *Reviews on Environmental Health* **24**: 303–309.

Alavanja, M.C., Sandler, D.P., McMaster, S.B., *et al.* (1996) The Agricultural Health Study. *Environmental Health Perspectives* **104**: 362–369.

Alavanja, M.C., Ross, M.K. and Bonner, M.R. (2013) Increased cancer burden among pesticide applicators and others due to pesticide exposure. *CA: A Cancer Journal for Clinicians* **63**: 120–142.

American Cancer Society (ACS) (2016) Lifetime Risk of Developing or Dying from Cancer. Available at: www.cancer.org/cancer/cancerbasics/lifetime-probability-of-developing-or-dying-from-cancer (accessed 7 March 2017).

Baltazar, M.T., Dinis-Oliveira, R.J., de Lourdes Bastos, M., Tsatsakis, A.M., Duarte, J.A. and Carvalho, F. (2014) Pesticides exposure as etiological factors of Parkinson's disease and other neurodegenerative diseases – a mechanistic approach. *Toxicology Letters* **230**: 85–103.

Beard, J.D., Umbach, D.M., Hoppin, J.A., *et al.* (2011) Suicide and pesticide use among pesticide applicators and their spouses in the agricultural health study. *Environmental Health Perspectives* **119**: 1610–1615.

Beard, J.D., Hoppin, J.A., Richards, M., *et al.* (2013) Pesticide exposure and self-reported incident depression among wives in the Agricultural Health Study. *Environmental Research* **126**: 31–42.

Beard, J.D., Umbach, D.M., Hoppin, J.A., *et al.* (2014) Pesticide exposure and depression among male private pesticide applicators in the Agricultural Health Study. *Environmental Health Perspectives* **122**: 984–991.

Berry, C., La Vecchia, C. and Nicotera, P. (2010) Paraquat and Parkinson's disease. *Cell Death and Differerentiation* **17**: 1115–1125.

Bertolote, J.M., Fleischmann, A., Eddleston, M. and Gunnell, D. (2006) Deaths from pesticide poisoning: a global response. *British Journal of Psychiatry* **189**: 201–203.

Boers, D., van Amelsvoort, L., Colosio, C., *et al.* (2008) Asthmatic symptoms after exposure to ethylenebisdithiocarbamates and other pesticides in the Europit field studies. *Human and Experimental Toxicology* **27**: 721–727.

Bradman, A., Quiros-Alcala, L., Castorina, R., *et al.* (2015) Effect of organic diet intervention on pesticide exposures in young children living in low-income urban and agricultural communities. *Environmental Health Perspectives* **123**: 1086–1093.

Brouwer, M., Schinasi, L., Beane Freeman, L.E., *et al.* (2016) Assessment of occupational exposure to pesticides in a pooled analysis of agricultural cohorts within the AGRICOH consortium. *Occupational and Environmental Medicine* in press.

Brown, T.P., Rumsby, P.C., Capleton, A.C., Rushton, L. and Levy, L.S. (2006) Pesticides and Parkinson's disease – is there a link? *Environmental Health Perspectives* **114**: 156–164.

Calvert, G.M., Beckman, J., Prado, J.B., *et al.* (2013) Acute occupational pesticide-related illness and injury – United States, 2007–2010. *Morbidity and Mortality Weekly Report* **62**: 5–9.

Centers for Disease Control and Prevention (CDC) (2014) National Diabetes Statistics Report, 2014. Available at: www.cdc.gov/diabetes/pubs/statsreport14/national-diabetes-report-web.pdf (accessed 7 March 2017).

Chakraborty, S., Mukherjee, S., Roychoudhury, S., Siddique, S., Lahiri, T. and Ray, M.R. (2009) Chronic exposures to cholinesterase-inhibiting pesticides adversely affect respiratory health of agricultural workers in India. *Journal of Occupational Health* **51**: 488–497.

Chatzi, L., Alegakis, A., Tzanakis, N., Siafakas, N., Kogevinas, M. and Lionis, C. (2007) Association of allergic rhinitis with pesticide use among grape farmers in Crete, Greece. *Occupational and Environmental Medicine* **64**: 417–421.

Chen, M., Chang, C.H., Tao, L. and Lu, C. (2015) Residential exposure to pesticide during childhood and childhood cancers: a meta-analysis. *Pediatrics* **136**: 719–729.

Chitra, G.A., Muraleedharan, V.R., Swaminathan, T. and Veeraraghavan, D. (2006) Use of pesticides and its impact on health of farmers in South India. *International Journal of Occupational and Environmental Health* **12**: 228–233.

Cho, Y.S., Oh, S.Y. and Zhu, Z. (2008) Tyrosine phosphatase SHP-1 in oxidative stress and development of allergic airway inflammation. *American Journal of Respiratory Cell and Molecular Biology* **39**: 412–419.

Colosio, C., Corsini, E., Barcellini, W. and Maroni, M. (1999) Immune parameters in biological monitoring of pesticide exposure: current knowledge and perspectives. *Toxicology Letters* **108**: 285–295.

Costello, S., Cockburn, M., Bronstein, J., Zhang, X. and Ritz, B. (2009) Parkinson's disease and residential exposure to maneb and paraquat from agricultural applications in the central valley of California. *American Journal of Epidemiology* **169**: 919–926.

Curl, C.L., Fenske, R.A., Kissel, J.C., *et al.* (2002) Evaluation of take-home organophosphorus pesticide exposure among agricultural workers and their children. *Environmental Health Perspectives* **110**: A787–A792.

Dalvie, M.A., White, N., Raine, R., *et al.* (1999) Long-term respiratory health effects of the herbicide, paraquat, among workers in the Western Cape. *Occupational and Environmental Medicine* **56**: 391–396.

Dayton, S.B., Sandler, D.P., Blair, A., Alavanja, M., Beane Freeman, L.E. and Hoppin, J.A. (2010) Pesticide use and myocardial infarction incidence among farm women in the agricultural health study. *Journal of Occupational and Environmental Medicine* **52**: 693–697.

Dinis-Oliveira, R.J., Remiao, F., Carmo, H., *et al.* (2006) Paraquat exposure as an etiological factor of Parkinson's disease. *Neurotoxicology* **27**: 1110–1122.

Donham, K. and Thelin, A. (2006) *Agricultural Medicine: Occupational and Environmental Health For The Health Professions*. Wiley-Blackwell, Ames, IA, USA.

Draper, A., Cullinan, P., Campbell, C., Jones, M. and Newman Taylor, A. (2003) Occupational asthma from fungicides fluazinam and chlorothalonil. *Occupational and Environmental Medicine* **60**: 76–77.

Elbaz, A. and Tranchant, C. (2007) Epidemiologic studies of environmental exposures in Parkinson's disease. *Journal of Neurological Sciences* **262**: 37–44.

Fieten, K.B., Kromhout, H., Heederik, D. and van Wendel de Joode, B. (2009) Pesticide exposure and respiratory health of indigenous women in Costa Rica. *American Journal of Epidemiology* **169**: 1500–1506.

Freire, C. and Koifman, S. (2012) Pesticide exposure and Parkinson's disease: epidemiological evidence of association. *Neurotoxicology* **33**: 947–971.

Fukuyama, T., Tajima, Y., Ueda, H., *et al.* (2009) Allergic reaction induced by dermal and/or respiratory exposure to low-dose phenoxyacetic acid, organophosphorus, and carbamate pesticides. *Toxicology* **261**: 152–161.

Garry, V.F. (2004) Pesticides and children. *Toxicology and Applied Pharmacology* **198**: 152–163.

Gonzalez-Alzaga, B., Lacasana, M., Aguilar-Garduno, C., *et al.* (2014) A systematic review of neurodevelopmental effects of prenatal and postnatal organophosphate pesticide exposure. *Toxicology Letters* **230**: 104–121.

Gunnell, D., Eddleston, M., Phillips, M.R. and Konradsen, F. (2007) The global distribution of fatal pesticide self-poisoning: systematic review. *BMC Public Health* **7**: 357.

Guyton, K.Z., Loomis, D., Grosse, Y., *et al.* (2015) Carcinogenicity of tetrachlorvinphos, parathion, malathion, diazinon, and glyphosate. *Lancet Oncology* **16**: 490–491.

Hatcher, J.M., Pennell, K.D. and Miller, G.W. (2008) Parkinson's disease and pesticides: a toxicological perspective. *Trends in Pharmacological Sciences* **29**: 322–329.

Hernandez, A.F., Casado, I., Pena, G., Gil, F., Villanueva, E. and Pla, A. (2008) Low level of exposure to pesticides leads to lung dysfunction in occupationally exposed subjects. *Inhalation Toxicology* **20**: 839–849.

Hodgson, E. (2010) *A Textbook of Modern Toxicology*. John Wiley and Sons, Hoboken, NJ, USA.

Honda, I., Kohrogi, H., Ando, M., *et al.* (1992) Occupational asthma induced by the fungicide tetrachloroisophthalonitrile. *Thorax* **47**: 760–761.

Hoppin, J.A., Umbach, D.M., London, S.J., Alavanja, M.C.R. and Sandler, D.P. (2002) Chemical predictors of wheeze among farmer pesticide applicators in the agricultural health study. *American Journal of Respiratory and Critical Care Medicine* **165**: 683–689.

Hoppin, J.A., Umbach, D.M., London, S.J., Lynch, C.F., Alavanja, M.C. and Sandler, D.P. (2006) Pesticides associated with wheeze among commercial pesticide applicators in the Agricultural Health Study. *American Journal of Epidemiology* **163**: 1129–1137.

Hoppin, J.A., Umbach, D.M., Kullman, G.J., *et al.* (2007) Pesticides and other agricultural factors associated with self-reported farmer's lung among farm residents in the Agricultural Health Study. *Occupational and Environmental Medicine* **64**: 334–341.

Hoppin, J.A., Umbach, D.M., London, S.J., *et al.* (2008) Pesticides and atopic and nonatopic asthma among farm women in the Agricultural Health Study. *American Journal of Respiratory and Critical Care Medicine* **177**: 11–18.

Hoppin, J.A., Umbach, D.M., London, S.J., *et al.* (2009) Pesticide use and adult-onset asthma among male farmers in the Agricultural Health Study. *European Respiratory Journal* **34**: 1296–1303.

Hoppin, J.A., Umbach, D.M., Long, S., *et al.* (2014) Respiratory disease in United States farmers. *Occupational and Environmental Medicine* **71**: 484–489.

Infante-Rivard, C. and Weichenthal, S. (2007) Pesticides and childhood cancer: an update of Zahm and Ward's 1998 review. *Journal of Toxicology and Environmental Health, Part B: Critical Reviews* **10**: 81–99.

International Agency for Research on Cancer (IARC) (1991) *Occupational Exposures in Insecticide Application, and Some Pesticides*. IARC, Lyon, France.

Jaacks, L.M. and Staimez, L.R. (2015) Association of persistent organic pollutants and non-persistent pesticides with diabetes and diabetes-related health outcomes in Asia: a systematic review. *Environment International* **76**: 57–70.

Jeyaratnam, J. (1985) Health problems of pesticide usage in the Third World. *British Journal of Industrial Medicine* **42**: 505–506.

Jeyaratnam, J. (1990) Acute pesticide poisoning: a major global health problem. *World Health Statistics Quarterly* **43**: 139–144.

Jones, D.R., Sutton, A.J., Abrams, K.R., Fenty, J., Warren, F. and Rushton, L. (2009) Systematic review and meta-analysis of mortality in crop protection product manufacturing workers. *Occupational and Environmental Medicine* **66**: 7–15.

Kamel, F. and Hoppin, J.A. (2004) Association of pesticide exposure with neurologic dysfunction and disease. *Environmental Health Perspectives* **112**: 950–958.

Kamel, F., Umbach, D.M., Bedlack, R.S., *et al.* (2012) Pesticide exposure and amyotrophic lateral sclerosis. *Neurotoxicology* **33**: 457–462.

Korrick, S.A. and Sagiv, S.K. (2008) Polychlorinated biphenyls, organochlorine pesticides and neurodevelopment. *Current Opinion in Pediatrics* **20**: 198–204.
Koutros, S., Alavanja, M.C., Lubin, J.H., *et al.* (2010) An update of cancer incidence in the Agricultural Health Study. *Journal of Occupational and Environmental Medicine* **52**: 1098–1105.
Krieger, R. (2010) *Hayes' Handbook of Pesticide Toxicology*. Academic Press, Burlington, MA, USA.
Langley, R.L. and Mort, S.A. (2012) Human exposures to pesticides in the United States. *Journal of Agromedicine* **17**: 300–315.
Lessenger, J.E. (1992) Five office workers inadvertently exposed to cypermethrin. *Journal of Toxicology and Environmental Health* **35**: 261–267.
Liu, B., Jung, K.H., Horton, M.K., *et al.* (2012) Prenatal exposure to pesticide ingredient piperonyl butoxide and childhood cough in an urban cohort. *Environment International* **48**: 156–161.
London, L., Beseler, C., Bouchard, M.F., *et al.* (2012) Neurobehavioral and neurodevelopmental effects of pesticide exposures. *Neurotoxicology* **33**: 887–896.
Loomis, D., Guyton, K., Grosse, Y., *et al.* (2015) Carcinogenicity of lindane, DDT, and 2,4-dichlorophenoxyacetic acid. *Lancet Oncology* **16**: 891–892.
Lu, C., Toepel, K., Irish, R., Fenske, R.A., Barr, D.B. and Bravo, R. (2006) Organic diets significantly lower children's dietary exposure to organophosphorus pesticides. *Environmental Health Perspectives* **114**: 260–263.
Magliano, D.J., Loh, V.H., Harding, J.L., Botton, J. and Shaw, J.E. (2014) Persistent organic pollutants and diabetes: a review of the epidemiological evidence. *Diabetes and Metabolism* **40**: 1–14.
Manthripragada, A.D., Costello, S., Cockburn, M.G., Bronstein, J.M. and Ritz, B. (2010) Paraoxonase 1, agricultural organophosphate exposure, and Parkinson disease. *Epidemiology* **21**: 87–94.
Merhi, M., Raynal, H., Cahuzac, E., Vinson, F., Cravedi, J.P. and Gamet-Payrastre, L. (2007) Occupational exposure to pesticides and risk of hematopoietic cancers: meta-analysis of case-control studies. *Cancer Causes and Control* **18**: 1209–1226.
Mills, K.T., Blair, A., Freeman, L.E., Sandler, D.P. and Hoppin, J.A. (2009) Pesticides and myocardial infarction incidence and mortality among male pesticide applicators in the Agricultural Health Study. *American Journal of Epidemiology* **170**: 892–900.
Montgomery, M.P., Kamel, F., Saldana, T.M., Alavanja, M.C. and Sandler, D.P. (2008) Incident diabetes and pesticide exposure among licensed pesticide applicators: Agricultural Health Study, 1993–2003. *American Journal of Epidemiology* **167**: 1235–1246.
Munoz-Quezada, M.T., Lucero, B.A., Barr, D.B., *et al.* (2013) Neurodevelopmental effects in children associated with exposure to organophosphate pesticides: a systematic review. *Neurotoxicology* **39**: 158–168.
Ndlovu, V., Dalvie, M.A. and Jeebhay, M.F. (2014) Asthma associated with pesticide exposure among women in rural Western Cape of South Africa. *American Journal of Industrial Medicine* **57**: 1331–1343.
Newton, J.G. and Breslin, A.B. (1983) Asthmatic reactions to a commonly used aerosol insect killer. *Medical Journal of Australia* **1**: 378–380.
Ohayo-Mitoko, G.J.A., Kromhout, H., Simwa, J.M., Boleij, J.S.M. and Heederik, D. (2000) Self reported symptoms and inhibition of acetylcholinesterase activity among Kenyan agricultural workers. *Occupational and Environmental Medicine* **57**: 195–200.

Perera, F.P., Rauh, V., Whyatt, R.M., *et al.* (2005) A summary of recent findings on birth outcomes and developmental effects of prenatal ETS, PAH, and pesticide exposures. *Neurotoxicology* **26**: 573–587.

Proskocil, B.J., Bruun, D.A., Lorton, J.K., *et al.* (2008) Antigen sensitization influences organophosphorus pesticide-induced airway hyperreactivity. *Environmental Health Perspectives* **116**: 381–388.

Proskocil, B.J., Bruun, D.A., Jacoby, D.B., van Rooijen, N., Lein, P.J. and Fryer, A.D. (2013) Macrophage TNF-alpha mediates parathion-induced airway hyperreactivity in guinea pigs. *American Journal of Physiology: Lung Cellular and Molecular Physiology* **304**: L519–L529.

Raanan, R., Harley, K.G., Balmes, J.R., Bradman, A., Lipsett, M. and Eskenazi, B. (2015) Early-life exposure to organophosphate pesticides and pediatric respiratory symptoms in the CHAMACOS cohort. *Environmental Health Perspectives* **123**: 179–185.

Raanan, R., Balmes, J.R., Harley, K.G., *et al.* (2016) Decreased lung function in 7-year-old children with early-life organophosphate exposure. *Thorax* **71**: 148–153.

Reardon, A.M., Perzanowski, M.S., Whyatt, R.M., Chew, G.L., Perera, F.P. and Miller, R.L. (2009) Associations between prenatal pesticide exposure and cough, wheeze, and IgE in early childhood. *Journal of Allergy and Clinical Immunology* **124**: 852–854.

Reigart, J. and Roberts, J. (2013) *Recognition and Management of Pesticide Poisonings.* United States Environmental Protection Agency, Office of Pesticide Programs, Washington, DC, USA.

Rinsky, J.L., Hoppin, J.A., Blair, A., He, K., Beane Freeman, L.E. and Chen, H. (2013) Agricultural exposures and stroke mortality in the Agricultural Health Study. *Journal of Toxicology and Environmental Health, Part A* **76**: 798–814.

Salam, M.T., Li, Y.F., Langholz, B. and Gilliland, F.D. (2004) Early-life environmental risk factors for asthma: findings from the Children's Health Study. *Environmental Health Perspectives* **112**: 760–765.

Schenker, M.B., Stoecklin, M., Lee, K., *et al.* (2004) Pulmonary function and exercise-associated changes with chronic low-level paraquat exposure. *American Journal of Respiratory and Critical Care Medicine* **170**: 773–779.

Schinasi, L. and Leon, M.E. (2014) Non-Hodgkin lymphoma and occupational exposure to agricultural pesticide chemical groups and active ingredients: a systematic review and meta-analysis. *International Journal of Environmental Research and Public Health* **11**: 4449–4527.

Shelton, D., Urch, B. and Tarlo, S.M. (1992) Occupational asthma induced by a carpet fungicide – tributyl tin oxide. *Journal of Allergy and Clinical Immunology* **90**: 274–275.

Shelton, J.F., Geraghty, E.M., Tancredi, D.J., *et al.* (2014) Neurodevelopmental disorders and prenatal residential proximity to agricultural pesticides: the CHARGE study. *Environmental Health Perspectives* **122**: 1103–1109.

Shirangi, A., Nieuwenhuijsen, M., Vienneau, D. and Holman, C.D.J. (2011) Living near agricultural pesticide applications and the risk of adverse reproductive outcomes: a review of the literature. *Paediatric and Perinatal Epidemiology* **25**: 172–191.

Starks, S.E., Gerr, F., Kamel, F., *et al.* (2012a) High pesticide exposure events and central nervous system function among pesticide applicators in the Agricultural Health Study. *International Archives of Occupational and Environmental Health* **85**: 505–515.

Starks, S.E., Gerr, F., Kamel, F., *et al.* (2012b) Neurobehavioral function and organophosphate insecticide use among pesticide applicators in the Agricultural Health Study. *Neurotoxicology and Teratology* **34**: 168–176.

Starks, S.E., Hoppin, J.A., Kamel, F., *et al.* (2012c) Peripheral nervous system function and organophosphate pesticide use among licensed pesticide applicators in the Agricultural Health Study. *Environmental Health Perspectives* **120**: 515–520.

Starling, A.P. and Hoppin, J.A. (2015) Environmental chemical risk factors for Type 2 diabetes: an update. *Diabetes Management* **5**: 285–299.

Starling, A.P., Umbach, D.M., Kamel, F., Long, S., Sandler, D.P. and Hoppin, J.A. (2014) Pesticide use and incident diabetes among wives of farmers in the Agricultural Health Study. *Occupational and Environmental Medicine* **71**: 629–635.

Sunyer, J., Torrent, M., Munoz-Ortiz, L., *et al.* (2005) Prenatal dichlorodiphenyldichloroethylene (DDE) and asthma in children. *Environmental Health Perspectives* **113**: 1787–1790.

Tanner, C.M., Kamel, F., Ross, G.W., *et al.* (2011) Rotenone, paraquat, and Parkinson's disease. *Environmental Health Perspectives* **119**: 866–872.

Trojsi, F., Monsurro, M.R. and Tedeschi, G. (2013) Exposure to environmental toxicants and pathogenesis of amyotrophic lateral sclerosis: state of the art and research perspectives. *International Journal of Molecular Sciences* **14**: 15286–15311.

United States Environmental Protection Agency (USEPA) (2002) Series 870 – Health Effects Test Guidelines. Group A 870.1000 Acute Toxicity Testing – Background. Available at: www.epa.gov/test-guidelines-pesticides-and-toxic-substances/series-870-health-effects-test-guidelines (accessed 7 March 2017).

United States Environmental Protection Agency (USEPA) (2015) Label Review Manual. Available at: www.epa.gov/pesticide-registration/label-review-manual (accessed 7 March 2017).

United States National Toxicology Program (NTP) (2014) 13th Report on Carcinogens. Available at: https://ntp.niehs.nih.gov/annualreport/2015/glance/roc/index.html (accessed 7 march 2017).

Van Maele-Fabry, G., Lantin, A.C., Hoet, P. and Lison, D. (2011) Residential exposure to pesticides and childhood leukaemia: a systematic review and meta-analysis. *Environmental International* **37**: 280–291.

Van Maele-Fabry, G., Hoet, P. and Lison, D. (2013) Parental occupational exposure to pesticides as risk factor for brain tumors in children and young adults: a systematic review and meta-analysis. *Environmental International* **56**: 19–31.

Waggoner, J.K., Kullman, G.J., Henneberger, P.K., *et al.* (2011) Mortality in the agricultural health study, 1993–2007. *American Journal of Epidemiology* **173**: 71–83.

Wagner, S.L. (2000) Fatal asthma in a child after use of an animal shampoo containing pyrethrin. *Western Journal of Medicine* **173**: 86–87.

Wang, A., Costello, S., Cockburn, M., Zhang, X., Bronstein, J. and Ritz, B. (2011) Parkinson's disease risk from ambient exposure to pesticides. *European Journal of Epidemiology* **26**: 547–555.

Wang, A., Cockburn, M., Ly, T.T., Bronstein, J.M. and Ritz, B. (2014) The association between ambient exposure to organophosphates and Parkinson's disease risk. *Occupational and Environmental Medicine* **71**: 275–281.

Weichenthal, S., Moase, C. and Chan, P. (2010) A review of pesticide exposure and cancer incidence in the Agricultural Health Study cohort. *Environmental Health Perspectives* **118**: 1117–1125.

Weselak, M., Arbuckle, T.E. and Foster, W. (2007) Pesticide exposures and developmental outcomes: the epidemiological evidence. *Journal of Toxicology and Environmental Health, Part B: Critical Reviews* **10**: 41–80.

Wigle, D.T., Turner, M.C. and Krewski, D. (2009) A systematic review and meta-analysis of childhood leukemia and parental occupational pesticide exposure. *Environmental Health Perspectives* **117**: 1505–1513.

World Health Organization (WHO) (2010) WHO Recommended Classification of Pesticides by Hazard and Guidelines to Classification 2009. Available at: www.who.int/ipcs/publications/pesticides_hazard_2009.pdf (accessed 7 March 2017).

12

Human Health Concerns Related to the Consumption of Foods from Genetically Modified Crops

Javier Magaña-Gómez and Ana Maria Calderón de la Barca

12.1 History of GM Foods and Associated Food Safety Concerns

Genetic engineering includes a large number of biotechnologies that are used for different purposes in crops, livestock, forestry, fisheries, aquaculture and agro-industry. The use of recombinant DNA technology makes it possible to modify genetic material in a way that does not occur naturally by mating or through natural recombination. It is now possible to modify endogenous genes or mobilize and express genes from unrelated species, called 'transgenes', into other organisms to obtain specific and desirable traits. Organisms created by these approaches are referred to as genetically modified (GM) organisms. When modifications are applied to either plants or animals for the purpose of human consumption, these are called 'genetically modified foods' (GM foods). GM foods can be classified into two categories (James 2014): those that are tissues of GM crops (potatoes, tomatoes, soya, maize, sunflowers, rice, pumpkins, melons, rape, etc.) and derivative products from GM crops (starch, oil, sugar, amino acids, vitamins, etc.).

In addition to the food industry, plant genetic modification has widespread applications in other areas such as agricultural, biological and medical research, pharmaceutical drug production and experimental nutrition. In accordance with the goal of genetic modification, GM plants are also classified into three generations. First generation refers to crop modifications conferring resistance to herbicides or pests, with benefits mainly for producers. Depending on legislation and approval granted in each country, these crops are currently grown for commercial or experimental use. Second generation refers to crops which are modified to improve their nutritional content, with direct benefits for consumers. Currently, these crops are not grown for commercial purposes but are still at the development stage. Third generation refers to plants engineered to produce pharmaceuticals, for example vaccines, antibodies and proteins to treat human or animal diseases, and industrial products such as biodegradable plastics, fibrous proteins, adhesives and synthetic proteins. Their usage state is similar to that of second-generation plants.

Consumers choose foodstuffs assuming there are minimal health risks based on their history of consumption and protective health regulations. This is also assumed for GM foods that do not differ in appearance from the corresponding non-GM foods. However,

Environmental Pest Management: Challenges for Agronomists, Ecologists, Economists and Policymakers, First Edition. Edited by Moshe Coll and Eric Wajnberg.

various reports about possible risks associated with GM food intake have spread through the media, thereby influencing consumers' decisions on whether to consume them or not.

One of the most well-known concerns is that GM foods may include some molecules that have no natural history of human consumption or that the GM processing has changed other food properties. However, one of the most widely perceived risks is related to the effects of GM crops on biodiversity, the environment and the economy. These are discussed elsewhere in this volume (see Chapter 6). Furthermore, the attitude toward GM crops and the foods produced from them is not ubiquitously negative because advantages such as improving food production and the availability of biomolecules with industrial or therapeutic properties are also realized (Ricroch and Hénard-Damave 2016; WHO/FAO 2009).

Historically, the rejection of GM foods could not be attributed only to the 'technical' aspects of genetic engineering, but also to circumstances at the time they were encountered in the market. At the beginning of their commercialization, GM foodstuffs such as tomato paste and dairy products made from milk obtained from cows treated with recombinant hormones were clearly labelled as being made with genetically modified tomatoes or recombinant hormones. Whether to buy these products or not was entirely up to the consumer, who usually was not well informed about the new biotechnological developments (Frewer *et al.* 2004; Uzogara 2000).

In the summer of 1998, in the middle of the scare in England over adverse food effects caused by mad cow disease and dioxin contamination in animal feeds, Dr Pusztai, from the Rowett Institute in Scotland, explained in a television interview that he had found adverse effects in the immune system of mice fed with GM potatoes (Ewen and Pusztai 1999). Some days later, GM soybean was detected in 60% of soy-containing products without full disclosure on the label. These events led to reduced confidence in the safety of GM foods and drew attention to the lack of effective regulatory frameworks in food production practices, thereby negatively affecting the acceptance of GM foods. Consumers believed that the industry was disregarding health risks in order to protect economic interests and this created a crisis of confidence. Regulators were quick to respond, banning GM foods from Great Britain and thereafter from Europe and other countries.

Several determinants have been found to be important in the acceptance of technology-based food innovations. These include characteristics of the innovation, the consumer and the social system, perceived cost–benefit ratio considerations, perceptions of risk and uncertainty, social norms and how much control consumers perceive that they have (Ronteltap *et al.* 2007). Clearly, no single study could cover all these aspects.

Unacceptability of GM foods could be exacerbated because people perceive that the main beneficiaries are biotech companies and crop growers, not consumers or society. On the other hand, consumers are seen as the ones exposed to the potential health risks. Even worse is the belief that genetic engineering alters the natural order or has unintended and unpredictable effects that could be hidden by producers or regulators. This perception may differ greatly between developed or developing countries due to differences in socioeconomic and political structures. Finally, limited public access to scientific information may further aggravate societal perception of government regulation.

12.2 Status and Commercial Traits Regarding Genetically Modified Organisms

12.2.1 Global Perspective on Genetically Modified Crops

Production data related to GM crops reveal their current importance. Despite the controversy, GM crops have been adopted rapidly. According to James (2014), GM crop hectares increased 100-fold between 1996 and 2014, from 1.7 to 181.5 million ha worldwide. Initially, the planting was concentrated in developed countries but in 2014, 20 countries that planted GM crops were developing countries and only eight were developed. Production is mainly concentrated in the USA, Brazil, Argentina, Canada and India (with 73.1, 42.2, 24.3, 11.6 and 11.6 million ha, respectively).

The most popular GM-derived traits are those that confer herbicide tolerance (59% of the total GM crops), insect resistance (15%) or both (26%) (Table 12.1) (see also Chapter 6). Each phenotypic or commercial trait is produced by specific genetic changes or 'events'. According to the International Service for the Acquisition of Agri-biotech Applications GM Approval Database (ISAAA 2015), the commercial trait with the largest number of events is herbicide tolerance, followed by insect resistance (see Table 12.1). Nearly all the GM crops are soybean (47%), maize (32%), cotton (15%) and canola (5%).

Currently, 40 countries have regulatory approval of GM crops for food, feed or cultivation use. Japan has the highest number of changes or events approved, followed by the USA, Canada and Mexico (Table 12.2). Maize has the biggest number of approved events, followed by cotton, potatoes and canola and soybean (Table 12.3). The herbicide-tolerant soybean event GTS-40-3-2 has the biggest number of regulatory approvals, followed by the herbicide-tolerant maize event NK603, insect-resistant maize events MON810, Bt11 and TC1507, herbicide-tolerant maize event GA21, insect-resistant cotton event MON531, and insect-resistant maize event MON89034 (James 2014).

Table 12.1 Introduced commercial traits through genetic modification of plants. Data taken from ISAAA (2015).

Trait	GM plant species/cultivar	Events*
Abiotic stress tolerance	3	8
Altered growth/yield	3	3
Disease resistance	7	26
Herbicide tolerance	15	239
Insect resistance	8	198
Modified product quality	13	79
Pollination control system	3	25

* Number of phenotypic or commercial traits that are being produced by specific genetic changes. An event could include modifications. Two or more events together are known as a 'stacked' event.

Table 12.2 Number of approved genetically modified events in different countries (alphabetical order). Data taken from ISAAA (2015).

Country	Approved events*	GM plants
Argentina	40	3
Australia	106	11
Bangladesh	1	1
Bolivia	1	1
Brazil	50	5
Burkina Faso	1	1
Canada	164	14
Chile	3	3
China	60	11
Colombia	73	9
Costa Rica	15	2
Cuba	1	1
Egypt	1	1
European Union	86	7
Honduras	8	2
India	11	2
Indonesia	15	3
Iran	1	1
Japan	214	11
Malaysia	22	3
Mexico	158	9
Myanmar	1	1
New Zealand	91	1
Norway	11	1
Pakistan	2	1
Panama	1	1
Paraguay	20	3
Philippines	88	8
Russian Federation	23	5
Singapore	23	6
South Africa	67	5
South Korea	137	7
Sudan	1	1
Switzerland	4	2
Taiwan	113	5
Thailand	15	2
Turkey	24	2
Uruguay	17	2
USA	189	20
Vietnam	6	1

* Number of phenotypic or commercial traits that are produced by specific genetic changes.

Table 12.3 Genetically modified plant species with approved events (alphabetical order). Data taken from ISAAA (2015).

Plant	Events*
Alfalfa (*Medicago sativa*)	5
Apple (*Malus x domestica*)	2
Bean (*Phaseolus vulgaris*)	1
Canola (*Brassica napus*)	32
Carnation (*Dianthus caryophyllus*)	19
Chicory (*Cichorium intybus*)	3
Cotton (*Gossypium hirsutum L.*)	56
Creeping bentgrass (*Agrostis stolonifera*)	1
Eggplant (*Solanum melongena*)	1
Eucalyptus (*Eucalyptus* sp.)	1
Flax (*Linum usitatissumum L.*)	1
Maize (*Zea mays L.*)	143
Melon (*Cucumis melo*)	2
Papaya (*Carica papaya*)	4
Petunia (*Petunia hybrida*)	1
Plum (*Prunus domestica*)	1
Polish canola (*Brassica rapa*)	4
Poplar (*Populus* sp.)	2
Potato (*Solanum tuberosum L.*)	44
Rice (*Oryza sativa L.*)	7
Rose (*Rosa hybrida*)	2
Soybean (*Glycine max L.*)	32
Squash (*Cucurbita pepo*)	2
Sugar Beet (*Beta vulgaris*)	3
Sugarcane (*Saccharum* sp.)	3
Sweet pepper (*Capsicum annuum*)	1
Tobacco (*Nicotiana tabacum L.*)	2
Tomato (*Lycopersicon esculentum*)	11
Wheat (*Triticum aestivum*)	1

* Number of phenotypic or commercial traits that are produced by specific genetic changes.

Regulatory approval should not be interpreted as an indication that the product is in commercial production. There are products that were granted regulatory approval but were never commercialized, or if they were, have been subsequently discontinued (Cressey 2013). Therefore, despite widespread controversy, the research and production of new generations of genetic modifications are steadily rising, and so does the importance of GM crops.

12.2.2 Risk Assessment of GM Crop-provided Foods in Developing Countries

Over the past several decades, the world grain supply has consistently outpaced demographic growth. At the same time, the number of people experiencing food insecurity has steadily risen to over 1 billion (FAO 2015; Shattuck and Holt-Giménez 2010) (see also Chapter 15). A situation has emerged where, despite an adequate supply, food remains unaffordable and inaccessible to the poorest and most vulnerable, the vast majority of people living in rural areas in developing countries, and relying on small subsistence farming for their livelihood. Therefore, the challenge is not simply to produce more food but to empower the poorest producers, particularly smallholders (Kaphengst and Smith 2013).

After a 15-year effort to achieve one of the goals of the Millennium Declaration, to eradicate extreme poverty and hunger, about 800 million people still live in extreme poverty and suffer from hunger (UN 2015). Therefore, GM crops should be accessible and economically available for those living in developing countries. However, as in the developed countries, the socioeconomic, environmental and health risks associated with GM crops and derived foods have been controversial also in developing countries (Herrera-Estrella and Álvares-Morales 2000). In addition, these countries have different agricultural practices, infrastructure, dietary traditions, nutritional deficiencies and climates. These differences do not allow developing countries to directly adopt practices and regulations from other regions or take a well-informed, independent position concerning GM food production and consumption (Adenle 2011).

Differences among countries in regulating new GM foods for human health safety and environmental risks complicate things further. Therefore, it is necessary to standardize evaluation procedures and criteria. The risk assessment process should not only consider the GM food itself, but also determine whether the cost–benefit ratio brings some advantages to the people in each country (Azadi and Ho 2010).

Results for Argentina, China, South Africa and Mexico reveal that both advantages and disadvantages of GM crops are highly country specific (Kaphengst and Smith 2013). The best distribution of benefits and improved incomes was observed in countries with functional and effective regulatory institutions, while those lacking institutional support experienced negative outcomes, such as increased inequality between farmers and other rural actors (Kaphengst and Smith 2013).

In Mexico, for example, several research groups have developed GM crops with different traits. Others have been working on methods for the detection and quantification of GM materials in locally produced maize. However, in this country very few studies are available on the health safety of GM foods consumption. Developing countries such as Mexico require collaborative networks for technology and data sharing to establish the safety of GM food consumption because people living in such high population density countries are most likely the main consumers of GM foods. A programme must also be implemented to educate policymakers about the uncertainty and complexity involved, so that, eventually, science-based decisions can be made. The previously described situation in Mexico possibly applies to other developing countries with similar epidemiological, economic and social characteristics.

12.3 The Bases for Unintended Health Risks

With regard to risk evaluation procedures of GM foodstuffs, it is first necessary to analyse whether there is any possibility of an unexpected or unintended component attributable to the production process of the corresponding GM crop (Chassy *et al.* 2008; EFSA 2008; WHO/FAO 2009). This task requires knowledge of the biotechnological transformation methodologies that continuously undergo changes and improvements, with new technologies being developed all the time.

Innovative traits in crops can be obtained using the entire spectrum of plant biotechnology: RNAi, transgenes isolated from a crossable donor plant, from the same or another species. Although these have been the most widely used techniques, other technologies such as gene editing tools through clustered regularly interspaced short palindromic repeats, oligonucleotide-directed mutagenesis, transcription activator-like effector nucleases and zinc finger nucleases are also under development (Hu *et al.* 2014; Pourcel *et al.* 2013; Ruiz-Lopez *et al.* 2015; Suen *et al.* 2014).

Currently, GM traits are obtained through basic techniques that incorporate DNA that encodes for the new desirable trait at random sites in the plant genome. For this, vehicles such as bacteria (*Agrobacterium tumefaciens* is the most widely used), viruses, biolistics bombardment, electroporation or chemical methods could be used (Dhar *et al.* 2011). As a result, deleterious effects may occur, such as, but not limited to, DNA that may be physically inserted into a transcriptionally active site and therefore inactivate a host gene or alter the control of its expression. Also, the expressed product may interact with a gene product or metabolite in a deleterious way (Conner and Jacobs 1999; FDA 2001). Unexpected changes from genetic transformation have the potential to result in loss, acquisition or underexpression of important traits. With *Agrobacterium*, small and large-scale deletions, rearrangements and insertion of superfluous DNA may occur. Using biolistics, random insertion occurs and sites appear to be associated with genome disruption, rearrangements and/or superfluous DNA. In any case, it may generate mRNA coding for fusion proteins – called 'intractable proteins' – which could modify the phenotype of the host organism (Verma *et al.* 2011).

One example of unintended changes is that of the Roundup Ready soybean, which was found to have an extra 250 bp fragment of the *epsps* gene localized at the 3' end of the introduced nopaline synthase transcription terminator (*nos*-T) (Rang *et al.* 2005; Windels *et al.* 2001). From this, a 150 bp fragment is transcribed as a consequence of failed transcription termination by the *nos*-T, resulting in four different RNA variants with the *nos*-T region completely deleted. This might express fusion proteins containing the *epsps* gene (Ricroch and Hénard-Damave 2016). Another example is a truncation event at the 3' end of the cryI(A)b gene in MON810 maize, leading to loss of the *nos*-T. An *in silico* analysis identified recombinant proteins that did not show homology to any known protein domains, confirming that DNA integration in the maize genome caused a recombination event (Klümper and Qaim 2014).

Another critical point to consider is the presence of genetic mutations as a result of the transformation process (Ben *et al.* 2014; Lambirth *et al.* 2015). There are also unintended changes not attributable to genomic alterations, such as the differences in the ratio of glycan variants between transgenic and non-transgenic alpha-amylase inhibitor (αAI) expressed in peas, although the DNA sequences were similar (Prescott *et al.* 2005).

Problems such as those described above are currently being overcome. The next generation of GM crops will be produced through high-precision techniques by editing the genome of the individual plant. However, these results highlight the current need to examine potential alterations of gene expression and protein expression, taking into consideration translational and posttranslational modifications. It is also necessary to analyse the metabolite content of new GM products to provide a more comprehensive search of causes and effects for human nutrition and health (Guyton *et al.* 2015).

As for unintended molecular changes, there are environmental aspects that may directly represent a potential health risk (Landrigan and Benbrook 2015) (see also Chapter 6). Among new traits introduced in GM crops are resistance to pests avoiding insecticide use, or herbicide resistance (glyphosate-tolerant crops) that allows chemical control of weeds without damaging the GM crop. Therefore, the deployment of GM crops can modify agrochemical applications. In the first years of GM crop use, herbicide use was not massive. However, poor management such as the repetitive culture of glyphosate-tolerant crops has selected for glyphosate-resistant weeds. Consequently, higher quantities of glyphosate and other herbicides are now required in order to kill the previously susceptible weeds (Bonny 2016). For example, between 1996 and 2011 insecticide sales in the USA decreased by 56×10^6 kg because of the use of insect-resistant GM crops, while herbicides increased by 239×10^6 kg due to the employment of herbicide-tolerant GM crops. Therefore, overall pesticide use increased by an estimated 183×10^6 kg, or about 7% (Benbrook 2012).

Concerns about the use of herbicides are not limited to the environmental aspect but also cover their potential effects on human health. Recently, the International Agency for Research on Cancer (IARC) classified glyphosate as a 'probable human carcinogen' (Guyton *et al.* 2015). More than 90% of the maize and soybean in the USA is Roundup Ready, which is glyphosate tolerant (USDA 2015), and residues of the herbicide are still detected in soybean flour after harvesting (Bohn *et al.* 2014). Therefore, additional, more complex impacts of GM food consumption should be thoroughly considered for their risk assessment. These assessments should integrate previously used as well as new indicators, such as a comparison of GM to conventionally pesticide-treated crops, the evaluation of low-dose or endocrine-mediated and epigenetic effects, potential health effects in children and the elderly, and more.

12.4 Guidelines and Approaches Used for Risk Assessment of GM Foods

There are three key international guidelines relevant to the issues of biosafety regulations:

- United Nations guidelines for consumer protection (United Nations 2016)
- Convention on Biological Diversity and Cartagena Protocol on Biosafety (SCBD 2000)
- Codex Alimentarius Principles and Guidelines on Food Derived from Biotechnology (WHO/FAO 2009).

Additionally, each country has its own regulatory framework for the production, commercialization and consumption of GM foods.

Risk assessment of GM crops and foods aims at identifying characteristics that may cause adverse effects, their potential consequences, assessment of the likelihood of occurrence and estimation of the risk caused by each characteristic. Guidelines indicate the items to be evaluated, such as the molecular characterization of DNA (Chassy *et al.* 2008; EFSA 2008; WHO/FAO 2009). However, the guidelines do not specify the techniques or models to be used for performing analyses, which is one of their main weaknesses.

The risk assessment involves the study of effects on human health and nutrition, with both *in vitro* and *in vivo* methods. Because the *in vitro* models do not suffice to study the effects of GM foods, *in vivo* models represent a useful experimental system for evaluating their immediate and long-term effects (Klümper and Qaim 2014). Although *in vivo* good practices for evaluation of GM foods have been published (Hartnell *et al.* 2007), there are still drawbacks for their adequate application. In addition, different animal models, assay periods, and biochemical, clinical, anthropometrical, anatomical and histopathological parameters are still being employed.

If the GM modification implies changes in proteins, carbohydrates or lipids, the guidelines recommend evaluating the performance of the test animals (feed intake, weight gain, feed efficiency, milk production, egg production, etc.), taking into consideration diet formulation. However, animal performance experiments may fail to show differences because growth rate and feed efficiency are not sensitive enough as indicators of vitamin or mineral adequacy. Therefore, more specific experiments are required.

In the earliest studies on the health risk of GM foods consumption, only animal performance was analysed. Interestingly, once these studies included histological evaluations, additional differences were found between experimental and control groups (Magaña-Gómez and Calderón de la Barca 2009; Pryme and Lembcke 2003). The relevance of microscopy studies is that they enable elucidation of the molecular events and basic mechanisms leading to pathological conditions (Pellicciari and Malatesta 2011). Therefore, this could be the beginning of detecting early relevant indicators. In fact, the pathophysiology of chronic diseases is due to slight molecular and cellular changes whose progression leads to metabolic disorders. One drawback is that, although current studies include histopathological examinations, there is a lack of references or standardization of protocols. Therefore, results are not considered biologically significant, although sometimes adverse effects are shown (Seralini *et al.* 2014).

12.5 Recent Research on *in vivo* Evaluation of GM Foods Consumption

The experimental design of studies on the risks to human health of GM foods consumption has progressively changed over the years. In the first decade GM foods became available, the majority of the studies were short term (2–4 weeks) and the measured indicators were linked to the model animal's capacity for biotransformation of nutrients in biomolecules. The main markers were weight gain, carcass composition, feeding efficiency and relative weight of organs (Bushey *et al.* 2014). However, after a while, microscopic, ultramicroscopic and molecular studies found some adverse effects in experimental animals fed with some GM foods. This was a landmark in the *in vivo* protocols, creating a new trend in both experimental design and expert opinion.

Table 12.4 summarizes 19 studies published from 2010 to 2015. All were conducted using *in vivo* tests based on feeding evaluation of GM whole foods (i.e. not extracts or isolated molecules). The assay periods were less than 30 days in four of the studies,

Table 12.4 Key studies on health risk evaluation of GM foods consumption.

GM crop	Trait	Animal model	Assay time	Parameters evaluated	Principal findings and conclusions
Maize (El-Shamei *et al.* 2012)	Insect resistant	Male rats	91 days	Histopathology of different organs	Adverse effects on hepatocytes, blood vessels and renal tubules. Activation of mucous glands and necrosis of intestinal villi
Maize MON 810 and RR soybean (Reichert *et al.* 2012)	Insect/herbicide resistant	Broiler chickens, hens, pigs and calves	42–210 days	Histopathology of different organs	No differences
Maize (multivitamin of inbred M37W) (Arjó *et al.* 2012)	B-carotene ascorbate and folate-enriched	Albino BALB/c mice	28–90 days	Body weight, biochemical tests, haematology and histopathology	No differences
Maize DP-004114-3 (Delaney *et al.* 2013)	Insect resistant	CD IGS rats	90 days	Haematology, clinical chemistry, urinary exams, histopathology	No differences
Maize MON810 (Gu *et al.* 2013)	Insect resistant	Atlantic salmon *Salmo salar*	97 days	Haematology, plasma chemistry, histology, digestive enzyme activity	Affected metabolism and increased IFN-gamma expression, potentiation of oxidative stress and immune affectation
Maize (Zhu *et al.* 2013)	Herbicide resistant	Sprague–Dawley rats	90 days	Body weight, haematology and serum chemistry, and pathology	Serum proteins, white cells and platelet volume augmented, not related to intake
Maize NK603 (Seralini *et al.* 2014)	Herbicide resistant	Sprague–Dawley rats	Two years	Body weight, blood and urine analysis, faecal microbes	Early deaths, mammary tumours and nephropathies. Deleterious effects attributable to GM maize and/or glyphosate
Maize BT799 (Guo *et al.* 2015)	Insect resistant	Wistar rats	90 days	Body and organ weights, haematology, hormones and histopathology examinations	No differences
Maize (Song *et al.* 2014)	Insect resistant	BALB/c mice	30 days	Immunopathology, body and organ weights, haematology and histopathology	No adverse immune-toxicological effects

Papaya 2210, 823 and 823–2210 (Lin *et al.* 2013)	Virus resistant	ICR strain mice, albino rats	28 days	Haematology, clinical chemistry, urine analysis and histopathology	Effects in white blood cells and lymphocytes, albumin, aminotransferase, cholesterol and triglycerides, without biological significance
Rice T1C-1 (Tang *et al.* 2012)	Insect resistant	Sprague–Dawley rats	90 days	Behaviour, weight, haematology, biochemistry and histopathology	Alterations in total serum protein, creatinine and cholesterol in female, not biologically significant
Rice T1-19 (Cao *et al.* 2012)	Insect/herbicide resistant	Sprague–Dawley rats	90 days	Urine analysis, faecal microbes	No differences
Rice (Zhou *et al.* 2014)	High-amylose and resistant starch	Sprague–Dawley rats	Three generations	Growth, reproduction, pathology and histopathology	Differences in blood parameters and liver enzymes and lipoproteins, within normal limits
Rice (Wang *et al.* 2014)	Insect resistant	Wistar rats	Two generations reproduction study	Body and organs weight, feed consumption, reproductive data, histopathology and haematology	Aspartate aminotransferase lower in F2 rats after transgenic TT51 rice intake, but differences not biologically significant or related to exposure to the transgenic rice
Soybean DP-305423-1 (Mejia *et al.* 2010)	Elevated oleic acid	Hy-line W-36 single-comb Leghorn hens	84 days	Body and eggs weight, feeding efficiency and production	No differences
Soybean DAS-68416-4 (Herman *et al.* 2011)	Herbicide resistant	Broiler chickens	42 days	Growth, body weight, survival rate, feed intake/weight gain	No differences
Soybean 305423 × 40-3-2 (Qi *et al.* 2012)	High oleic acid and herbicide resistant	Sprague–Dawley rats	90 days	Clinical pathology, haematology and serum chemistry	Increased mean platelet volume in males and count in females, low serum phosphorus content. No diet-related differences
Soybean 40-3-2 (Cirnatu *et al.* 2011)	Herbicide resistant	Ross broiler	42 days	Growth, histology and histochemistry	Liver lesions, muscle hypertrophy, necrosis of kidney cells, ulceration of bowel and pancreas dystrophies. This might be due to other causes
Wheat (Liang *et al.* 2013)	Salt and drought tolerant	BALB/c mice	30 days	Body weight, histopathology, toxicology and haematology, immunological parameters	No immune-toxicological effects

42 days in another two and more than 90 days in the remaining 13. Only two studies analysed gross indicators such as animal performance, and another two analysed gut microbiota. In the 15 remaining studies, haematological, biochemical and histopathological indicators were considered. No differences were found between animals fed GM and non-GM foods in eight studies. Although seven studies reported some effects, they were classified as 'not biologically significant' or 'considered not treatment related'. Therefore, only four studies clearly attributed adverse changes in the model animals to the consumption of GM food.

The results of studies that found adverse effects of GM foods on model animals (presented in Table 12.4) are not directly applicable to human beings. This is because diets formulated for animals' subchronic intake have unique sources of nutrients that do not change throughout the test, while in human diets, nutrients change from day to day and even within each eating session. However, the results of these studies may be informative for the analysis of human dietary consumption of GM foods.

An additional problem in the risk evaluation of GM foods is that some negative or adverse results of performed studies remain unpublished. This publication bias is common in many scientific fields when statistically non-significant results and even results that are contrary to expectations remain unpublished (Fanelli 2011; Matosin *et al.* 2014).

To save time and resources looking for rapid and reliable progress in the complex field of risk assessment of GM foods for human consumption, international consortia could be formed. In this respect, differences between developed and developing countries should be considered. It is widely believed that developing countries will be the foremost users of GM foods because of their dense populations and the need for more and better foods. Therefore, it is important to have well-established regulatory frameworks, global trade relationships and scientific collaborations and share know-how between developed and developing countries.

12.6 Shortcomings and Research Needs in the Risk Assessment of Genetically Modified Foods

12.6.1 Substantial Equivalence for Testing Safety of GM Foods

The Codex Alimentarius is a standard that considers substantial equivalence as a starting point for assessing the safety and nutritional value of a food or ingredient that has been modified by modern biotechnological methods (WHO/FAO 2009). Yet it is not a safety assessment *per se*. Instead, it implies the safety of a GM crop by comparing it to its closest conventional counterpart. The goal is to identify similarities and differences, using existing food sources as references. If the novel food is found to be substantially equivalent to its conventional counterpart, it can be treated in the same manner with respect to safety. If it is different, it should be investigated further (OECD 1993).

The use of substantial equivalence as part of the evaluation procedure for GM foods has supporters and detractors (Konig *et al.* 2004; Kuiper *et al.* 2002). Both pros and cons are based on:

- the availability of an appropriate comparator
- the choice of parameters in the single-constituent compound analyses

- the ability to discriminate between differences in the GM food and the comparator that result from the genetic modification and differences attributed to somaclonal variation introduced during tissue culture and environmental or cultivation conditions (Konig *et al.* 2004).

The concept of substantial equivalence might be improved through consensus on the appropriate components (e.g. key nutrients, key toxicants and antinutritional compounds) on a crop-by-crop basis. Key nutrients typically include proximal composition, amino acids, fatty acids, calcium, phosphorus, antinutrients and toxicants that are harmful to health (OECD 2015). However, few compositional data in terms of food safety assessment are available. One explanation is that documents cannot be easily updated to reflect current data, even though many varieties and cultivars of crops have been developed (Kitta 2013). Useful databases include those developed by the OECD (OECD 2016), the US Department of Agriculture (USDA 2016), the International Life Science Institute (ILSI 2014) and the US National Research Council (NRC 1982).

The argument that GM crops are essentially unchanged, except for the intended additions, is not generalizable (Bushey *et al.* 2014). In fact, some genetic modifications intentionally modify the chemical composition of the GM product, such as cultivars with improved content of iron, folate, ascorbate, oleic acid, omega-3, amylose or anthocyanins (Chen and Lin 2013). Therefore, special assessment protocols and substantial equivalence principles should be proposed for such cases. Yet, even if substantial equivalence is demonstrated, it does not imply that consumption is safe and that no additional studies are required. Likewise, undesirable differences between GM and non-GM foods are not necessarily indicative of health risks. The new GM food could replace conventional food and nutritional consequences of the consumption need to be evaluated in terms of the direct or indirect, immediate or cumulative effects (Malatesta 2009).

Currently, guidelines recommend that the following parameters are determined: the number of copies of the inserted gene and the insertion site, the absence of gene disruption, the presence of the marker gene at the same locus, integrated vector sequences and the stability of the transgene. Phenotypic analysis should include the evaluation of agronomic characteristics, chemical composition, antiphysiological factors, expressed proteins and potential changes in plant constituents of the GM crop compared to the conventional one (Bushey *et al.* 2014). For this approach, exploratory techniques based on 'omics' such as transcriptomics, proteomics and metabolomics have been used to compare GM with conventional crops with broader and deeper information about them (García-Cañas *et al.* 2011; Ouakfaoui and Miki 2005; Ruebelt *et al.* 2006; Simo *et al.* 2014). However, these exploratory techniques cannot yet be applied as large-scale methods because of the lack of standardization and certification.

12.6.2 A Nutritional Genomic Approach for *in vivo* Risk Assessment of GM Foods

Nowadays, 'omics' techniques are limited mainly to compositional analysis of GM crops rather than assessment of their effects on *in vivo* models. It is very important to evaluate the effects on the right model. For instance, if a GM crop is modified to produce a food with improved nutritional quality for any deficiency, the receptor animal model must present this deficiency. If the model immune system is compromised, the response may be quite different from that of healthy models. Different physiological processes are regulated by negative feedback mechanisms in humans and mammals in general. For example, specific nutrient deficiency may lead to compensatory increase in its absorption rate. So, the *in vivo* risk assessment of GM foods intake requires a complex experimental design to evaluate such classes of physiological conditions.

In addition to the right evaluation model, the concept of the model itself should be considered. An experimental model (*in vivo*, *in vitro* or *in silico*) just represents part of a given phenomenon and does not provide a complete and conclusive explanation. Phenomena such as allergenicity, cytotoxicity, mutagenicity, toxicology, reproduction, development and nutrition involve multiple molecular and cellular mechanisms. Therefore, it would be almost impossible to analyse the intake effects of a new GM food as a whole with a single experimental model (EFSA 2008; Malatesta 2009). For instance, different experimental models that were used to evaluate the intake effect of the same GM food yielded highly variable conclusions, from no adverse effects to negative ones (Brake and Evenson 2004; Gu *et al.* 2013; Magaña-Gomez *et al.* 2008; Malatesta *et al.* 2002). Although data analysis could be different, the measured biological or biochemical indicators were not the same either, further demonstrating the complexity of these evaluation tasks.

The methods of analysis of biological effects in contrast to chemical or compositional characterization have limitations. According to the Institute of Medicine National Research Council (2004), advances in analytical chemistry have exceeded the ability to interpret the consequences for human health of changes in food composition. Therefore, further development of analytical technologies for health evaluations and their interpretation is needed to overcome these limitations of risk evaluation of GM food intake.

One critical problem in the risk assessment of GM foods consumption is precisely the concept of 'food'. From the beginning, when assaying possible toxicants or antinutrients, risk evaluation has been based on protocols for testing specific molecules. These are low molecular weight chemicals, pharmaceutical products, industrial chemicals, pesticides and food additives, which are included at different concentrations in diets of animal models for testing. Before the development of GM crops, there were no assessments of the effects of a whole food, which involve interactions between nutrients, the variability of diet composition and the physiological state of the receptor. Therefore, studies should evolve to include aspects that were not previously considered.

Among the most difficult aspects to evaluate is the effect of chronic consumption of a GM food on human health. After 20 years of some GM foods such as glyphosate-tolerant soybeans being available on the market, we may be able to evaluate the health risk associated with their intake. For this task, it is necessary to develop strategies to obtain epidemiological and dietary information from several populations to identify possible health changes associated with consumption of these GM foods. However, the challenge would be to distinguish between the effects of GM foods and other environmental factors. At the same time, research efforts should also be directed at developing profiling techniques that relate dietary metabolites to altered gene expression in relevant experimental models. In this respect, toxicological evaluations of whole foods and complex mixtures, including microarray analysis, proteomics and metabolomics, should be developed and applied (EFSA 2008; Institute of Medicine National Research Council 2004; Malatesta 2009).

The molecular characterization of foods and feeds, and environmental risk/safety assessment of GM crops, are very important (Chassy *et al.* 2004; Kuiper *et al.* 2003). In the initial stages of the risk assessment of GM crops, many necessary techniques were not easily available and therefore not applied by regulatory agencies in risk/safety assessment. In the future, after sufficient development and validation, these techniques will be used to obtain GM crop safety assessments based on integral examinations (Davies 2010; Malatesta 2009).

Currently, there is a growing effort in the field of nutrition to understand the relationship between dietary components and genome, conducted in two areas of knowledge: nutrigenomics and nutrigenetics (Norheim *et al.* 2012). The aim of nutrigenomics is to determine the influence of common dietary ingredients on the genome, and to understand their effect on metabolic pathways and homeostatic control. Nutrigenetics, on the other hand, is applied to understand how the genetic make-up of a person co-ordinates response to diet. The aim of both disciplines is to unravel diet–genome interactions based on technologies that can provide information about several expressed components. Some studies of nutritional genomics use transcriptomics, proteomics, metabolomics and systems biology by microarray, RNA sequencing, protein separation and mathematical modelling (Muller and Kersten 2003; Mutch *et al.* 2005; Norheim *et al.* 2012). However, these methods are currently not suitable for large-scale studies because of the lack of complete databases.

Although assessment of GM foods safety is not a subject for nutrigenomics, understanding the scope of this discipline can be useful for breaking some paradigms about the assessment of health risks involved in GM foods intake. In fact, a few studies have already used these techniques for characterizing the health risks of GM foods (Fenech *et al.* 2011; Isaak and Siow 2013). Evidence that both known and unknown nutrients are able to interact with molecular mechanisms underlying the physiological functions of organisms and to regulate them is increasing. It is possible to establish enlightening relationships between molecules and metabolic changes using techniques and experimental designs, data analysis and interpretation coming from nutrigenomics. A bioactive dietary component is able to affect cellular response to a stimulus, either by directly activating nuclear receptors and transcription factors or indirectly as metabolites from different biochemical pathways (Serrano *et al.* 2015). For example, vitamins and microelements, acting as co-factors or regulators, can regulate basic biological processes such as DNA synthesis and DNA repair.

There are also intrinsic causes, which may modify the body's physiological responses. For instance, genetic polymorphisms could produce modified proteins with unknown functional properties. Because proteins have diverse functions as signalling molecules, hormones, receptors, transporters, transcription factors or enzymes, the complexity of the interactions in an organism is obvious, and its evaluation is therefore an overwhelming task (Mutch *et al.* 2005).

Health status depends not only on the amount of dietary carbohydrates, proteins and lipids, but also on other minor bioactive nutrients, as has been stated through nutrigenomics. In nutrition, the main implications could be the need for a new way of analysing the relationship between diet and health, taking into consideration other disciplines and innovative research strategies rather than just epidemiological data. For risk assessment, this could be a way to study issues not previously considered.

The reductionist approach, in which GM foods safety is defined by differences or similarities in some agronomic parameters between plants or gross biological indicators in animal models, must change. The idea that substantial equivalence between GM and non-GM foods is enough to assume safe consumption should lead to redefinition of key nutrients, experimental techniques, reference standards and a more comprehensive interpretation. Each foodstuff includes macronutrients, micronutrients and other beneficial or antiphysiological compounds in different concentrations with synergistic or antagonistic relationships between them (Sapone *et al.* 2012). Thus, although good progress has been made in the identification of compounds and compositional changes in GM foods, the effects on human health should be studied with the whole GM food available and the most recently developed techniques (Gu *et al.* 2013).

12.7 Conclusion

The health risk assessment of the consumption of GM foods is still a topic of debate. There are technical limitations making it difficult to demonstrate their safe consumption, which must be compensated for by adequate design and analysis of experiments. It has been demonstrated that some genetic modifications on food crops could produce unexpected changes such as modified proteins or levels of metabolites. Furthermore, molecules themselves or metabolic interactions may modify the body homeostasis after consumption. The safety of GM foods should be demonstrated beyond substantial equivalence made by chemical analyses and animal performance indicators or haematological and biochemical parameters, in the best of cases. To date, only a few studies have used ultramicroscopy or metabolomics for GM foods intake evaluation. In the near future, with the second-generation GM foods modified for nutrients, 'omics'-supported studies will be required to gain the best insight into the body's response to food intake. Therefore, we agree with a recent National Academies of Sciences, Engineering, and Medicine (2016) report that states that 'In the case of foods, including GM foods, it can be reasonably argued that even a small adverse chronic effect should be guarded against, given that billions of people could be consuming the foods.'

References

Adenle, A.A. (2011) Global capture of crop biotechnology in developing world over a decade. *Journal of Genetic Engineering and Biotechnology* **9**: 83–95.

Arjó, G., Capell, T., Matias-Guiu, X., Zhu, C., Christou, P. and Piñol, C. (2012) Mice fed on a diet enriched with genetically engineered multivitamin corn show no sub-acute toxic effects and no sub-chronic toxicity. *Plant Biotechnology Journal* **10**: 1026–1034.

Azadi, H. and Ho, P. (2010) Genetically modified and organic crops in developing countries: a review of options for food security. *Biotechnology Advances* **28**: 160–168.

Ben Ali, S.E., Madi, Z.E., Hochegger, R., *et al.* (2014) Mutation scanning in a single and a stacked genetically modified (GM) event by real-time PCR and high resolution melting (HRM) analysis. *International Journal of Molecular Sciences* **15**: 19898–19923.

Benbrook, C. (2012) Impacts of genetically engineered crops on pesticide use in the U.S. – the first sixteen years. *Environmental Sciences Europe* **24**: 24.

Bohn, T., Cuhra, M., Traavik, T., Sanden, M., Fagan, J. and Primicerio, R. (2014) Compositional differences in soybeans on the market: glyphosate accumulates in Roundup Ready GM soybeans. *Food Chemistry* **153**: 207–215.

Bonny, S. (2016) Genetically modified herbicide-tolerant crops, weeds, and herbicides: overview and impact. *Environmental Management* **57**: 31–48.

Brake, D.G. and Evenson, D.P. (2004) A generational study of glyphosate-tolerant soybeans on mouse fetal, postnatal, pubertal and adult testicular development. *Food and Chemical Toxicology* **42**: 29–36.

Bushey, D.F., Bannon, G.A., Delaney, B.F., *et al.* (2014) Characteristics and safety assessment of intractable proteins in genetically modified crops. *Regulatory Toxicology and Pharmacology* **69**: 154–170.

Cao, S., He, X., Xu, W., *et al.* (2012) Safety assessment of transgenic *Bacillus thuringiensis* rice T1c-19 in Sprague-Dawley rats from metabonomics and bacterial profile perspectives. *IUBMB Life* **64**: 242–250.

Chassy, B.M., Parrott, W.A. and Roush, R. (2004) Nutritional and safety assessments of foods and feeds nutritionally improved through biotechnology: an executive summary a task force report by the International Life Sciences Institute, Washington, DC, USA. *Comprehensive Reviews in Food Science and Food Safety* **3**: 35–104.

Chassy, B., Egnin, M., Gao, Y., *et al.* (2008) Chapter 2: Recent developments in the safety and nutritional assessment of nutritionally improved foods and feeds. Prepared by a task force of the ILSI International Food Biotechnology Committee. *Comprehensive Reviews in Food Science and Food Safety* **7**: 65–74.

Chen, H. and Lin, Y. (2013) Promise and issues of genetically modified crops. *Current Opinion in Plant Biology* **16**: 255–260.

Cirnatu, D., Jompan, A., Sin, A.I. and Zugravu, C.A. (2011) Multiple organ histopathological changes in broiler chickens fed on genetically modified organism. *Romanian Journal of Morphology and Embryology* **52**: 475–480.

Conner, A.J. and Jacobs, J.M. (1999) Genetic engineering of crops as potential source of genetic hazard in the human diet. *Mutation Research* **443**: 223–234.

Cressey, D. (2013) Transgenics: a new breed. *Nature* **497**: 27–29.

Davies, H. (2010) A role for 'omics' technologies in food safety assessment. *Food Control* **21**: 1601–1610.

Delaney, B., Karaman, S., Roper, J., *et al.* (2013) Thirteen week rodent feeding study with grain from molecular stacked trait lepidopteran and coleopteran protected (DP-ØØ4114-3) maize. *Food and Chemical Toxicology* **53**: 417–427.

Dhar, M.K., Kaul, S. and Kour, J. (2011) Towards the development of better crops by genetic transformation using engineered plant chromosomes. *Plant Cell Reports* **30**: 799–806.

EFSA (2008) Safety and nutritional assessment of GM plants and derived food and feed: the role of animal feeding trials. *Food and Chemical Toxicology* **46**: S2–S70.

El-Shamei, Z., Gab-Alla, A., Shatta, A., Moussa, E. and Rayan, A. (2012) Histopathological changes in some organs of male rats fed on genetically modified corn (Ajeeb YG). *Journal of American Science* **8**: 684–696.

Ewen, S.W. and Pusztai, A. (1999) Effect of diets containing genetically modified potatoes expressing *Galanthus nivalis* lectin on rat small intestine. *Lancet* **354**: 1353–1354.

Fanelli, D. (2011) Negative results are disappearing from most disciplines and countries. *Scientometrics* **90**: 891–904.

FAO (2015) Crop prospects and food situation. *Global Overview* **4**: 6–11.

FDA (2001) Premarket notice concerning bioengineered foods. *Federal Registry* **66**: 4706–4738.

Fenech, M., El-Sohemy, A., Cahill, L., *et al.* (2011) Nutrigenetics and nutrigenomics: viewpoints on the current status and applications in nutrition research and practice. *Journal of Nutrigenetics and Nutrigenomics* **4**: 68–89.

Frewer, L., Lassen, J., Kettlitz, B., Scholderer, J., Beekman, V. and Berdal, K.G. (2004) Societal aspects of genetically modified foods. *Food and Chemical Toxicology* **42**: 1181–1193.

García-Cañas, V., Simó, C., León, C., Ibáñez, E. and Cifuentes, A. (2011) MS-based analytical methodologies to characterize genetically modified crops. *Mass Spectrometry Reviews* **30**: 396–416.

Gu, J., Krogdahl, Å., Sissener, N.H., *et al.* (2013) Effects of oral Bt-maize (MON810) exposure on growth and health parameters in normal and sensitised Atlantic salmon, *Salmo salar* L. *British Journal of Nutrition* **109**: 1408–1423.

Guo, Q., He, L., Zhu, H., *et al.* (2015) Effects of 90-day feeding of transgenic maize BT799 on the reproductive system in male wistar rats. *International Journal of Environmental Research and Public Health* **12**: 15309–15320.

Guyton, K.Z., Loomis, D., Grosse, Y., *et al.* (2015) Carcinogenicity of tetrachlorvinphos, parathion, malathion, diazinon, and glyphosate. *Lancet Oncology* **16**: 490–491.

Hartnell, G., Cromwell, G., Dana, G., *et al.* (2007) *Best Practices for the Conduct of Animal Studies to Evaluate Crops Genetically Modified for Output Traits.* ILSI, Washington, DC, USA.

Herman, R.A., Dunville, C.M., Juberg, D.R., Fletcher, D.W. and Cromwell, G.L. (2011) Performance of broiler chickens fed diets containing DAS-68416-4 soybean meal. *GM Crops* **2**: 169–175.

Herrera-Estrella, L.R. and Álvares-Morales, A. (2000) Genetically modified crops and developing countries. *Plant Physiology* **124**: 923–926.

Hu, T., Zeng, H., Hu, Z., Qv, X. and Chen, G. (2014) Simultaneous silencing of five lipoxygenase genes increases the contents of alpha-linolenic and linoleic acids in tomato (*Solanum lycopersicum* L.) fruits. *Journal of Agricultural and Food Chemistry* **62**: 11988–11993.

ILSI (2014) International Life Sciences Institute Crop Composition Database, Version 6. Available at: www.cropcomposition.org/query/index.html (accessed 7 March 2017).

Institute of Medicine National Research Council (2004) *Safety of Genetically Engineered Foods: Approaches to Assessing Unintended Health Effects.* The National Academies Press Washington, DC, USA.

ISAAA (2015) GM Approval Database. Available at: www.isaaa.org/gmapprovaldatabase/ (accessed 7 March 2017).

Isaak, C.K. and Siow, Y.L. (2013) The evolution of nutrition research. *Canadian Journal of Physiology and Pharmacology* **91**: 257–267.

James, C. (2014) *Global Status of Commercialized Biotech/GM Crops: 2014.* ISAAA Brief No. 49. ISAAA, Ithaca, NY, USA.

Kaphengst, T. and Smith, L. (2013) *The Impact of Biotechnology on Developing Countries.* Framework Contract Development Policy for the European Parliament, Ecologic Institute, Berlin, Germany.

Kitta, K. (2013) Availability and utility of crop composition data. *Journal of Agricultural and Food Chemistry* **61**: 8304–8311.

Klümper, W. and Qaim, M. (2014) A meta-analysis of the impacts of genetically modified crops. *PLoS ONE* **9(11)**: e111629.

Konig, A., Cockburn, A., Crevel, R.W., *et al.* (2004) Assessment of the safety of foods derived from genetically modified (GM) crops. *Food and Chemical Toxicology* **42**: 1047–1088.

Kuiper, H.A., Kleter, G.A., Noteborn, H.P.J.M. and Kok, E.J. (2002) Substantial equivalence – an appropriate paradigm for the safety assessment of genetically modified foods? *Toxicology* **181–182**: 427–431.

Kuiper, H.A., Kok, E.J. and Engel, K.H. (2003) Exploitation of molecular profiling techniques for GM food safety assessment. *Current Opinion in Biotechnology* **14**: 238–243.

Lambirth, K.C., Whaley, A.M., Blakley, I.C., *et al.* (2015) A comparison of transgenic and wild type soybean seeds: analysis of transcriptome profiles using RNA-Seq. *BMC Biotechnology.* **15**: 89.

Landrigan, P.J. and Benbrook, C. (2015) GMOs, herbicides, and public health. *New England Journal of Medicine* **373**: 693–695.

Liang, C.L., Zhang, X.P., Song, Y. and Jia, X.D. (2013) Immunotoxicological evaluation of wheat genetically modified with TaDREB4 gene on BALB/c mice. *Biomedical and Environmental Sciences* **26**: 663–670.

Lin, H.T., Yen, G.C., Huang, T.T., *et al.* (2013) Toxicity assessment of transgenic papaya ringspot virus of 823-2210 line papaya fruits. *Journal of Agricultural and Food Chemistry* **61**: 1585–1596.

Magaña-Gómez, J.A. and Calderón de la Barca, A.M. (2009) Risk assessment of genetically modified crops for nutrition and health. *Nutrition Reviews* **67**: 1–16.

Magaña-Gomez, J.A., Cervantes, G.L., Yepiz-Plascencia, G. and Calderón de la Barca, A.M. (2008) Pancreatic response of rats fed genetically modified soybean. *Journal of Applied Toxicology* **28**: 217–226.

Malatesta, M. (2009) Animal feeding trials for assessing GMO safety: answers and questions. *CAB Reviews: Perspectives in Agriculture, Veterinary Science, Nutrition and Natural Resources* **4**: 1–13.

Malatesta, M., Caporaloni, C., Rossi, L., *et al.* (2002) Ultrastructural analysis of pancreatic acinar cells from mice fed on genetically modified soybean. *Journal of Anatomy* **201**: 409–415.

Matosin, N., Frank, E., Engel, M., Lum, J.S. and Newell, K.A. (2014) Negativity towards negative results: a discussion of the disconnect between scientific worth and scientific culture. *Disease Models and Mechanisms* **7**: 171–173.

Mejia, L., Jacobs, C.M., Utterback, P.L., *et al.* (2010) Evaluation of the nutritional equivalency of soybean meal with the genetically modified trait DP-3Ø5423-1 when fed to laying hens. *Poultry Science* **89**: 2634–2639.

Muller, M. and Kersten, S. (2003) Nutrigenomics: goals and strategies. *Nature Reviews Genetics* **4**: 315–322.

Mutch, D.M., Wahli, W. and Williamson, G. (2005) Nutrigenomics and nutrigenetics: the emerging faces of nutrition. *FASEB Journal* **19**: 1602–1616.

National Academies of Sciences, Engineering, and Medicine (2016) *Genetically Engineered Crops: Experiences and Prospects*. National Academies Press, Washington, DC, USA.

Norheim, F., Gjelstad, I.M., Hjorth, M., *et al.* (2012) Molecular nutrition research: the modern way of performing nutritional science. *Nutrients* **4**: 1898–1944.

NRC (1982) *United States-Canadian Tables of Feed Composition: Nutritional Data for United States and Canadian Feeds, Third Revision*. National Academies Press, Washington, DC, USA.

OECD (1993) *Safety Evaluation of Foods Derived by Modern Biotechnology: Concepts and Principles*. OECD, Paris.

OECD (2015) *Safety Assessment of Foods and Feeds Derived from Transgenic Crops, Volume 1, Novel Food and Feed Safety*. OECD, Paris.

OECD (2016) Consensus Documents for the Work on the Safety of Novel Foods and Feeds, 2016. Available at: www.oecd.org/science/biotrack/consensusdocumentsfortheworkonthesafetyofnovelfoodsandfeeds.htm (accessed 7 March 2017).

Ouakfaoui, S.E. and Miki, B. (2005) The stability of the *Arabidopsis* transcriptome in transgenic plants expressing the marker genes nptII and uidA. *Plant Journal* **41**: 791–800.

Pellicciari, C. and Malatesta, M. (2011) Identifying pathological biomarkers: histochemistry still ranks high in the omics era. *European Journal of Histochemistry* **55**: 235–238.

Pourcel, L., Moulin, M. and Fitzpatrick, T.B. (2013) Examining strategies to facilitate vitamin B1 biofortification of plants by genetic engineering. *Frontiers in Plant Science* **4**: 1–8.

Prescott, V.E., Campbell, P.M., Moore, A., *et al.* (2005) Transgenic expression of bean alpha-amylase inhibitor in peas results in altered structure and immunogenicity. *Journal of Agricultural and Food Chemistry* **53**: 9023–9030.

Pryme, I.F. and Lembcke, R. (2003) *In vivo* studies on possible health consequences of genetically modified food and feed – with particular regard to ingredients consisting of genetically modified plant materials. *Nutrition and Health* **17**: 1–8.

Qi, X., He, X., Luo, Y., *et al.* (2012) Subchronic feeding study of stacked trait genetically-modified soybean (3Ø5423 × 40-3-2) in Sprague–Dawley rats. *Food and Chemical Toxicology* **50**: 3256–3263.

Rang, A., Linke, B. and Jansen, B. (2005) Detection of RNA variants transcribed from the transgene in Roundup Ready soybean. *European Food Research and Technology* **220**: 438–443.

Reichert, M., Kozaczyński, W., Karpińska, T.A., *et al.* (2012) Histopathology of internal organs of farm animals fed genetically modified corn and soybean meal. *Bulletin of the Veterinary Institute in Pulawy* **56**: 617–622.

Ricroch, A.E. and Hénard-Damave, M.C. (2016) Next biotech plants: new traits, crops, developers and technologies for addressing global challenges. *Critical Reviews in Biotechnology* **36**: 675–690.

Ronteltap, A., van Trijp, J.C., Renes, R.J. and Frewer, L.J. (2007) Consumer acceptance of technology-based food innovations: lessons for the future of nutrigenomics. *Appetite* **49**: 1–17.

Ruebelt, M.C., Leimgruber, N.K., Lipp, M., *et al.* (2006) Application of two-dimensional gel electrophoresis to interrogate alterations in the proteome of genetically modified crops. 1. Assessing analytical validation. *Journal of Agricultural and Food Chemistry* **54**: 2154–2161.

Ruiz-Lopez, N., Usher, S., Sayanova, O.V., Napier, J.A. and Haslam, R.P. (2015) Modifying the lipid content and composition of plant seeds: engineering the production of LC-PUFA. *Applied Microbiology and Biotechnology* **99**: 143–154.

Sapone, A., Canistro, D., Melega, S., Moles, R., Vivarelli, F. and Paolini, M. (2012) On enzyme-based anticancer molecular dietary manipulations. *Journal of Biomedicine and Biotechnology* **2012**: 790987.

SCBD (2000) *Cartagena Protocol on Biosafety to the Convention on Biological Diversity: Text and Annexes*. Secretariat of the Convention on Biological Diversity, Montreal, Canada.

Seralini, G.E., Clair, E., Mesnage, R., *et al.* (2014) Republished study: long-term toxicity of a Roundup herbicide and a Roundup-tolerant genetically modified maize. *Environmental Sciences Europe* **26**: 14.

Serrano, J.C., Jove, M., Gonzalo, H., Pamplona, R. and Portero-Otin, M. (2015) Nutridynamics: mechanism(s) of action of bioactive compounds and their effects. *International Journal of Food Sciences and Nutrition* **66**: S22–S30.

Shattuck, A. and Holt-Giménez, E. (2010) Moving from food crisis to food sovereignty. *Yale Human Rights and Development Journal* **13**: 421–434.

Simo, C., Ibanez, C., Valdes, A., Cifuentes, A. and Garcia-Canas, V. (2014) Metabolomics of genetically modified crops. *International Journal of Molecular Sciences* **15**: 18941–18966.

Song, Y., Liang, C., Wang, W., *et al.* (2014) Immunotoxicological evaluation of corn genetically modified with *Bacillus thuringiensis* Cry1Ah gene by a 30-day feeding study in BALB/c mice. *PLoS ONE* **9(2)**: e78566.

Suen, Y.L., Tang, H., Huang, J. and Chen, F. (2014) Enhanced production of fatty acids and astaxanthin in *Aurantiochytrium* sp. by the expression of *Vitreoscilla* hemoglobin. *Journal of Agricultural and Food Chemistry* **62**: 12392–12398.

Tang, X., Han, F., Zhao, K., *et al.* (2012) A 90-day dietary toxicity study of genetically modified rice T1C-1 expressing Cry1C protein in Sprague Dawley rats. *PLoS ONE* **7(12)**: e52507.

United Nations (2015) The Millennium Development Goals Report 2015. Available at: www.un.org/millenniumgoals/2015_MDG_Report/pdf/MDG%202015%20Summary%20web_english.pdf (accessed 7 March 2017).

United Nations (2016) *The United Nations Guidelines for Consumer Protection*. Document A/RES/70/186. United Nations, Geneva, Switzerland.

USDA (2015) *Adoption of Genetically Engineered Crops in the U.S.* Department of Agriculture, Economic Research Service, Washington, DC, USA.

USDA (2016) USDA National Nutrient Database for Standard Reference, Release 28. Version Current: September 2015, slightly revised May 2016. Available at: https://ndb.nal.usda.gov/ndb/ (accessed 7 March 2017).

Uzogara, S.G. (2000) The impact of genetic modification of human foods in the 21st century: a review. *Biotechnology Advances* **18**: 179–206.

Verma, C., Nanda, S., Singh, R.K., Singh, R.B. and Mishra, S. (2011) A review on impacts of genetically modified food on human health. *Open Nutraceuticals Journal* **4**: 3–11.

Wang, E.H., Yu, Z., Hu, J., Jia, X.D. and Xu, H.B. (2014) A two-generation reproduction study with transgenic Bt rice TT51 in Wistar rats. *Food and Chemical Toxicology* **65**: 312–320.

WHO/FAO (2009) *Codex Alimentarius: foods derived from modern biotechnology*, 2nd edn. World Health Organization, Food and Agriculture Organization of the United Nations, Rome, Italy. Available at: ftp://ftp.fao.org/codex/Publications/Booklets/Biotech/Biotech_2009e.pdf (accessed 7 March 2017).

Windels, P., Taverniers, I., Depicker, A., van Bockstaele, E. and De Loose. M. (2001) Characterisation of the Roundup Ready soybean insert. *European Food Research and Technology* **213**: 107–112.

Zhou, X.H., Dong, Y., Zhao, Y.S., *et al.* (2014) A three generation reproduction study with Sprague–Dawley rats consuming high-amylose transgenic rice. *Food and Chemical Toxicology* **74**: 20–27.

Zhu, Y., He, X., Luo, Y., *et al.* (2013) A 90-day feeding study of glyphosate-tolerant maize with the G2-aroA gene in Sprague-Dawley rats. *Food and Chemical Toxicology* **51**: 280–287.

Part VI

Policies Related to Environmental Pest Management

13

Effectiveness of Pesticide Policies: Experiences from Danish Pesticide Regulation 1986–2015

Anders Branth Pedersen and Helle Ørsted Nielsen

13.1 Introduction

The European Union (EU)'s Sustainable Use of Pesticides Directive (Directive 2009/128/EC) was approved in 2009. As part of the implementation, the EU expected that:

> *Member States shall adopt National Action Plans to set up their quantitative objectives, targets, measures and timetables to reduce risks and impacts of pesticide use on human health and the environment and to encourage the development and introduction of integrated pest management and of alternative approaches or techniques in order to reduce dependency on the use of pesticides [...]. By 26 November 2012, Member States shall communicate their National Action Plans to the Commission and to other Member States.*
>
> (Directive 2009/128/EC, article 4)

When Pesticide Action Network Europe (2013) assessed the Member State implementation in 2013, the results were disappointing. Only one of the 24 Member States that had published national action plans had an overall, clear, quantifiable objective for the pesticide policy as demanded in article 4: Denmark. Denmark has, in fact, a long track record in developing both national pesticide action plans and quantifiable objectives, starting with its first pesticide action plan in 1986. This makes it relevant to analyse the Danish pesticide policy design in a historical perspective to detect potential lessons for other countries considering the introduction of quantitative objectives for pesticide reduction.

Denmark is one of Europe's most intensively farmed countries – agricultural land amounts to 62% of the land area in 2012 (World Bank 2015) – and is among Europe's pesticide policy pioneers. Denmark has learned during three decades of pesticide policies that meeting ambitious objectives can prove difficult. Ambitious objectives must be matched by strong policy instruments directed at curbing agricultural pesticide use, and Denmark has applied carrots, sticks and sermons (Vedung 1998) to prompt farmers to reduce their use of pesticides, but with mixed results. Consequently, there are lessons to be learned, both positive and negative, for countries trying to reduce agricultural pesticide use today.

Environmental Pest Management: Challenges for Agronomists, Ecologists, Economists and Policymakers, First Edition. Edited by Moshe Coll and Eric Wajnberg.

In this chapter, we first describe the different Danish national action plans, including the most important policy instruments directed towards agricultural pesticide use (section 13.2) – Danish farmers use 93% of the pesticides sold in Denmark, while the remaining 7% is used by forestry, nurseries, municipalities, golf courses and households (Danish Government 2013). In the following section (13.3), we assess the overall effects of the plans on agricultural pesticide use. Section 13.4 contains a short comparison of pesticide use across the EU, including some considerations on the possibility of policy transfer. Finally, we conclude on the main policy lessons in section 13.5.

The chapter is based on an analysis of official documents on the Danish pesticide action plans, the development of pesticide use and other indicators, a review of comparisons of pesticide use in EU countries, and a review of existing knowledge on the effects of Danish pesticide action plans. Furthermore, the chapter uses survey data gathered in two research projects in which the authors of this chapter participated (Christensen *et al.* 2011; Pedersen *et al.* 2011, 2012, 2014).

13.2 Denmark – a Pioneer in Pesticide Policies

The first Danish regulation on the general use of toxics dates back to the end of the 18th century, but the first act specifically regulating pesticides was not introduced until 1948. In 1980, the regulations were merged into a new Act on Chemicals, and the jurisdiction was moved from the Ministry of Agriculture to the Ministry of Environment (Bichel Committee 1998). In 1986, a first pesticide action plan was introduced. In the subsequent three decades, Denmark has applied a wide range of different pesticide policy instruments. Below is a description of the aims and policy instruments targeting the agricultural sector in the pesticide action plans.

13.2.1 First Plan: 1986–2000

Denmark became one of Europe's pioneers in 1986, when a first pesticide action plan was presented (Ministry of Environment 1986). Having largely avoided burdening environmental regulation until the mid-1980s, Danish agriculture was suddenly challenged by a very unusual situation in the Danish parliament, where the centre right minority government faced a so-called 'alternative green majority coalition' consisting of the Social Democrats, the Social-Liberal Party and other left-wing parties (Andersen and Hansen 1991; Pedersen 2010). This coalition forced the government to implement several environmental regulations of agriculture (Andersen and Hansen 1991; Pedersen 2010). Among these was the demand to develop a pesticide action plan (Danish Parliament 1986). The focus on regulation of agricultural use of pesticides was underpinned in part by a strong norm in Danish society for maintaining unpolluted groundwater as a source of untreated drinking water (Danish Water and Wastewater Association 2011; Hasler *et al.* 2007; Ministry of Taxation 2011).

The action plan aimed at a 50% reduction in pesticide use, measured as the amount of active ingredients. This objective was to be achieved in two stages: a 25% reduction from 1986 to 1990 (baseline was an average of the years 1981–1985) and a further 25% before 1997. Besides reducing active ingredients the plan also set a target of reducing the so-called treatment frequency index (TFI) by the same amount (25% by 1990, 50%

by 1997) (Ministry of Environment 1986). The TFI is a standard indicator for pesticide use, calculated as the number of pesticide applications on cultivated areas per calendar year in conventional farming, assuming use of a fixed standard dose (based on sales data) (Danish Environmental Protection Agency 2012).

The reason for focusing on both active ingredients and the TFI was that there is no direct correlation between the amount of active ingredients in a pesticide and its environmental load. Some pesticides are biologically active in very small quantities and can be used at lower dosages. Consequently, it is possible simultaneously to decrease the amount of active ingredients and increase the environmental load (Ministry of Taxation *et al.* 2001), and therefore Danish experts considered the TFI to be the best indicator for environmental load at the time (Bichel Committee 1998).

To achieve these targets, the plan contained a range of policy instruments. First and foremost, it introduced an information effort aimed at farmers, primarily through the agricultural consultancies, who were expected to include environmental effects in their advisory services, but also directly through government information to farmers. The plan also included intensified research (an information-based instrument too), for example in new resistant crops, crop rotation, Integrated Pest Management (IPM), etc. Additionally, the Act on Chemicals was changed to tighten up the approval procedure for pesticides (Ministry of Environment 1986), and some further possible policy instruments were suggested and later implemented, including mandatory spraying certificates for professional users of pesticides (Ministry of Environment 1990), mandatory spraying journals for professional users (implemented in 1994) (Plantedirektoratet 1994) and reform of the pesticide tax (implemented in 1996, see below). Additionally, a new act on education for professional users of pesticides was implemented in 1993 (Bichel Committee 1998).

Years earlier, Denmark had introduced economic instruments directed towards pesticide use: an approval fee (introduced in 1972) at 3% of the wholesale price and a 1982 tax amounting to 20% of the wholesale price of pesticides approved before 1982 for pesticides in packaging up to 1 litre or 1 kilogram. The tax was directed towards household consumption only and had no effect on agricultural pesticide use, since pesticides for agriculture normally were sold in packages larger than 1 kilo or 1 litre. Moreover, many of the agricultural pesticides at the time were introduced, and therefore approved, after 1982 (Lovtidende 1982; Ministry of Taxation *et al.* 2001).

In 1994, it was determined that the set of policy instruments described above would be adequate to reach the objective for reduction of active ingredients, but not for the TFI. Meanwhile, the Danish government, led by the Social Democratic Party, made a general move in the 1990s towards a green tax reform, shifting the tax burden from income taxes to environmental taxes (Ministry of Taxation 2001). Consequently, a pesticide tax covering all types of pesticide use (also agricultural) and providing (expectedly) stronger incentives to reduce pesticide use was introduced in 1996 (Pedersen *et al.* 2015) (Table 13.1). The tax was levied on the sales price of pesticides and, in combination with other policy instruments, it aimed to reduce the use of approved pesticides by 50%. The tax revenue was fully reimbursed to the agricultural sector, primarily through a reduction of land taxes (Ministry of Taxation *et al.* 2001). Consequently, farmers who substantially reduced their use of pesticides would gain a net benefit through a relatively low pesticide tax and lower land tax. An *ex ante* impact assessment estimated that the tax would reduce the use of pesticides by 8%, assuming

Table 13.1 Danish pesticide tax 1996–2013 (% of retail price, excluding VAT and other taxes).

	Tax rates	
Pesticide type	**1996–1998**	**1998–2013**
Insecticides	37	54
Fungicides	15	33
Herbicides	15	33
Growth regulators	15	33

Source: Ministry of Taxation (1998) and Pedersen *et al.* (2015).

a price increase of 15% and a price elasticity of demand of –0.5 (Ministry of Taxation 1995). There is no argument in the official documents for using these exact estimates, but it appears to be roughly comparable to estimates cited in the scientific literature which generally suggest that the demand for pesticide among European farmers is relatively inelastic (Falconer and Hodge 2000). It soon became clear that the tax was not as effective as predicted, which led the Danish government to double (on average) the tax rates in 1998 (see Table 13.1).

13.2.2 Second Plan: 2000–2004

In 2000, following failure to reach the 1997 aim of reducing TFI by 50% and based on an expert evaluation of the Danish pesticide policy (Bichel Committee 1998), the government (Social Democratic Party and Social-Liberal Party) introduced a new pesticide action plan, which contained the following main elements:

- an objective to decrease pesticide use measured through the TFI as much as possible. The first partial aim was to reach a TFI of 2.0 before the end of 2002 and decreasing further to a TFI of 1.4–1.7 within 5–10 years. This level could be reached without significant economic losses for the farmers and the Danish economy, according to the expert evaluation committee
- an increase in pesticide-free buffer zones along watercourses and lakes and in vulnerable areas (50 000 ha around watercourses and lakes for the whole period, 20 000 ha before end of 2002)
- an increase in the share of organic farmland (170 000 ha increase before the end of 2003)
- revision of the pesticide approval procedure.

The main policy instruments laid out in the plan were:

- increased advisory of farmers on reduction of pesticide use
- establishment of demonstration farms and knowledge exchange groups in agricultural counselling
- increased use of farmer decision-making tools and pest monitoring systems
- information campaign from agricultural organizations directed at farmers

- continuing education of farmers and agricultural consultants
- development of target figures for TFI in each crop
- subsidies (supported by the EU Common Agricultural Policy) for buffer zones along watercourses and lakes
- a more restrictive approval procedure for pesticides constituting a risk to ground water
- more research on organic farming
- development activities for organic farming.

It was further mentioned in the plan that if the aims were not reached by 2002, the government would consider increasing the pesticide tax or introducing a pesticide quota system (Ministry of Environment and Energy and Ministry of Food, Agriculture and Fishery 2000).

13.2.3 Third Plan: 2004–2009

Coming within the reach of achieving the target for 2002 of a TFI at 2.0, the Danish government (Liberal Party and Conservative People's Party) maintained the overall aim of a TFI at 1.7 in the third plan, an aim that was set to be reached no later than the end of 2009. It was expected that it would be possible to reach even lower levels of TFI after 2009. The key policy instruments were advice to farmers and research on farmer decision-making systems, precision spraying and other technology-based measures in combination with the tax (Ministry of Environment and Ministry of Food, undated). As in the second plan, a second aim was to increase the amount of organic farming and, additionally, increase pesticide-free farming on conventional farms. This was to be achieved through subsidies for conversion to organic farming or pesticide-free farming (financed by the Danish state and the Common Agricultural Policy). Third, the plan aimed at continuing the development of a restrictive approval system. This objective was supported by more research and more consultancy for farmers about the importance of not violating the instructions for using the pesticides. Finally, the plan aimed at increasing the amount of pesticide-free buffer zones from 8000 to 25 000 ha before the end of 2009 through increased information and advice on the subsidies available for such zones.

While advisory services continued to constitute a core element in pesticide policy, it is noteworthy that the direct subsidy from the Danish state to farmers using agricultural advisors was abolished shortly after the plan was introduced. As part of the budget negotiations, the Ministry of Food decided to abolish, from 2006, subsidies to farmers contracting with agricultural extension services or private agricultural advisors. These subsidies had been in place since 1999. Instead, the farmers received a reduction in the land tax as compensation (Ministry of Food 2005). Although the direct subsidy for hiring an advisor was removed, agricultural consultancies were still economically supported through a foundation that channels part of the pesticide tax revenue back into agricultural uses (Ministry of Food, Agriculture and Fishery 2004; Promilleafgiftsfonden for landbrug undated; Ugebrevet A4 2011). The foundation is administered by a board consisting of six representatives from different agricultural organizations and five representatives from different public interests (Promilleafgiftsfonden for landbrug 2016). During some periods since then, there have been subsidies for specific advisory activities, for example for advice on IPM (see below), but no general scheme.

13.2.4 Fourth Plan (Green Growth): 2009–2013

In 2009, the Liberal-Conservative government, supported by the Danish People's Party, adopted a 'Green Growth Plan' including a fourth pesticide action plan. The plan concluded that the objectives of the previous plan of reaching a TFI of 1.7 and 25 000 ha of pesticide-free buffer zones along watercourses and lakes had not been reached and that, therefore, it was necessary to change and strengthen the policy to achieve a significant reduction of the damaging effects of pesticides (Danish Government 2009). In the plan, a new 'pesticide load indicator' (PLI) was introduced as a replacement for the TFI. The load indicator measures environmental load on all Danish farmland, whereas the TFI included no exact load estimation and only covered conventionally cultivated fields.

The PLI consists of three main categories of load: (1) human health (measures the degree of exposure to pesticides of the spray operator), (2) environmental fate (a measure of the degradation time of the pesticides in soil and their potential for accumulation in food chains and for transport from soil to ground water), (3) environmental toxicity (a measure of the toxicity of the pesticide to non-target organisms in the field and adjacent nature) (Danish Environmental Protection Agency 2012). The new indicator corrects what some have pointed out as a paradox, i.e. that an increase in the size of organic farmland or other areas not sprayed with pesticides did not count as a reduction in the TFI. The aim set for pesticide reduction measured by the new indicator was to reach a PLI of 1.4 (a level that was estimated to be equal to a TFI of 1.7). At that time, the TFI had increased to 2.5. Regarding pesticide free-buffer zones along watercourses and lakes, the aim was to reach 50 000 ha by the end of 2012.

Some of the most important new instruments to reach these targets were as follows.

- A planned reform of the pesticide tax, changing the tax base from sales price to the environmental load of each specific pesticide (the lowest tax rates were applied to the 'greenest' pesticides, based on a calculation of PLI for each pesticide) and an increase in the average tax level. The increased revenue would be reimbursed to the agricultural sector through reduced land tax. However, due to design difficulties and the EU approval process, the reform of the pesticide tax was not implemented until the end of the plan period (July 2013).
- A mandatory establishment of buffer zones along all watercourses and lakes, amounting to a total of 50 000 ha.
- Establishment of 25 m buffer zones around public water supplies.

Furthermore, as part of the plan to introduce more market-based instruments, the policy package included a farmer subsidy for advisory on IPM (2010–2015), among other measures (Ministry of Food, Agriculture and Fishery 2010). The subsidy covered 80% of the costs for IPM consultancy. According to Kudsk and Jensen (2014), it was estimated that approximately 15% of Danish agricultural land would be covered by this type of advice during the period 2010–2015. It is noteworthy that the Green Growth Plan is the first pesticide action plan explicitly mentioning 'integrated pest management' (IPM) (Kudsk and Jensen 2014). IPM was included in the plan as a direct response to EU Directive 2009/128/EC on sustainable use of pesticides, which demands, for instance, that professional users will have to apply the general principles of IPM (from 1 January 2014 onwards). The directive lists eight general principles of IPM in the EU (Directive 2009/128/EC, annex III).

13.2.5 Fifth Plan: 2013–2015 (2016)

In 2013, the fifth pesticide action plan was presented by the government (Social Democratic Party, Social-Liberal Party, Socialist People's Party). The stated premise of the plan was that since the use of pesticides had increased by 35% since 2007, tough measures were necessary. The overall aim of the plan was to reduce the environmental load of pesticides by 40% between 2011 and 2015 and likewise the health load from particularly problematic pesticides. Moreover, the plan maintained the aims of lowest possible amount of pesticide residues in Danish food and no approved pesticides in ground water above limit values. Finally, all commercial use of pesticides had to follow IPM principles. According to the government, the plan would secure an implementation of the EU's directive on Sustainable Use of Pesticides (Directive 2009/128/EC) and add to Denmark's status as a pioneer country in reducing pesticide load in the environment (Danish Government 2013). In May 2015, the plan period was extended by 1 year, to the end of 2016.

In this fifth plan, the most important policy instrument was the revised pesticide tax, according to the government, but in addition to the tax there were a number of initiatives, including a more restrictive approval system; subsidies for alternative pesticides (low risk); information; investigation of the possibility of introducing mandatory use of drift-reducing sprinklers; research; more control directed towards, for example, illegal import, stricter sanctions for illegal import; focused IPM advisory.

13.2.6 Summing Up

Danish pesticide policies over the last three decades, as implemented through successive pesticide action plans, have been centred around the following main objectives.

- A significant reduction in the use of approved pesticides (50% reduction of active ingredients and TFI; TFI of 1.7; PLI of 1.4; PLI reduced by 40%).
- Pesticide-free zones adjacent to vulnerable areas (i.e. watercourses, lakes, public water supply).
- An increase in organic farmland.
- Tightening of the pesticide approval system.
- No pesticide leaches to ground water above limit values.

Overall, the plans can be characterized as being quite ambitious (especially concerning the aim of significant reductions in the use of approved pesticides). As previously mentioned, the level of ambition of the plans may be driven by the strong norm in the Danish population and among politicians for maintaining a system of untreated drinking (ground) water (Danish Water and Wastewater Association 2011; Hasler *et al.* 2007; Ministry of Taxation 2011), and the fact that Denmark is among Europe's most intensively farmed countries (World Bank 2015), which, all things being equal, leads to a higher overall pesticide load than in less farmed countries. Regarding political support, we observe that two of the plans were presented by Social Democratic-led governments, one by a Conservative-led government, and two by governments led by the Liberal Party, which signals a relatively broad political support for reducing the use of pesticides. A second observation is that Denmark has been using a broad suite of different policy instruments, including:

- economic instruments (pesticide tax; subsidies for pesticide-free buffer zones/organic farming/alternative pesticides/IPM advisory; economic support to targeted research)

- command-and-control instruments (approval system of pesticides, mandatory spraying certificates, mandatory spraying journals, mandatory buffer zones along watercourses, lakes and public water supplies, more control and punishment, etc.)
- information, advisory, education, etc. (for a detailed analysis of this element, see Kudsk and Jensen 2014).

The pesticide tax has attracted much attention over the years, and has been buttressed by other instruments. Table 13.2 sums up the main aims and policy instruments in the different plans, and the next section addresses the effects of these instruments.

13.3 Effects

In the following, we discuss the effects of the pesticide action plans on the five main types of aims presented above.

13.3.1 A Large Reduction in the Use of Approved Pesticides

As described above, during the last three decades Denmark has set a number of objectives for reductions in the use of approved pesticides. The objectives have evolved from a 50% reduction of active ingredients and of the TFI (introduced in 1986) to a TFI at 1.4–1.7 (2000) and a 40% reduction of PLI in the fifth plan.

An expert committee, known as the Bichel Committee for its chairman, set up to evaluate the first pesticide action plan, concluded that this first plan had led to a 40% reduction of active ingredients in the period 1986–1996, below the targeted 50% (Bichel Committee 1998). Yet, eventually the focus shifted from the aim regarding active ingredients to the TFI aim since, as mentioned above, this indicator was considered a better proxy for environmental load than 'amount of active ingredients'. Therefore, we will focus on this indicator below.

In the period between 1986 and 2015, Denmark had never come very close to reaching the overall aim of a TFI of 1.7 (Figure 13.1). However, between 2000 and 2002 the TFI was only slightly above 2.0, but started rising in the years after. In the years 2011–2013 the TFI was above 3, but then it declined below 3 in 2014.

The general trend over the years has been that pesticide use, as measured by the TFI, has oscillated around 2.5, albeit with a general upward tendency since the mid-2000s. A few years show spikes in pesticide use, followed by significant drops. The spikes in 1995, 1997 and 2012–2013, at least, appear to be related to the introduction of the pesticide tax (1996), its doubling (1998) and its redesign (mid-2013), probably reflecting hoarding behaviour. Anticipating tax-induced price hikes, farmers would buy larger amounts of pesticides and then use them in the following years, leading to lower sales directly after the tax took effect. Thus, the TFI increased from 2.51 (1994) to 3.49 (1995) and then fell to 1.92 in 1996 when the tax was introduced. In 2011, the TFI was at 3.22, rose to 3.96 in 2012, but only fell to 3.76 in 2013 when the tax was introduced in July. In 2014, the TFI fell sharply. This is not surprising since the TFI is based on sales figures, and, due to the hoarding effect in 2012–2013, lower sales were to be expected in 2014.

We have no way of knowing how pesticide use would have developed without the introduced policy instruments, which makes any assessment of the effects measured by

Table 13.2 Important policy instruments directed towards farmers in Danish pesticide action plans (1986–2015).

Year	Main objectives	Main policy instruments in addition to those introduced earlier
1986–2000	• 50% reduction of total amount of active ingredients by 1997 • 50% reduction of TFI* (to 1.34) by 1997	• Information and advisory • Research • Tighter regulation on approval of pesticides • Mandatory spraying certificates • Mandatory spraying journals (from 1994) • Pesticide tax (from 1996, doubled in 1998) • Education of users (1993)
2000–2004	• TFI of 2.0 before the end of 2002; TFI of 1.4–1.7 for the period 2005–2010 • 20 000 ha pesticide-free buffer zones along watercourses and lakes before the end of 2002 and later 50 000 ha • Protect vulnerable areas • Increase organic farmland to 170 000 ha in the period 2000–2003 • Revision of the approval procedure (e.g. to protect ground water)	• Increased advisory and information • Establishment of demonstration farms and knowledge exchange groups in agricultural advisory • Increased use of decision support and monitoring systems • Continuing education of farmers and consultants • Subsidies for pesticide-free buffer zones • More targeted use of EU's set-aside regulation • More restrictive approval procedure for pesticides constituting a risk for the ground water • Research on organic farming • Development activities for organic farming • Specific targets for pesticide use in different crops
2004–2009	• TFI of 1.7 before the end of 2009 • Increase amount of pesticide-free farmland • Continuing restrictive approval procedure • 25 000 ha pesticide-free buffer zones along watercourses and lakes before the end of 2009 • No leaching of approved pesticides to the ground water	• Focused advisory and information • Research • Subsidies for organic and other pesticide-free farming • Continue warning system • New regulation on filling up pesticides and washing of sprinklers

(*Continued*)

Table 13.2 (Continued)

Year	Main objectives	Main policy instruments in addition to those introduced earlier
2009–2013	• PLI** of 1.4 (similar to TFI of 1.7) before the end of 2013 • Reach 50 000 ha of pesticide, fertilizer and no-cultivation buffer zones along watercourses and lakes before the end of 2012 • Double amount of organic farmland in 2020 • Lowest possible amount of pesticide residues in Danish food • No approved pesticides in ground water above the 0.1 μg limit	• Change the pesticide tax to tax environmental load instead of price • Mandatory establishment of 50 000 ha buffer zones before the end of 2012 • Research • Tight approval procedure of pesticides and easier approval of non-chemical products, etc. • Promote pesticide-free agriculture through market-based instruments (e.g. subsidy for IPM advisory) • Strengthen advisory on IPM • Spray journals • Regular control of sprinklers • Pesticide-free buffer zones (25 m) around public water supplies
2013–2015(2016)	• Reduce the environmental load of pesticides by 40% for the period 2011–2015 • Lowest possible amount of pesticide residues in Danish food • No approved pesticides in ground water above the limits • 40% reduction in health load from particularly problematic pests by 2011–2015 • Use of pesticides must follow IPM principles	• Implement reformed pesticide tax • Tighter approval system • Subsidies for alternative pesticides (low risk) • Information • (Maybe) mandatory use of drift-reducing sprinklers • Research • More control directed towards illegal import • Stricter sanctions for illegal import • Focused IPM advisory

* TFI, treatment frequency index. The TFI is a standard indicator for pesticide use, calculated as the number of pesticide applications on cultivated areas per calendar year in conventional farming assuming use of a fixed standard dose (Danish Environmental Protection Agency 2012).

** PLI, pesticide load indicator. The PLI measures environmental load on all Danish farmland, not only, as for the TFI, conventionally cultivated fields (Danish Environmental Protection Agency 2012).

TFI slightly speculative. For instance, fluctuating grain prices, changes in crop composition, climate change effects, etc., may also have an effect on pesticide use. Taking into account these external factors, Ørum *et al.* (2008) calculated that the economically optimal level of pesticide use for the average farmer was 2.08 in 2007, and not 1.70 as expected by the Bichel Committee (see above). Still, it would probably be difficult to explain those very high numbers for TFI in recent years by purely external factors. It seems reasonable to conclude that the pesticide policy package has had some effects, but it has not been able to deliver the desired TFI in the past three decades.

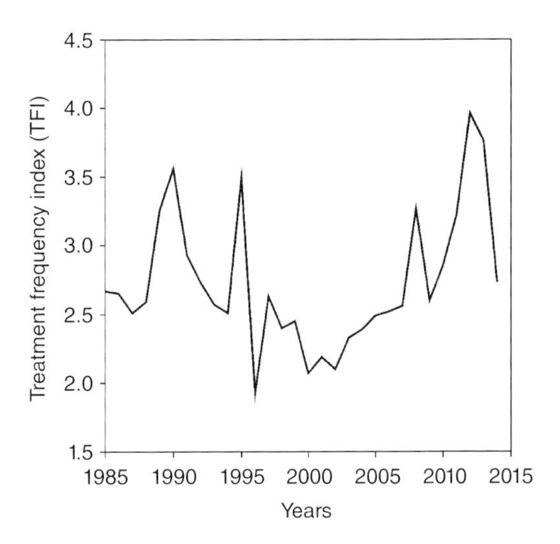

Figure 13.1 Treatment frequency index (TFI) for Denmark 1985–2014. The TFI is a standard indicator for pesticide use, calculated as the number of pesticide applications on cultivated areas per calendar year in conventional farming, assuming the use of a fixed standard dose and based on sales numbers (Danish Environmental Protection Agency 2012). The figure for 1985 is an average of the years 1981–1985 (Danish Environmental Protection Agency 1998). For the years 1997–2013, the numbers reflect the Danish Environmental Protection Agency's so-called 'new method' for calculating TFI. The change in calculation methods in the late 1990s meant that the TFI figure calculated was a bit higher (in the interval 0.07–0.27) compared to when the old method was used (Pedersen *et al.* 2015). Sources: Index made by Christina Bøje (Danish Environmental Protection Agency) based on the agency's annual reports. The years 2007–2013 are corrected with the newest figures from the Danish Environmental Protection Agency (2014, 2015). Adapted from Pedersen *et al.* (2012).

Regarding the new PLI, it is still too early to assess its development, since the redesigned tax has only been working since July 2013. Moreover, there appears to have been some hoarding in 2012 and 2013 before the tax was introduced in July 2013, and, conversely, lower sales in 2014 when farmers were able to use pesticides bought in the previous years. The development in the PLI based on sales numbers (2007–2014) is described in Figure 13.2. Meanwhile, an analysis of pesticide use based on the mandatory reporting of each farm's use shows that the actual pesticide load has not changed much between 2011 and 2014 (see Figure 13.2). In fact, the load has increased a bit from 2.22 in 2011 to 2.37 in 2014 (Ørum 2015). This development fits expectations as the large purchases of pesticides prior to the introduction of the tax in July 2013 allowed farmers to maintain their habitual pesticide use in the first years after the new tax. However, as Danish farmers exhaust their supply of stored pesticides in the coming years, the 'real' effects of the tax will probably emerge as they will have to purchase pesticides affected by the new taxation.

In general, the tax can provide both producers and users of pesticides with incentives to minimize the use of the most damaging pesticides. Producers have an incentive to phase out the most damaging pesticides since demand will probably decrease as a result of these pesticides being hit with the greatest price increases, as well as an incentive to invent new products with lower environmental load. Farmers, at the same time, will have an incentive to buy pesticides with lower environmental load (or switch to organic

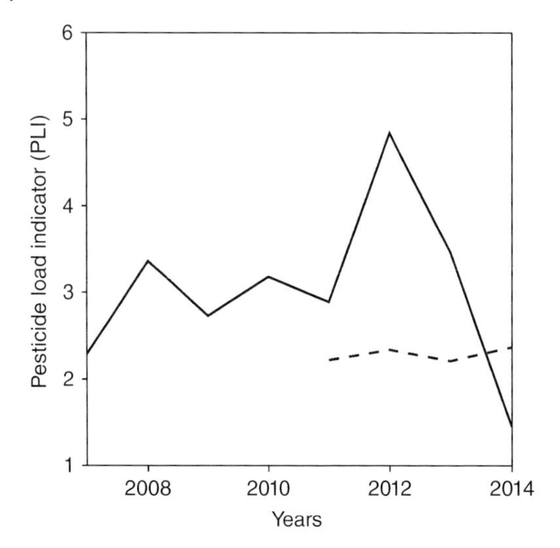

Figure 13.2 Pesticide load indicator (PLI), based on sales (*solid line*) or on use (*dashed line*) for Denmark, 2007–2014. The PLI consists of three main categories of load: (1) human health (measures the degree of exposure to pesticides of the spray operator), (2) environmental fate (a measure of the degradation time of the pesticides in soil and their potential for accumulation in food chains and for transport from soil to ground water), (3) environmental toxicity (a measure of the toxicity of the pesticide to non-target organisms in the field and surrounding nature) (Danish Environmental Protection Agency 2012). The sales figures in the table are based on sales data from the companies and estimated by Copenhagen University in December 2015 (Ørum 2015). These data are more up to date than the data in official statistics from the Danish Environmental Protection Agency. Data on PLI based on use are not available before 2011. Source: Adapted from Ørum (2015).

farming) if there are substitutes that will minimize expenses. The policy objective is a 40% reduction in the PLI between 2011 and 2015.

13.3.2 Pesticide-free Buffer Zones Adjacent to Vulnerable Bodies of Water

The aim for pesticide-free buffer zones along watercourses, lakes and public water supplies has been moved several times within the interval 20 000 to 50 000 ha. With the 2009–2013 plan, the aim was reformulated to 50 000 ha of pesticide-free, fertilizer-free and no-cultivation buffer zones along watercourses and lakes before the end of 2012. An evaluation conducted by a consultancy in 2008 found that, in 2002, 8300 ha of pesticide-free buffer zones had been established (the aim was 20 000 ha) (Rambøll Management 2008). In 2006, the area was 12 000 ha and it was assessed that it would be impossible to reach 25 000 ha in 2009 (an aim in the third plan) (Rambøll Management 2008). Consequently, in 2011 the Danish government, led by the Liberal Party, chose to make mandatory 10 m pesticide-, fertilizer- and cultivation-free buffer zones adjacent to watercourses and lakes (Ministry of Food 2011). The regulation was met with widespread protests from farmers and agricultural organizations – one organization directly encouraged farmers to break the law (Jydske Vestkysten 2012) – leading to a political agreement between the government and the right wing opposition to scale down the measure (Ministry of Environment and Food 2015). However, following the return to

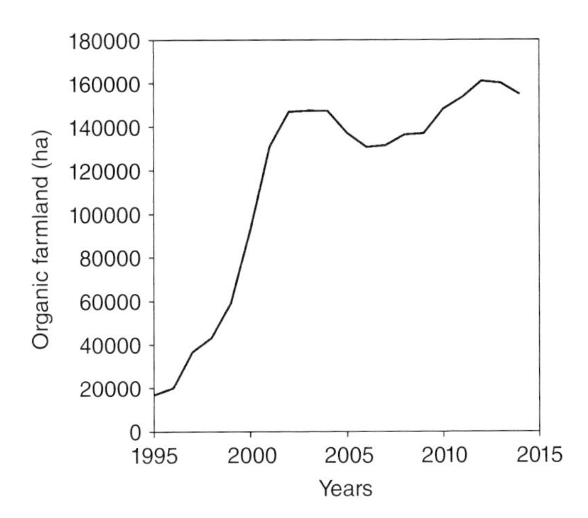

Figure 13.3 Organic farmland in Denmark, 1995–2014. The figure shows amount of farmland which is fully organic. Source: Statistics Denmark (2015).

government of the Liberal Party after 4 years in opposition, the Minister of Environment and Food presented, in November 2015, a proposal to abolish the regulation entirely, arguing that economic growth, jobs and welfare and cost-effective solutions constitute preconditions for the protection of the environment (Ministry of Environment and Food 2015).

A rule requiring 25 m buffer zones around public groundwater extraction sources, which was implemented in 2011, remains in force (before 2011, generally, there were only 10 m pesticide-free buffer zones). About half of the 10 000 Danish extraction sources are located in rural areas with pesticide use (Ministry of Environment, undated).

13.3.3 Increase in Organic Farmland

The development in the amount of agricultural land (ha) certified for organic production is described in Figure 13.3. Denmark experienced a large increase in organic farmland in the period 1995–2002. Between 2002 and 2014, the amount of organic farmland has been relatively stable. The aim to double the amount of organic farmland to 280 000 ha by 2020, which was set in 2009 when the Liberal Party last held office, has been dropped by the current Liberal government (Altinget 2015). It maintains an unquantified aim of increasing conversion to organic farming.

13.3.4 Restrictive Pesticide Approval System

Ten years after the first pesticide action plan was created, an expert committee assessed whether or not a strict approval system of pesticides had been reached (Bichel Committee 1998). At the time, 213 active substances had been reassessed. Of these, 105 had been removed from the Danish market either because producers/importers did not attempt to get them approved due to lack of documentation of absent harmful effects or because they were withdrawn by the applicant. Of the remaining 108 substances, 78 were approved and 30 were banned or strongly restricted (Bichel Committee 1998).

When the pesticide action plan was evaluated in 2008, it was concluded that Denmark, for a long period, had had a restrictive approval procedure compared to many other European countries. However, at the same time, the evaluation concluded that it was difficult to assess this aim more precisely (Rambøll Management 2008). According to Pesticide Action Network Europe (2013), only 80 active substances are approved in Denmark while, in many other European countries, 300–400 substances are approved for use.

However, the approval system experienced significant implementation difficulties during the period 2000–2011, specifically in terms of delays in the reassessment of substances, according to an evaluation undertaken by the Danish National Audit Office, an independent agency under the Danish Parliament Auditor General (Rigsrevisionen 2011). Thus, in 2011, a report from the National Audit Office concluded that some pesticides that should have been either banned or limited had in fact remained on the market unrestricted for up to 5 years (Rigsrevisionen 2011). A follow-up audit in 2014 concluded that the problems had been solved after the Ministry of Environment allocated more funding to the approval procedure (Rigsrevisionen 2014).

One concern about a restrictive approval system is that it may lead to illegal import of pesticides. Current Danish levels of illegal imports are difficult to estimate, but some illegal pesticide transports have been uncovered by the authorities (Pedersen *et al.* 2015). In December 2015, the Danish Broadcasting Corporation (DR) reported that 17% of the 2014 inspections on Danish farms had revealed illegally imported pesticides (Danish Broadcasting Corporation 2015). Based on data from the government's inspection reports obtained through a freedom-of-information request, the report also showed that the percentage of detected violations had actually increased from 10% in 2012. However, only 1% of the violations involved illegally imported pesticides, according to a representative for SEGES, a knowledge and research organization connected to the Danish Agriculture and Food Council. The representative questioned the interpretation that the violations indicate use of illegal pesticides (Landbrugsavisen 2015). He pointed out that the great majority of the violations involved possession of illegal products that had been approved for use in Denmark until 2012, when the rules were changed, suggesting that farmers simply had not yet disposed of the products. Inspections are carried out on a sample of slightly fewer than 3% of farms and nurseries (Danish Broadcasting Corporation 2015).

It is worth noting that tests for pesticide residues in agricultural crops have consistently shown pesticide residues in significantly fewer samples of Danish-grown fruits, vegetables and cereals compared with imported products from the EU or from other countries (Ministry for Food, Agriculture and Fishery 2014). For example, in 2013, 18% of Danish vegetable samples contained pesticide residues compared with 48% of the EU products tested. Fruit was more likely to contain pesticide residues and as many as 48% of the sampled products contained pesticides, compared to 70% of the EU samples.

13.3.5 No Pesticide Leaching to the Ground Water Above Limit Values

A 1998 expert committee found that there were some violations of the limit values in groundwater samples (Bichel Committee 1998). The lowest number of violations since 1995 was registered in 2005 when pesticides were found in 24% of the test drillings and limit values showed a 4% excess. For the period 2006–2013, there were still samples exceeding the limit values. However, the pesticides most frequently detected are generally those that have been off the market or have been illegal for the last 6–15 years. Often, water extracted for drinking purposes has been stored in aquifers for more than 15 years (Thorling *et al.* 2015). Hence, many finds might be characterized as 'sins of the past' (Danish Environmental Protection Agency 2010). Consequently, while the aim has not been achieved, the number of violations has been decreasing over the years and it is expected that it will decrease further due to the restrictive pesticide approval procedure.

13.3.6 Why did the Danish Pesticide Action Plans not Fulfil Their Aims?

In brief, it has been difficult to fulfil several of the aims in the Danish pesticide action plans. Denmark has implemented and maintained a relatively restrictive approval system of pesticides, and it is therefore fair to conclude that this particular aim has been met. However, the objectives relating to reduction of the use of approved pesticides have proved much more difficult to reach; Denmark has not come near reducing the TFI to 1.7 (or 1.4). Apart from protection buffer zones around groundwater drillings, it has proved difficult to protect the vulnerable buffer zones surrounding lakes and watercourses from pesticide use. There was a large increase in organic farmland during the years around the millennium but in subsequent years, the conversion rate has stalled, so the quantitative target for conversion to organics has also not been achieved. Finally, pesticides are still being detected above the limit values in ground water for drinking, although the number of samples above the limits is decreasing.

Thus, although Denmark has implemented a broad scheme of instruments, covering the three main types of policy instruments, including command-and-control instruments, economic instruments and information/advice, its policy performance has been mixed and, importantly, has fallen short of achieving aims to reduce pesticide use. This raises the question of whether the incentives in the policies have been too weak to generate the intended effect on farmers' decision making and, by implication, whether target groups have behaved differently than anticipated in *ex ante* policy assessments. We will briefly discuss this for each of the main instrument types.

In a stable country with relatively strong institutions such as Denmark, it is expected that citizens generally obey the law (Winter and Nielsen 2010). However, as demonstrated when the Danish government decided to make it mandatory for farmers to establish 10 m buffer zones along watercourses and lakes, there are exceptions where command-and-control policy instruments meet active resistance from the target group (Winter and Nielsen 2010), particularly when the policy is considered unfair. While some farmers publicly announced in a show of civil disobedience that they would not obey the regulation (Politiken 2013), there is no indication as to how many farmers actually chose not to implement the buffer zone regulation. Furthermore, as mentioned above, some illegal pesticide imports have been detected in Denmark, but despite such evidence of non-compliance, Winter and May (2001) concluded in a study that large percentages of farmers in Denmark comply with the agro-environmental regulations, although some vary with regard to the circumstances under which they meet the demands and when they do not.

Denmark has many years of experience with economic instruments in its pesticide policy, in particular with taxation and subsidies (e.g. for buffer zones and organic farming). Although the Danish pesticide tax is probably the highest in the world (Pedersen *et al.* 2015), it has not offered adequate incentives for farmers to reduce their application of pesticides to a TFI of 1.7 (it still remains, though, to see the results of the reformed tax). Milder winters and an increase in winter crops may partially account for the failure to reach the 1.7 target, but it still cannot entirely explain why Denmark has been so far away from fulfilling the aims (Ørum *et al.* 2008; Pedersen *et al.* 2015).

A study of the decision patterns among Danish farmers suggests that it is worth looking at the assumptions built into *ex ante* policy analyses that model farmers as economic optimizers (Pedersen *et al.* 2012). The study, which is based on a survey of more than 1100 Danish farmers, shows that farmers may be grouped into segments based on the degree to which they weigh economic rationales versus more professionally oriented rationales (e.g. growing as large a crop as possible) (Pedersen *et al.* 2012). The study shows that about half of the farmers are primarily economically motivated and place greater emphasis on the price of pesticides than other variables when making decisions regarding pesticide use. However, approximately one-third of the farmers are very focused on optimizing crop yield and pay relatively little attention to both expense and crop prices when making decisions regarding pesticides. The study demonstrates that this last group is not as responsive to three types of economic instrument (tax increase, tradeable pesticide quota and increase in subsidy schemes for reduced pesticide consumption) – as the group of more economically motivated farmers (Pedersen *et al.* 2012). These findings seem to indicate that the demand for pesticides is more inelastic than anticipated by the Ministry of Taxation when the tax was first introduced.

Regarding the difficulties in making subsidies work as well as intended (for buffer zones along watercourses and lakes, and for organic farming), Christensen *et al.* (2011) found that there are fundamental barriers to increasing farmers' interests in signing up for agri-environmental schemes. Importantly, the study shows that it is very important for farmers to maintain flexibility in the scheme and to avoid bureaucratic red tape and that payment above and beyond direct costs, in general, is a necessary condition for farmers to show an interest in a subsidy scheme. This latter finding conflicts with EU regulation that does not allow for schemes offering compensation levels significantly above direct cost. Consequently, if *ex ante* analyses of economic instruments assume that farmers generally fit the model of 'economic man', this could be one explanation for the gap between policy aims and policy effects.

As shown in Table 13.2, the Danish pesticide action plans have included a substantial number of information and advisory instruments, including education (spraying certificates, etc.), research, development of monitoring and decision support systems, information campaigns and subsidies for advisory. Kudsk and Jensen (2014) conclude that the early start with Danish pesticide action plans with quantitative targets for three decades means that Danish advisory services and research institutions have been focusing on reducing pesticide use. Consequently, many of the 'easier' reductions such as optimized pesticide dosages and using disease-resistant varieties have already been implemented and therefore it can be a challenge to meet new demands in national and EU regulations. Jørgensen and Jensen (2011) found that intense Danish research on, for instance, reduced dosages, decision support systems and independent advisory services has contributed to a reduction in pesticide use.

But it is clear that these soft instruments have not fulfilled the aims in the pesticide action plans either, given the failure to decrease use of approved pesticides to the 'economically optimal' level of 1.7. First, as mentioned above, a significant share of farmers do not behave as 'economic man' and therefore do not necessarily react to advice based on economic arguments (Nielsen 2010). Second, many advisors as well as farmers express the conviction that pesticides, if approved by environmental authorities and if used in amounts that do not exceed the recommended dosages, are not harmful to the environment (Pedersen *et al.* 2012, 2014). Consequently, these farmers perceive no environmental incentive to reduce pesticide use. Finally, in the past, group-based advisory was used by many farmers and provided good fora for participatory learning and competition on minimizing pesticide use. But increasing farm size (the average size of managed units has quadrupled over the last two decades) has caused many farmers to skip the group-based advisory in order to use their limited amount of time solely on their own farm (Kudsk and Jensen 2014).

There exists no baseline scenario indicating the counterfactual trajectory of pesticide use in Denmark for the period 1986–2015 since no described policy instruments have been implemented. It seems fair to conclude, however, that, over the 30-year perspective, the policy has had some effects. For instance, the relatively restrictive approval system has weeded out some harmful products and the amount of organic farmland has increased. It is also conceivable that the TFI would have been even higher if no pesticide tax had been in place. However, the incentives incorporated in the different policy instruments have not been strong enough to achieve the ambitious aims in the five Danish pesticide action plans. Still, based on the existence of the broad spectrum of policy instruments, we would expect Denmark to perform better in relation to pesticide reduction than other European countries.

13.4 Comparing Denmark to the EU and Internationally

In this section we provide a tentative comparison of Danish pesticide policies in an international context, beginning with the European countries which are subject to the same overall pesticide and agricultural policies as Denmark. However, surprisingly, comparison between the performance of Danish pesticide policies and the policies in other European countries is hampered by a lack of good data.

One comparative study of five EU Member States (Denmark, Germany, the Netherlands, France and UK) shows that there has been a tendency to move away from volume/use reduction targets towards risk/impact reductions (Barzman and Dachbrodt-Saaydeh 2011). As described above, Denmark continues along these lines with the introduction of the PLI and the redesigned pesticide tax. What appears to be lacking in the international literature is an analysis of the effectiveness of different policy instruments in different settings. Below, we compare pesticide use across the EU with the purpose of identifying whether or not being among the pesticide policy pioneers has made a comparative difference for Danish pesticide use.

As mentioned above, only 80 active substances are approved in Denmark, while in many other European countries 300–400 substances are approved, according to Pesticide Action Network Europe (2013). It seems fair to assume that part of the explanation for this is the restrictive Danish approval system which is, for instance, focused on avoiding pesticides in drinking water.

Regarding the use of approved pesticides, Eurostat currently measures only pesticide sales in the Member States (Table 13.3). However, a new 'main' indicator that measures application rates of different pesticide categories is under development (November 2015)

Table 13.3 Pesticide sales 1999–2008 in Europe, tonnes of active ingredient (index: year 2000 = 100). Countries are listed in alphabetical order.

Country/year	1999	2000	2001	2002	2003	2004	2005	2006	2007	2008
Austria	96	100	88	86	95	93	96	na	na	na
Belgium	96	100	89	92	89	92	98	na	na	na
Denmark	101	100	108	101	104	102	114	113	116	140
Estonia	60	100	108	108	105	117	128	153	150	na
Finland	100	100	124	141	145	130	125	144	na	na
France	127	100	105	87	79	80	83	76	82	na
Germany	100	100	92	97	99	95	97	105	108	114
Greece	91	100	100	na	na	na	na	na	na	na
Hungary	106	100	118	150	159	182	177	211	204	221
Ireland	99	100	117	131	137	146	130	135	na	na
Italy	103	100	96	119	109	106	107	102	na	na
Latvia	na	100	130	119	147	210	258	788	370	na
Luxembourg	na	na	na	na	na	na	na	na	na	na
Malta	na	100	118	121	132	na	na	na	na	na
Netherlands	106	100	83	84	81	94	96	97	111	na
Norway	211	100	137	216	174	218	135	183	190	na
Poland	96	100	100	117	81	99	181	193	173	na
Portugal	100	100	100	113	110	109	106	102	108	110
Slovenia	na	100	95	79	93	106	94	87	na	na
Spain	97	100	103	na	na	na	na	na	na	na
Sweden	103	100	105	104	124	57	92	103	na	na
United Kingdom	107	100	100	100	96	99	100	90	na	na

Source: Eurostat 2015a. Note from Eurostat 2015a: 'Most of the Member States refer to the definition of plant protection product given in Directive 91/414/EEC to delimit the scope of this indicator. Nevertheless, there is no common definition adopted by all Member States and there can be significant differences in the range of products used in different countries, so that comparability is limited. Additional information on the situation in specific countries is required for any detailed assessment'.
na, data not available.

(Eurostat 2015b). The table shows that Denmark has increased sales of active ingredient (in tonnes) by 40% in the period 2000–2008. Several other countries (Estonia, Ireland, Latvia, Hungary, Poland and Finland) have experienced large increases as well, while some countries (France, Slovenia, UK) have experienced large decreases. However, the table does not offer a solid basis for comparative conclusions regarding pesticide load since:

- the table covers a relatively short period – the Danish action plans had already been working for 13 years in 1999 when the available Eurostat data series started, and Denmark had already achieved a 40% reduction in use of active ingredients between 1986 and 1996 (Bichel Committee 1998)

Table 13.4 Treatment frequency index (TFI) in wheat and yield in wheat (2006–2007) (Jørgensen and Jensen 2011).

	UK (2006)	France (2006)	Germany (2007)	Denmark (2007)
TFI in wheat	6.74	4.1	5.8*	2.62
Wheat yield, tonnes ha^{-1}	8.0	6.9	7.3	7.3

* Snail pesticides not included.

- 'tonnes of active ingredients' is not a good indicator for pesticide load since different types of active ingredients constitute quite different levels of environmental threat
- the Member States' definitions of pesticides differ (see the footnote to Table 13.3).

Reservations concerning the validity of 'active ingredients' as an indicator for pesticide consumption was the reason why, decades ago, Danish authorities chose to focus more on the TFI as an indicator for pesticide use (and today has moved to the PLI). Unfortunately, there is a lack of comparable European data of this type. However, a comparison of treatment frequency levels in a single year in four countries with large agricultural sectors can be found in Table 13.4. From this, it is apparent that, in 2006–2007, Denmark had a TFI for wheat which was substantially lower than those applying to other large agricultural countries in Western Europe, such as France, Germany and the UK. Jørgensen and Jensen (2011) point to several explanations for these differences. For instance, the UK, France and Germany use 3–4 times as many fungicides as Danish farmers. Danish farmers have fewer problems with fungi than their colleagues in these countries (partly due to tougher winters) as well as more experience in using reduced doses and lower marginal profit for using fungicides. Similarly, Danish farmers use fewer growth regulators, insecticides and herbicides compared to farmers in the three other countries. In addition to climatic variations and variations in pest attack levels, these differences may be explained by the knowledge gained from experiments in the field, independent advisory services and good forecasting systems for pest attacks, according to Jørgensen and Jensen (2011). This comparison tentatively suggests, then, that the lower Danish pesticide use in wheat is due to its active pesticide policy as well as to different natural and economic factors that condition the need for pesticides.

Like Denmark, other European countries have also been relatively early adopters of pesticide action plans: Sweden (1986), the Netherlands (1990), Germany (2004) and Belgium (2005) (Jørgensen and Kudsk 2008). Today, all 28 EU Member States have presented national action plans as part of the implementation of Directive 2009/128/EC, article 4 (European Commission 2016). However, as the assessment by Pesticide Action Network Europe (2013) demonstrated, Denmark was the only country out of the 24 Member States having published national action plans at the time that included an overall, clear, quantifiable objective for the pesticide policy, as demanded in article 4. Some of these Member States might choose to copy parts of the Danish policy design by either introducing similar quantitative targets and/or by copying the policy instruments. Needless to say, subsidy schemes and pesticide taxes, among other things, need to be adjusted to the particular national context, but there would probably not be any barriers as such for these types of instruments. However, as the Danish experiences demonstrate, when trying to construct an effective policy design, it is very important to have knowledge of factors that motivate different forms of behaviour within the target group, i.e. the farmers. As the evidence from Denmark demonstrates, there are no easy shortcuts.

Comparing policy effects beyond the European context should be undertaken with some caution due to differences in the character and structures of the agricultural sectors and political and social cultures. Yet, the international literature offers some comparative analyses of pesticide policies, for example Schreinemachers *et al.* (2015) comparison of pesticide regulation in South-East Asia. This study aims at identifying challenges and entry points for governments in South-East Asia and elsewhere to reduce the risk from agricultural pesticides by comparing levels of pesticide use, pesticide regulation and farm-level practices in Cambodia, Laos, Thailand and Vietnam. The authors identify three challenges in South-East Asia.

- There is a rapid increase in pesticide trade and, meanwhile, government capacity to enforce regulation has not expanded.
- Farmers' demand for and satisfaction with synthetic pesticides is high. Farmers are aware of adverse effects, but they give high priority to stable crop yields and are unaware/sceptical about alternative pest management methods.
- There is a lack of national systems to monitor pesticide risk regularly, and if data are collected, they are often not made publicly available (Schreinemachers *et al.* 2015).

In Denmark, enforcement of the regulation on approved/illegal pesticides is probably relatively good today, although, as described above, Denmark has also experienced some problems with use of illegal pesticides. Regarding demand for synthetic pesticides, Danish conventional farmers, like the farmers in South-East Asia, are satisfied with synthetic pesticides, and many do not think that there are high environmental risks connected to the use of pesticides (Pedersen *et al.* 2012). This may be one of the main reasons why it has proven difficult to reduce the use of approved pesticides in Denmark. Finally, Denmark does have a relatively good system for monitoring pesticide risk. Consequently, approval/banning and control of pesticides seem to work relatively well in the Danish regulatory system, although there have been some breaches of the law.

Finally, Pelaez *et al.* (2013) compared three different representative regulatory frameworks for pesticides (Brazil, USA and the EU). They found that all three systems have brought 'significant progress by incorporating social regulation in attempts to compensate for the long-standing hegemony of production-based rationales defended by vested interests in agriculture and the chemical industry'. However, they also noticed some differences. For instance, the EU regulatory framework appears to have institutionalized the principle of 'precaution' more explicitly than is the case in the USA regulatory framework. The authors also conclude that 'there is plenty of room for manoeuvre by regulatory agents as precaution is often used as discretionary measures in risk assessments'. They also noted that some aspects of pesticide regulation need time for institutional transformation.

Consequently, policy learning from frontrunner countries such as Denmark may become important – what are the 'do's and don'ts' when designing pesticide policies and institutions? What seems to be lacking in particular in the international literature is analyses on the effectiveness of different pesticide policy instruments in different settings.

13.5 Conclusion

Denmark may be considered one of Europe's pioneers when it comes to pesticide policies. The country was among the first to adopt a pesticide action plan, introducing a broad range of policy instruments. The action plan and follow-up plans have applied a two-pronged approach. One concerned implementation of a restrictive approval system that banned the use of pesticides known to have unwanted environmental or health effects. The other approach concerned the objective to significantly reduce the use of approved pesticides. A broad range of policy instruments have been introduced covering the three main types of instruments: command-and-control, economic and information/advice. While the policy has been repeatedly evaluated and reformed, it has fallen short of achieving most of its quantitative objectives. In fact, pesticide consumption has increased steadily since the early 2000s and policy measures targeting land use, such as buffer zones, have been taken up by farmers to a much smaller extent than anticipated.

It is reasonable to conclude that since the introduction of the first pesticide action plan in 1986, the policy has had some effects. The relatively restrictive approval system has taken some harmful products off the Danish market, the amount of organic farmland has increased, and it is even possible that the treatment frequency would have been higher if none of the policy instruments had been in place. However, the incentives incorporated in the different policy instruments appear to have been too weak to encourage sufficient changes in pesticide use for achieving the ambitious aims in the five Danish pesticide action plans. One key problem may be that a substantial number of farmers do not react to economic incentives as assumed by economic models (Pedersen *et al.* 2012). This implies that economic incentives need to be much stronger to motivate this group or that different types of policy measures that target the rationales of these farmers more directly should be implemented. It will be interesting to see whether the redesigned pesticide tax, which includes overall higher tax rates and a differentiation based on load, will be more effective.

Compared to other European countries, the sparse data solely based on pesticide use in wheat indicate that Danish farmers perform well – at least in 2007, which was a fairly average year for Denmark. The sparse amount of relevant comparable data for pesticide consumption across EU Member States demonstrates a need for the European Commission to initiate this type of data collection. Better indicators than sales figures for active ingredients, as for instance the Danish indicators on treatment frequency (TFI) and pesticide load (PLI), would make it possible to benchmark the Member States. If such measures were accompanied by more comparative analyses on farmer decision-making patterns, for example segmentation studies based on the degree to which farmers weigh economic rationales versus more professionally oriented rationales, knowledge on effective policy designs for reduction of pesticide use in the European Union would significantly increase.

References

Altinget (2015) Regeringen dropper egen målsætning for økologi. Available at: www.altinget.dk/artikel/regeringen-dropper-egen-maalsaetning-for-oekologi (accessed 8 March 2017).

Andersen, M.S. and Hansen, M.W. (1991) *Vandmiljøplanen – Fra Forhandling Til Symbol.* Niche, Harlev, Denmark.

Barzman, M., and Dachbrodt-Saaydeh, S. (2011) Comparative analysis of pesticide action plans in five European countries. *Pest Management Science* **67**: 1481–1485.
Bichel Committee (1998) *Rapport Fra Hovedudvalget*, Ministry of Environment, Copenhagen. Available at: http://mst.dk/service/publikationer/publikationsarkiv/1999/maj/rapport-fra-hovedudvalget/ (accessed 8 March 2017).
Christensen, T., Pedersen, A.B., Nielsen, H.Ø. and Hasler, B. (2011) Determinants of farmers' willingness to participate in subsidy schemes for pesticide-free buffer zones. *Ecological Economics* **70**: 1558–1564.
Danish Broadcasting Corporation (2015) Hver sjette kontrollerede landmand har ulovlige sprøjtegifte. Available at: www.dr.dk/nyheder/indland/hver-sjette-kontrollerede-landmand-har-ulovlige-sproejtegifte (accessed 8 March 2017).
Danish Environmental Protection Agency (1998) *Bekæmpelsesmiddelstatistik 1997*. Danish EPA, Copenhagen. Available at: http://mst.dk/service/publikationer/publikationsarkiv/1998/jul/bekaempelsesmiddelstatistik-1997/ (accessed 8 March 2017).
Danish Environmental Protection Agency (2010) *Fortidens synder kan give pesticider i grundvandet*. Danish EPA, Copenhagen.
Danish Environmental Protection Agency (2012) *The Agricultural Pesticide Load in Denmark 2007–2010, Environmental Review no. 2, 2012*. Danish EPA, Copenhagen. Available at: http://www2.mst.dk/Udgiv/publikationer/2012/03/978-87-92779-96-0.pdf (accessed 8 March 2017).
Danish Environmental Protection Agency (2014) *Bekæmpelsesmiddelstatistik 2013 – Behandlingshyppighed og Pesticidbelastning, Baseret på Salgsstatistik og Sprøjtejournaldata*. Danish EPA, Copenhagen. Available at: http://www2.mst.dk/Udgiv/publikationer/2014/12/978-87-93283-33-6.pdf (accessed 8 March 2017).
Danish Environmental Protection Agency (2015) *Bekæmpelsesmiddelstatistik 2014 – Behandlingshyppighed og Pesticidbelastning, Baseret på Salgsstatistik og Sprøjtejournaldata Orientering fra Miljøstyrelsen nr. 13, 2015*. Danish EPA, Copenhagen. Available at: http://www2.mst.dk/Udgiv/publikationer/2015/12/978-87-93435-00-1.pdf (accessed 8 March 2017).
Danish Government (2009) *Grøn Vækst*. Available at: www.stm.dk/multimedia/Groen_vaekst.pdf (accessed 8 March 2017).
Danish Government (2013) *Beskyt Vand, Natur og Sundhed – Sprøjtemiddelstrategi 2013–2015*. Available at: http://mst.dk/virksomhed-myndighed/bekaempelsesmidler/sproejtemidler/sproejtemiddelstrategi-2013-2015/ (accessed 8 March 2017).
Danish Parliament (1986) F 26 (oversigt): Forespørgsel til miljøministeren om forbruget af bekæmpelsesmidler. Available at: http://webarkiv.ft.dk/?/samling/19851/forespoergsel/f26.htm (accessed 8 March 2017).
Danish Water and Wastewater Association (2011) Employee (anonymous), interviewed November 2011 (unpublished).
Department of Agriculture, Forestry and Fishery (2010) Pesticide management policy for South Africa. Available at: www.nda.agric.za/docs/Policy/PesticideManag.pdf (accessed 8 March 2017).
Directive 2009/128/EC. Sustainable use of pesticides. Available at: http://eur-lex.europa.eu/legal-content/EN/TXT/?uri=CELEX%3A32009L0128 (accessed 8 March 2017).
Directive 91/414/EEC. Concerning the placing of plant protection products on the market. Available at: http://eur-lex.europa.eu/legal-content/en/ALL/?uri=CELEX:31991L0414 (accessed 8 March 2017).

European Commission (2016) National action plans. Available at: http://ec.europa.eu/food/plant/pesticides/sustainable_use_pesticides/nap/index_en.htm (accessed 8 March 2017).
Eurostat (2015a) Pesticide sales 1997–2008. Available at: http://ec.europa.eu/eurostat/ (accessed 8 March 2017).
Eurostat (2015b) Agri-environmental indicators – Farm management practices – AEI6 Consumption of pesticides. Available at: http://ec.europa.eu/eurostat/ (accessed 8 March 2017).
Falconer, K. and Hodge, I. (2000) Using economic incentives for pesticide usage reductions: responsiveness to input taxation and agricultural systems. *Agricultural Systems* **63**: 175–194.
Hasler, B., Lundhede, T. and Martinsen, L. (2007) Protection versus purification – assessing the benefits of drinking water quality. *Nordic Hydrology* **38**: 373–386.
Jørgensen, L.N. and Jensen, J.E. (2011) Strategier for planteværn i Danmark og i vore nabolande. In: *Plantekongres 2011: Sammendrag af Indlæg*. Aarhus University, Aarhus, pp. 77–79.
Jørgensen, L.N. and Kudsk, P. (2008) Pesticidhandlingsplaner i andre lande. In: *Plantekongres 2008, D3 1. Pesticidhandlingsplan 2004–2009 (I)*. Aarhus University, Aarhus, pp 82–85.
Jydske Vestkysten (2012) Opfordrer til boykot af randzoner. Available at: www.jv.dk/artikel/1387222:Indland--Opfordrer-til-boykot-af-randzoner (accessed 8 March 2017).
Kudsk, P. and Jensen, J.E. (2014) Experiences with implementation and adoption of Integrated Pest Management in Denmark. In: Peshin, R. and Pimentel, D. (eds) *Integrated Pest Management, Global Overview*, Vol.4. Springer, Dordrecht, pp. 467–485.
Landbrugsavisen (2015) Landskonsulent: Misvisende brug af tal for kontrol af sprøjtemidler (16 December 2015). Available at: http://landbrugsavisen.dk/mark/landskonsulent-misvisende-brug-af-tal-kontrol-af-spr%C3%B8jtemidler (accessed 8 March 2017).
Lovtidende A. (1982) Lov om ændring af lov om forbrugsbegrænsende foranstaltninger (act no. 259, 09.06.1982). Schultz, Copenhagen, Denmark.
Ministry of Environment (1986) Miljøministerens handlingsplan for nedsættelse af forbruget af bekæmpelsesmidler (Bilag til bet. o. lovf. vedr. kemiske stoffer og produkter – Bilag IV, December 1986).
Ministry of Environment (1990) Bekendtgørelse om undervisning for erhvervsmæssige brugere af bekæmpelsesmidler m.v. (BEK no. 731 05 November 1990).
Ministry of Environment (undated) Vejledning om 25 Meters Beskyttelseszone Omkring Indvindingsboringer, jf. Miljøbeskyttelseslovens § 21 b, jf. § 64 c. Available at: http://naturstyrelsen.dk/media/nst/64930/Vejledning%20om%2025%20meters%20beskyttelseszone%20omkring%20indvindingsboringer%20%201%20juli%202013.pdf (accessed 8 March 2017).
Ministry of Environment and Energy and Ministry of Food, Agriculture and Fishery (2000) Pesticidhandlingsplan II. Available at: www.google.dk/url?sa=tandrct=jandq=andesrc=sandfrm=1andsource=webandcd=1andcad=rjaanduact=8andved=0ahUKEwjYiOjV9YPLAhWBXCwKHXfTCYsQFggdMAAandurl=http%3A%2F%2Fmst.dk%2Fmedia%2Fmst%2F67444%2F03020400.docandusg=AFQjCNHAcVB6p3xgHM-9qIaPbJDx0IfJAgandsig2=tccnpPanmI5Q4LH6i7qBMw (accessed 8 March 2017).
Ministry of Environment and Food (2015) Forslag til Lov om ophævelse af lov om randzoner. Available at: www.ft.dk/RIpdf/samling/20151/lovforslag/L59/20151_L59_som_fremsat.pdf (accessed 8 March 2017).

Ministry of Environment and Ministry of Food (undated) Pesticidplan 2004–2009 for nedsættelse af pesticidanvendelsen og pesticidbelastningen. Available at: www.google.dk/url?sa=tandrct=jandq=andesrc=sandfrm=1andsource=webandcd=1andcad=rjaanduact=8andved=0ahUKEwj2xfaG9oPLAhXI2ywKHVTbBocQFggbMAAandurl=http%3A%2F%2Fmst.dk%2Fmedia%2Fmst%2F9108849%2Fpesticidplan_2004.docxandusg=AFQjCNHG9xn6JBud7Wu4chxo6KbskdjQYAandsig2=z1sYIrKCxvOsnk_cMycYUQ (accessed 8 March 2017).

Ministry of Food (2005) Forslag til Lov om ophævelse af lov om statsgaranti og ydelsestilskud vedrørende høsttabslån til tørkeramte jordbrugere og lov om tilskud til jordbrugets konsulentvirksomhed. Available at: www.retsinformation.dk/Forms/R0710.aspx?id=88789 (accessed 8 March 2017).

Ministry of Food (2011) Lov om randzoner (LOV no. 591 14 June 2011). Available at: www.retsinformation.dk/forms/r0710.aspx?id=137429 (accessed 8 March 2017).

Ministry of Food, Agriculture and Fishery (2004) Bekendtgørelse af lov om administration af Det Europæiske Fællesskabs forordninger om markedsordninger for landbrugsvarer m.v. (LBK nr. 297 af 28 April 2004). Available at: www.retsinformation.dk/Forms/R0710.aspx?id=7987 (accessed 8 March 2017).

Ministry of Food, Agriculture and Fishery (2010) Bekendtgørelse om tilskud til fremme af dyrkning efter retningslinjerne for integreret plantebeskyttelse (IPM) (BEK 409 21 April 2010). Available at: www.retsinformation.dk/Forms/R0710.aspx?id=131581 (accessed 8 March 2017).

Ministry of Food, Agriculture and Fishery (2014) Pesticidrester I danske fødevarer 2013. Resultater fra den danske pesticidkontrol. Glostrup: Fødevarestyrelsen. Available at: www.foedevarestyrelsen.dk/SiteCollectionDocuments/Pressemeddelelser/2014/Rapport-om-pesticider-i-foedevarer-2013.pdf (accessed 8 March 2017).

Ministry of Taxation (1995) Forslag til lov om bekæmpelsesmidler. Lovforslag nr. L 186. Fremsat den 7 marts 1995 af skatteministeren (Carsten Koch). Available at: www.retsinformation.dk/forms/R0710.aspx?id=111748 (accessed 8 March 2017).

Ministry of Taxation (1998) Forslag til Lov om ændring af lov om afgift af bekæmpelsesmidler og lov om ændring af forskellige punktafgiftslove (Afgiftsforhøjelse m.v.) (LSF 44, approved June 23 1998). Available at: www.retsinformation.dk/Forms/R0710.aspx?id=102594 (accessed 8 March 2017).

Ministry of Taxation (2001) Indkomstopgørelser i Ikke-Erhvervsmæssig Virksomhed – Hvorfor Skatteministeret Afviste Udlevering af Visse Dokumenter i 10-Mandsprojektsagerne – Omlægning af Personskatterne 1987–2002. Ministry of Taxation, Copenhagen.

Ministry of Taxation (2011) Interview with an expert (anonymous) who participated in the 1995–1996 negotiations on the pesticide tax. Interviewed November 2011 (unpublished).

Ministry of Taxation, Ministry of Food, Environmental Protection Agency, Danish Plant Protection, Association of Danish Professional Gardeners, Association of Danish Farmers, Danish Family Farmers, Danish Agricultural Council, Danish Institute of Agricultural Sciences (2001) Rapport om Muligheden for at Omlægge Pesticidafgiften til en Afgift på Behandlingshyppighed. Available at: www.skm.dk/media/128369/pesticid.pdf (accessed 8 March 2017).

Nielsen, H.Ø. (2010) *Bounded Rationality in Decision-Making: How Cognitive Shortcuts and Professional Values May Interfere With Market-Based Regulation*. Manchester University Press, Manchester, UK.

Ørum, J.E. (2015) *Behandlingshyppighed og Pesticidbelastning 2007-2014 - Supplerende Bilag til Bekæmpelsesmiddelstatistik 2014 (notat)*. Department of Food and Ressource Economics, Copenhagen. Available at: http://curis.ku.dk/ws/files/156091426/IFRO_Dokumentation_2016_1.pdf (accessed 8 March 2017).

Ørum, J.E., Boesen, M.V., Jørgensen, L.N. and Kudsk, P. (2008) *Opdateret Analyse af de Driftsøkonomiske Muligheder for en Reduceret Pesticidanvendelse i Dansk Landbrug – en Beskrivelse af Udviklingen fra 2003–2008*. Institute for Food Economics, Copenhagen. Available at: http://mfvm.dk/footermenu/publikationer/publikation/pub/hent-fil/publication/opdateret-analyse-af-de-driftsoekonomiske-muligheder-for-en-reduceret-pesticidanvendelse-i-dansk-landbrug-en-beskrivelse-af-udviklingen-fra-2003-2008-1/ (accessed 8 March 2017).

Pedersen, A.B. (2010) The fight over Danish nature: explaining policy network change and policy change. *Public Administration* **88**: 346–363.

Pedersen, A.B., Christensen, T., Nielsen, H.Ø. and Hasler, B. (2011) *Barrierer i Landmændenes Beslutningsmønstre Vedrørende Ændret Pesticidanvendelse (Bekæmpelsesmiddelforskning fra Miljøstyrelsen)*. Danish Environmental Protection Agency, Copenhagen. Available at: http://www2.mst.dk/udgiv/publikationer/2011/11/978-87-92779-18-2/978-87-92779-18-2.pdf (accessed 8 March 2017).

Pedersen, A.B., Nielsen, H.Ø., Christensen, T. and Hasler, B. (2012) Optimising the effect of policy instruments: a study of farmers' decision rationales and how they match the incentives in Danish pesticide policy. *Journal of Environmental Planning and Management* **55**: 1094–1110.

Pedersen, A.B., Nielsen, H.Ø., Christensen, T., Martinsen, L. and Ørum, J.E. (2014) *Konsulenters Rådgivning om Bekæmpelsesmidler – Muligheder og Barrierer for at Reducere Forbruget (Bekæmpelsesmiddelforskning fra Miljøstyrelsen nr. 157 2014)*. Danish Environmental Protection Agency, Copenhagen. Available at: http://mst.dk/service/publikationer/publikationsarkiv/2014/apr/konsulenters-raadgivning-om-bekaempelsesmidler---muligheder-og-barrierer-for-at-reducere-forbruget/ (accessed 8 March 2017).

Pedersen, A.B., Nielsen, H.Ø. and Andersen, M.S. (2015) The Danish pesticide tax. In: Lago, M., Mysiak, J., Gómez, C.M., Delacámara, G. and Maziotis, A. (eds) *Use of Economic Instruments in Water Policy: Insights from International Experience*. Springer, Dordrecht, pp. 73–87.

Pelaez, V., Rodrigus da Silva, L. and Araújo, E.B. (2013) Regulation of pesticides: a comparative analysis. *Science and Public Policy* **40**: 644–656.

Pesticide Action Network Europe (2013) Reducing pesticide use across the EU. Available at: www.pan-europe.info/old/Resources/Reports/PANE%20-%202013%20-%20Reducing%20pesticide%20use%20across%20the%20EU.pdf (accessed 8 March 2017).

Plantedirektoratet (1994) Bekendtgørelse om sprøjtejournaler og eftersyn af sprøjteudstyr i jordbruget (BEK nr.492 07.06.1994). Available at: www.retsinformation.dk/Forms/R0710.aspx?id=77676 (accessed 8 March 2017).

Politiken (2013) Landmænd blæser på omstridt miljølov. Available at: http://politiken.dk/indland/ECE2134785/landmaend-blaeser-paa-omstridt-miljoelov/ (accessed 8 March 2017).

Promilleafgiftsfonden for landbrug (2016) Bestyrelsen. Available at: www.promilleafgiftsfonden.dk/om-fonden/bestyrelsen (accessed 8 March 2017).

Promilleafgiftsfonden for landbrug (undated) Promilleafgiftsfonden for landbrug – Oversigt over bevillinger for 2014. Available at: http://promilleafgiftsfonden.dk/%7E/media/promilleafgiftsfonden/det-har-fonden-stoettet/fondens-bevillinger/fondens-bevillinger-2014.pdf?la=da (accessed 8 March 2017).

Rambøll Management (2008) *Evaluering af Målopfyldelse og Virkemidler i Pesticidplan 2004–09*. Ministry of Environment. Available at: http://mst.dk/service/publikationer/publikationsarkiv/2008/sep/evaluering-af-maalopfyldelse-og-virkemidler-i-pesticidplan-2004-09/ (accessed 8 March 2017).

Rigsrevisionen (2011) Beretning til statsrevisorerne om statens sikring af grundvandet mod pesticider, no. 4/2011. Available at: www.ft.dk/statsrevisor/20111/beretning/SB4/1057106.PDF (accessed 8 March 2017).

Rigsrevisionen (2014) Notat til statsrevisorerne om beretning om statens sikring af grundvandet mod pesticider. Available at: www.rigsrevisionen.dk/media/2010106/706-14.pdf (accessed 8 March 2017).

Schreinemachers, P., Afari-Sefa, V., Heng, C.H., Dung, P.T.M., Praneetvatakul, S. and Srinivasan, R. (2015) Safe and sustainable crop protection in Southeast Asia: status, challenges and policy options. *Environmental Science and Policy* **54**: 357–366.

Statistics Denmark (2015) OEKO1: Økologiske arealer efter afgrøde. Available at: www.statistikbanken.dk (accessed 8 March 2017).

Thorling, L., Brüsch, W., Ernstsen, V., *et al.* (2015) *Grundvand. Status og Udvikling 1989–2013 (Teknisk rapport)*. GEUS.

Ugebrevet A4 (2011) Landbruget tildeler sig selv en halv milliard kroner. Available at: www.ugebreveta4.dk/landbruget-tildeler-sig-selv-en-halv-milliard-skattek_14641.aspx (accessed 8 March 2017).

Vedung, E. (1998) Policy instruments: typologies and theories. In: Bemelmans-Videc, M.L., Rist, R.C. and Vedung, E. (eds) *Carrots, Sticks and Sermons*. Transaction Publishers, New Brunswick, pp. 21–58.

Winter, S.C. and May, P. (2001) Motivation for compliance with environmental regulations. *Journal of Policy Analysis and Management* **20**: 675–698.

Winter, S.C. and Nielsen, V.L. (2010) *Implementering af politik*. Academica Frederiksberg.

World Bank (2015) Agricultural land (% of land area). Available at: http://data.worldbank.org/indicator/AG.LND.AGRI.ZS/countries?display=default (accessed 8 March 2017).

14

Impacts of Exotic Biological Control Agents on Non-target Species and Biodiversity: Evidence, Policy and Implications

Barbara I.P. Barratt and Clark A.C. Ehlers

14.1 Environmental Safety of Biological Control

Biological control can include: the enhancement of naturally occurring herbivores, predators, parasitoids and pathogens (conservation biological control); the release of natural enemies which are not expected to establish, but to control pests during favourable seasons or in protected environments (augmentative biological control); and the importation of natural enemies introduced from the area of origin of the weed or pest and which are expected to establish and spread (classical biological control). Here we focus on the latter, since it is the predicted impact of exotic organisms imported for pest management that is subject to legislation and regulatory policy before release into a new environment can be approved. A range of policies have been put in place in many countries to regulate the introduction of biological control agents. While each country has approached this in a slightly different way, in general policy has developed that requires a risk assessment to be carried out to predict as accurately as possible the environmental safety of the proposed biocontrol agent.

14.1.1 Is Biological Control Safe?

Reservations about the environmental safety of biological control have been expressed over many years when observations, particularly in Hawaii, indicated that adverse impacts had resulted from biological control introductions (Swezey 1931; Zimmerman 1958) (see also Chapter 5). However, from the 1980s the debate became more polarized, with Howarth (1991), Simberloff and Stiling (1996), Louda *et al.* (1997) and Lockwood (2000) calling for more consideration of non-target impacts before release of imported biological agents. In response, Hopper (1995), Thomas and Willis (1998) and others put forward the view that biological control practice could be severely curtailed if risk assessments need to be carried out for all possible non-target species, and that the adverse impact of doing nothing was equally risky. The controversy, at times quite heated (Barratt *et al.* 2010), was eventually tempered by the acceptance that some biological control practices (e.g. the shotgun approach) were probably not appropriate, and that indeed, there was opportunity for good ecological research to be incorporated into biological control programmes that could lead to improved beneficial impacts while reducing potentially adverse impacts (Waage and Greathead 1988).

Environmental Pest Management: Challenges for Agronomists, Ecologists, Economists and Policymakers, First Edition. Edited by Moshe Coll and Eric Wajnberg.

Increased availability of funds for research in this area and developing regulatory attention also contributed to greater agreement about the need for risk assessment for biological control agent imports to new areas. It is now generally accepted that regulation of biological control introductions should be subject to public policy legislation because of the irreversibility of biological control releases and the ability of organisms to disperse.

Biological control practitioners need to be able to provide regulators and stakeholders with information about the likely ecological risk from releasing a new natural enemy into the environment. Given the complexity of food webs in communities into which the new organism will be introduced, it is challenging to be able to predict impacts with a high degree of certainty. However, recent research using mathematical modelling of data that can be acquired from laboratory experiments, knowledge of the biology of the organisms, and taking into account factors such as apparent competition (two hosts sharing the same natural enemy), has shown potential for efficacy and risk to be linked in a way that could provide useful information for regulatory decision makers (Kaser and Heimpel 2015). Similarly, food web analyses have shown potential for predicting non-target effects (Tylianakis and Binzer 2013), and so sophisticated modelling approaches might in the future provide important decision support for regulators.

14.1.2 What are the Risks?

The risks identified for biological control introductions can be either direct or indirect (Barratt *et al.* 2006; Wajnberg *et al.* 2001). The former includes unintended impacts that a biological control agent might have on organisms in the same trophic level as the target in the receiving environment (Stiling and Simberloff 2000), including effects on native non-target species, beneficial or valued exotic species (Murray *et al.* 2002) or other pests (McNeill *et al.* 2002), sometimes known as 'fortuitous biological control'. Indirect effects can result from a wide range of factors and trigger food web perturbations that can be very difficult to predict (Messing *et al.* 2006). Species such as other natural enemies in the same trophic level as the biological control agent can be at risk from competition with, or displacement by, the introduced biocontrol agent (Bennett 1993; Wang and Messing 2002). Furthermore, there is potential for hybridization between species in the same genus which might compromise biocontrol efficacy (Hopper *et al.* 2006). Goldson *et al.* (2003) found that when a European biotype of the parasitoid *Microctonus aethiopoides* (Loan), intended for control of clover root weevil (*Sitona obsoletus* Gmelin), was crossed with an existing Moroccan biotype introduced earlier for lucerne weevil (*S. discoideus* Gyllenhal), the result indicated that there would be compromised efficacy in both biocontrol programmes. This was avoided by releasing a parthenogenetic strain of *M. aethiopoides* from Ireland (Gerard *et al.* 2006).

Concern has at times been expressed about the risk of habitat change that could result from an effective weed biological control agent, for example soil erosion from rapid plant cover loss or exacerbation of alternative weed species. However, there are few examples of this, and rarely do biocontrol agents precipitate such dramatic or rapid change. Sheppard *et al.* (2003) listed some examples of hazards (and benefits) identified in applications to the Environmental Risk Management Authority New Zealand (now replaced by the Environmental Protection Authority) for weed biological control agents.

Although generalist predators can be effective biological control agents, it is generally accepted that they are unlikely to meet biosafety standards required by regulators because of the risk to non-target species (Elkinton *et al.* 2006) and, as a result, generalist natural enemies, predators or herbivores, are rarely considered for biological control. Also, in the future, climate change might introduce a new and potentially unpredictable element of risk as a result of 'uncoupling' of biological control/host/prey relationships, resulting from a change in fitness, phenology or distribution of organisms (Gerard *et al.* 2010; Lu *et al.* 2015; Thomson *et al.* 2010) (see also Chapter 9), allowing for changes in host/prey distribution and host/prey range expansion of natural enemies (Evans *et al.* 2011).

14.2 Legislation and Regulation of Biological Control

14.2.1 General Summary

A number of countries have adopted legislation that regulates the introduction of exotic biological control agents. A brief comparative review of the relevant biological control-related policies in different countries follows with the emphasis on procedures used for the assessment of risks posed by exotic biocontrol agents.

There are about 25 countries that have implemented policy concerning the importation of exotic biological control agents (Lockwood *et al.* 2001). Some countries, such as Australia, Canada, New Zealand, South Africa, Switzerland and the USA, have significant experience in the application of biological control using exotic arthropod and microbial pathogen agents while other countries have minimal or no experience, such as some nations within the European Union (EFSA Panel on Plant Health 2015). More countries are considering the development of policy that will regulate the introduction of biological control agents and this is thought to be a direct consequence of the Convention on Biological Diversity (CBD). Biological control is increasingly being recognized as an important tool in an Integrated Pest Management strategy to prevent the spread and proliferation of invasive exotic species. Sensible policy that is cognisant of the hazards and advantages of biocontrol practice can achieve introduction of safe biocontrol agents.

An overview of international organizations that have adopted regulations, guidelines and codes of conduct relevant to biological control is presented in Table 14.1. Australia

Table 14.1 Summary of international organizations that regulate or advise on the release of biocontrol agents. Source: Adapted from van Lenteren *et al.* (2006a).

Organization	Environmental regulations	Authority	Evaluation; non-target effects considered
European and Mediterranean Plant Protection Organization	Standards Safe Use Biological Control 1999, 2000	Member States	Pest risk assessment; biology, ecology, environment
FAO: International Plant Protection Convention	Code of Conduct 1996, ISPM No. 3 2005	Member States	Pest risk assessment; health, environmental impact
OECD	Guidance Invertebrate Biological Control Agents 2002	Member States	Health, biology, ecology, non-target effects
European Community	Habitat Directive 1992, Plant Health Directive 2000	Member States	Protection habitat, endangered species, biodiversity

has the Biological Control Act, a piece of legislation that is specific to biological control. Also relevant to biocontrol are the Environmental Protection and Biodiversity Conservation Act and the Quarantine Act. Canada regulates the import and release of biocontrol agents under its Plant Protection Act. New Zealand, on the other hand, regulates the introduction of all new organisms, including exotic biocontrol agents, under the Hazardous Substances and New Organisms Act. In the USA, the use of biological control is regulated by state laws. However, the Plant Protection Act is the federal US statute relating to plant pests and noxious weeds.

14.2.1.1 Australia

There are three steps an applicant must take. First, an application is made to obtain a permit to import the agent into quarantine which is assessed by the Department of Agriculture, Fisheries and Forestry-Bio-security Australia and the Department of the Environment and Heritage. Both these agencies must approve the application which is granted by the Australian Quarantine Inspection Service.

During the second step, the list of species that will be tested to establish specificity of the agent must be submitted for approval by the Department of Agriculture, Fisheries and Forestry-Bio-security Australia. The test list is reviewed by a total of 21 members of a co-operative, an autonomous association of stakeholders who make the decision to approve or request amendments to the list of test species. The applicant then moves to the third stage where an application is submitted to release the agent. Again, the 21-member co-operative reviews the application and the Australian public are also notified of the intention to release a new biocontrol agent. Members of the public are invited to submit any concerns they have about the proposed introduction and once the co-operative is satisfied that all concerns have been dealt with and the potential risks are acceptable, the Department of the Environment and Heritage and Australian Quarantine Inspection Service issue an approval.

14.2.1.2 Canada

The Canadian Food Inspection Authority (CFIA) receives petitions for the introduction of new biocontrol agents. The CFIA is responsible for issuing release permits. An application is reviewed by a biocontrol review committee which consists of taxonomists, ecologists, specialists in federal and provincial governments, a technical advisory group from the USA and representatives from Mexico. A recommendation is then made by the committee of regulatory experts at the CFIA to the director of the plant health division of the CFIA, who will then make the decision on whether or not to issue a permit.

14.2.1.3 New Zealand

Applicants seeking release of new biocontrol agents in New Zealand apply to the Environmental Protection Authority (EPA). The EPA conducts a full evaluation and review of an application which includes an assessment of benefits and risks associated with a candidate agent. The process also includes engagement with the Māori, New Zealand's indigenous people. The application is publically notified and the people of New Zealand are invited to submit their concerns or support for the candidate agent. A decision on whether or not to approve the agent is made by an independent committee following a public hearing.

14.2.1.4 USA

Exotic biocontrol agents are assessed by a technical advisory group (TAG) of the Animal and Plant Health Inspection Service which is part of the United States Department of Agriculture (USDA). The TAG assesses the lists of hosts or prey that will be tested, the import into containment and release of the candidate agent. The TAG includes representatives of five USDA agencies, six Department of Interior agencies, the Environmental Protection Agency and the Department of Defense, as well as delegates from Canada and Mexico.

14.2.2 Other International Obligations

Almost 200 countries are now signatories to the Convention on Biological Diversity (CBD), and about a third of these countries have ratified the Nagoya Protocol which is the instrument for the implementation of the Access and Benefit-Sharing (ABS) provision of the CBD. The principles of ABS are that access to genetic resources (GRs) such as potential biological control agents from another sovereign state are subject to 'prior informed consent' and granted subject to 'mutually agreed terms'. A certificate of compliance will need to be obtained as evidence that the country has agreed to make GRs available under ABS requirements of that country. ABS was developed with the main aim of reducing the opportunity for financial gain to be made from GRs from another country with no benefit to be returned to the donor country, but concerns have been expressed that there might be negative consequences of ABS for biological control (Cock *et al.* 2010; Coutinot *et al.* 2013). The International Organisation for Biological Control (IOBC), with support from the CABI (Centre for Agriculture and Biosciences International), was invited to report to the FAO Commission on GRs for Food and Agriculture. The IOBC made the case that biological control is generally not a profit-making exercise, there are public good benefits from biocontrol, and that the practice has worked well in reciprocal relationships between countries for many years (Cock *et al.* 2009). The full extent of the impact of ABS on biological control will become clear within the next few years as countries develop their own legislation.

14.3 Risk Assessment

14.3.1 Purpose of a Risk Assessment for Biological Control

Risk assessments are carried out as part of the decision-making process for regulators receiving applications from practitioners wanting to introduce a new biological control agent. Regulators are generally charged with avoiding adverse environmental, social and economic impacts which might result from their decisions, and for biological control, a risk assessment is conducted both to identify all possible hazards associated with a biological control release and to assess the likelihood of each hazard occurring. An analysis of these factors then provides the evidence that will be used to inform the decision. In some countries, benefits of the biocontrol introduction are also taken into account and balanced against risk (e.g. in New Zealand), while in other countries the benefits are not considered.

By and large, benefits are considered to be positive effects that may be associated with a new biocontrol agent. This can include effects on the environment, human health, market economy, communities and so on when a biological control agent successfully suppresses its target pest (see section 14.3.7 regarding the assumptions that risk assessors make). The benefits are considered in light of the status quo and future pest management scenarios (e.g. use of chemical and/or physical control) in lieu of the biological control agent. Benefits can also be evaluated on a likelihood scale, and thereby weighed up against hazards.

Risk assessments for biological control are mostly qualitative, and only when balancing costs with benefits for economic impacts can quantitative assessment be undertaken. For this reason, a qualitative risk assessment framework including a scale of the magnitude of the hazard (minimal to massive) and likelihood of occurrence (highly improbable to highly likely) is often constructed as part of the risk analysis. This equips decision makers with greater clarity on the likelihood of any adverse effects and benefits occurring, and the magnitude of those effects when a new biocontrol agent is considered for release. The identification of the adverse effects is prescribed by legislation in some cases; for example, the Hazardous Substances and New Organisms Act (HSNO Act 1996) in New Zealand compels applicants to take into account the effects of exotic biocontrol agents on the environment, the market economy, public health and the relationship of Māori with their culture and traditions. In other cases, guidelines are available to assist with this task. For example, the European Union-funded programme on Evaluating Environmental Risks of Biological Control Introductions into Europe (ERBIC) (van Lenteren *et al.* 2003) and the European and Mediterranean Plant Protection Organization (EPPO 2014) developed guidelines to assist biocontrol practitioners preparing their dossier of supporting data and for regulators performing environmental risk assessments on new biological control agents.

A risk assessment for biological control can comprise several elements: (1) host specificity in its natural (native) range, (2) information available from other countries where the same biocontrol agent has been used, and (3) host specificity testing in quarantine. The former two aspects can be completed before the proposed biocontrol agent is brought into the new country, whereas quarantine testing usually requires that the agent is in secure containment in the receiving country.

The required outcome for risk assessment is two-fold: to reduce the likelihood of releasing an organism that will cause unacceptable levels of environmental or economic damage, but at the same time to avoid unnecessarily rejecting potentially useful and effective biological control agents.

14.3.2 Preimportation Evidence

Exploration for a biological control agent usually takes place in the area of origin of the pest. The main purpose is to identify potential biological control agents, investigate their efficacy against the target, and to narrow the candidates down to a small number of organisms worthy of further investigation. Part of this selection process can include the collection and consideration of natural host range data to further reduce the 'shortlist' down to those that have a narrow host range or that are specific to the intended host.

It has become increasingly evident that strains/biotypes of biocontrol agents need to be considered in selecting candidates. Host range and other biological characteristics of organisms may vary from one population to another across their native range, so it is important that the agent finally released, if approval is given, is the same biotype or from the same population as that tested in quarantine. Furthermore, additional imports of organisms for release should then also be restricted to organisms from the same biotype. For example, host range testing of gorse pod moth *Cydia succedana* (Denis and Schiffermüller) sourced from the UK showed that it was highly specific to gorse, and it was released in New Zealand. However, the offspring of Portuguese moths were also released in New Zealand and were later determined to have a different host range to the UK populations that included several exotic Genisteae and Loteae species, although no adverse impacts have been reported (Paynter *et al.* 2008). Approvals for biological control introductions in New Zealand are now often conditional on the same biotype or source provenance of insects being released as those that were tested in quarantine.

A Moroccan biotype of the braconid parasitoid *Microctonus aethiopoides* Loan was introduced for biological control of *Sitona discoideus* (Gyllenhal) (Coleoptera: Curculionidae), a lucerne pest in New Zealand, and postrelease studies have shown that it has a number of non-target and native weevil hosts. Interestingly, when *Sitona obsoletus* Gmelin arrived in New Zealand, it was expected that *M. aethiopoides* would readily attack this new pest, but this was found not to be the case (Barratt *et al.* 1997). Further exploration revealed that another biotype of the same parasitoid species from Ireland was an effective agent (Goldson *et al.* 2005). Although the regulatory agency (Environmental Risk Management Authority New Zealand at the time) did not require risk assessment to be carried out for taxa below the species level, the researchers' advice on the different host range characteristics of the biotypes convinced the regulator that a new application should be prepared.

In the event that a biological control agent has been used previously in a different country for a particular pest, useful data on biosafety (as well as efficacy) can be collected as part of a dossier for risk assessment. Clearly, this can be little more than a guide, given intended exposure to a different environment and biota, but should experience elsewhere show that a species imported for biological control has a wider physiological or ecological host range than anticipated, then this can provide some prior warning.

Biocontrol practitioners also need to consider the ability of candidate agents to establish in a new environment. It is expected that practitioners will submit applications to release new agents only if those agents are predicted to establish successfully, while data that underpin establishment may also be important to determine dispersal of candidate agents in new environments. Classic biocontrol programmes rely on agents building large populations to be successful against pests. For example, biocontrol practitioners should consider if the climates in the area of origin and area of release match (van Lenteren *et al.* 2003). Climate matching is a useful guide that will allow practitioners to gain an early indication of whether an agent will successfully establish in its new environment. Sensitivity to temperature is considered key to determine survivability, voltinism and likely dispersal of agents – parameters that are fundamental to predict a candidate agent's success in a new territory (van Lenteren *et al.* 2006a). There are also a number of biotic factors that should be considered, including, for example, the ability of a parasitoid to synchronize its life cycle with that of the host.

14.3.3 Testing in Quarantine

Host/prey range testing in quarantine provides one of the most important datasets which inform the decision-making process. It is this series of tests where for the first time the proposed biological control agent can be exposed to species from the receiving environment. Quarantine tests are considered to be conservative and, depending upon how they are conducted, provide, for example, evidence of physiological host range. If a non-target species is successfully attacked, that is, the biocontrol agent develops successfully, then this provides evidence that the non-target species is a potential field host, but provides little evidence that the non-target species would actually be attacked in the field. Further information on distribution, phenology, ecoclimatic tolerance, etc. is required to determine the likelihood that the non-target species sits within the ecological host/prey range of the biocontrol agent.

The approach to quarantine testing for proposed weed and insect biological control agents is similar in principle but quite different in practice, for two reasons. First, the number of potential non-target species to be considered is much smaller for plants than insects simply as a function of species richness. Second, for weed biocontrol there are two trophic levels to consider – the host plant and the agent (herbivore or plant pathogen). For insect biological control, there is a third trophic level, the natural enemy of the herbivore, whether it is a predator, parasitoid or a pathogen.

Weed biocontrol practitioners have generally adopted the principles of the 'centrifugal phylogenetic testing' method and subsequent improvements to that practice were developed by Wapshere (1974, 1989). Testing is carried out on plants closely related to the target, and then successively more distantly related species until a point is reached where a profile of the host range breadth of the proposed biological control agent can be established. This process can be used to determine the physiological host range of weed biological control agents. There are no records of field attack of plants outside the physiological host range as determined in laboratory tests (Andreas *et al.* 2008). More recently, host range testing has evolved to include technological developments such as molecular or DNA-based methods to compile non-target test species lists that better represent phylogenetic relationships. The access to molecular methods has led to new understanding of the degree of relatedness between species, and of organism behaviour and ecological relationships that drive host use by specialist (as opposed to generalist) insects (Briese 2005).

Quarantine host or prey range testing for entomophagous biological control agents attempts to follow the same principle, selecting for testing those species with taxonomic (phylogenetic) affiliations with the target host, as well as those with ecological affinities. However, selection of insect biocontrol agents is complicated by factors such as limited data on arthropod phylogeny, dispersal rates in new environments and similar/dissimilar feeding niches compared to native habitats (van Lenteren *et al.* 2006b). It was also pointed out that tests for predators which tend to be more oligophagous need to be considered carefully, and van Lenteren *et al.* (2006b) suggested that host size and geographical distribution might be more meaningful than taxonomic relatedness to the target species. Furthermore, a wider range of prey might need to be tested for predator biosafety testing, and adult and larval predators often have different prey ranges which should be considered (van Lenteren *et al.* 2006b).

For both weed and insect biocontrol, species of economic or iconic value are often included in lists of test species (Babendreier *et al.* 2006; Kuhlmann *et al.* 2006). Species that have endangered or threatened status, or are recognized as being keystone species in ecological function should also be considered during test list development (Kuhlmann *et al.* 2006).

Contemporary thinking recognizes the limitations of using phylogenetic/taxonomic affinities for selection of non-target species for testing. Biocontrol practitioners are encouraged also to consider biogeographical factors, such as overlapping distribution patterns of biological control agents and possible hosts or prey, and ecological features of the agent in its native range and of species in its receiving environment when they compile species lists for host or prey specificity testing of candidate agents.

More recently, a method known as PRONTI (Priority Ranking on Non-Target Invertebrates) has been developed for test species selection of entomophagous biocontrol agents (Barratt *et al.* 2016; Todd *et al.* 2015) whereby a model is applied to a reasonably comprehensive dataset of invertebrate species from the receiving environment. The model takes a set of predetermined selection criteria, applies them to each species in the dataset and produces a PRONTI score for each. These scores are then used to rank the species in order of likely risk and exposure to the biocontrol agent, including consideration of mechanisms whereby organisms can reduce their risk (resilience), estimates of ecological and anthropocentric value, and the practicality of including them as a test species (testability).

14.3.4 Providing Robust Data

Quarantine testing is often criticized for overestimating host or prey range, largely as a result of the artificiality of the quarantine environment such as the proximity and numbers of the test species. No-choice tests where only the target or non-target species is present are conservative, as mentioned above, although a negative result is comparatively informative. Choice tests where the target and non-target species are both present provide information on host preference. However, clearly neither test is representative of natural field conditions. There are also a number of other test designs that address different elements of biocontrol agent behaviour, such as sequential testing (van Driesche and Murray 2004; Withers and Mansfield 2005).

Experimental design with positive controls (to test the efficacy of the biocontrol agents on the target) and negative controls (unexposed hosts or prey to show that the test species are viable in the absence of the biological control agent) are important. Needless to say, adequate replication and appropriate statistical analyses are required to present robust evidence to regulators for evaluation. Hoffmeister *et al.* (2006) argued that it is not possible to statistically demonstrate that there will be no non-target effects, but when no significant non-target effects are found, the power of the test dictates the level of confidence in the data. Withers *et al.* (2013) compared statistical tools to analyse data on quarantine testing of *Cleopus japonicus* (Wingelmüller) (Coleoptera: Curculionidae) proposed for biological control of *Buddleja davidii* Franchet (Scrophulariaceae). They concluded that, to achieve sufficient power in statistical tests, it was important to predefine effect size based on a biological relevance, maximize sample size as far as possible, understand the statistical implications of the test, and carefully select an appropriate experimental design consistent with the statistical test to be used.

Depending upon the organisms involved, there are many variables that can be measured in quarantine tests (Barratt *et al.* 2007; van Driesche and Murray 2004) and clearly these need to be determined case by case to provide data that will best inform the analysis selected. The selection of test species and the testing method used in host range testing should ultimately incorporate biological and physical characteristics that determines an agent's life cycle and behaviour, but recognition should be given to unique ecological and biophysical features of the territory where it is to be introduced. The testing regime should maximize the chance of detecting non-target effects, and follow-up tests should be completed to describe the mechanisms of interaction between the agent and non-targets (Barratt *et al.* 2010). Selection of test species and, to a large extent, selection of testing methods should be supported by objective and justified reasoning that will underpin robust environmental risk assessments.

14.3.5 Identification and Assessment of Risk and Benefits

Risk assessment can be informed by data collected before importation, and from quarantine tests, usually provided by the applicant. The ultimate goal is to predict postrelease impacts with the highest level of certainty possible. Depending upon how risk averse the policy requires the regulator to be, and the threshold for a precautionary approach to be taken, a decision is made.

14.3.5.1 Non-target Effects

As mentioned earlier, assessment of the magnitude and likelihood of adverse impacts on non-target species that could result from the release of a natural enemy for biological control is the most obvious focus of concern.

14.3.5.2 Biodiversity/Ecosystem Effects

Indirect effects of a new organism in the environment, particularly one that is specifically selected to have a negative population impact on the target species and a high probability of successful establishment, may share the qualities of an invasive species with potential to threaten biodiversity. A well-known example of this is the harlequin ladybird *Harmonia axyridis* (Pallas) (Coleoptera: Coccinellidae) released as a biological control agent for aphids. The species was very slow to establish but eventually provided successful biological control. However, it was then discovered to be invading natural ecosystems, attacking native aphids and other invertebrates, as well as other coccinelid species. It is now considered to be a serious invasive species capable of spreading its range rapidly and widely (Roy and Brown 2015). While this is probably an extreme example, and predatory insects used for biological control are treated with particular caution, the likelihood of adverse food web and ecosystem effects of any proposed biological control agent need to be considered.

14.3.5.3 Risk of Doing Nothing or Continuing Current Practices

Risk assessment should incorporate consideration of the risk of either doing nothing (i.e. an application is declined) or continuing to manage a pest using potentially more hazardous or less cost-effective methods. Pesticides can be a hazard for applicators, the environment and consumers (see Chapter 4). Land management methods used to control pests such as cultivation and burning can also be environmentally damaging, possibly leading to soil erosion. Any adverse impacts need to be balanced against those that might have been identified for a candidate biological control agent. Many would say that when faced with a severely damaging pest, doing nothing is not an option, and might represent a greater risk. In Integrated Pest Management systems, biocontrol can often considerably reduce the need for other, much more hazardous practices (Liu *et al.* 2014; Urbaneja *et al.* 2015; Walker *et al.* 2010).

14.3.5.4 Assessment of Benefits

A risk assessment is often balanced with an analysis of benefits, and a framework of magnitude and likelihood can also be used to evaluate benefits such as treatment cost savings, crop yield increases (see Chapter 3), health benefits, better biodiversity and conservation values, time saving, reduced requirement for withholding periods, market access, etc. An extensive analysis of weed biological control programmes in Australia showed an average benefit–cost ratio of 23:1, with some individual programmes showing massive ratios of over 100:1 (Page and Lacey 2006). Data on environmental and social benefits were not easily obtained and hence were largely precluded from this analysis. While risk was not included in that analysis, Seier *et al.* (2013) compared examples of rust fungi being considered for weed biological control programmes in Australia. They reported that the authorities regarded the risk posed to three non-target *Jatropha* (Euphorbiacaea) species by the rust *Phakopsora jatrophicola* Cummins was outweighed by the benefit of potential impact on the major weed, *J. gossypifolia* L. (bellyache bush). However, when the rust *Ravenelia acaciae-arabicae* Mundkur and Thirumalacha, proposed for control of the invasive weed prickly acacia, was found to sporulate on the native species *A. sutherlandii* (F. Mueller), the risk was considered unacceptable regardless of benefits.

14.3.6 Public, Stakeholder and Technical Expert Input

Public and stakeholder input may play an important role in application and decision-making processes when a new biocontrol agent is considered for release. Information obtained from members of the public, stakeholders and technical experts can be considered in the risk assessment. Stakeholder agencies can articulate the advantages and disadvantages of a candidate biocontrol agent and its target species. In particular, they can voice their concerns or support for a biocontrol programme's potential conservation value or its value to economically significant industries. For example, a proposal to introduce a weed biological control agent might draw a response from a beekeeping industry noting that the 'weed' is in fact an important resource for bees.

Public opinion, which is elicited through a notification process required by legislation in some countries, can deliver important information from a grassroots level which is not always available in widely accessible formats such as published literature. A decision-making authority may also call for expert input to an application to release a new biocontrol agent. For example, the HSNO Act that regulates the introduction of new organisms to New Zealand, including biological control agents, prescribes in section 58 that the EPA may commission a report or seek advice from any person on any matters raised in relation to an application, including a review of the information provided by the applicant party (HSNO Act 1996).

14.3.7 Decision Making and Predicting Outcomes

The regulatory agency responsible for making a decision allowing an applicant to release a new biological control agent has to make a number of assumptions when it conducts its risk assessment. First, for the hazards associated with the introduction of an exotic organism to a new territory to occur, the biocontrol agent will need to successfully establish in its new environment and develop self-sustaining populations. It is also expected that, to be beneficial, any population of a biocontrol agent will need to reach sufficiently high numbers that will significantly suppress pest populations. If an agent is approved for release and subsequently does not establish or establishes but does not build to large population densities, the risks are diminished. This means that any hazards but also benefits analysed in a risk assessment may not eventuate, and the likelihood and magnitude of effects may be significantly lower than determined in the risk assessment. The decision maker therefore has to assume that a candidate agent will fully establish and build large populations to achieve maximum effect when a risk assessment is conducted. In this way, it makes it easier for a decision maker to determine if the identified risks will be significant and if the benefits truly outweigh the risks where positive effects are taken into account.

When an applicant applies to the appropriate authority for the release of a new biological control agent, it is expected that they will identify the potential hazards of the agent. In addition, it is likely that they will need to determine the likelihood of each hazard occurring and the consequences that may follow once an agent establishes in the environment based on data submitted to support their application. A regulator will evaluate the adverse effects and proceed to conduct a risk assessment of those effects, and any others not identified by an applicant that they consider may have significant impacts. Generally, any effects that are considered speculative or where the likelihood of those effects and their consequences occurring is low are not addressed. Biological control environmental risk assessments focus on the key effects and their anticipated outcomes when decisions are made. For instance, an agent may have economic costs associated with the loss of both target and non-target species if they have monetary value or are beneficial to particular industries. A decision maker may also consider the cultural or aesthetic value of the species, or the ecosystem functions that they have, to facilitate robust and informed decision making that should be independent from political influence or generic public opinion regarding the introduction of exotic organisms.

Where benefits are considered in a risk assessment, the decision maker will weigh up the benefits against the hazards. Where an agent is shown to attack non-target species in quarantine testing, the adverse effects would need to be considered in light of its beneficial effects against the pest species. One example where a regulatory agency approved the release of a biocontrol agent in spite of adverse effects shown in quarantine tests is the release of blackberry rust fungus (*Phragmidium violaceum* (Schultz) G. Winter 1880) to control European blackberry in Australia. The regulator approved the release of the agent even though testing indicated that native Australian *Rubus* species would be at risk of attack by the fungus. This decision was based on the fact that potential benefits were seen to outweigh the hazards, since European blackberry itself was perceived to be a threat to native *Rubus* species (Barton 2012).

Once an agent is approved for release and successfully controls the target, reductions in pest populations and collateral effects on non-targets may ultimately improve biodiversity values by restoring native ecological processes and reducing costs of pest management. Conversely, the intrinsic value of ecosystems might be reduced as a result of displacing valued fauna and disturbing ecosystem interactions. These ultimate measures of impact are difficult to predict. Population-level impacts on targets and non-targets are rarely estimated and relatively few biocontrol programmes obtain pre- and postrelease data in the field to assess effects. Biocontrol practitioners are starting to recognize the value of postrelease assessments to make confident predictions of the impacts of candidate agents. For regulators making decisions to approve or reject a new biocontrol agent, postrelease case studies can help to signal the impacts that predicted adverse effects might have. This is discussed in more detail below.

Regulators routinely deal with uncertainty when it comes to making a decision to approve the introduction of a new biocontrol agent. Ecosystem and other indirect effects are much more difficult to predict than direct effects on target and non-target species. One way of dealing with uncertainty is to add a weighting factor to the adverse effect that may follow the release of an agent in a risk assessment (van Lenteren *et al.* 2003). Another approach is to consider the benefits of releasing the biocontrol agent in light of any uncertainties. In New Zealand, for example, the benefits of a candidate agent are carefully weighed up against its adverse effects. The Environmental Risk Management Authority New Zealand (now replaced by the Environmental Protection Authority) concluded in its consideration of an application to introduce *Ceratapion onopordi* (W. Kirby) (Coleoptera: Brentidae) and *Cassida rubiginosa* (Mueller) (Coleoptera: Chrysomelidae) to control Californian thistle (*Cirsium arvense*) that any unintended adverse effects on other plants, including globe artichoke and cornflower, was outweighed by the beneficial effects of thistle control. The non-target hosts were of limited economic interest in New Zealand which contributed to that decision. Where multiple biocontrol agents are considered for release against a pest, incompatibility issues between the agents may be dealt with by providing information on the interaction between agents by describing the temporal and spatial activities of the agents on the target.

Another mechanism to deal with uncertainty in decision making is to apply a condition to a release approval that aims to reduce possible adverse effects. In the *Microctonus aethiopoides* example described above, the Environmental Risk Management Authority New Zealand approved release of *M. aethiopoides* to control clover root weevil subject to the condition that only a parthenogenetic strain collected from a particular area in Ireland may be introduced. The condition aimed to reduce uncertainty regarding the incompatibility of the new insect with the closely related Moroccan strain of *M. aethiopoides* that was already established in New Zealand to control lucerne weevil. While a conditional release approval may reduce uncertainty in some cases, restrictions on the use of an agent should be carefully considered so as not to obviate the purpose of a biocontrol programme: establishment of self-sustaining populations that will disperse freely in their new environment to target the pest.

14.4 Postrelease Validation of Predicted Outcomes

Biological control practitioners often have insufficient resources to carry out postrelease studies to monitor efficacy of the biological control agent, let alone non-target impacts. Regulators also rarely have means to support such studies although the logic of follow-up evaluations has long been recognized (Stanley and Julien 1998) and retrospective assessment of their decisions would clearly be informative. In New Zealand, the EPA, in implementing the legislation regulating biological control agents (HSNO Act 1996), is unable to make it a condition of approval that postrelease monitoring of biological control releases is carried out, since 'controls' applied to approvals are imposed only to reduce risk. However, the regulator encourages postrelease monitoring where possible because such information can inform future decisions. Nevertheless, the EPA does have a requirement to monitor effectiveness of the HSNO Act in reducing adverse effects on the environment or people from new organism introductions and so can have a strong advocacy role.

Sheppard *et al.* (2003) noted that the EPPO, in considering improvements to the IPPC Code of Conduct (IPPC 2005), suggested that postrelease monitoring should be included in the guidelines. However, the recently revised guidelines did not go that far, but do recommend that the organization undertaking a release should take account of 'any problems encountered in postrelease monitoring' (OEPP/EPPO 2014). Despite the scarcity of resources to fund postrelease monitoring, a number of studies have been carried out (see Barratt *et al.* 2006 and references therein).

14.4.1 Retrospective Case Studies and Evidence for Impacts

Weed biological control has been claimed (and generally proven) to be environmentally safe, with few examples of non-target impacts that were not predicted (Fowler *et al.* 2004; Paynter *et al.* 2004; Willis *et al.* 2003). An exception often quoted is the attack on native North American thistles by *Rhinocyllus conicus* Oke (Coleoptera: Curculionidae) introduced to control exotic thistles (Louda 2000), one of few examples of a population impact demonstrated on a non-target species. In fact, this non-target attack could have been predicted since *R. conicus* was known not to be host specific (Gassmann and Louda 2001) and poor decision making was more at fault than poor risk assessment. Using a different approach which analysed community responses, a meta-analysis by Clewley *et al.* (2012) showed that the non-target plant diversity and abundance increased after the release of biological control agents for reasons that were not entirely clear.

In New Zealand, the release of *Microctonus aethiopoides* for control of the lucerne weevil, *S. discoideus*, was approved in 1982 when the regulatory legislation was less stringent than it became once the HSNO Act was enacted in 1996. The release was approved after quarantine testing with a number of beneficial weed biocontrol agents (M. Stufkens, personal communication), one of which was *R. conicus*. This species has subsequently been shown to be attacked in the field by *M. aethiopoides*. Quarantine tests, however, were carried out at a time when *R. conicus* was likely to be in aestivation, and inactive. *M. aethiopoides* requires an active host in order to achieve successful oviposition (Phillips 2002), and as this did not occur, *R. conicus* was consequently discounted as a potential host.

The weevil *Mogulones cruciger* (Herbst) was released in Canada to control hound's tongue, (*Cynoglossum officinale*, Boraginaceae), but not approved for release in the USA where its physiological host range was considered too broad (Andreas *et al.* 2008). The authors found that, indeed, non-target plants were attacked in the field in Canada to a lesser extent than the target, but as a result of this research they suggested that measuring response to plant volatiles in quarantine tests could be a useful tool in understanding host selection.

14.4.2 Modelling Population Impact *versus* Attack Rates

Although there are many examples of studies on non-target attack in the field by biological control agents, it was pointed out by Hopper (1998) that few studies have demonstrated population impact. Theoretical studies by Holt and Hochberg (2001) confirmed the view that attack rates mean very little, and that knowledge of the intrinsic rate of increase (r) of a species can assist in determining impact on population density. Mills and Kean (2010) suggested that modelling can enable theory and empirical observation to help predict quantitative impacts on target and non-target populations. Barlow *et al.* (2004) showed that population impact can be estimated from a given attack rate of a parasitoid, using knowledge of the host life cycle and the intrinsic rate of increase of the non-target population, parameters that can be estimated relatively easily. Non-target species will generally have lower r values than target pests and weeds, and so they are likely to be impacted more by a similar attack rate (Barratt *et al.* 2010). In the example discussed above, population impact has been calculated from attack rates for *M. aethiopoides* and modelling suggests that at lower altitudes, where the intrinsic rate of increase for native weevils is higher, a particular attack rate translates to a lower population reduction than the same attack rate at higher altitude, where r is lower (Barlow *et al.* 2004). Some of these principles developed from postrelease studies could be relatively easy for regulators to take into account in future decision making.

14.4.3 Value for Regulators

Regulators are likely to benefit from information on postrelease monitoring on target and non-target species of biological control programmes, but this is particularly true in cases where this is relevant to their own previous decisions. It is rarely possible to reverse a decision once biocontrol organisms have been released into the environment, particularly given that undesirable non-target impacts are very unlikely to be detectable within a time frame where eradication would be remotely feasible. However, information on unpredicted non-target impacts can feed back into decision making in the future. All decisions have to be made case by case, and since experiences in another country using the same biocontrol agent can be very valuable, principles or generic factors that have been reported from postrelease field studies can provide useful guidance for future deliberations. Benefits from successful biological control programmes beyond those anticipated at the application stage can also benefit regulators. Unpredicted benefits can accrue, which might be applicable to future decision making.

14.5 Implications of Biological Control Regulation Policy: What has it Meant for Biological Control Practice?

Increased policy requirements for environmental risk assessment for biological control agent applications has meant that the time frames for obtaining approval to release have become longer and the costs have become higher for the practitioners (Fowler *et al.* 2000). For example, there are often fees for regulators, costs involved in quarantine testing and report preparation, quarantine maintenance of colonies, consultation with stakeholders, etc. Highly risk-averse policy might mean that some biological control agents are rejected for release which might result in lost opportunities for cost-effective pest management (Groenteman *et al.* 2011). Practitioners have expressed frustration at very long timelines for obtaining approval under some jurisdictions (Messing 2005) while other jurisdictions have capped the period for processing applications. The policy in New Zealand, for example, is to process applications within 100 days from official acceptance of the application. Exceptions can be made if, for example, further information is required from applicants.

It can, however, be strongly argued that legislative requirement for environmental risk assessment has brought many benefits to biological control practice. First, targets for biological control programmes are now more carefully selected so that benefits will offset costs of the programme. Second, success rates of biological control programmes would be expected to increase given that more research effort is likely to be given to each programme, and more information is required to make the case for release. Biological control has become less of a technical exercise, with a 'shotgun' approach which characterized some programmes in the past, and more of a carefully researched ecological science. Indeed, Waage (2001), the then Director of CABI Bioscience, pointed out that the increasing interest in biosafety of biocontrol provides researchers with some outstanding opportunities to carry out research that will make a substantial contribution to ecological theory. More recently, Fowler *et al.* (2012) have demonstrated how important ecological research has been to weed biocontrol success. Finally, the regulatory process has in general provided a platform upon which the interaction between researchers and stakeholders has been able to provide mutual benefit and exchange of perspective.

14.6 The Future for Biological Control Regulation

It seems likely that most countries will eventually adopt policy relating to biological control agent introductions. However, compliance with costly, protracted or challenging legislation will encourage practitioners to investigate alternative, less constraining pest management approaches such as using conservation biological control for enhancing the impact of naturally occurring generalist predators, for example, or habitat manipulation to encourage biodiversity and associated ecosystem services (see Chapter 7). Conservation biological control research is gaining traction and the benefits for developing countries were recently reviewed by Wyckhuys *et al.* (2013). These authors noted that pesticides are heavily overused in many developing countries, and that the majority of pests that are resistant to insecticides are not being considered for new biological control programmes or enhancement of existing natural enemies. IPM incorporating natural enemies from the area of origin of the pest is likely to continue to be the best solution in many situations, and we would like to optimistically predict that classic weed and pest biological control using the suite of agents at our disposal will result in safe, effective and enduring solutions. Modelling approaches to understanding natural enemy performance in the environment and interactions between species in complex communities are likely in the future to incrementally improve certainty of predictions made before release. The benefit of postrelease validation of predictions will also contribute to building an enhanced capacity for decision support in the future.

Acknowledgements

We would like to thank colleagues in AgResearch and the Better Border Biosecurity (B3) Research Collaboration for valuable discussion during the preparation of this chapter. BIPB was supported by AgResearch Core funding as part of the B3 Research Collaboration.

References

Andreas, J.E., Schwarzlander, M., Ding, H. and Eigenbrode, S.D. (2008) Post-release non-target monitoring of *Mogulones cruciger*, a biological control agent released to control *Cynoglossum officinale* in Canada. In: Proceedings of the XII International Symposium on Biological Control of Weeds, La Grande Motte, France, pp. 75–82.

Babendreier, D., Bigler, F. and Kuhlmann, U. (2006) Current status and constraints in the assessment of non-target effects. In: Bigler, F., Babendreier, D. and Kuhlmann, U. (eds) *Environmental Impact of Invertebrates for Biological Control of Arthropods – Methods and Risk Assessment*. CABI Publishing, Wallingford, UK, pp. 1–13.

Barlow, N.D., Barratt, B.I.P., Ferguson, C.M. and Barron, M.C. (2004) Using models to estimate parasitoid impacts on non-target host abundance. *Environmental Entomology* **33**: 941–948.

Barratt, B.I.P., Evans, A.A. and Ferguson, C.M. (1997) Potential for control of *Sitona lepidus* Gyllenhal by *Microctonus* spp. *New Zealand Plant Protection* **50**: 37–40.

Barratt, B.I.P., Blossey, B. and Hokkanen, H.M.T. (2006) Post-release evaluation of non-target effects of biological control agents. In: Bigler, F., Babendreier, D. and Kuhlmann, U. (eds) *Environmental Impact of Arthropod Biological Control – Methods and Risk Assessment*. CABI Publishing, Wallingford, UK, pp. 166–186.

Barratt, B.I.P., Berndt, L.A., Dodd, S.L., *et al.* (2007) BIREA – Biocontrol Information Resource for EPA Applicants. Available at: http://www.b3nz.org/birea/ (accessed 9 March 2017).

Barratt, B.I.P., Howarth, F., Withers, T., Kean, J. and Ridley, G. (2010) Progress in risk assessment for classical biological control. *Biological Control* **52**: 245–254.

Barratt, B.I.P., Todd, J. and Malone, L.A. (2016) Selecting non-target species for arthropod biological control agent host range testing: evaluation of a novel method. *Biological Control* **93**: 84–92.

Barton, J. (2012) Predictability of pathogen host range in classical biological control of weeds: an update. *BioControl* **57**: 289–305.

Bennett, F.D. (1993) Do introduced parasitoids displace native ones? *Florida Entomologist* **76**: 54–63.

Briese, D. (2005) Translating host-specificity test results into the real world: the need to harmonize the yin and yang of current testing procedures. *Biological Control* **35**: 208–214.

Clewley, G.D., Eschen, R., Shaw, R.H. and Wright, D.J. (2012) The effectiveness of classical biological control of invasive plants. *Journal of Applied Ecology* **49**: 1287–1295.

Cock, M.J.W., van Lenteren, J.C., Brodeur, J., *et al.* (2009) *The Use and Exchange of Biological Control Agents for Food and Agriculture*. Food and Agriculture Organization, Delémont, Switzerland.

Cock, M.J.W., van Lenteren, J.C., Brodeur, J., *et al.* (2010) Do new access and benefit sharing procedures under the Convention on Biological Diversity threaten the future of biological control? *BioControl* **55**: 199–218.

Coutinot, D., Briano, J., Parra, J.R.P., de Sa, L.A.N. and Cônsoli, F.L. (2013) Exchange of natural enemies for biological control: is it a rocky road? The road in the Euro-Mediterranean region and the South American common market. *Neotropical Entomology* **42**: 1–14.

EFSA Panel on Plant Health (2015) Statement on the assessment of the risk posed to plant health in the EU territory by the intentional release of biological control agents of invasive alien plant species. *EFSA Journal* **13**: 4134.

Elkinton, J.S., Parry, D. and Boettner, G.H. (2006) Implicating an introduced generalist parasitoid in the invasive browntail moth's enigmatic demise. *Ecology* **87**: 2664–2672.

EPPO (2014) Safe use of biological control: PM6/2(3) Import and release of non-indigenous biological control agents. *EPPO Bulletin* **44**: 320–329.

Evans, E.W., Comont, R. and Rabitsch, W. (2011) Alien arthropod predators and parasitoids: interactions with the environment. *BioControl* **56**: 395–407.

Fowler, S.V., Syrett, P. and Jarvis, P. (2000) Will expected and unexpected non-target effects, and the New Hazardous Substances and New Organisms Act, cause biological control of broom to fail in New Zealand? In: Spencer, N.R. (ed.) Proceedings of the X International Symposium on Biological Control of Weeds, Bozeman, Montana, USA, 4–14 July, 1999. Montana State University, Bozeman, MT, USA, pp. 173–186.

Fowler, S.V., Gourlay, A.H., Hill, R.H. and Withers, T. (2004) Safety in New Zealand weed biocontrol: a retrospective analysis of host-specificity testing and the predictability of impacts on non-target plants. In: Cullen, J.M., Briese, D.T., Kriticos, D.J., Lonsdale, W.M., Morin, L. and Scott, J.K. (eds) Proceedings of the XI International Symposium on Biological Control of Weeds. CSIRO Entomology, Canberra, Australia, pp. 265–270.

Fowler, S.V., Paynter, Q., Dodd, S. and Groenteman, R. (2012) How can ecologists help practitioners minimize non-target effects in weed biocontrol? *Journal of Applied Ecology* **49**: 307–310.

Gassmann, A. and Louda, S.M. (2001) *Rhinocyllus conicus*: initial evaluation and subsequent ecological impacts in North America. In: Wajnberg, E., Scott, J.K. and Quimby, P.C. (eds) *Evaluating Indirect Ecological Effects of Biological Control*. CABI Publishing, Wallingford, UK, pp. 147–183.

Gerard, P.J., McNeill, M.R., Barratt, B.I.P. and Whiteman, S.A. (2006) Rationale for release of the Irish strain of *Microctonus aethiopoides* for biocontrol of clover root weevil. *New Zealand Plant Protection* **59**: 285–289.

Gerard, P.J., Kean, J.M., Phillips, C.B., *et al.* (2010) *Possible Impacts of Climate Change on Biocontrol Systems in New Zealand*. Report for MAF Pol Project 0910-11689, AgResearch, Hamilton, New Zealand, p. 64.

Goldson, S.L., McNeill, M.R. and Proffitt, J.R. (2003) Negative effects of strain hybridisation on the biocontrol agent *Microctonus aethiopoides*. *New Zealand Plant Protection* **56**: 138–142.

Goldson, S.L., McNeill, M.R., Proffitt, J.R. and Barratt, B.I.P. (2005) Host specificity testing and suitability of a European biotype of the braconid parasitoid *Microctonus aethiopoides* Loan as a biological control agent against *Sitona lepidus* (Coleoptera: Curculionidae) in New Zealand. *Biocontrol Science and Technology* **15**: 791–813.

Groenteman, R., Fowler, S.V. and Sullivan, J.J. (2011) St. John's wort beetles would not have been introduced to New Zealand now: a retrospective host range test of New Zealand's most successful weed biocontrol agents. *Biological Control* **57**: 50–58.

Hoffmeister, T.S., Babendreier, D. and Wajnberg, E. (2006) Statistical tools to improve the quality of experiments and data analysis for assessing non-target effects. In: Bigler, F. Babendreier, D. and Kuhlmann, U. (eds) *Environmental Impact of Invertebrates for Biological Control of Arthropods – Methods and Risk Assessment.* CABI Publishing, Wallingford, UK, pp. 222–240.

Holt, R.D. and Hochberg, M.E. (2001) Indirect interactions, community modules and biological control: a theoretical perspective. In: Wajnberg, E., Scott, J.K. and Quimby, P.C. (eds) *Evaluating Indirect Ecological Effects of Biological Control.* CABI Publishing, Wallingford, UK, pp. 13–37.

Hopper, K.R. (1995) Potential impacts on threatened and endangered insect species in the United States from introductions of parasitic Hymenoptera for the control of insect pests. In: Hokkanen, H.M.T. and Lynch, J.M. (eds) *Biological Control: Benefits and Risks.* Cambridge University Press, Cambridge, UK, pp. 64–74.

Hopper, K.R. (1998) Is biological control safe – or much ado about nothing? In: Zalucki, M.P., Drew, R.A.I. and White, G.G. (eds) Pest Management – Future Challenges. Proceedings of the 6th Australasian Applied Entomological Research Conference. Brisbane, Australia, pp. 501–509.

Hopper, K.R., Britch, S.C. and Wajnberg, E. (2006) Risks of interbreeding between species used in biological control and native species, and methods for evaluating their occurrence and impact. In: Bigler, F., Babendreier, D. and Kuhlmann, U. (eds) *Environmental Impact of Arthropod Biological Control – Methods and Risk Assessment.* CABI Publishing, Wallingford, UK, pp. 78–97.

Howarth, F.G. (1991) Environmental impacts of classical biological control. *Annual Review of Entomology* **36**: 489–509.

HSNO Act (1996) *Hazardous Substances and New Organisms Act, Vol.* **2015**. New Zealand Legislation Parliamentary Counsel Office, Wellington, New Zealand.

IPPC (2005) *Guidelines for the Export, Shipment, Import and Release of Biological Control Agents and Other Beneficial Organisms.* Secretariat of the International Plant Protection Convention, FAO, Rome, p. 32.

Kaser, J.M. and Heimpel, G.E. (2015) Linking risk and efficacy in biological control host-parasitoid models. *Biological Control* **90**: 49–60.

Kuhlmann, U., Schaffner, U. and Mason, P.G. (2006) Selection of non-target species for host specificity testing. In: Bigler, F., Babendreier, D. and Kuhlmann, U. (eds) *Environmental Impact of Invertebrates for Biological Control of Arthropods – Methods and Risk Assessment.* CABI Publishing, Wallingford, UK, pp. 15–37.

Liu, Y., Shi, Z., Zalucki, M.P. and Liu, S. (2014) Conservation biological control and IPM practices in Brassica vegetable crops in China. *Biological Control* **68**: 37–46.

Lockwood, J.A. (2000) Nontarget effects of biological control: what are we trying to miss? In: Follett, P.A. and Duan, J.J. (eds) *Nontarget Effects of Biological Control Introductions.* Kluwer Academic Publishers, Norwell, MA, USA, pp. 15–30.

Lockwood, J.A., Howarth, F.G. and Purcell, M.F. (2001) Balancing nature: assessing the impact of importing non-native biological control agents. In: Proceedings of the Entomological Society of America. Thomas Say Publications in Entomology, Langham, MD, p. 130.

Louda, S.M. (2000) Negative ecological effects of the musk thistle biological control agent, *Rhinocyllus conicus.* In: Follett, P.A. and Duan, J.J. (eds) *Nontarget Effects of Biological Control Introductions.* Kluwer Academic Publishers, Norwell, MA, USA, pp. 213–243.

Louda, S.M., Kendall, D., Connor, J. and Simberloff, D. (1997) Ecological effects of an insect introduced for the biological control of weeds. *Science* **277**: 1088–1090.

Lu, X., Siemann, E., He, M., Wei, H., Shao, X. and Ding, J. (2015) Climate warming increases biological control agent impact on a non-target species. *Ecology Letters* **18**: 48–56.

McNeill, M.R., Kean, J.M. and Goldson, S.L. (2002) Parasitism by *Microctonus aethiopoides* on a novel host, *Listronotus bonariensis*, in Canterbury pastures. *New Zealand Plant Protection* **55**: 280–286.

Messing, R. (2005) Hawaii as a role model for comprehensive U.S. biocontrol legislation: the best and the worst of it. In: Hoddle, M.S. (ed.) Second International Symposium on Biological Control of Arthropods, Davos, Switzerland. USDA Forest Service Publication FHTET-2005-08, Washington, DC, pp. 686–691.

Messing, R.H., Roitberg, B.D. and Brodeur, J. (2006) Measuring and predicting indirect impacts of biological control, competition, displacement and secondary interactions. In: Bigler, F., Babendreier, D. and Kuhlmann, U. (eds) *Environmental Impact of Invertebrates for Biological Control of Arthropods – Methods and Risk Assessment.* CABI Publishing, Wallingford, UK, pp. 64–77.

Mills, N.J. and Kean, J.M. (2010) Behavioral studies, molecular approaches, and modeling: methodological contributions to biological control success. *Biological Control* **52**: 255–262.

Murray, T.J., Barratt, B.I.P. and Ferguson, C.M. (2002) Field parasitism of *Rhinocyllus conicus* Froehlich (Coleoptera: Curculionidae) by *Microctonus aethiopoides* Loan (Hymenoptera: Braconidae) in Otago and South Canterbury. *New Zealand Plant Protection* **55**: 263–266.

OEPP/EPPO (2014) PM 6/2 (3) Import and release of non-indigenous biological control agents. *Bulletin OEPP/EPPO Bulletin* **44**: 320–329.

Page, A.R. and Lacey, K.L. (2006) *Economic Impact Assessment of Australian Weed Biological Control.* Prepared by the AEC group for the CRC for Australian Weed Management, p. 149.

Paynter, Q.E., Fowler, S.V., Gourlay, A.H., *et al.* (2004) Safety in New Zealand weed biocontrol: a nationwide survey for impacts on non-target plants. *New Zealand Plant Protection* **57**: 102–107.

Paynter, Q., Gourlay A.H., Oboyski, P.T., *et al.* (2008) Why did specificity testing fail to predict the field host-range of the gorse pod moth in New Zealand? *Biological Control* **46**: 453–462.

Phillips, C.B. (2002) Observations of oviposition behaviour of *Microctonus hyperodae* Loan and *M. aethiopoides* Loan (Hymenoptera: Braconidae: Euphorinae). *Journal of Hymenoptera Research* **11**: 326–337.

Roy, H.E. and Brown, P.M.J. (2015) Ten years of invasion: *Harmonia axyridis* (Pallas) (Coleoptera: Coccinellidae) in Britain. *Ecological Entomology* **40**: 336–348.

Seier, M.K., Ellison, C.A., Cortat, G., Day, M. and Dhileepan, K. (2013) How specific is specific enough? Case studies of three rust species under evaluation for weed biological control in Australia. In: Proceedings of the XIII International Symposium on Biological Control of Weeds, Waikoloa, Hawaii, USA, 11-16 September, 2011. USDA Forest Service, Pacific Southwest Research Station, Institute of Pacific Islands Forestry, Hilo, HI, USA.

Sheppard, A.W., Hill, R.L., DeClerck-Floate, R.A., *et al.* (2003) A global review of risk-benefit-cost analysis for the introduction of classical weed biological control agents against weeds: a crisis in the making? *Biocontrol News and Information* **24**: 91N–108N.

Simberloff, D. and Stiling, P. (1996) How risky is biological control? *Ecology* **77**: 1965–1974.
Stanley, J.N. and Julien, M.H. (1998) The need for post-release studies to improve risk assessments and decision making in classical biological control, Vol. Addendum: Pest management –future challenges. In: Zalucki, M., Drew, R. and White, G. (eds) Proceedings of the 6th Australasian Applied Entomological Research Conference. Cooperative Research Centre for Tropical Pest Management, Brisbane, Australia, pp. 561–564.
Stiling, P. and Simberloff, D. (2000) The frequency and strength of nontarget effects of invertebrate biological control agents of plant pests and weeds. In: Follett, P.A. and Duan, J.J. (eds) *Nontarget Effects of Biological Control Introductions*. Kluwer Academic Publishers, Norwell, MA, USA, pp. 31–43.
Swezey, O.H. (1931) Records of introduction of beneficial insects into the Hawaiian Islands. In: Williams, F.X. (ed.) *Handbook of Insects of Hawaiian Sugar Cane Field*. Advertiser, Honolulu, HI, USA, pp. 368–389.
Thomas, M.B. and Willis, A.J. (1998) Biocontrol – risky but necessary? *Trends in Evolution and Ecology* **13**: 325–329.
Thomson, L.J., Macfadyen, S. and Hoffmann, A.A. (2010) Predicting the effects of climate change on natural enemies of agricultural pests. *Biological Control* **52**: 296–306.
Todd, J., Barratt, B.I.P., Tooman, L., Beggs, J.R. and Malone, L.A. (2015) Selecting non-target species for risk assessment of entomophagous biological control agents: evaluation of the PRONTI decision-support tool. *Biological Control* **80**: 77–88.
Tylianakis, J.M. and Binzer, A. (2013) Effects of global environmental changes on parasitoid–host food webs and biological control. *Biological Control* **75**: 77–86.
Urbaneja, A., Tena, A., Jacas, J.A. and Monzo, C. (2015) IPM in Spanish citrus: current status of biological control. In: Sabater-Munro, B., Moreno, P. and Navarro, L. (eds) XII International Citrus Congress, Valencia, Spain. *Acta Horticulturae* **1065**: 1075–1082.
Van Driesche, R. and Murray, T.J. (2004) Overview of testing schemes and designs used to estimate host ranges. In: van Driesche, R. and Reardon, R. (eds) *Assessing Host Ranges for Parasitoids and Predators Used for Classical Biological Control: a Guide to Best Practice*. Forest Health Technology Enterprise Team, Morgantown, VA, USA, pp. 68–89.
Van Lenteren, J., Babendreier, D., Bigler, F., *et al.* (2003) Environmental risk assessment of exotic natural enemies used in inundative biological control. *BioControl* **48**: 3–38.
Van Lenteren, J., Bale, J., Bigler, F., Hokkanen, H. and Loomans, A. (2006a) Assessing risks of releasing exotic biological control agents of arthropod pests. *Annual Review of Entomology* **51**: 609–634.
Van Lenteren, J.C., Cock, M.J.W., Hoffmeister, T.S. and Sands, D.P.A. (2006b) Host specificity in arthropod biological control, methods for testing and interpreting the data. In: Bigler, F., Babendreier, D. and Kuhlmann, U. (eds) *Environmental Impact of Invertebrates for Biological Control of Arthropods – Methods and Risk Assessment*. CABI Publishing, Wallingford, UK, pp. 38–63.
Waage, J.K. (2001) Indirect ecological effects of biological control: the challenge and the opportunity: In: Wajnberg, E., Scott, J.K. and Quimby, P.C. (eds) *Evaluating Indirect Ecological Effects of Biological Control*. CABI Publishing, Wallingford, UK, pp. 1–12.
Waage, J.K. and Greathead, D.J. (1988) Biological control: challenges and opportunities. *Philosophical Transactions of the Royal Society of London, Series B, Biological Sciences* **318**: 111–128.
Wajnberg, E., Scott, J.K. and Quimby, P.C. (2001) *Evaluating Indirect Ecological Effects of Biological Control*. CABI Publishing, Wallingford, UK.

Walker, G.P., Herman, T.J.B., Kale, A.J. and Wallace, A.R. (2010) An adjustable action threshold using larval parasitism of *Helicoverpa armigera* (Lepidoptera: Noctuidae) in IPM for processing tomatoes. *Biological Control* **52**: 30–36.

Wang, X.G. and Messing, R.H. (2002) Newly imported larval parasitoids pose minimal competitive risk to extant egg-larval parasitoid of tephritid fruit flies in Hawaii. *Bulletin of Entomological Research* **92**: 423–429.

Wapshere, A.J. (1974) A strategy for evaluating the safety of organisms for biological weed control. *Annals of Applied Biology* 77: 201–211.

Wapshere, A.J. (1989) A testing sequence for reducing rejection of potential biological control agents for weeds. *Annals of Applied Biology* **114**: 515–526.

Willis, A.J., Kilby, M.J., McMaster, K., Cullen, J.M. and Groves, R.H. (2003) Predictability and acceptability. In: Spafford, J.H. and Briese, D.T. (eds) *Potential for Damage to Nontarget Native Plant Species by Biological Control Agents For Weeds, Vol. 7: Improving the Selection, Testing and Evaluation of Weed Biological Control Agents*. Proceedings of the CRC for Australian Weed Management Biological Control of Weeds Symposium and Workshop, 13 September 2002. CRC for Australian Weed Management, Glen Osmond, Australia.

Withers, T.M. and Mansfield, S. (2005) Choice or no-choice tests? Effects of experimental design on the expression of host range. In: Hoddle, M. (ed.) Second International Symposium on Biological Control of Arthropods, Davos, Switzerland. USDA Forest Service Bulletin, Morgantown, VA, USA, pp. 620–633.

Withers, T.M., Carlson, C.A. and Gresham, B.A. (2013) Statistical tools to interpret risks that arise from rare events in host specificity testing. *Biological Control* **64**: 177–185.

Wyckhuys, K.A.G., Lu, Y., Morales, H., *et al.* (2013) Current status and potential of conservation biological control for agriculture in the developing world. *Biological Control* **65**: 152–167.

Zimmermann, E.C. (1958) *Insects of Hawaii, Volume 7 Macrolepidoptera*. University of Hawaii Press, Honolulu, HI, USA.

15

Pesticides in Food Safety *versus* Food Security

Pieter Spanoghe

15.1 Introduction

Plant protection products, or pesticides, are not only used to protect fruit and vegetables against insects, diseases and weeds, but are also considered to ensure a good harvest and related income. Their widespread use in global food production is further enhanced by the demand for high cosmetic quality (colour, shape, defects) in export markets for fresh fruit and vegetables (Okello and Swinton 2011). In general, the use of pesticides emphasizes the economic goal of maximum productivity at minimum costs, resulting in an intensification of agricultural production. This intensification is seen as the solution for food security concerns. On the other hand, questions arise when pesticides used for securing food production jeopardize food safety.

A recent definition of food security is: 'Food security exists when all people, at all times, have access to sufficient, safe, and nutritious food to meet their dietary needs and food preferences for an active and healthy life' (FAO 2016b). The concept of food security developed over time. Looking back to the 1960s and 1970s, 'food production and availability' (physical presence of enough food), rather than 'food accessibility' (physical presence and ability to buy), were the main concerns (Schoonbeek *et al.* 2013). Also, food can never be entirely safe. Food safety is threatened by numerous pathogens that cause a variety of food-borne diseases, including algal toxins that cause mostly acute disease and fungal toxins that may be acutely toxic but may also have chronic sequelae, such as teratogenic, immunotoxic, nephrotoxic and oestrogenic effects. Industrial activities of the last century have resulted in massive increases in our exposure to toxic metals such as lead, cadmium, mercury and arsenic, which now are present in the entire food chain. Industrial processes also released chemicals that, although banned a long time ago, persist in the environment and contaminate our food. Other food contaminants arise from the treatment of animals with veterinary drugs or the spraying of food crops, which may leave residues. Numerous chemical contaminants are also formed during the processing and cooking of foods (Borchers *et al.* 2010).

Every year, millions of people fall ill and many die as a result of eating unsafe food. The World Health Organization defines food safety as all actions aimed at ensuring that all food is as safe as possible. Food safety policies and actions need to cover the entire food chain, from production to consumption (WHO 2016).

Environmental Pest Management: Challenges for Agronomists, Ecologists, Economists and Policymakers, First Edition. Edited by Moshe Coll and Eric Wajnberg.

Pesticides became widely used in many parts of the world during the 20th century. In the 1940s, inorganic pesticides (based on, for example, copper or sulphur) gave way to synthetic pesticides derived from organic chemistry (e.g. organochlorine, organophosphorus and pyrethroid pesticides). Subsequently, numerous kinds of pesticides have been developed and released globally into agricultural fields and thus the environment (Galt 2008b). At present, pest management is primarily accomplished through the use of pesticides that induce resistance development in pest populations and affect environmental (Matson *et al.* 1997) and human health (Konradsen *et al.* 2003) (see also Chapters 11 and 4).

In the European Union (EU), pesticide residues in food and animal feed have a negative public perception due to reported health (e.g. Parkinson's disease; Elbaz and Moisan 2016) (see Chapter 11) and environmental problems (e.g. DDT; Carson 1962). Closely related to this are concerns about pesticide use in intensive farming systems. Pesticides are viewed as a production system-based risk, and the presence of pesticide residues in fruit and vegetables is one manifestation of this (Tait and Bruce 2001). However, historical risks related to pesticide use could not be compared with the post-1991 situation. Currently authorized products in the EU are significantly safer from both toxicological and environmental aspects. Today's EU products are evaluated under Council Directives 79/117/EEC and 91/414/EEC, and will be re-evaluated under Regulation 1107/2009 (EC 2009) through a science-based risk assessment of their properties and adverse effects for human health and the environment (see also Chapter 13). This safety assessment considers sensitive consumer groups (young, old, pregnant women, immunodeficient people) as well as environmental effects. Products that do not meet the set requirements are removed from the authorized product list for use in the EU or on products consumed in the EU.

Currently, world population is growing at an annual rate of 1.2%, i.e. 77 million new mouths to feed per year. Six countries account for half of this growth: India, China, Pakistan, Nigeria, Bangladesh and Indonesia (Carvalho 2006). In addition, a high percentage of people in many developed countries are obese. This situation stands in sharp contrast to the conditions in many developing countries where currently 925 million people are suffering from undernourishment (Rosenthal and Ort 2012). The combination of population and economic growth with rising incomes is predicted to double worldwide calorie consumption over the next 25 years. The Food and Agriculture Organization of the United Nations (FAO 2016b) expects that, by 2050, an additional 1 billion tons of cereals and 200 million tons of meat will be needed annually to satisfy the growing food demand. Therefore, in developing countries, it is clear that concerns for food safety are far less of an issue compared to food security issues.

15.2 Use of Plant Protection Products in Farming Systems

Between production and consumption, fruit and vegetables are attacked by diseases and pests, which often lead to substantial losses for farmers, traders and consumers (Galt 2008b). The use of pesticides is one of the responses to the issue of food insecurity due to production losses. This section focuses on current challenges in the use, choice and availability of pesticides as crop protection agents. In other farming practices, such as sustainable cropping systems, organic farming and Integrated Pest Management programmes, the same challenges are being addressed with fewer or no synthetic pesticides. The contribution of such practices to food safety is discussed in the latter part of the section.

15.2.1 Pesticide Misuse

15.2.1.1 Pesticide Mismatch with Targeted Damage-Causing Agents

Proper identification of yield-reducing agents is important whether you are dealing with an insect, weed, plant disease or vertebrate. Misidentification and lack of information about a pest cause people to choose the wrong control method or to apply the control at the wrong time. Plants are also damaged by a lack of nutrients or water and by non-living agents, such as extreme weather, air pollutants, road salt and inadequate or excessive fertilization. Sometimes the damage is mistaken for that caused by living pests or diseases. Applying pesticides in such situations leads to unnecessary contamination with toxic residues.

One of the major challenges in crop protection is identifying and controlling plant diseases. Disease symptoms of plants include necrosis, over- or underdeveloped tissues, discoloration and wilt. These symptoms can be caused by several pathogens. It is obvious that, to treat fungi, fungicides have to be applied, bacteria are treated with antibiotics, viruses with viricides and nematodes with nematicides. Knowing that plant pathogens are taxonomically classified into Protozoa (e.g. *Plasmodiophoromycota*), Chromista (e.g. *Oomycota*) or Fungi (e.g. *Chytridiomycota, Zygomycota, Ascomycota, Basidiomycota*), the biodiversity between the different species is much more pronounced compared to the kingdoms of insects or plants. No broad-spectrum fungicides really exist. They are not effective against all plant pathogens at once. If the right pathogen is not correctly identified, the wrong products may be applied. One solution to overcome this problem is to apply mixtures of fungicides. However, applying several pesticides may result in cocktails of pesticide residues in crops, contributing to another food safety issue.

15.2.1.2 Non-optimal Timing of Pesticide Application

Timing of pesticide application is important. Little is known about pesticide use patterns for different pests, weeds or diseases in each crop. For instance, the life cycle and infection process of fungi necessitate contact fungicides to be applied preventively before disease symptoms occur. Once the plant is externally infected, it is usually too late to heal the crop. Eradication is thus the only solution. Contact products only protect the sprayed parts of the plants so after growth of new plant parts, other pesticide treatments are necessary. This explains the high application rates of fungicides on an almost weekly basis.

15.2.1.3 Non-optimal Frequency of Application

The frequency of pesticide applications depends on the persistence and corresponding biological action of the pesticides used. Application technology, formulation and climatic conditions such as rainfall and temperature further influence the deposit needed to sustain plant protection.

15.2.1.4 Non-optimal Application Dose

Due to limited knowledge of pest occurrence, pest control and pesticide use, farmers may overuse pesticides to control some pests and underuse pesticides for other pests. Both over- and underuse of pesticides may have side effects. Pesticide overuse is known to induce pest resistance, threaten food safety, damage human health and pollute the environment. It will also kill many beneficial organisms and natural enemies. The negative effects of pesticide underuse have often been ignored in the existing literature. Besides low crop yield caused by the inefficient control of some crop pests, the other negative externalities of pesticide underuse are poorly understood. Since crop pests can cross the physical boundaries of farm fields, pesticide underuse may put pressure on neighbouring farmers to increase their pesticide use to achieve effective control. Moreover, it is also known that even small amounts of pesticides may be sufficient to kill some beneficial organisms and natural enemies. Thus, pesticide underuse may promote the proliferation of crop pests. In addition, insufficient usage of pesticides is associated with the development of pest resistance (Zhang *et al.* 2015a).

15.2.1.5 Non-optimal Pesticide Formulation

A pesticide is rarely used or applied in its pure form. After manufacture, the technical-grade compound must be formulated, whether it is a herbicide, insecticide, fungicide or another classification. It is processed into a usable form for direct application or for dilution. Formulating a pesticide improves its properties of storage, handling, application, effectiveness or safety. Tank-mix adjuvants are added to active ingredients of pesticides to make them even more effective for application. Ryckaert *et al.* (2007) showed, for instance, that the use of tank-mix adjuvants better prevented mycotoxin-producing *Fusarium* species and optimized the formulation. For example, when no tank-mix adjuvant was used, the lower part of the ear was reached five times less by the propiconazole spray than the upper part of the ear. When the tank-mix adjuvant was combined with the propiconazole formulation, an increase in residue on both the upper and lower part of the ear was observed.

15.2.1.6 Non-optimal Spatial Application

Residues of crop protection products on foodstuffs in samples taken from field and monitoring samples are shown to exhibit great variability, with coefficients of variation of 80–110% being common. Variability in residues among individual samples is inevitable, partly because it is impossible to achieve uniform deposition of pesticides and partly because variables influencing dissipation processes, such as microclimate and crop growth, are inherently heterogeneous. Application technique, canopy architecture and growth stage have all been shown to affect variability in initial deposit. Significant loss of residues may result from wash-off due to rain, although the exact relationship of residue losses with the amount of rainfall may vary between pesticides (Xu *et al.* 2008).

15.2.2 Pesticide Resistance Development

Pesticide resistance, the ability of an organism to withstand a poison, is a predictable consequence of repeated pesticide use. It is defined as 'any heritable decrease in sensitivity to a chemical within a pest population' (Brent 1986). Resistant organisms are simply following the rules of evolution: the best-adapted individuals survive and pass their resistance traits on to their offspring. In many cases, pesticide resistance has resulted in more frequent spraying, influencing food safety as farmers and residential pest control operators scramble to destroy the resilient organisms, followed by increasing resistance and escalating crop losses or food insecurity.

15.2.3 Availability of New Compounds

15.2.3.1 Costs and Selection of New Compounds

New pesticide chemistry modifies/improves existing compounds or formulates new ones (Whitford *et al.* 2016). The development of new pesticides is an extremely demanding business. It is risky: development cycles are long and uncertain, and market changes might occur during the process. It takes 8–10 years and millions of euros to test, develop and register a new pesticide once a promising compound is created. During this lengthy cycle, customer needs and attitudes fluctuate, new regulatory requirements may come into effect, competitive products may pre-empt opportunities and market prices often change (Whitford *et al.* 2016). Less than one in 140 000 molecules tested for pesticide properties ever becomes a registered pesticide product.

There are European and global patent laws to protect owners of products and use patents. These laws protect the owners' rights for a period of 17–20 years, which allows time for the owners to develop and market their products and recover investment costs (Whitford *et al.* 2016). New guidelines for authorization of new plant protection products increasingly demand more studies to demonstrate human and environmental safety. These demands extend even further the time until the product can be released to the market, which is now more than 10 years. This results in a shorter period for marketing before patent protection expires, which is translated into higher prices for new products. From a food safety viewpoint, the side effect of a newer compound being more expensive is the continued use of old products that are less expensive but probably put human and environmental health at greater risks.

15.2.3.2 Agro-chemical Companies

In 2016, 70% of the €51 billion agro-chemical market was in the hands of five companies – and the top seven players own about 95% of the market. The key players in the agro-chemical market dropped down from 13 in the 1990s to five in 2016 (C. Faitz, Kepler Cheuvreux, personal communication). Although this ongoing trend of merging companies may benefit shareholders in the short term, there is the risk that, later on, research departments will be merged or closed, which would hamper research for a broader spectrum of new plant protection products with different modes of actions and low (eco-)toxicological profile. These capabilities are of the utmost importance for overcoming pesticide resistance and enhancing food safety and security.

15.2.4 Use of Genetically Modified Crops

Agro-chemical companies make large investments in the development of genetically modified (GM) crops. In the context of pesticides, GM crops, like soybeans, corn or potatoes, engineered with Bt-, glyphosate- or pathogen-resistant enzyme systems are on the market. Genetic plant breeding is aimed at reducing pesticide use as well as the improvement of food safety by minimizing pesticide residues (Dias and Ortiz 2013). GM crops have been and continue to be the subject of controversy despite their rapid adoption by farmers when approved (Bonny 2016) (see also Chapter 12). Countries vary in their market acceptance of transgenic crops. GM labelling is mandatory in the EU, whereas in the USA, this practice is not imposed.

For the last two decades, an important matter of debate has been the impact of GM crops on pesticide use, particularly for herbicide-tolerant crops. Some claim that these crops bring about a decrease in herbicide use, while others claim the opposite. In all cases, it became clear that genetic engineering is not the only solution to solve the issue of crop threats. Weeds, insects and fungi also develop resistance to GM crops, meaning that these crops will never be a standalone solution. Other crop protection methods such as pesticides will remain necessary to secure GM food production. Biotechnology products will only be successful if clear advantages and safety are demonstrated to both growers and consumers.

15.2.5 Farming Systems: Integrated Pest Management, Organic Farming, Rational Pesticide use, Intensive Sustainable Farming

In Europe and the USA, for example, the postwar 'Green Revolution' saw ancient soil management methods replaced by fertilizers, while herbicides provided an alternative to the old crop rotations as a means of controlling weeds. Industrial farming systems based on large-scale monocultures enabled a substantial increase in yields and a drop in food prices, ending centuries of food insecurity. However, with more and more land being cultivated, it is now recognized that production gains were accompanied by negative human and environmental impacts, jeopardizing the very future of agriculture as intensive farming methods constrain the natural resources upon which they rely. Alternative farming systems are likely to contribute to a more sustainable way of food production (Euractiv 2016).

The oldest alternative to conventional farming, characterized by high use of pesticides, is organic farming. The role of organic farming, how this model of agriculture affects farmers and whether or not it can actually feed the global population is often discussed (Todhunter 2016). Although exclusion of mineral fertilizers, synthetic pesticides and GM crops are the principal differences between organic and conventional farming, these systems often also differ in terms of crop rotation, nutrient supply from manure or other organic amendments, weed control, soil management and crop protection. Authors suggest that organic crops tend to provide farmers with a higher net income compared to their conventional counterparts due to lower production costs (Todhunter 2016). Evidence has shown that organic farming has a positive influence on smallholder food security and livelihoods. This is important because smallholder agriculture is key to food production in developing countries, where food insecurity is most prevalent (Todhunter 2016). As synthetic pesticides may not be used, it is believed that these farming systems also enhance food safety. However, the output per hectare is perceived to be considerably lower in organic than conventional systems in the developed world (Kirchmann *et al.* 2016). Furthermore, pesticides from natural origins that are used in organic crops may also exert toxic effects. Finally, organic farming usually consumes more water and, overall, may not be risk free.

During the latter part of the 1980s, rational pesticide use (RPU) was put forward as another factor in sustainability (Brent and Atkin 1987). RPU combines warning systems to identify the right time of application. This application is done with precision technology, applying the active substance in the right place using new, highly efficient (at low dose, e.g. seed coating, pheromones, etc.) and selective compounds. Advantages of RPU are operator safety, lower or equal costs and a lower impact on the environment due to its selectivity and precision technology. Thus, the combination of or alternation between selective products with different modes of action undermines pesticide resistance development.

The principles of RPU are included in Integrated Pest Management (IPM) which started to gain popularity after the 1970s. The United Nations (UN)'s Food and Agriculture Organization defines IPM as:

> *... the careful consideration of all available pest control techniques and subsequent integration of appropriate measures that discourage the development of pest populations and keep pesticides and other interventions to levels that are economically justified and reduce or minimize risks to human health and the environment. IPM emphasizes the growth of a healthy crop with the least possible disruption to agro-ecosystems and encourages natural pest control mechanisms.*
>
> (UN 2016) (see also Chapter 2)

This plant protection approach became mandatory in Europe in January 2014 (European Sustainable Use Directive).

Intensive sustainable farming (ISF) is the most recent farming idea. The concept here is 'producing more with less'. Intensive agriculture uses high levels of complementary inputs such as irrigation, chemical fertilizers and plant protection products to achieve maximum yields at the lowest possible cost. ISF looks for ways to make conventional farming methods sustainable (Euractiv 2016).

With a growing world population and shrinking arable land, some argue that cities of the future must generate their own food supply. If traditional farming practices are not modified, new land will be needed to grow enough food to feed the growing human population. People see a future in vertical farming such as urban farming and multi-floor farming. The concept of vertical farming is based on LED grow lighting, robotics, aeroponics and hydroponics, and rooftop greenhouse crop production without pesticide use (Despommier 2011).

15.3 Food Security in a Changing World

15.3.1 Climate Change

There is widespread agreement that anthropogenic greenhouse gas emissions are leading to climate change (Miraglia *et al.* 2009). This seems to have a number of impacts, including changes in food production and supply. Without sufficient adaptation, climate change will have negative impacts on at least some of the crops that are important to large food-insecure human populations in South Asia and southern Africa, such as wheat, rice, maize and rapeseed. Furthermore, it is projected that the impacts of climate change on food security will be significant, causing an estimated 5–170 million additional people to be at risk of hunger by 2080 (Schoonbeek *et al.* 2013).

Agricultural yields strongly depend on crop protection measures. In a changing climate, not only crop yields but also pesticide use are expected to be affected. However, the direction of this effect is uncertain and has not yet been thoroughly investigated. Climate change has a substantial effect on the environmental fate of pesticides by altering fundamental mechanisms of partitioning between the environmental compartments (see Chapter 4). For example, global warming may lead to lower pesticide residues on crops that would cause greater vulnerability of crops to pests, weeds and diseases. Thus, in the future, farmers may need to apply higher amounts of pesticides during the growing season. A higher pest, weed or disease pressure will also enhance application frequencies and active ingredient dose. As a consequence, detected pesticide residue concentrations on crops might double for some products, while other plant protection products will degrade faster in the environment and thus will not be detectable on harvested foods. Yet elevated ambient temperatures are expected to lead to overall higher volumes and more frequent application of pesticides. This, in turn, will result in greater consumer exposure to these pesticides (Delcour *et al.* 2015).

Harmful insects, plant diseases and weeds are an ongoing challenge to agricultural producers. An average of 35% of potential crop yield is lost to preharvest pests (Zhang *et al.* 2015b). Temperature, light and water are the key elements that control the growth and development of organisms. Consequently, biodiversity responses to climate change characterized by these environmental elements are expected. Climatic variation may influence the physiology and phenology of the host species, host resistance and growth, all possibly disrupting the synchrony between host and parasite. In contrast, the most severe and least predictable disease outbreaks may occur when altered geographic ranges cause formerly disjunctive species and populations to converge. Due to cropping intensification, crop rotation reductions, increased areas of perennial crops, introduction of new species or varieties and autumn sowing, the rural landscape is always changing. These changes influence the location and availability of host plants for pest species and provide a green bridge for insects and fungi during winter (Delcour *et al.* 2015).

15.3.2 Food Waste

Policies to improve food security do not only involve actions to increase agricultural production by means of appropriate crop protection tools. These days, the focus is oriented more and more towards minimizing food waste. A third of all food produced globally for human consumption is lost or wasted: around 1.3 billion tons per year. Around 100 million tons of food are wasted annually in the EU alone (EC 2016a). Food loss and waste in industrialized countries are as high as in developing countries, but their distribution is different. In developing countries, over 40% of food losses happen after harvest and during storage and processing. In industrialized countries, over 40% occurs at the retail and consumer levels. Therefore, food is lost or wasted along the whole food supply chain: on the farm, in storage, processing and manufacture, in shops, in restaurants and canteens, and at home (EC 2016a). While pesticides are said to enhance crop yields, questions arise over whether these agro-chemicals are really necessary when food waste could be reduced.

15.3.3 Water Scarcity

To grow crops, access to water is obviously important. The more farmland that is used to grow crops, the more water is needed. According to the UN, water scarcity is among the main problems to be faced by many societies in the 21st century (FAO Water

Reports 2012). The UN states that water scarcity already affects every continent. Around 1.2 billion people, or almost one-fifth of the world's population, live in areas of poor water access, and 500 million people are approaching this situation. Another 1.6 billion people, or almost one-quarter of the world's population, face economic water shortage (where countries lack the necessary infrastructure to take water from rivers and aquifers).

Water scarcity is both a natural and a human-made phenomenon. Demographic pressure, the rate of economic development, urbanization and pollution are all putting unprecedented pressure on such renewable but finite resources, particularly in semi-arid and arid regions. Of all economic sectors, agriculture is the sector where water scarcity has the greatest role (FAO Water Reports 2012).

15.3.4 Land Degradation and Restricted Land Use

The World Bank (2011) estimates that approximately 90% of the required increase in food production must come from yield increases on existing farmland. Actually, the current trend is a decrease of agriculture land (ha per inhabitant) in all regions of the globe (Carvalho 2006).

As with demographic and economic transitions, societies appear to follow a dynamic sequence of different land use regimes: from presettlement natural vegetation to frontier clearing, then to subsistence agriculture and small-scale farms, and finally to small areas of intensive agriculture combined with urban areas, and protected recreational lands. Different parts of the world are in different transition stages, depending on their history, social and economic conditions, and ecological context (Foley *et al.* 2005).

The continued conversion of agricultural areas to urbanized developed land (residences, institutions, commerce, industry, urban recreation, and urban and rural transportation uses) in the most productive regions in the world may interfere with the long-term ability to produce food and fibre for much of the world's population. As a solution, new land is cultivated by sacrificing forest areas which are often classified as ecological reserves and natural parks. In this way, a lot of uncultivated farmland seems available in developing countries. However, large-scale expansion of cultivated areas poses significant risks, especially if they are not well managed (see Chapter 8). Without access to technology, traffic infrastructure and market access, large gaps will remain between potential and actual yields (World Bank 2011). Expanding agricultural lands will not necessarily improve the situation.

Land degradation due to acute and chronic droughts or erratic heavy rainfall resulting in soil erosion further reduces the capacity of the land to provide ecosystem goods and services and to ensure its functions over a period of time. Land degradation already affects large areas in the world, especially in dry regions (Huang *et al.* 2016). The removal of the soil crust to reduce competition for water and nutrients, ploughing, heavy grazing and deforestation all leave the soil highly vulnerable to wind erosion, particularly during severe droughts. The FAO (2016a) estimates that land degradation costs approximately US$40 billion annually worldwide, without taking into account hidden costs of increased fertilizer use, loss of biodiversity and loss of unique landscapes. The consequences of land degradation are reduced land productivity, socioeconomic problems, including uncertainty in food security, migration, limited development and damage to ecosystems (FAO 2016a).

15.3.5 Food Price

Food supply and demand in Europe is driven by the EU's commitment to support long-term food supply and meet European and growing world food demand. As a result of Common Agricultural Policy reforms and rising incomes, the share of European household expenditure on food has also been steadily declining over the years (Cupak *et al.* 2015). In the past, the whole family budget was spent on food. In some developing countries and regions worldwide, this is still the case. Data from the USA show that, 30 years ago, an average household spent about 17% of its income on food. Today, this value is estimated to be 11% (Thompson 2013). Food prices influence food security; increase in the price of foods leads to a reduction in food consumption (Green *et al.* 2013). It cannot be ignored that, next to the contributions of modern plant breeding and extensive application of synthetic fertilizers, the use of pesticides played a major role in modern farm production. Pesticides maximize crop yields per hectare and prevent postharvest losses. High food production leads to lower cost of products entering the market.

15.4 Food Safety and Pesticides in a Global Market

15.4.1 Food Safety Risk Linked to Pesticide Residues on Crops

Different governments worldwide define a maximum residue level (MRL), referring to the permitted residue level on a commodity. This level is based on the highest residue amount that can be found on a crop when the pesticide is used according to standard agricultural practices, also called Good Agricultural Practices or GAP (EFSA 2010). MRLs are not safety barriers, implying that a stricter MRL does not necessarily indicate a safer food product. Hence, when detected residue levels exceed the crop's MRL, the human health safety risk has to be assessed case by case. Safety limits consider the Acceptable Daily Intake (ADI) for long-term exposure and the Acute Reference Dose (ARfD) for short-term exposure. Furthermore, if a pesticide's MRL is not considerably lower than the public health safety limit (ADI), the pesticide is banned during the authorization process.

In developed countries, pesticide residues in fruits and vegetables are a major concern to consumers due to their possible negative health effects. Although such residues have also been found in processed products, several studies have found that food processing reduces pesticide residues (Keikothlaile *et al.* 2010). Awareness of pesticide residues in agricultural produce is also increasing in developing countries.

Monitoring of pesticide residues in food may be confusing to consumers because the majority of samples contain no detectable levels of pesticide residues. Therefore, and counterintuitively, the recitation of findings from regulatory monitoring programmes is of little value in terms of assessing the potential health risks posed by the consumption of the tested foods. This is primarily due to the fact that the allowable residue levels are not indicators of safety but rather reflect enforcement tools to assess whether Good Agricultural Practices have been followed (Winter 1992). As such, excessive residue levels often indicate breaches of Good Agricultural Practices but only on very rare circumstances represent cases of health concern (Winter and Jara 2015).

According to the Codex Alimentarius, a risk is defined by a function of the probability of an adverse health effect and the severity of that effect, consequential to hazards in food. The potential health risks posed by pesticide residues in foods can best be assessed by developing estimates of dietary exposure to pesticides and comparing exposure estimates to toxicological indicators of health concern such as the ARfD or ADI. An accurate estimation of dietary pesticide exposure requires data on specific levels of pesticide residues detected (not just whether the residues were legal or excessive) as well as estimations of consumption amounts of all foods for which residues are detected.

Several findings indicate that both the probability (due to wrong use, inappropriate equipment) and severity (e.g. banned, adulterated pesticides) of pesticide hazards for consumers seem higher in developing countries than, for example, in EU member states. However, an accurate assessment of pesticide food safety in developing countries is difficult because these calculations require data on pesticide residues in the food products combined with knowledge of food consumption. These data are available in only a few developing countries (FAO/WHO 2016).

Dietary risk assessment has traditionally been done for an individual compound in a single crop. However, in real life, humans are usually exposed to multiple compounds in their diet. This combined exposure can have toxicological effects that can be independent, dose additive or interactive (i.e. synergistic or antagonistic). Until cumulative risk assessment tools become available, it will remain difficult to determine whether continuously lowering residue levels towards zero makes sense from the public health, agricultural and economical points of view.

In developed countries, risk assessments are frequently performed to investigate where measures should be taken to lower health risk. An example is the Belgian risk assessment study (Claeys *et al.* 2011) which used official monitoring data on pesticide residues and food consumption data. This study suggests that pesticide residues on fruit and vegetables do not pose major public health risks. However, it was pointed out that although exposure of the adult population to pesticide residues appeared to be under control, a high consumption of fruit and vegetables by young children may be exceeding the ADI levels. This indicates that risks of pesticide use and the acceptable residue level depend on the context. Furthermore, to further reduce coincidental risks, the safety of pesticides is currently being assessed separately in Europe for several pesticide–food commodity combinations.

15.4.2 International *versus* Local Production

15.4.2.1 Pesticide Use in EU *versus* Developing World

In the EU, plant protection products cannot be placed on the market or used without prior authorization. A two-tier system is in place, in which the European Commission evaluates active substances used in plant protection products and Member States evaluate and authorize the products at the national level (active ingredient plus additives and/or adjuvants). The authorization of products occurs according to Regulation (EC) 1107/2009 (EC 2009) after a peer review by the risk assessment body of the European Food Safety Authority (EFSA). Each new regulation implies the removal of a (large) number of active substances from the market due to decreasing risk tolerance.

The use of pesticides in the EU is further restricted because of environmental reasons. Pesticide contamination in surface water and ground water is addressed by several directives, such as the Pesticides Framework Directive (Directive 2009/128/EC) and the Water Framework Directive (WFD) (2000/60/EC). Other EU measures are established to strictly control pesticides in the environment. The Thematic Strategy on the Sustainable Use of Pesticides (COM/2006/0372 final) includes a number of measures to encourage the sustainable use of pesticides that were later transformed into Directive 2009/128/EC, establishing a framework for community action to achieve the sustainable use of pesticides, and Regulation (EC) 1107/2009 concerning the placing of plant protection products on the market and repealing Council Directives 79/117/EEC and 91/414/EEC (EC 2016b).

Harmonization in the framework of EU Directive 91/414 has led to a reduction in the number of pesticides authorized for use in Europe. As a consequence of the high costs of authorization, pesticide producers do not tend to invest in the authorization of pesticides for small crops because of insufficient return. Consequently, the number of authorized pesticides for small crops is limited, which may lead to illegal use of pesticides and a higher exposure risk, comparable to what is observed in developing countries. It may also cause insufficient protection and accordingly possible growth of spoilage-causing or mycotoxin-producing fungi or pests.

Apart from the use of illegal pesticides, the EU market today faces a number of new threats, such as generics, rising internet sales and pirate companies copying the packaging of big international brands so that the distinction between the original and the replacement is difficult to make without the right analytical equipment (B. Schiffers, personal communication, 2012). Questions arise about the purity of generic active substances and the additives and adjuvants used to improve efficiency. The use of these sometimes unauthorized unknown molecules is motivated by the fact that they may not be fully included in the scope of monitoring programmes, and that no analytical detection methods exist yet.

In some developing countries such as Cameroon (Amouh 2011), there is a market for the recycling of old pesticides from obsolete stocks. Obsolescence arises because a product has been deregistered locally or banned internationally. More commonly, a stock of pesticides becomes obsolete because of long-term storage (in third countries) during which the product and/or packaging degrades (FAO 1995). Reuse must be strictly controlled. If not, a food safety issue occurs. This may also affect food security because the dated material is no longer effective for pest control.

To overcome food safety issues in developing countries characterized by smallholder farms, access to biodegradable, low-cost and low-risk pesticides should be improved. This is important since it is expected that these countries may suffer first from climate change. The risk today is that some countries may even reintroduce or increase the use of banned or restricted pesticides (Delcour *et al.* 2015).

15.4.2.2 Harmonization of the Maximum Residue Limit (MRL)

In the EU, the EFSA and the European Commission are responsible for setting the MRLs. However, MRLs are part of the legal requirements in most countries. If national or regional limits are not available, the MRLs of internationally recognized bodies such as the WHO Codex Alimentarius Commission (2016) can be used as guidance. Furthermore, MRL harmonization, having the same settings across countries, is an emerging trend which is highly supported by international organizations such as the FAO, WHO, Codex Committee on Pesticide Residues (CCPR) and

Organization for Economic Co-operation and Development (OECD). For example, in the EU, MRL harmonization started in 2008 under Regulation (EC) 396/2005 (OECD 2010). In some developing regions, efforts to harmonize MRLs were initiated by the Global MRL Harmonization Initiative (an Africa Project supported by the US Department of Agriculture, the Foreign Service, IR-4 Project and US Environmental Protection Agency). This project showed that most African countries have adopted the Codex Alimentarius MRLs, while only South Africa established MRLs of its own (Keikotlhaile 2011).

Although in some developing countries, pesticide legislation enhancing food safety is in place, it is often not efficiently implemented which affects, for example, pesticide monitoring (Ecobichon 2001). Amouh (2011) indicated a shortage of trained personnel to enforce legislation and to monitor the use of pesticides and residue levels in food and the environment. This shortage is caused by a lack of knowledge and the cost of laboratory equipment. However, other developing countries, often with a high level of agricultural export, succeed in organizing pesticide monitoring, because their international trade forces them to achieve a higher level of food safety assurance.

15.4.2.3 Guidelines for Trade

Historically, most fruit and vegetables were produced for local consumption. However, trade has rapidly grown since the 1980s, allowing the virtual elimination of seasonality and geographical distances between production and consumption (Huang 2004). The EU is both the leading importer and exporter in the global trade of fruit and vegetables (Huang 2004). In 2011, the EU imported 11.6 million tons of fruit and 1.8 million tons of vegetables. Export accounted for 3.6 million tons of fruit and 1.6 million tons of vegetables. The trade in these products also implies the trade of pesticide residue risks.

Many African countries have ratified international legal instruments related to the management of chemicals (the FAO Code of Conduct, the Rotterdam Convention on Prior Informed Consent (PIC), the Stockholm Convention on Persistent Organic Pollutants (POPs), the Strategic Approach to International Chemical Management (SAICM), the Basel Convention, the Bamako Convention, and the Common Regulations of the Comité Permanent Inter-états de Lutte contre la Sécheresse dans le Sahel – Permanent Inter states Committee for Drought Control in the Sahel – CILSS). One of these agreements, the POP Convention adopted in May 2001, calls for an outright banning and destruction of some of the world's most dangerous chemicals. Moreover, the FAO worked out a code for the distribution and use of pesticides, which describes the responsibilities and standards of conduct for all public and private entities engaged in, or affecting, the distribution and use of pesticides, particularly where there is little or no adequate national law to regulate pesticides. These specifications also provide quality standards for buying and selling pesticides, guidelines for the official approval of pesticides, and support to manufacturers dealing with national and other specifications. The code covers guidelines for vendors to protect themselves against inferior products, and describes the link between biological efficacy and specification requirements (FAO 1995).

In the global context of trade of foods and frequent human travel, the food safety impact is not restricted to the local situation. In Europe, the safety of locally grown foods and products imported from developing countries seems to be guaranteed at a high level. However, in developing countries, the local population and tourists or people travelling for business consume foods from local markets and restaurants, and thus may be at a much higher risk of exposure to pesticide residues through food consumption.

15.4.2.4 Implementation of Guidelines

A study in Tanzania on fruit and vegetables (Ngowi and Partanen 2002) and a study in Nigeria (Ivbijaro 1990) on cocoa and other related crops (Asogwa and Dongo 2009) show that the absence of strong enforcement policies with regard to pesticide use represents a human health problem for farmers (via direct exposure to the pesticides during application) and likely consumers (indirectly through the intake of contaminated food). A case in point is the agricultural use of organochlorine insecticides which have been banned or severely restricted in many developed nations and some developing countries due to their adverse effects on human health and the environment. Regrettably, due to a combination of factors including inadequate regulation and management, trade, weak import controls, illegal use and lack of logistics to monitor these pesticides, the ban or restrictions on these pesticides may be ineffective, as there is evidence for their continued application to crops, vegetables and fruits (Okoffo *et al.* 2016). Also, the pattern of herbicide usage in developing countries has changed. In many areas, a change from multiple soil cultivation for weed control to reduced tillage to prevent soil erosion has led to greater dependence on herbicides such as glyphosate. Associated factors include the wide availability of cheap herbicides and an increase in the cost of hand weeding resulting from socioeconomic changes (Okoffo *et al.* 2016). The continuous use and even greater dependence on these sometimes problematic pesticides is a result of their low cost, their versatility against various pests, weeds and diseases and their availability (Okoffo *et al.* 2016).

Although the implementation of international and national agreements needs to be substantially improved, they can be seen as a first step towards sustainable production and towards controlling the risk associated with the use of pesticides. This can be illustrated by the case of Senegal, which was the first developing country to notify the population of highly hazardous pesticides under the Prior Informed Consent (PIC) Convention, and to alert decision makers in other countries to the risk posed by these particular products (Williamson 2003).

Aiming for safe exposure, the FAO also worked out a code on the distribution and use of pesticides. This code describes responsibilities and standards of conduct for all public and private entities engaged in or affecting the distribution and use of pesticides, particularly where there is little or no adequate national law to regulate pesticides. The specifications also provide quality standards for buying and selling pesticides. They provide guidelines for the official approval and acceptance of pesticides and support manufacturers on how to deal with national and other specifications. Finally, the code includes guidelines on ways in which purchasers can protect themselves against inferior products and describes the link between biological efficacy and specification requirements (FAO 1995).

15.4.2.5 Border Control

In recent years, pesticide residues in food expressed as MRLs have become a focus for food safety and trade. Quarantine regulations sometimes require pesticide treatment of food shipments to prevent establishment of exotic pests. Nonetheless, local consumers and international trading partners increasingly demand food that is free from unsafe pesticide residues. Therefore, many countries have initiated programmes to monitor pesticide residues in food (Zhang *et al.* 2015b). In Europe, the Rapid Alert System for Food and Feed (RASFF) was created to allow food and feed control authorities to exchange information about measures taken in response to serious risks detected in food or feed products. This exchange of information helps Member States to act more rapidly and in a co-ordinated manner in response to a health threat caused by food or feed.

Border rejections are one type of RASFF notification concerning food and feed consignments that have been tested and rejected at the external borders of the EU (and the European Economic Area) when a health risk has been found. Galt (2009) raised the question as to what extent residue testing methods used on imported produce correspond to the pesticides used on crops. He reported that the US Food and Drug Administration (FDA)'s residue testing failed to identify residues of the majority of pesticides used on crops imported into the USA, and suggested that FDA residue testing on imported produce is inadequate in its coverage. A similar situation occurs in the EU, where the total number of pesticides identified in analytical laboratories significantly differs between countries and even between laboratories. Moreover, the number of identified pesticides increases annually: the average number of pesticides labelled in 2006, 2007 and 2008 increased from 209 to 218 and up to 235, respectively (EFSA 2010). Food considered as (pesticide) safe or of low risk in one country or by one laboratory or in one year may be considered not safe in another place or a couple of years later.

Border rejections are directly related to the safety of food and feed products, but, at the same time, they indirectly affect food security. Food produced for export in developing countries replaces the local food production which is needed to contribute to local food security. It is obvious that, when exported food is rejected at the border, the financial return for this replacement to produce local food is totally lost. One of the solutions is perhaps monitoring food and feed products before shipment. However, when this control shows that products do not comply with EU guidelines, moral and ethical questions arise when these products are then sent to a less stringent, often local, market.

In Europe, it seems that harmonization across the borders of EU countries has not been fully realized. In fact, there exists a (regionally) different authorization/registration procedure among European Member States. In certain countries, some product formulations of active ingredients (marketed plant protection products) are not allowed, although the active ingredient is authorized at the EU level. This raises the issue of the implementation of harmonization and risk management among EU Member States; what is considered to be acceptable or safe in one EU country seems not to be so in another. For example, crops containing a given pesticide can be exported to every European country due to freedom of trade. A country can accept crops containing residues of that pesticide, while its own farmers are not allowed to use it to cultivate that same crop, which seems to create a paradox.

15.4.2.6 Secondary Standards

Export crops are considered to be pesticide intensive, while local and national crops are regarded as environmentally benign. This discrepancy has fuelled debates concerning developing countries' dependence on agricultural export and the impact on development, equity and the environment (Galt 2008a). However, in a globalizing economy, the organization of the food chain and its requirements is not only influenced by countries' legislation but also by voluntary private, also called secondary, standards, defining a food safety approach to pesticide residues. The standards are based on established regulatory levels but go beyond the legal requirements, for example, they may be a percentage of the MRL or the ARfD. The secondary standards are set or benchmarked (e.g. by GLOBALG.A.P. certification, available for three scopes of production: Crops, Livestock, Aquaculture or the Global Food Safety Initiative as collective recognized food safety schemes, supported predominantly by European and USA retailers or international food companies) or set as private standards on an individual basis per retailer or manufacturer. Although these standards are not legally binding, farmers are virtually obliged to comply with what has become an unofficial licence for national and international trade. The stringent demands of course have an impact on market access for farmers worldwide (Henson *et al.* 2011; Jaffee 2012).

15.5 Towards Sustainability

15.5.1 Judicious Use of Pesticides

Here, sustainability is referred to in the context of lowering pesticide risks to avoid human and environmental exposure. A study by Mokhele (2011) showed that farm workers in Lesotho were exposed to a greater health risk when there was lack of training on the use of pesticides. Pesticide companies, large corporate international food companies and universities organize such trainings on pesticide use. A system of 'stewardship', where teachers train farmers, is used to reduce the hazard of operator exposure. Education on the correct products to be applied against a certain pest, disease or weed is an ongoing activity. Many students move to developed countries to get a higher education, which they can implement back home. Since 2015, farmers in the EU need to be licensed to spray pesticides, by proving that they have relevant education or sufficient experience. In order to keep this certificate, they have to attend several training courses. These can be provided by the government or information services and cover personal protection, spraying techniques and equipment and integrated pest management.

Failure to adhere to trade standards can result in loss of revenue for farmers supported by the food industry (Okello and Swinton 2010). This is illustrated by Kenya's green bean farmers, who implemented pesticide standards from developed countries because of retailer requirements to show evidence of compliance with UK pesticide legislation (Okello and Swinton 2010). The same study noted that Kenya had become one of the leading producers of green beans since the 1990s, and an important supplier for developed countries. The study by Mokhele (2011) also demonstrated the benefits of reduced pesticide use by comparing the occurrence of illnesses and acute symptoms of pesticide exposure in monitored *versus* unmonitored farmers. This benefit was attributed to farmers' education on the use and handling of pesticides and the implementation of protective measures. Although the Kenya story seems positive, the question arises whether the impact of globalization in trade has really spread across the entire country or whether it is restricted to the regions where green beans are being produced for export.

Unlike the Kenya bean farmers' situation, banana production in Cameroon shows no real benefit from globalization (Amouh 2011). Nevertheless, another study illustrated the impact of globalization on the banana trade by the multinational Chiquita which introduced sustainability principles more than 20 years ago. According to the Rainforest Alliance (RA), a non-profit organization dedicated to protecting tropical forests, this banana company made significant strides. In the early 1990s, the two organizations started talking about reducing pesticide use, recycling, eliminating deforestation and respecting workers' rights. In 1994, the RA began to certify Chiquita's plantations when they met its social and environmental standards. By 2005, Chiquita was selling bananas in Europe with the rainforest-safe label, i.e. the 'green frog'. Now, all Chiquita farms and most of its independent suppliers are certified by the RA group. One negative point of Chiquita's tactics was that they made consumers believe that they were producing 'green' (ecological) bananas, which is not the case.

Similarly, in the EU, the use of certain pesticides from natural origins such as copper, sulphur or the synergist piperonylbutoxide (sometimes in very high amounts) in organic farming is questioned because it might negatively influence food safety. Using these active substances is even more toxic compared to synthetic ones. Organic farming is in fact not equal to 'zero risk' farming but is a more philosophical way of farming of which people believe that it results in healthier/safer or more sustainable products. Hoefkens *et al.* (2009) showed no real differences in food safety and food quality between organic and conventional produce. However, if products of organic farming are indeed proved to be healthier, the question remains whether it is ethical that people have to pay more for these 'low-risk' products (i.e. poor people would have to consume high-risk food).

Besides organic farming, products obtained directly from the farm are currently gaining popularity. It should be noted that although all farmers must comply with legal requirements and are open to inspection by the authorities, farmers delivering crops directly to the market are exempt from additional controls and demands from private standards occurring in business-to-business sales in the conventional (long) supply chain. It remains to be seen whether less controlled fresh produce implies less safe produce. Since this seems to be the case in developing countries, it is likely that EU farm products (home sales) are also not necessarily healthier than products in the retail market.

15.5.2 Low-residue Farming

In contrast to natural contaminants produced by micro-organisms or fungi, pesticide residues in food can be controlled by human actions. Highly toxic crop protection chemicals can, for instance, be replaced by agro-ecological, less dangerous or more human and environmental friendly alternatives.

Winter (2012) analysed pesticide residue data and confirmed that, while pesticide residues are frequently detected in a variety of food products, typical dietary exposure to pesticides continues to be at levels far lower than levels considered to be of health concern. Consumer fears about pesticide residues provide the potential for consumers to reduce their consumption of fruits, vegetables and grains, negating the positive health benefits attributed to consumption of large amounts of such foods. Consumers of conventionally produced food are typically exposed to a very low level of pesticide residues. Findings also indicate that further reducing one's exposure to pesticide residues through purchase of organic foods may not provide any appreciable health benefit. Furthermore, organic foods have been shown to contain pesticide residues as well, although at lower frequency than their conventional counterparts (Winter 2012).

Agricultural farming today can be roughly divided into two main groups: conventional farming, which is now changed to integrated farming in Europe (IPM), and organic farming. Both apply pesticides or plant protection products to maximize crop quality and ensure crop yields. Both methods have drawbacks. Organic agriculture, applying biopesticides, is more labour intensive and water consuming and has lower crop yields and also contains residues, while it is believed that synthetic crop protection products pose potentially higher risks for consumers and the environment.

An upcoming farming system in Belgium is low-residue farming. As biopesticides and chemically synthesized pesticides may have a human and environmental impact, low-residue cropping seeks to find a compromise between these two worlds. In low-residue cropping, the fate of crop protection products in plants is modelled over time. Dissipation of pesticides in plants depends on plant growth and external factors like sun photodegradation, evaporation, metabolism and/or chemical breakdown. Farmers are allowed to use biological or chemical crop protection products to protect their crops from pests or diseases, and are also guaranteed that the crop will have no detectable residue of said products at harvest. This alleviates concern for consumers' health, without compromising crop yield.

15.6 Conclusion

Under the joint pressure of population growth and changes in dietary habits, food consumption is increasing in most regions of the world. Food security is influenced by external factors like climate change, water scarcity, land degradation and farm land restriction, and pest, disease or weed pressure.

Although a legal framework is present in both the developed and developing world to secure food safety, implementation and follow-up by inspection and monitoring are lacking in developing countries, which may lead to an abuse of chemicals. The primary risk factors for human health in developing countries are the application of chemicals and the awareness of pesticide residues. In addition, assessments of pesticide risks cannot be conducted because of a lack of data.

In developed countries, like those in Europe, problems related to the practical use of pesticides no longer require urgent identification because of enhanced capacity, spraying licences and farm inspections in line with certification systems. However, a high food safety risk due to pesticide residues on fruit and vegetables is perceived by European consumers. EU retailers sometimes even require more stringent multiple residue limits than the legislation. However, food safety concerns related to pesticide residues are not confirmed by experts or identified in current risk assessment studies.

References

Amouh, C.N. (2011) A case study of health risk estimate for pesticide-users of fruits and vegetable farmers in Cameroon. Master in Bioscience Engineering, Ghent University, Belgium.

Asogwa, E.U. and Dongo, L.N. (2009) Problems associated with pesticide usage and application in Nigerian cocoa production: a review. *African Journal of Agricultural Research* **4**: 675–683.

Bonny, S. (2016) Genetically modified herbicide-tolerant crops, weeds, and herbicides: overview and impact. *Environmental Management* **57**: 31–48.
Borchers, A., Teuber, S.S., Keen, C.L. and Gershwin, M.E. (2010) Food safety. *Clinical Review in Allergy and Immunology* **39**: 95–141.
Brent, K.J. (1986) Detection and monitoring of resistant forms: an overview. In: National Research Council (ed.) *Pesticide Resistance: Strategies and Tactics for Management*. National Academy Press, Washington, DC, USA, pp. 298–312.
Brent, K.J. and Atkin, R.K. (1987) *Rational Pesticide Use*. Cambridge University Press, Cambridge, UK.
Carson, R. (1962) *Silent Spring*. Houghton Mifflin, New York, USA.
Carvalho, F.P. (2006) Agriculture, pesticides, food security and food safety. *Environmental Science and Policy* **9**: 685–692.
Claeys, W.L., Schmit, J.F., Bragard, C., Maghuin-Rogister, G., Pussemier, L. and Schiffers, B. (2011) Exposure of several Belgian consumer groups to pesticide residues through fresh fruit and vegetable consumption. *Food Control* **22**: 508–516.
Cupak, A., Pokrivcak, J. and Rizov, M. (2015) Food demand and consumption patterns in the new EU Member States: the case of Slovakia. *Ekonomicky Casopis* **63**: 339–358.
Delcour, I., Spanoghe, P. and Uyttendaele, M. (2015) Literature review: impact of climate change on pesticide use. *Food Research International* **68**: 7–15.
Despommier, D. (2011) *The Vertical Farm: feeding the world in the 21st century*. St Martin's Press, New York, USA.
Dias, J.S. and Ortiz, R. (2013) Transgenic vegetables for 21st century horticulture. II Genetically modified organisms in horticulture symposium. *Acta Horticulturae* **974**: 15–30.
EC (2009) *Regulation (EC) N° 1107/2009 of the European Parliament and of the Council of 21 October 2009 Concerning the Placing of Plant Protection Products on the Market*. Official Journal of the European Union 309.
EC (2016a) Stop food waste. Available at: http://ec.europa.eu/food/safety/food_waste/stop/index_en.htm (accessed 10 March 2017).
EC (2016b) Agri-environmental indicator – pesticide pollution of water. Available at: http://ec.europa.eu/eurostat/statistics-explained/index.php/Agri-environmental_indicator_-_pesticide_pollution_of_water (accessed 10 March 2017).
Ecobichon, D.J. (2001) Pesticide use in developing countries. *Toxicology* **160**: 27–33.
EFSA (2010) *2008 Annual Report on Pesticide Residues According to article 32 of Regulation (EC) N° 396/2005*. EFSA Journal.
Elbaz, A. and Moisan, F. (2016) The scientific bases to consider Parkinson's disease an occupational disease in agriculture professionals exposed to pesticides in France. *Journal of Epidemiology and Community Health* **70**: 319–321.
Euractiv (2016) Intensive farming: Ecologically sustainable? Available at: www.euractiv.com/section/agriculture-food/linksdossier/intensive-farming-ecologically-sustainable/#ea-accordion-background (accessed 10 March 2017).
FAO (1995) Prevention of accumulation of obsolete pesticide stocks. Available at: www.fao.org/docrep/v7460e/v7460e00.htm (accessed 10 March 2017).
FAO (2016a) Land degradation assessment in drylands. Available at: http://www.fao.org/3/a-i3241e.pdf (accessed 10 March 2017).
FAO (2016b) How to Feed the World in 2050. Available at: www.fao.org/fileadmin/templates/wsfs/docs/expert_paper/How_to_Feed_the_World_in_2050.pdf (accessed 10 March 2017).

FAO Water Reports (2012) *Coping with Water Scarcity. An Action Framework for Agriculture and Food Security*. Food and Agriculture Organization of the United Nations, Rome, Italy, pp. 38–78.

FAO/WHO (2016) *Assuring Food Safety and Quality: Guidelines for Strengthening National Control Systems*. Food and Agriculture Organization of the United Nations and World Health Organization. Available at: http://www.wpro.who.int/foodsafety/documents/docs/English_Guidelines_Food_control.pdf (accessed 10 March 2017).

Foley, J.A., de Fries, R., Asner, G.P., *et al.* (2005) Global consequences of land use. *Science* **309**: 570–574.

Galt, R.E. (2008a) Pesticides in export and domestic agriculture: reconsidering market orientation and pesticide use in Costa Rica. *Geoforum* **39**: 1378–1392.

Galt, R.E. (2008b) Beyond the circle of poison: significant shifts in the global pesticide complex, 1976–2008. *Global Environmental Change* **18**: 786–799.

Galt, R.E. (2009) Overlap of US FDA residue tests and pesticides used on imported vegetables: empirical findings and policy recommendations. *Food Policy* **34**: 468–476.

GLOBALG.A.P. (2016) www.globalgap.org/uk_en/what-we-do/globalg.a.p.-certification (accessed 10 March 2017).

Green, R., Cornelsen, L., Dangour, A.D., *et al.* (2013) The effect of rising food prices on food consumption: systematic review with meta-regression. *British Medical Journal* **346**: f3703.

Henson, S., Masakure, O. and Cranfield, J. (2011) Do fresh produce exporters in sub-saharan Africa benefit from GlobalGAP certification? *World Development* **39**: 375–386.

Hoefkens, C., Verbeke, W., Aertsens, J., Mondelaers, K. and van Camp, J. (2009) The nutritional and toxicological value of organic vegetables: consumer perception versus scientific evidence. *British Food Journal* **111**: 1062–1077.

Huang, J.P., Yu, H.P., Guan, X.D., Wang, G.Y. and Guo, R.X. (2016) Accelerated dryland expansion under climate change. *Nature Climate Change* **6**: 166–171.

Huang, S.W. (2004) Global trade patterns in fruits and vegetables. USDA electronic outlook REPORT from the Economic Research Service. Available at: http://citeseerx.ist.psu.edu/viewdoc/download?doi=10.1.1.200.6306&rep=rep1&type=pdf (accessed 10 March 2017). Ivbijaro, M.F.A. (1990) *Natural Pesticides: Role and Production Potential in Nigeria*. National Workshop on the Pesticide Industry in Nigeria, University of Ibadan, Nigeria.

Jaffee, S. (2012) From challenge to opportunity. Transforming Kenya's fresh vegetable trade in the context of emerging food safety and other standards in Europe. Agriculture and Rural Development Discussion Paper 2. Available at: http://documents.worldbank.org/curated/en/598771468753012002/From-challenge-to-opportunity-transforming-Kenyas-fresh-vegetable-trade-in-the-context-of-emerging-food-safety-and-other-standards-in-Europe (accessed 10 March 2017).

Keikotlhaile, B.M. (2011) Effects of food processing on pesticide residues in fruits and vegetables. PhD thesis. Crop Protection Department, Ghent University, Ghent, Belgium.

Keikotlhaile, B.M., Spanoghe, P. and Steurbaut, W. (2010) Effects of food processing on pesticide residues in fruits and vegetables: a meta-analysis approach. *Food and Chemical Toxicology* **48**: 1–6.

Kirchmann, H., Katterer, T., Bergstrom, L., Borjesson, G. and Bolinder, M.A. (2016) Flaws and criteria for design and evaluation of comparative organic and conventional cropping systems. *Field Crops Research* **186**: 99–106.

Konradsen, F., van der Hoek, W., Cole, D.C., *et al.* (2003) Reducing acute poisoning in developing countries – options for restricting the availability of pesticides. *Toxicology* **192**: 249–261.

Matson, P.A., Parton, W.J., Power, A.G. and Swift, M.J. (1997) Agricultural intensification and ecosystem properties. *Science* **277**: 504–509.

Miraglia, M., Marvin, H.J.P., Kleter, G.A., *et al.* (2009) Climate change and food safety: an emerging issue with special focus on Europe. *Food and Chemical Toxicology* **47**: 1009–1021.

Mokhele, T.A. (2011) Potential health effects of pesticide use on farm workers in Lesotho, South Africa. *South African Journal of Science* **107**: 509.

OECD (2010) OECD survey on pesticide maximum residue limit (MRL) policies: survey results. Series on Pesticides No. 51, Environment Directorate Joint Meeting of the Chemicals Committee and the Working Party on Chemicals, Pesticides and Biotechnology, Paris, France.

Ngowi, A.V.F. and Partanen, T. (2002) Treatment of pesticide poisoning: a problem for health care workers in Tanzania. *African Newsletter on Occupational Health and Safety* **12**: 71.

Okello, J.J. and Swinton, S.M. (2010) From circle of poison to circle of virtue: pesticides, export standards and Kenya's green bean farmers. *Journal of Agricultural Economics* **61**: 209–224.

Okello, J.J. and Swinton, S.M. (2011) International food safety standards and the use of pesticides in fresh export vegetable production in developing countries: implications for farmer health and the environment, In: Stoytcheva, P.M. (ed.) *Pesticides – Formulations, Effects, Fate*. InTech, Rijeka, Croatia, pp. 183–198.

Okoffo, E.D., Fosu-Mensah, B.Y. and Gordon, C. (2016) Persistent organochlorine pesticide residues in cocoa beans from Ghana, a concern for public health. *International Journal of Food Contamination* **3**: 5.

Rosenthal, D.M. and Ort, D.R. (2012) Examining cassava's potential to enhance food security under climate change. *Tropical Plant Biology* **5**: 30–38.

Ryckaert, B., Spanoghe, P., Haesaert, G., Heremans, B., Isebaert, S. and Steurbaut, W. (2007) Quantitative determination of the influence of adjuvants on foliar fungicide residues. *Crop Protection* **26**: 1589–1594.

Schoonbeek, S., Azadi, H., Mahmoudi, H., Derudder, B., de Maeyer, P. and Witlox, F. (2013) Organic agriculture and undernourishment in developing countries: main potentials and challenges. *Critical Reviews in Food Science and Nutrition* **53**: 917–928.

Tait, J. and Bruce, A. (2001) Globalisation and transboundary risk regulation: pesticides and genetically modified crops. *Health, Risk and Society* **3**: 99–112.

Thompson, D. (2013) Cheap eats: how America spends money on food. Available at: www.theatlantic.com/business/archive/2013/03/cheap-eats-how-america-spends-money-on-food/273811/ (accessed 10 March 2017).

Todhunter, C. (2016) Global agribusiness, dependency and the marginalisation of self-sufficiency, organic farming and agroecology. Available at: www.counterpunch.org/2016/03/29/global-agribusiness-dependency-and-the-marginalisation-of-self-sufficiency-organic-farming-and-agroecology/ (accessed 10 March 2017).

UN (2016) AGP - Integrated Pest Management. Available at: www.fao.org/agriculture/crops/core-themes/theme/pests/ipm/en/ (accessed 10 March 2017).

Whitford, F., Pike, D., Burroughs, F., *et al.* (2016) The pesticide marketplace, discovering and developing new products. Purdue University. Available at: www.extension.purdue.edu/extmedia/PPP/PPP-71.pdf (accessed 10 March 2017).

WHO (2016) Food safety. Available at: www.who.int/topics/food_safety/en/ (accessed 10 March 2017).

WHO Codex Alimentarius Commission (2016) www.fao.org/fao-who-codexalimentarius/standards/pestres/pesticides/en/ (accessed 10 March 2017).

Williamson, S. (2003) Pesticide provision in liberalised Africa: out of control? Agricultural Research and Extension Network. Agricultural Research and Extension Network Newsletter No. 47.

Winter, C.K. (1992) Pesticide tolerances and their relevance to safety standards. *Regulatory Toxicology and Pharmacology* **15**: 137–150.

Winter, C.K. (2012) Pesticide residues in imported, organic, and 'suspect' fruits and vegetables. *Journal of Agricultural and Food Chemistry* **60**: 4425–4429.

Winter, C.K. and Jara, E.A. (2015) Pesticide food safety standards as companions to tolerances and maximum residue limits. *Journal of Integrative Agriculture* **14**: 2358–2364.

World Bank (2011) *The World Bank Rising Global Interest in Farmland. Can It Yield Sustainable and Equitable Benefits?* International Bank for Reconstruction and Development/World Bank, Washington, DC, USA.

Xu, X.M., Huo, R., Salazar, J.D. and Hyder, K. (2008) Investigating factors affecting unit-to-unit variability in non-systemic pesticide residues by stochastic simulation modelling. *Human and Ecological Risk Assessment* **14**: 992–1006.

Zhang, C., Hua, R., Shi, G., Jin, Y., Robson, M.G. and Huang, X. (2015a) Overuse or underuse? An observation of pesticide use in China. *Science of the Total Environment* **538**: 1–6.

Zhang, M., Zeiss, M.R. and Geng, S. (2015b) Agricultural pesticide use and food safety: California's model. *Journal of Integrative Agriculture* **14**: 2340–2357.

16

External Costs of Food Production: Environmental and Human Health Costs of Pest Management

Nir Becker

16.1 Introduction: Pesticide Externalities

While pest control increases both the productivity and the yield of various crops, it also has widespread impact on public health, the environment and the productivity of other farms (Carlson 1989) (see also Chapters 4, 11, 13 and 15). The economic costs of these impacts affect not only the parties conducting agribusiness (see Chapter 3), but third parties as well. Hence, the full costs are not internalized into the cost structure of either party to the transaction. With respect to chemical pest control, for example, neither pesticide manufacturers nor pesticide users (i.e. farmers) account for these costs when calculating their financial bottom lines. The costs are paid for by society as a whole, by consumers, other farmers, neighbouring communities and the general public who benefit from environmental services. These 'social costs', 'external costs' or 'externalities' (terms used here interchangeably) are considered to be negative in this chapter, although positive externalities that create benefit also exist in other cases in which actions of one party create some external diffusion of benefits to another party. However, this is not the case in pesticides use.

Since externalities are not internalized by market entities such as companies and other organizations, they are excluded from the production planning process, probably resulting in a market failure. This, in turn, prevents markets from optimizing social welfare and reaching equilibrium. Therefore, in order to maximize social welfare, governments should intervene in the markets to impose the economic burden created by externalities on the organizations that created them, through the application of tax and subsidy-related environmental policies. Such rewards should incentivize companies to allocate more resources to reducing negative externalities and increasing positive ones.

In order to implement a scheme of taxes and subsidies, we must first estimate the externalities in monetary terms, a rather complicated task. It is challenging primarily due to the nature of externalities; their non-market factors pose a great challenge to putting a 'price tag' on damages. As a gross figure, Pimentel (2009) estimated that US$ 10 billion in pest control saves about US$ 40 billion of crop value. However, it creates external effects costing US$ 9 billion. The main components of this gross estimate include groundwater contamination (US$ 2 billion), public health (US$ 1.1 billion), pesticide resistance in pests (US$ 1.5 billion), crop losses due to pesticide use (US$ 1.1 billion) and bird losses (US$ 2.2 billion).

Environmental Pest Management: Challenges for Agronomists, Ecologists, Economists and Policymakers, First Edition. Edited by Moshe Coll and Eric Wajnberg.

It should be emphasized, however, that price incentive mechanisms are most effective when consumers or producers can adjust their behaviour to the price change. That is, that the supply or demand of the specific good has a meaningful elasticity. If this is not the case and the production or consumption of the good is constant, it may be more efficient to use quantity versus price mechanisms. The reason for this is that price needs to change dramatically when elasticity of demand is low in order to affect demand. In some extreme situations, it might be that the good should be banned altogether.

The remainder of the chapter addresses the economic assessment of the externalities associated with the use of pesticides. Similar principles should be applied to other pest management practices that harbour external costs, such as biological control (see Chapter 5) and the use of transgenic crops (see Chapters 6 and 12). In this chapter, I first present the associated damages and externalities, together with the challenges associated with the estimation of these externalities. The economic methods to cope with these challenges and to estimate the true cost of externalities will then be reviewed. Lastly, an overview of studies that attempted to quantify externalities from pesticide use will be presented, followed by introduction of the notion of Integrated Pest Management (IPM) (see also Chapter 2).

16.2 Background: The Impact of Pesticide Use

Health impacts from pesticides affect farmers, surrounding communities and consumers. They include air pollution, water and soil contamination and food contamination, which can lead to acute and chronic illnesses (see Chapter 4). There is also environmental damage associated with externalities in the form of threats to natural ecosystems and to flora and fauna biodiversity. Pesticide use also contributes, in the long term, to increased pest resistance, as explained by Waterfield and Zilberman (2012):

> *By its nature, pest control imposes selection pressure on traits that enable particular individuals within a targeted pest population to survive treatment. Individuals with that trait thus become an increasingly large fraction of the pest population within the relevant region, and the efficacy of the treatment is expected to decline.*

These costs are considered externalities because in addition to the negative impact on the pesticide user's future crop productivity, there is an impact on the future productivity of other farmers (who are not directly involved in the transaction but who suffer from the impact nonetheless).

Pesticide use may entail some positive externalities as well. This is due, in part, to positive spillover effects and to the contribution of pest control to food safety in ways unaccounted for in the producers' objective function (hence, an externality). Waterfield and Zilberman (2012) explained that:

> *The benefits of pest management to food safety are not incorporated into the producer's objective function, except to the extent that food safety is related to product quality and the price the farmer can obtain. Prices may not, however, fully reflect the degree of contamination from mycotoxins, which are toxic and carcinogenic chemicals produced by fungal mold in a number of crops, primarily*

grains and nuts. The prevalence of these compounds increases as a result of insect damage, so pest control reduces the likelihood of harmful contamination.

Pimentel and Burgess (2014) provide a list of adverse effects of pesticide use (see also Chapter 4, this volume).

- *Soil contamination*: pesticides easily find their way into soils where they may be toxic. Small organisms are vital to ecosystems because they dominate both the structure and function of ecosystems (Pimentel *et al.* 1992). In particular, pesticides reduce species diversity in the soil. Although microbes and invertebrates are essential to the vital structure and function of both natural and agricultural ecosystems, to date, no relevant quantitative data on the value of microbe and invertebrate destruction could be found.
- *Bird and mammal poisoning*: pesticides harm and destroy birds and mammals. Effects on wildlife include death from direct exposure to pesticides or secondary poisoning from consuming contaminated food, and habitat reduction caused by the elimination of food resources and refuges. The full extent of wildlife kills is difficult to determine because birds and mammals are mobile and live in dense grass, shrubs and trees. Studies of the effects of pesticides often obtain low estimates of bird and mammal mortality (Mineau *et al.* 1999). Low estimates result because bird and small mammal carcasses disappear quickly, well before they can be found and counted. In addition, studies seldom account for birds that die a distance from treated areas. Although gross values for wildlife are not available, wildlife-related expenditures made by humans are one measure of the monetary output. The estimated cost for bird watching is about US$ 0.4 per bird and for hunting US$ 216 per bird (Pimentel 2009). In addition, the estimated cost of replacing a bird of an affected species to the wild (the recovery cost), as in the case of the Exxon Valdez oil spill, was US$ 800 US per bird (Dobbins 1986).
- *Impact on aquatic systems*: pesticides are washed into aquatic ecosystems by water run-off and soil erosion (Unnevehr *et al.* 2003). Pesticides also can drift during application and contaminate aquatic systems. High pesticide concentrations in water kill fish directly, causing fishery losses. In addition, because government safety restrictions ban the catching or sale of fish contaminated with pesticide residues, such fish are unmarketable and an economic loss.
- *Water contamination*: pesticides applied at recommended dosages to crops eventually may end up in ground and surface waters. The three most common pesticides found in ground water are aldicarb, alachlor and atrazine (Trautmann and Porter 2012). Costs to sample and monitor well and ground water for pesticide residues are in addition to the costs of cleaning and treating the contaminated water. The issue of water pollution is particularly important because pesticide residues remain for long periods of time.
- *Yield reduction*: pesticides are applied to protect crops from pests in order to increase yields, but sometimes the crops are damaged by the pesticide treatments, especially when farmers apply excessive dosage of pesticides which leads to phytotoxic effects. Crop losses translate into financial losses. Ultimately, the consumer pays for these losses through higher marketplace prices and the producer by reduced profits. Damage to crops may occur even when recommended dosages are applied to crops

under normal environmental conditions. These costs do not take into account other crop losses, nor do they include major events causing large-scale losses. Additional losses are incurred when food crops are disposed of because they exceed regulation pesticide residue levels. Special investigations and testing for pesticide contamination comprise another segment associated with that damage.

- *Pollinator poisoning*: honeybees and wild bees are vital for pollination of fruits, vegetables and other crops. Most agricultural pesticides are toxic to bees, resulting in reduced honey production, the moving of colonies and the loss of suitable crop-growing locations. In addition to these direct losses caused by the destruction of bees, many crops are lost due to lack of pollination. Mussen (1990) indicated that poor pollination also reduces the quality of some crops, such as melon and fruits.
- *Selection for pest resistance*: extensive use of pesticides has often resulted in evolution of pesticide resistance in insect pests. About 520 insect and mite species, a total of nearly 150 plant pathogen species and about 273 weed species are reported to be resistant to pesticides (Bates *et al.* 2005). Increased pesticide resistance in pest populations frequently results in the need for several additional applications of commonly used pesticides to maintain crop yields. These additional pesticide applications compound the problem by increasing environmental selection for resistance.
- *Poisoning of biological control agents*: in many natural and agricultural ecosystems, many species suppress plant-feeding arthropod populations. With the parasites and predators keeping plant-feeding populations at low levels, only a relatively small amount of plant biomass is removed each growing season by arthropods (Hairston *et al.* 1960). Beneficial natural enemies and biodiversity are adversely affected by pesticides.
- *Poisoning of domestic animals*: pesticides not only pose a threat to humans; domestic animals are often accidentally poisoned by pesticides. Poisoning of dogs and cats is common because they usually wander freely and therefore have greater contact with pesticides than other domesticated animals. Additional economic losses related to domestic animals occur when meat, milk and eggs are contaminated with pesticides. Similar to animal carcasses, pesticide-contaminated milk cannot be sold and must be disposed of.
- *Human poisoning*: human pesticide poisoning is a significant external cost of pesticide use. Worldwide, the application of 3 million tons of pesticides results in more than 26 million cases of non-fatal pesticide poisonings annually (Richter, 2002) (see also Chapter 4). In addition, about 3 million individuals are hospitalized, there are 220 000 fatalities and 750 000 people suffer from chronic illnesses every year (Hart and Pimentel 2002). The latter include damage to the neurological, respiratory and reproductive systems (Colborn *et al.* 1996). There is some evidence that pesticides can cause sensory disturbances as well as cognitive damage such as memory loss, language problems and learning impairment (Hart and Pimentel 2002). Evidence also exists for a carcinogenic threat stemming from pesticide use. Several studies have shown that the risks of certain types of cancers are higher in farm workers and pesticide applicators, who have greater exposure to pesticides than the general public (Pimentel and Hart 2001). Certain pesticides have been shown to induce tumours in laboratory animals and there is some evidence suggesting similar effects occurring in humans (Colborn *et al.* 1996). Many pesticides are also oestrogenic: they mimic or interact with the hormone oestrogen, linking the pesticides to an increase in breast

cancer (Colborn *et al.* 1996). The negative health effects of pesticides can be far more significant in children than adults, for several reasons. First, children have higher metabolic rates than adults, so their ability to activate toxic pesticides differs from adults. Also, children consume more pesticides per unit of weight than adults. According to the Environmental Protection Agency (EPA), babies and toddlers are 10 times more at risk for cancer than adults (Hebert 2003).

Although no one can place a precise monetary value on a human life, the economic 'costs' of human pesticide poisonings have been estimated. For our assessment, we use the EPA value of US$ 3.7 million per human life (Kaiser 2003).

- *Pesticide residue in supermarket produce*: the majority of produce purchased in supermarkets has detectable levels of pesticide residues. Pesticide residues were found in 90% of apples, peaches, pears, strawberries and celery (Baker *et al.* 2002).

16.3 The Challenge in Estimating Externalities from Pesticide Use

16.3.1 Market Values

Out of the above 11 suggested external effects from pesticide use, some are relatively easy to estimate. For example, calculating the actual costs resulting from crop damage is fairly straightforward. Actual costs represent that part of the damage that is accounted and paid for in cash money. The actual costs amount is usually based on an agreement between consumers and producers. Additional examples are the money spent by government on cleaning a contaminated water source. We can classify these straightforward costs into three broad categories: lost revenues, increased costs and replacement/restoration costs (if indeed the need arises to engage in such an activity). These actual costs, however, represent only a portion of the true costs of the externalities associated with pesticide use. While actual costs are fairly easy to measure, other types of externalities are more difficult to estimate.

16.3.2 Health Issues

Externalities from pesticide use are expressed in the form of increased health costs caused by environmental degradation. Some of the damage can be calculated using the actual costs of treatment, as discussed above. Another component associated with health is missed work days, a cost which is borne by both employees and employers. Again, these two examples (i.e. treatment costs and loss of work days) are relatively easy to calculate since market data are available. The primary challenge here is to relate a specific health impact to specific pesticide use. It is nearly impossible to trace a direct linkage between a specific chronic poisoning health effect and a specific pesticide.

However, some components of public health are less tangible in terms of monetary value. It is very difficult, for example, if at all possible, to put a price tag on the aggravation a person feels when sick. But since the sick person would be willing to pay a certain amount to avoid the discomfort of being ill, it is still a cost. The premium this person is willing to pay to avoid illness is part of the cost imposed by externalities. Additionally, actual costs include only cases for which people reported sick and got treated for direct pesticide-related impact. In reality, the damage extends much farther than the formally registered cases.First, some people will not report sick and will not seek treatment, and the costs associated with their illness will not be included. Second, the often long-term damage may be difficult to identify and, therefore, to consider. Third, with respect to health impacts and harms, the relation between pesticide use and health damage might be indirect, so the link between the polluter and the sickness will not be made. For all the reasons described above, the impact of pesticide use on public health may be underestimated.

Another complexity derived from human health issues is a chronic effect such as cancer. The complexity of a causal relationship between exposure to pesticide and the occurrence of cancer makes it extremely difficult to assign a precise price tag to a unit of pesticide.

The last issue concerning human health is the mortality effect. In other words, how much is a person's life worth? Of course, the question is not posed with regard to a specific person. Since pest control is probably beneficial from society's point of view, the real trade-off is the treatment cost versus life saved. Is spending US$ 4 million to save a human life cost-effective? The answer to this question depends upon how society values an individual's life, which turns the question into a moral issue.

16.3.3 Ecological Damage

Pesticide use damages ecosystems. Ross and Birnbaum (2003) described the severity of these damages. They claimed that the continual increase of environmental pollution from both intentional and unintentional sources that release chemicals, including pesticides, has been identified as the single major source of various negative effects on humans, animals, crops and ecosystems. For a number of reasons, the estimation of such damages can be extremely difficult.

First, natural resources have, in addition to direct use values, indirect use and non-use values and these are quite complicated and challenging to estimate. Direct use values relate to active engagement within a given environment (for example, a visit to a national park and eating fruits or fishing there). Indirect use values relate to lesser involvement within a given environment (a visit that includes taking pictures only). Non-use values are the vaguest since they are related to a completely passive value from the resource (persons staying at home and enjoying the idea that some resources are pesticide free).

Second, sometimes the full extent of the long-term damage to the environment is unclear. Due to the complexity of ecosystems, science does not yet fully understand the extent of the long-term impact and resulting potential damage of associated externalities. The impact of pesticides on ecosystems might play a significant role in externalities, and therefore should be measured as well.

A good illustration of the challenges in estimating the true economic costs of damage to ecosystems can be found in the example of the honeybees' colony collapse disorder (CCD). This is a phenomenon that occurs when the majority of worker bees in a colony disappear and leave behind a queen, plenty of food and a few nurse bees to care for the remaining immature bees and the queen. A recent study by Spector (2014) implies a connection between pesticide use and the CCD phenomenon. This connection raises two questions with respect to externalities associated with pesticide use: (1) what would be the economic damage to society due to potential exacerbation of the CCD phenomenon, and (2) what is the contribution of pest control to the CCD phenomenon? Answering these two questions will enable estimation of the externalities associated with CCD that were created from pesticide use. However, the ecosystem may be so complex that the future increase in CCD, or its future economic impact, may not be known today with a high degree of confidence. Hence, the task of estimating the economic impact of externalities associated with CCD seems very challenging at best.

16.3.4 Impact on Neighbouring Farms

The negative impact of pesticide application on neighbouring farms is more complex to estimate. It may involve an increase in pesticide resistance that develops in pest populations that would negatively affect crop productivity. This calculation can be thought of as fairly simple because all required data are relatively available; market prices and projections of both crops and pesticides are known, and the correlation between pesticide use (application rates) and increase in pesticide resistance has been studied; see Praneetvatakul *et al.* (2013) for an example of this correlation in Thailand. Yet, an important variable is the spatial scale of this impact. It is difficult to determine the maximum distance of negative effect on neighbouring farms. The distance will vary with the extent of pesticide movement in the environment and pest ability to disperse. Some of these complex issues are treated in Tisdell (2015).

16.4 Externality Estimation Methods

Estimating externalities related to market values is straightforward; they are measured by accounting costs of market loss, cost increase, replacement costs and restoration costs. The more complicated areas concern health and ecological damage. With respect to health, the monetary values are usually derived by the cost of illness, in which costs of medication, loss of work and hospitalization are gathered from secondary sources and weighted by dose–response functions. In cases in which hospitalization is unnecessary, loss of work time for recovery and for family care provision should be estimated instead of hospitalization costs. Today, there exists a good database of the values of different symptoms, whether acute or chronic. With respect to human life, the EPA and the European Union agencies use either a value of an entire life or the value of a saved year of life. The latter calculation biases against elderly people. Hence, if pesticide affects elderly people more significantly, it creates a moral issue.

Ecological damages are usually associated with degradation of natural resources whose values consist of use values (direct and indirect) and non-use values. While the direct value of natural resources is relatively easy to measure, the indirect use and non-use values are generally responsible for creating the challenge in estimating the true cost of externalities. Estimation of ecological impact related to actual costs usually addresses only a portion of the direct use value, and always ignores both indirect use and non-use values which might account for a significant share of the full economic costs. A method that deals with such complexity by asking people about their willingness to pay for an improvement or for avoiding deterioration is the Contingent Valuation Method (CVM). The CVM is used to estimate economic values for a variety of ecosystems, natural resources and environmental services (Carson 2000). It can be used to estimate both use and non-use values, and it is the most widely used method for estimating non-use values (perhaps the only method that fully captures non-use values). The CVM employs a survey to gather data on how much respondents would be willing to pay for specific environmental services (Field and Field 2006). In some cases, people are asked the amount of compensation they would be willing to accept to give up a specific environmental service (Gallardo and Wang 2013). This method is called 'contingent' valuation, because respondents are asked to state their willingness to pay contingent on a specific hypothetical scenario and description of the environmental service. The CVM is referred to as a 'stated preference' method because it asks people to state their values directly, rather than inferring values from actual choices, as do 'revealed preference' methods. The fact that the CVM is based on what people say they would do, as opposed to what people are observed doing, is the source of its greatest strength but also of the controversy surrounding the method.

Contingent valuation is one of the only ways to assign monetary values to non-use values for natural resources (such as ecosystems), values that do not involve market purchases and may not involve direct participation (Jetter and Paine 2004). These values are sometimes referred to as 'passive use' values. They include everything from the basic life support functions associated with a natural resource to the enjoyment of its scenery or the right to bequeath those options to one's grandchildren. Non-use values also include the value people place on simply knowing that some ecosystems still exist. It is clear that people are willing to pay for non-use, or passive use, environmental benefits. However, these benefits are likely to be implicitly treated as zero unless their monetary value is somehow estimated. So, how much are they worth? Since people do not reveal their willingness to pay for them through their purchases or by their behaviour, the only option for estimating value is through asking people questions, and then using the answers to estimate their willingness to pay.

Implementing CVM research on pesticide use, by providing relevant stakeholders (e.g. communities, other farmers and consumers) with the potential tangible ramifications of the use of certain pesticides, would make it possible to examine their willingness to pay for avoiding this particular use. It would then be possible to provide a monetary estimation of the true cost associated with specific pesticide use.

Studies have applied the CVM to estimate consumer willingness to pay for reduced exposure to pesticides (Florax *et al.* 2005), in particular to pay for crops and urban plants protected by IPM technologies (Gallardo and Wang 2013; Jetter and Paine 2004). However, it was found that even if consumers are willing to pay premiums for using biological control, growers may not capture such premiums without some sort of inspection and labelling system providing verification to consumers (see section 16.7). CVM studies are also used to estimate farmer willingness to accept reduced profits resulting from reduced pesticide use in exchange for environmental benefits, including protection of beneficial insects (Lohr *et al.* 1999).

16.5 Overview of Existing Studies on Externalities of Pesticides

The topic of externalities and social costs of pesticide control has been discussed in the literature for nearly half a century. McCarl (1981) noted that the need to consider long-term spillover effects and social costs had been mentioned as early as the 1960s. However, research that attempts to quantify the external costs of pest control is still limited, and most of the studies conducted in this area failed in estimating the true economic costs. Major studies on damage costs related to pesticide use include Pimentel *et al.* (1992, 1999), Waibel *et al.* (1999), Davidson (2004), Tegtmeier and Duffy (2004),

Pretty and Waibel (2005), Leach and Mumford (2008), Rabl and Holland (2008), and Pimentel and Burgess (2014). However, these studies also underestimated the true costs of damages from pesticide use (Fantke 2012).

Other research dealing with more specific case studies provides a more concentrated estimation of the true social costs of pesticide use. They are presented in Table 16.1, and additional explanations are given as footnotes.

Table 16.1 Sample studies of pest control externality costs.

Reference	Region	Key results (costs, unless mentioned otherwise)
Cyuno *et al.* (2001)[a]	Philippines	Environmental benefits from IPM: • US$ 0.5–7.5 per person per cropping season • Aggregated benefits of US$ 150 000 for the 4600 local residents
Palumbi (2001)[b]	USA	Increased resistance in pest and diseases totalled at US$ 2–7 billion per year
Khan *et al.* (2002)[c]	Punjab, India	External costs of US$ 181 million per year, distributed as follows (only key costs): • US$ 86 million production losses due to resistance development • US$ 20 million: damages to domestic animals • US$ 15 million: human health • US$ 57 million: loss of biodiversity
Steiner *et al.* (1995)	USA	US$ 1.3–3.6 billion per year for the US economy
Waibel *et al.* (1999)	Germany	At least US$ 146 million per year for Germany
Houndekon *et al.* (2006)[d]	Niger	• Impact on human health: US$ 1.70 ha^{-1} • Livestock losses: US$ 0.33 ha^{-1}
Pretty and Waibel (2005), Williamson (2011)	China, Germany, UK, USA	An average of US$ 4.28 kg^{-1} of used pesticide active ingredient, including externalities
Leach *et al.* (2008)	Senegal	Externalities of over € 8.05 million, distributed as follows: • 2.75 million €: environmental costs • 2.5 million €: human health • 2.1 million €: agricultural production losses • 0.7 million €: damage prevention costs
Pimentel (2009)	USA	US$ 9 billion in externalities. Major costs are distributed as follows: • US$ 2 billion: ground water contamination • US$ 1.1 billion: impact on public health • US$ 1.5 billion: pesticide resistance in pests • US$ 1.1 billion: crop losses caused by pesticides • US$ 2.2 billion: bird losses due to pesticides

(*Continued*)

Table 16.1 (Continued)

Reference	Region	Key results (costs, unless mentioned otherwise)
Koleva and Schneider (2009)[e]	USA	Cost of US$ 50.5 for use of 1 kg ha^{-1}
Praneetvatakul *et al.* (2013)[f]	Thailand	• External costs for rice cultivation: US$ 19.29 ha^{-1} • External costs for intensive horticulture: US$ 105.75 ha^{-1} • Total external cost of pesticide use in Thailand was US$ 353 million in 2010

[a] Contingent valuation survey was used to assess the reduction in environmental risks to human health and to flora and fauna as a result of implementing IPM, and the public willingness to pay for the reduction of such risks.
[b] Costs to US farmers due to human-induced increased resistance in pesticide expenditures and yield losses.
[c] External cost of pesticide use in cotton.
[d] Based on farmers' interviews, with some additional indirect costs totalled at US$ 2.09 ha^{-1} to prevent further damage.
[e] In 2015 dollars.
[f] Based on Pesticides Environmental Accounting tools. Based on average expenditures on pesticides per ha data, the average external costs for each US$ 1 spent on pesticides was calculated to be US$ 0.66 for rice and US$ 0.23 for intensive horticulture. Using the market cost method, the external cost of pesticide use in Thailand was estimated to be US$ 353 million in 2010. Of this amount, costs related to health were US$ 0.134 million, based on registered cases of acute pesticide poisoning in 2010. Because registered cases underestimate the actual health costs, a cost transfer function was used based on health costs data among tangerine growers. This analysis yielded an estimated total health cost of US$ 2.79 million.

16.6 Integrated Pest Management

As can be seen from the above analysis, pest management is an issue of public as well as private interest. Thus, the term Integrated Pest Management (IPM) is used in order to reflect internalization of the external effects generated by pest control. The FAO (2012) defines IPM in this way:

> *... the careful consideration of all available pest control techniques and subsequent integration of appropriate measures that discourage the development of pest populations and keep pesticides and other interventions to levels that are economically justified and reduce or minimize risks to human health and the environment.*

Another definition is that of the University of California Statewide IPM programme (University of California, Agriculture and Natural Resources 2016):

> *Integrated Pest Management is an ecosystem-based strategy that focuses on long term prevention of pests or their damage through a combination of techniques such as biological control, habitat manipulation, and modification of cultural practices. Pesticides are used only after monitoring indicates they are needed, and pest control materials are selected and applied in a manner that minimizes risks to humans, non-target organisms, and the environment.*

Both definitions are quite vague and open to different interpretations, enabling emphasis on only a singular aspect of the strategy (economic profitability, health issues, etc.).

In practical terms, IPM is frequently reduced to several key points. The primary focus is on continuous monitoring (see also Chapter 2). The IPM approach also suggests that, while biological and ecological control ought to be employed preventively to keep pest populations low and stable, pesticides should be available for precise application in response to an observed population that would otherwise cause significant crop damage (Kos *et al.* 2009; Waterfield and Zilberman 2012). This strategy minimizes the externalities generated by pesticide use and, at the same time, prevents major economic losses to farmers. Precision agriculture tools also offer new support in achieving the goals of IPM (Gebbers and Adamchuk 2010).

Because the practical application of IPM is open to interpretation, estimating the impact of its adoption on pesticide use, yields and profit is difficult. In particular, comparing estimates across existing empirical evaluations of IPM is not so meaningful since different IPM programmes can entail different strategies. Case studies of the impacts of the IPM approach often find reduced pesticide use or increased yields and profits compared to conventional pest control approaches (Brumfield *et al.* 2000; Reddy and Guerrero 2000; Trumble *et al.* 1997). This result is not surprising, particularly regarding pesticide reduction, since IPM is designed to improve farmer profitability by minimized pesticide use.

The success of IPM also depends upon the success of the information-sharing mechanisms employed but less evidence is available on the success of such mechanisms. Goodhue *et al.* (2010) found that a grower education programme for California almond growers significantly reduced use of organophosphate pesticides. A number of studies examined the success of particular projects with controversial results (Epstein and Zhang 2014; Feder *et. al.* 2004; van der Berg and Jiggins 2007).

16.7 The Role of Information

In addition to external effects, the primary focus of this chapter, there are other issues concerning market failures in pest management. One is the asymmetric information available to producers and consumers. Generally, consumers are willing to pay for safer food but food safety is a non-market characteristic of the good itself and not something that can be purchased at a given price. Consumers are usually unable to determine the level of food safety risk both before and after the purchasing act since pesticide residues are not visible.

The idea of consumer sovereignty is well summarized by Korthals (2001):

> *According to the narrow liberal response ... with respect to food one should conceptualize consumer sovereignty as the right of the individual consumer to get information on food products and to make his or her own choice on the market of food products. In this conception, there is a very strong emphasis on rules and principles with respect to the autonomy of individuals.*

With regard to pesticides, clearly there is insufficient information to enable this concept to be realized. Furthermore, consumers cannot distinguish between true and false safety claims made by producers, so they have little reason to pay more for unverifiable claims of safer food. This obstacle creates a market failure since the production and marketing of food contains an aspect hidden from consumers. Caswell (1998) argued that information asymmetry is the most important issue in analysing labelling programmes. Here, since consumers cannot distinguish between products of different safety levels, producers of safer food cannot charge higher prices to cover their higher production costs. Thus, there is economic incentive to grow less safe products. A reliable set of standards for reporting food safety information is necessary before producers will have economic incentives to grow safer food. For example, this is the case for genetically modified foods (Hobbs and Plunkett 1999).

In conclusion, if there are no regulated standards for reporting product safety information, producers have little incentive to grow safer food. This asymmetry of food safety information between producers and consumers leads to a case of market failure (Buzby *et al.* 1998; Oger *et al.* 2001).

16.8 Conclusion

Investment in pesticide control significantly increases crop value (Pimentel and Burgess 2014). However, the indirect costs of pesticide use to the environment and public health need to be balanced against these benefits. Users of pesticides pay only a fraction of the true costs of pesticide resistance and destruction of natural enemies. Society eventually pays additional costs in environmental and public health damages. Our assessment of the environmental and health problems associated with pesticides is difficult due to the complexity of the issues and the scarcity of data. For example, it is difficult to calculate an acceptable monetary value for a human life lost or cancer illness due to pesticides. Equally difficult is placing a monetary value on wild birds and other wildlife killed, on the death of invertebrates or microbes lost, or on the price of contaminated food and ground water. In addition to the costs that cannot be measured accurately, there are many costs absent from the above list. If the full environmental, public health and social costs could be measured, the total cost might increase considerably. Such a complete and long-term cost–benefit ratio analysis of pesticide use would reduce the perceived profitability of pesticides. It should also be noted that estimating the benefits and costs (private and external) of pesticide use, despite the fact that such estimation passes a cost–benefit ratio test according to Pimentel and Burgess (2014), does not take into account any optimization. That is, the net social benefit could be enhanced by government regulations aimed at increasing the benefits of pesticide use together with reducing their overall burden to society.

The efforts of many scientists to devise ways to reduce pesticide use in crop production while still maintaining crop yields have helped but much more needs to be done. Sweden, for example, has reduced pesticide use by 68% without reducing crop yields and/or cosmetic standards (see also Chapter 13). At the same time, public pesticide poisonings have been reduced by 77%. Such reduction is the result of more potent chemistry (less active ingredient is needed) and fewer pesticide applications. IPM, therefore, should not be a mere slogan. It should be exercised worldwide in a manner that can be translated into price and quantity incentives that reflect the true cost of pest control.

References

Baker, B.P., Benbrook, C.M., Groth III, E. and Benbrook, K.L. (2002) Pesticide residues in conventional, integrated pest management (IPM)-grown and organic foods: insights from three US data sets. *Food Additives and Contaminants* **19**: 427–446.

Bates, S.L., Zhao, J.Z., Roush, R.T. and Shelton, A.M. (2005) Insect resistance management in GM crops: past, present and future. *Nature Biotechnology* **23**: 57–62.

Brumfield, R.G., Rimal, A. and Reiners, S. (2000) Comparative cost analyses of conventional, integrated crop management, and organic methods. *HortTechnology* **10**: 785–793.

Buzby, J.C., Fox, J.A., Ready, R.C. and Crutchfleld, S.R. (1998) Measuring consumer benefits of food safety risk reductions. *Journal of Agricultural and Applied Economics* **30**: 69–82.

Carlson, G.A. (1989) Externalities and research priorities in agricultural pest control. *American Journal of Agricultural Economics* **71**: 453–457.

Carson, R.T. (2000) Contingent valuation: a user's guide. *Environmental Science and Technology* **34**: 1413–1418.

Caswell, J. (1998) How labeling of safety and process attributes affects markets for food. *Agricultural and Resource Economics Review* **27**: 151–158.

Colborn, T., Dumanoski, D. and Meyers, J. (1996) *Our Stolen Future: how we are threatening our fertility, intelligence and survival – a scientific detective story.* Dutton-Signet, Auckland, New Zealand.

Cuyno, L., Norton, G.W. and Rola, A. (2001) Economic analysis of environmental benefits of integrated pest management: a Philippine case study. *Agricultural Economics* **25**: 227–233.

Davidson, C. (2004) Declining downwind: amphibian population declines in California and historical pesticide use. *Ecological Applications* **14**: 1892–1902.

Dobbins, J. (1986) *Resources Damage Assessment of the T/V Puerto-Rican Oil Spill Incident.* Unpublished report. James Dobbin Associates, Inc., Alexandria, VA, USA.

Epstein, L. and Zhang, M. (2014) The impact of integrated pest management programs on pesticide use in California, USA. In: Peshin, R. and Dhawan A.K. (eds) *Integrated Pest Management.* Springer, Dordrecht, The Netherlands, pp. 173–200.

Fantke, P. (2012) Health impact assessment of pesticide use in Europe. PhD thesis, University of Stuttgart.

FAO (2012) AGP – integrated pest management. Available at: www.fao.org/agriculture/crops/core-themes/theme/pests/ipm/en (accessed 10 March 2017).

Feder, G., Murgai, R. and Quizon, J.B. (2004) Sending farmers back to school: the impact of farmer field schools in Indonesia. *Applied Economic Perspectives and Policy* **26**: 45–62.

Field B.C. and Field, M.K. (2006). *Environmental Economics: An Introduction*, 4th edn. McGraw-Hill, Boston, MA, USA.

Florax, R.J., Travisi, C.M. and Nijkamp, P. (2005) A meta-analysis of the willingness to pay for reductions in pesticide risk exposure. *European Review of Agricultural Economics* **32**: 441–467.

Gallardo, R.K. and Wang, Q. (2013) Willingness to pay for pesticides' environmental features and social desirability bias: the case of apple and pear growers. *Journal of Agricultural and Resource Economics* **38**: 124–139.

Gebbers, R. and Adamchuk, V.I. (2010) Precision agriculture and food security. *Science* **327**: 828–831.

Goodhue, R.E., Klonsky, K. and Mohapatra, S. (2010) Can an education program be a substitute for a regulatory program that bans pesticides? Evidence from a panel selection model. *American Journal of Agricultural Economics* **92**: 956–971.

Hairston, N.G., Smith, F.E. and Slobodkin, L.B. (1960) Community structure, population control, and competition. *American Naturalist* **94**: 421–425.

Hart, K. and Pimentel, D. (2002) Public health and costs of pesticides. In: Pimentel, D. (ed.) *Encyclopedia of Pest Management*. Marcel Dekker, New York, USA, pp. 677–679.

Hebert, H.J. (2003) *EPA Guidelines Address Kids*, Cancer Risks. Detroit Free Press, Detroit, MI, USA.

Hobbs, J.E. and Plunkett, M.D. (1999) Genetically modified foods: consumer issues and the role of information asymmetry. *Canadian Journal of Agricultural Economics* **47**: 445–455.

Houndekon, V.A., de Groote, H. and Lomer, C. (2006) Health costs and externalities of pesticide use in the Sahel. *Outlook on Agriculture* **35**: 25–31.

Jetter, K. and Paine, T.D. (2004) Consumer preferences and willingness to pay for biological control in the urban landscape. *Biological Control* **30**: 312–322.

Kaiser, J. (2003) How much are human lives and health worth? *Science* **299**: 1836.

Khan, M.A., Iqbal, M., Ahmad, I., Soomro, M.H. and Chaudhary, M.A. (2002) Economic evaluation of pesticide use externalities in the cotton zones of Punjab, Pakistan. *Pakistan Development Review* **41**: 683–698.

Koleva, N.G. and Schneider, U.A. (2009) The impact of climate change on the external cost of pesticide applications in US agriculture. *International Journal of Agricultural Sustainability* 7: 203–216.

Korthals, M. (2001) Taking consumers seriously: two concepts of consumer sovereignty. *Journal of Agricultural and Environmental Ethics* **14**: 201–215.

Kos, M., van Loon, J.J., Dicke, M. and Vet, L.E. (2009) Transgenic plants as vital components of integrated pest management. *Trends in Biotechnology* **27**: 621–627.

Leach, A.W. and Mumford, J.D. (2008) Pesticide environmental accounting: a method for assessing the external costs of individual pesticide applications. *Environmental Pollution* **151**: 139–147.

Lohr, L., Park, T. and Higley, L. (1999) Farmer risk assessment for voluntary insecticide reduction. *Ecological Economics* **30**: 121–130.

McCarl, B.A. (1981) *Economics of Integrated Pest Management: an interpretive review of the literature*. Agricultural Experiment Station, International Plant Protection Center, and Department of Agricultural and Resource Economics, Oregon State University, Corvallis, OR, USA.

Mineau, P., Fletcher, M.R., Glaser, L.C., *et al.* (1999) Poisoning of raptors with organophosphorus and carbamate pesticides with emphasis on Canada, US and UK. *Journal of Raptor Research* **33**: 1–37.

Mussen, E. (1990) California crop pollination. *Gleanings in Bee Culture* **118**: 646–647.

Oger, R., Woods, T.A., Jean-Albert, P. and Allan, D. (2001) *Food Safety in the US Fruit and Vegetable Industry: awareness and management practices of producers in Kentucky* (No. 37867). Department of Agricultural Economics, University of Kentucky, Lexington, KY, USA.

Palumbi, S.R. (2001) Humans as the world's greatest evolutionary force. *Science* **293**: 1786–1790.

Pimentel, D. (2009) Pesticides and pest control. In: Peshin, R.P. and Dhawan, A.K. (eds) *Integrated Pest Management: innovation-development process*. Springer, Dordrecht, The Netherlands, pp. 83–87.
Pimentel, D. and Burgess, M. (2014) Environmental and economic costs of the application of pesticides primarily in the United States. In: Peshin, R. and Dhawan, A.K. (eds) *Integrated Pest Management*. Springer, Dordrecht, The Netherlands, pp. 47–71.
Pimentel, D. and Hart, K. (2001) Pesticide use: ethical, environmental, and public health implications. In: Galston, W. and Shurr, E. (eds) *New Dimensions in Bioethics*. Springer, New York, USA, pp. 79–108.
Pimentel, D., Stachow, U., Takacs, D.A., *et al.* (1992) Conserving biological diversity in agricultural/forestry systems. *BioScience* **42**: 354–362.
Pimentel, D., Bailey, O., Kim, P., *et al.* (1999) Will limits of the Earth's resources control human numbers? *Environment, Development and Sustainability* **1**: 19–39.
Praneetvatakul, S., Schreinemachers, P., Pananurak, P. and Tipraqsa, P. (2013) Pesticides, external costs and policy options for Thai agriculture. *Environmental Science and Policy* **27**: 103–113.
Pretty, J. and Waibel, H. (2005) Paying the price: the full cost of pesticides. In: Pretty, J. (ed.) *The Pesticide Detox: towards a more sustainable agriculture*. Earthscan, London, UK, pp. 39–54.
Rabl, A. and Holland, M. (2008) Environmental assessment framework for policy applications: life cycle assessment, external costs and multi-criteria analysis. *Journal of Environmental Planning and Management* **51**: 81–105.
Reddy, G.V.P. and Guerrero, A. (2000) Pheromone-based integrated pest management to control the diamondback moth *Plutella xylostella* in cabbage fields. *Pest Management Science* **56**: 882–888.
Richter, E.D. (2002) Acute human pesticide poisonings. In: Pimentel D. (ed.) *Encyclopedia of Pest Management*. Dekker, New York, USA, pp. 3–6.
Ross, P.S. and Birnbaum, L.S. (2003) Integrated human and ecological risk assessment: a case study of persistent organic pollutants (POPs) in humans and wildlife. *Human and Ecological Risk Assessment* **9**: 303–324.
Spector, D. (2014) Scientists may have finally pinpointed what's killing all the honeybees. *Business Insider* May 13. Available at: www.businessinsider.com/harvard-study-links-pesticides-to-colony-collapse-disorder-2014-5?IR=T (accessed 10 March 2017).
Steiner, R.A., McLaughlin, L., Faeth, P., *et al.* (1995) Incorporating externality costs into productivity measures: a case study using US agriculture. In: Barnett, V., Payne, R. and Steiner, R. (eds) *Agricultural Sustainability: economic, environmental and statistical considerations*. John Wiley, Chichester, UK, pp. 209–230.
Tegtmeier, E.M. and Duffy, M.D. (2004) External costs of agricultural production in the United States. *International Journal of Agricultural Sustainability* **2**: 1–20.
Tisdell, C.A. (2015) Valuing and sustaining natural ecosystem services: assumptions, estimates and public policies. In: Tisdell, C.A. (ed.) *Sustaining Biodiversity and Ecosystem Functions: economic issues*. Edward Elgar Publishing, Chichester, UK, pp. 363–385.
Trautmann, N.M. and Porter, K.S (2012) Pesticides and groundwater: a guide for the pesticide user. Available at: psep.cce.cornell.edu/facts-slides-self/facts/pest-gr-gud-grw89.aspx (accessed 10 March 2017).

Trumble, J.T., Carson, W.G. and Kund, G.S. (1997) Economics and environmental impact of a sustainable integrated pest management program in celery. *Journal of Economic Entomology* **90**: 139–146.

University of California Agriculture and Natural Resources (2016) State wide pest management program. Available at: www.ipm.ucdavis.edu/ (accessed 10 March 2017).

Unnevehr, L.J., Lowe, F.M., Pimentel, D., *et al.* (2003) *Frontier in Agricultural Research: food, health, environment, and communities.* National Academies of Science, Washington, DC, USA.

Van den Berg, H. and Jiggins, J. (2007) Investing in farmers – the impacts of farmer field schools in relation to integrated pest management. *World Development* **35**: 663–686.

Waibel, H., Fleischner, G. and Becker, H. (1999) The economic benefits of pesticides: a case study from Germany. Available at: www.gjae-online.de/inhaltsverzeichnisse/pages/protected/show.prl?params=recent%3D1%26type%3D2andid=41andcurrPage=andtype=2 (accessed 10 March 2017).

Waterfield, G. and Zilberman, D. (2012) Pest management in food systems: an economic perspective. *Annual Review of Environment and Resources* **37**: 223–245.

Wesseling, C., McConnell, R., Partanen, T. and Hogstedt, C. (1997) Agricultural pesticide use in developing countries: health effects and research needs. *International Journal of Health Services* **27**: 273–308.

Williamson, S. (2011) *Understanding the Full Costs of Pesticides: experience from the field, with a focus on Africa.* InTech, Rijeka, Croatia.

Wu, F. (2006) Bt corn's reduction of mycotoxins: regulatory decisions and public opinion. In: Just, R.E., Alston, J.M. and Zilberman, D. (eds) *Regulating Agricultural Biotechnology: economics and policy.* Springer, New York, USA, pp. 179–200.

17

The Role of Pest Management in Driving Agri-environment Schemes in Switzerland

Felix Herzog, Katja Jacot, Matthias Tschumi and Thomas Walter

17.1 Introduction

Agri-environment schemes (AES) are payments granted by governments to farmers and other landholders to address environmental problems and/or promote the provision of environmental amenities (OECD 2003). The farmers' participation in AES is voluntary, and the financial reward is calculated to compensate farmers either for income foregone or for additional efforts required. AES include the Conservation Reserve Program and the Wetland Reserve Program in the USA and the agri-environmental measures in European countries. Some schemes are competitive: for example, farmers can bid for participation in measures to create landscape features, and amongst the offers, the ones with the best cost–benefit ratio are selected. Most schemes are action oriented (e.g. late cut of grassland, restrictions in fertilizer and/or pesticide inputs, etc.). Some schemes are results oriented with bonus payments being granted for an environmental outcome such as the occurrence of specific farmland birds, mammals or plants (Burton and Schwarz 2013; Dobbs and Pretty 2004; Primdahl *et al.* 2010). Whilst AES are instruments of public authorities, label programmes are instruments of the private sector. Organic farming labels were amongst the first, and they originate from farmers who voluntarily subscribe to certain environmental standards. Various labels exist today, relating to production standards, social standards, animal welfare, etc. Often they are initiated by the food processing industry or by large retailers.

The so-called 'cross-compliance' mechanism requires farmers to comply with environmental standards in order to qualify for any type of subsidy, which is a strong incentive for farmers. Cross-compliance was first introduced in the USA in 1985, mainly to control soil erosion and prevent farmers from reclaiming wetlands (Hoag and Holloway 1991). In Switzerland, cross-compliance was introduced in 1998 (Aviron *et al.* 2009) and in the European Union (EU) in 2013 (http://ec.europa.eu/agriculture/direct-support/cross-compliance/index_en.htm). Cross-compliance requirements may be limited to simple compliance with laws and regulations but may also comprise additional standards (Figure 17.1).

In this chapter, we summarize the role of cross-compliance and AES with respect to the conservation of biodiversity and pest management in Switzerland. Switzerland is an interesting case because, although located geographically in the centre of Europe, it is not an EU Member State and therefore has its own agricultural policy independent of

Environmental Pest Management: Challenges for Agronomists, Ecologists, Economists and Policymakers, First Edition. Edited by Moshe Coll and Eric Wajnberg.

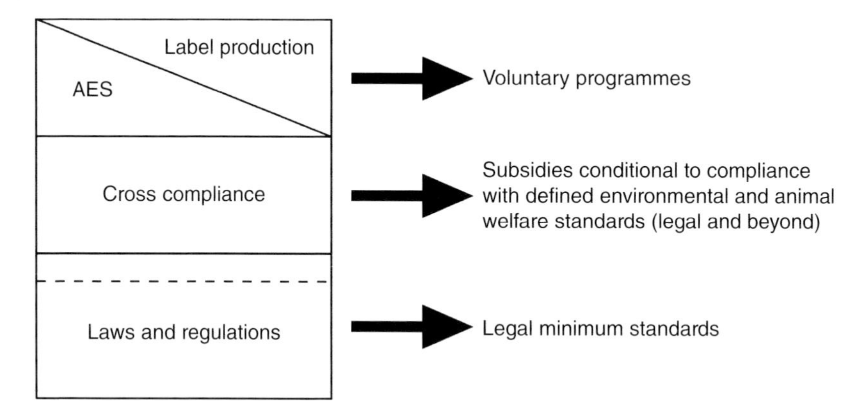

Figure 17.1 Schematic view of the role of cross-compliance mechanism, agri-environment schemes (AES) and label programmes in relation to laws and regulations.

the Common Agricultural Policy (CAP). At times, policy measures have been introduced in Switzerland which were later also adopted by the CAP, which makes Switzerland a kind of laboratory for agri-environmental policy development.

17.2 Policy Context of the Swiss Agricultural Sector

Switzerland is a landlocked industrial country dominated by mountains, except for the Central Plateau and a few other lowland areas (Figure 17.2). Situated in a transition area between oceanic and continental climates, Switzerland enjoys regular rainfall (essentially 700–1500 mm yr^{-1}) and moderate seasonal fluctuation of temperature (15–20 °C difference between January and July). It is not surprising that forest and permanent grassland are the dominating land cover types. Of the total surface area of 41 293 km^2, 28% is biologically unproductive areas (high mountains and built-up areas), 30% is forests, 4% lakes and rivers, 30% grasslands and 8% arable land, vineyards, orchards and special cultures (ARE 2016).

As agriculture occupies 35% of Switzerland's surface (summer grazing area included), it shapes large parts of the country's landscapes and provides an important habitat for many wild species. Farming is characterized by traditional mixed family farms, which are still relatively small (average farm size is slightly below 20 ha). The country's share of self-sufficiency with food is estimated at 60% (SBV 2012).

Switzerland has integrated the special feature of regular ballots in the system of political decisions. A rejected bill on sugar production in 1986 was the first big defeat for farmer production lobbies. In fact, since World War II, agricultural policy has supported farmers by ensuring guaranteed prices for agricultural commodities and taxing cheaper imports. This led to the overproduction of some goods (e.g. milk, some seasonal fruits and vegetables) and to environmental problems (e.g. eutrophication of surface waters, nitrate pollution of drinking water, decline in farmland biodiversity). It was a completely new and painful experience for farmer production lobbies to lose that vote and to realize that, in the future, they would depend on the goodwill of the population. Also, in subsequent referenda, voters repeatedly refused to approve support to agriculture without well-defined performance in ecological matters (Günter *et al.* 2002).

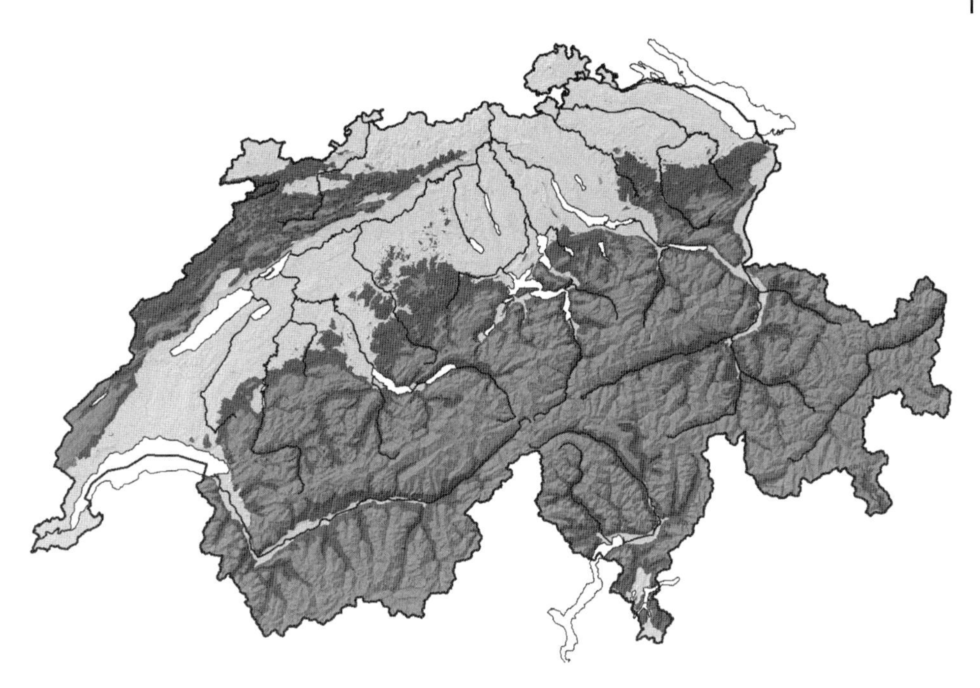

Figure 17.2 Agricultural production regions of Switzerland. Arable farming is mostly concentrated in the lowlands (*light grey*), which also include hilly parts. The mountain regions (*dark grey*) comprise the areas with permanent settlement and mostly grassland-based agriculture. The summer grazing areas (*medium grey*) are only seasonally used. Source: Agricultural zones in Switzerland (2014 © Federal Office for Agriculture); Water courses and relief (2014 © Federal Office of Topography).

This process was also linked to international treaties on trade, which aimed at reducing agricultural subsidies and linked them to environmental performance (so-called 'green box' payments). Within the Uruguay Round (1986–1993), the GATT (General Agreement on Tariffs and Trade, later World Trade Organization – WTO) set up new rules for global trade. These rules restricted export subventions and aimed at avoiding overproduction in agriculture. This objective was a major driving force to move the Swiss agricultural post-World War II policy from a one-sided product subvention and product protection strategy towards a multifunctional agriculture (Popp 2013).

The starting point of AES implementation in Switzerland was in 1992 and they became operational in the field in 1993, before the GATT's Uruguay Round was closed. These first schemes addressed the promotion of farmland biodiversity (Moser 2005), and later the application of pesticides, the balance of the nutrient budget, animal welfare, etc. were targeted. These AES gave Switzerland a small edge in moving agriculture towards multifunctionality, aiming at reduced environmental impacts. Since 1992, annual revisions of the legislation have had the objective of improving the environmental performance of agriculture. In 1996, a revised constitutional article gained a large (78%) majority in a popular referendum. Article 104 of the Swiss Constitution stipulates that the agricultural sector contributes to (1) the provision of the population with food, (2) the conservation of natural resources and the upkeep of the countryside, and (3) a decentralized population settlement in the country. The Constitution also ties direct subsidies to 'proof of compliance with ecological requirements'. The cross-compliance system, which was then developed, was based on experience with ongoing voluntary projects. Since its inception, the cross-compliance requirements have essentially remained the same (Bundesrat 2016), and are based on the following points.

- *Animal-friendly keeping of livestock.* Farm animals have to be kept according to legal requirements. In addition, two AES are available to further increase animal welfare.
- *Balanced use of nutrients.* The annual nitrogen and phosphorus balance of individual farms needs to be lower than 110% of crop requirements. Soil analysis in each plot is required every 10 years at least.
- *Regular crop rotation.* On an arable farm, there must be at least four different crops every year with a maximum of two-thirds of the arable land planted with a single crop.
- *Appropriate soil protection.* Erosion is to be avoided. If it occurs, the farmer has to demonstrate that appropriate soil protection measures had been in place.
- *Targeted application of pesticides.* No application in winter. Regular revision and testing of spraying equipment. Insecticide applications only when action threshold is reached.
- *Adequate share of Ecological Focus Areas (EFA).* At least 7% of the agricultural area of a farm needs to be managed as EFA (3.5% for farms specializing in horticultural production or vineyards).

Almost the entire farming community follows these regulations. Today, 90% of farmers manage 98% of the utilized agricultural area (UAA) (summer pastures not included) according to cross-compliance rules, including organic farming (FOAG 2015). The implementation of cross-compliance regulations is regularly controlled: about one-third of the farms are visited every year. Whilst levels of productivity were maintained, the regulations contributed, for example, to the reduction of nutrient surplus and erosion (Herzog *et al.* 2008). Animal welfare standards are similar to standards in Sweden and comparatively higher than in neighbouring countries.

The success of cross-compliance regulations with respect to pesticide applications is difficult to evaluate because statistical data on the actual application of pesticides are not available (only the total quantities sold, which have remained constant over the last 10 years) and the toxicity of individual pesticides differs (Poiger *et al.* 2005). The actual observation of action thresholds differs according to pesticide category. The percentage of farmers who decide on the application of a pesticide only after verifying in the field whether the action threshold has been reached is higher for insecticides and fungicides (40–90% of farmers, depending on the pest or disease) than for the application of herbicides (25%) (Ramseier *et al.* 2016b).

17.3 Ecological Focus Areas for Biodiversity Protection

17.3.1 Types of Ecological Focus Areas

Table 17.1 lists the most important types of EFA available to farmers in Switzerland. Farmers are free to choose the types and locations of EFA on their farm. Grassland EFA are the most popular type, together with traditional high-stem orchards. The regulations for their management correspond to the traditional management of extensively used grasslands and orchards (see Table 17.1). It was more difficult to propose EFA for arable land because such traditional examples were missing. Therefore, specific seed

Table 17.1 Management regulations and quality criteria for major types of Ecological Focus Areas (EFA) in Switzerland in 2016 (Bundesrat 2016), ordered according to their popularity with farmers, i.e. extensively used and low-input hay meadows make up the highest share of EFA (FOAG 2015). Q I: Quality level I, which means that basic rules for EFA management are respected (management-oriented approach); Q II: Quality level II, which means that certain indicators for ecological quality are actually present on the EFA (result-oriented approach; has not been defined for all types of EFA).

EFA types	Management regulations (Quality level Q I)	Criteria for ecological quality (if applicable, Quality level Q II)
Extensively used and low-input hay meadows	Meadows with minimum size of 0.05 ha, restrictions on fertilization and mowing (late cut, specific dates for agricultural production zones according to altitude). Since 1993	Required plant indicator species present in the plot core area (edge excluded). Since 2001
Extensively used pastures	Grazed at least once each year, no use of fertilizer, no additional feeding of grazing animals. Can contain up to 20% of unproductive rock, shrubs, etc. Since 1993	Same as for extensively used hay meadows. Since 2007
Litter meadows	Meadows with minimum size of 0.05 ha for traditional litter use, prescriptions on mowing, no use of fertilizer. Since 1993	Same as for extensively used hay meadows. Since 2001
Hedges, field and riverside woods	Hedges with grassland buffers of 3 m width on both sides. Since 1993	≥2 m width (excluding buffer), no invasive species, ≥5 shrub or tree species per 10 m length, ≥20% of thorny shrubs, alternatively one native tree every 30 m (stem perimeter ≥170 cm at 150 cm above ground). Since 2001
Traditional orchards	Standard fruit, walnut and chestnut trees, mostly on grassland. Since 1993	≥0.2 ha with ≥10 trees, 30–100 trees ha^{-1}, combination with another EFA within ecological effective distance (stipulated as 50 m in the implementation of the by-law). Since 2001
Wildflower strips (WS)	WS are 2–6 year fallows sown with local wildflowers and maintained in annual crops and fruit orchards. They are at least 3 m wide and do not receive fertilizer or pesticides. Since 1994	No quality criteria specified, but of high nature value quality for species richness according to Walter *et al.* (2013)
Rotational fallows (RF)	RF are special 1–3 year components within a crop rotation. They develop by sowing of locally adapted wildflowers. They do not receive fertilizer or pesticides. Since 1999	No quality criteria specified, but of high nature value quality for species richness according to Walter *et al.* (2013)
Improved field margins	Permanent field margins sown with local wildflower and grassland species. They are at least 3 m wide and do not receive fertilizer or pesticides. Cutting is restricted to alternate mowing of half of the strip no more than once a year. Since 2007	No quality criteria specified, but of high nature value quality for species richness according to Walter *et al.* (2013)

(*Continued*)

Table 17.1 (Continued)

EFA types	Management regulations (Quality level Q I)	Criteria for ecological quality (if applicable, Quality level Q II)
Conservation headland (CH)	CH is defined as the outer few metres of the crop and is harvested with the crop. It does not receive fertilizer or pesticides. Additional local/indigenous wildflowers can be sown. Since 1993	No quality criteria specified, but of high nature value quality for species richness according to Walter *et al.* (2013)
Flower strips for pollinators and other beneficials	Annual flower strips sown with indigenous ecotypes and cultivated plant species. Since 2015	No quality criteria specified

Additional EFA types not listed in the table because of only minor uptake as yet: shoreline pasture, forest pasture, species-rich summer grazing area, native single trees and alleys appropriate to the habitat, ditches, puddles, ponds, ruderal areas, stone heaps, dry stone walls, species-rich vineyards.

mixtures have been developed for wildflower strips since 1994 (Günter 2000). These seed mixtures were designed for perennial elements (6–8 years).

To facilitate the uptake of EFA by farmers, the administration initiated a platform for ecological compensation in agriculture. This consisted of 20 representatives of the major stakeholder groups: farmer associations, nature conservation lobby groups, regional administrations, farm advisory services, independent consultancies in charge of controlling the implementation of EFA on farms, etc. The platform was – and still is – managed by the Ministries for Agriculture and for Environment and became operational in 1997. It is an important player in the decision-making process and, especially when EFA were introduced, was instrumental in developing consensual solutions and rules for EFA implementation (Moser 2005). There was a rapid uptake of EFA management even before their implementation became conditional under the cross-compliance mechanism in 1998 (Figure 17.3). Then, the share of EFA stabilized at about 12% of farmland – with a lower share in the more productive lowland region and a higher share in the mountain region where farmers actually did not have to adapt grassland management significantly. There, the EFA scheme helps to maintain traditional, extensive farming and related biodiversity (Kampmann *et al.* 2012). More recently, there has been an additional increase in EFA due to policy reforms which even more strongly link public payments to public services.

To improve the quality of EFA, in 2001 better financial incentives for ecological quality (Level Q II in Table 17.1) were set up. Ecological quality was defined in several ordinances for low-input meadows, hedges and orchards and a few years later for extensively used pastures and vineyards. Additionally, the ordinance grants financial contributions if the farmers participate in a 'network project' which includes measures to improve the habitat network and quality for target species in high nature value farming areas. All these efforts are aimed at conducting agriculture in a way that is more sustainable and halting the loss of biodiversity.

From 1999 to 2005, the introduction of the cross-compliance mechanism was accompanied by an evaluation of the effectiveness of EFA for protecting farmland biodiversity. Its results suggested a moderately positive effect (Herzog and Walter 2005). In the lowland area (see Figure 17.2), the policy objective for the quantitative uptake of EFA had nearly

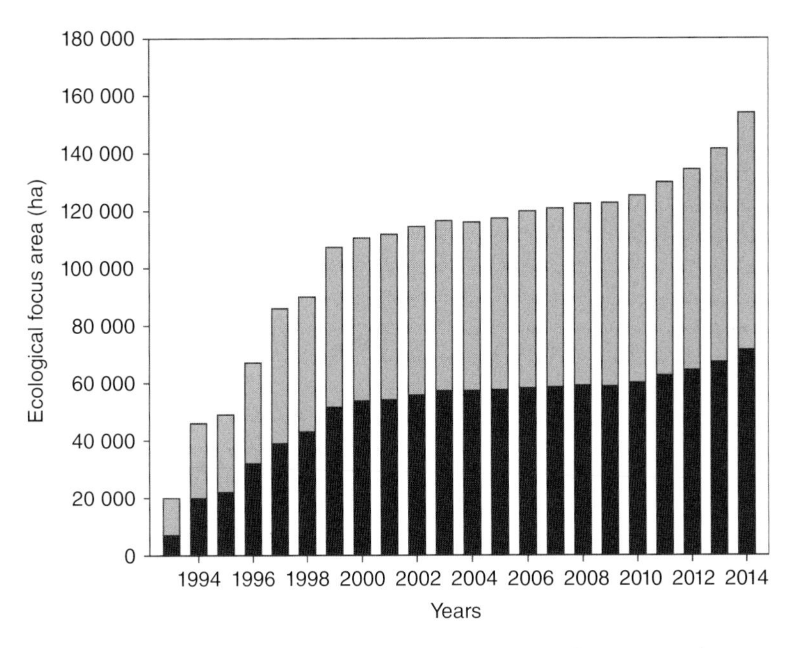

Figure 17.3 Development of the total surface of ecological focus areas in hectares within the Swiss utilized agricultural area (UAA) (without trees). Total UAA in 2014 was 1 051 183 ha. An additional 2 400 000 high-stem trees are also managed under the scheme. With an assumed average density of 100 trees ha^{-1}, this corresponds to another 24 000 hectares EFA. Lowland (*black*) and mountain regions (*grey*) are shown on Figure 17.2.

been achieved in 2005 and was later reached in 2010. Species richness was generally higher on EFA than on conventionally managed but otherwise comparable fields or grasslands (Aviron *et al.* 2007, 2009; Knop *et al.* 2006). However, the promotion of threatened species failed and actual biodiversity goals were still missing. To fill this gap, the federal government specified agri-environmental objectives for biodiversity and landscape, climate, air, water and soil (FOEN and FOAG 2008). For the first time also, agriculture was committed to preserving the ecosystem services of biodiversity, such as the maintenance of fertile soils, natural pest management and pollination (FOEN and FOAG 2008) (see also Chapter 7). Evidence-based quantitative objectives for habitats of Agriculture-related Environmental Objective (AEO) quality in Swiss agricultural zones and regions were elaborated, estimating the actual share (in 2010) of 'AEO-quality land' and proposing target shares for the various agricultural regions (see Figure 17.2). Whereas sufficient AEO-quality land is still available in the upper mountain regions and the summer grazing area, there is a shortfall in the lower mountain regions and the lowland, mainly in the arable area (Lachat *et al.* 2010; Walter *et al.* 2013).

17.3.2 Ecological Focus Areas on Arable Land

On arable land, the uptake of EFA is much lower than in grassland farming areas. In 2014, there were less than 3000 ha of sown EFA designed for the promotion of target species of arable farmland (Figure 17.4), which corresponds to only 0.28% of the UAA or about 0.7% of the arable areas. The predominant arable EFA types are wildflower

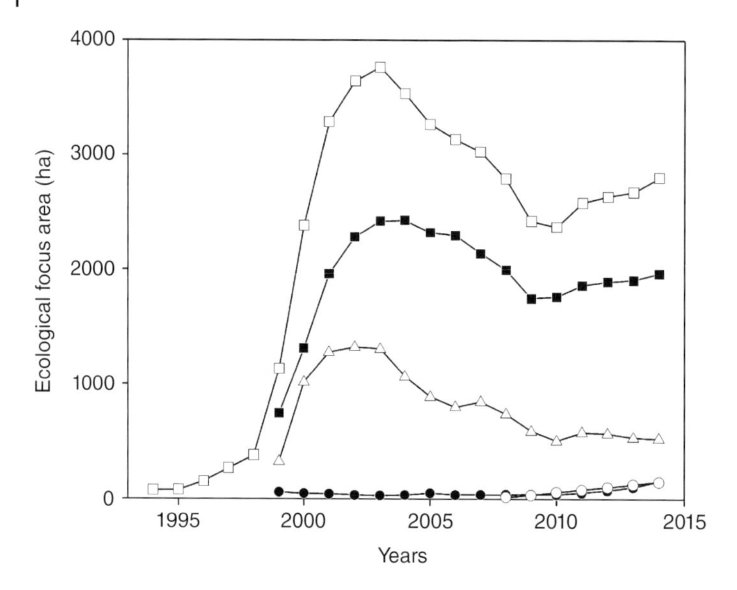

Figure 17.4 Evolution of the area of EFA in arable regions in Switzerland. Total EFA types for arable area – open squares; wildflower strip – closed squares; rotational fallow – open triangles; conservation headland – closed circles; improved field margin – open circles.

strips and rotational fallows, which must be sown with a recommended seed mixture containing buckwheat (*Fagopyrum esculentum*) and 17–37 indigenous plant species (Table 17.2). The presence of species with contrasting life cycles, both annual and perennial, leads to a marked succession and ensures a certain level of species richness over several years (Günter 2000). Rotational fallows are maintained for 2–3 years and wildflower strips for up to 8 years.

To increase the acceptance of EFA by arable farmers, novel EFA types were designed. Conservation headlands at the crop edge are neither fertilized with nitrogen nor treated with pesticides (Bundesrat 2016). The intention is to maintain the arable flora. If valuable plants are no longer present in the field, specific seed mixtures are available. Another type of seed mixture was developed for improved field margins (Jacot *et al.* 2007). This mixture differs from that for wildflower strips in containing grasses and perennial herbs. In contrast to wildflower strips, improved field margins are typical border structures, and they are permanent elements on arable soils.

All seed mixtures currently available on the market were developed to promote threatened arable plant species and did not focus on biological pest control. Despite the additional arable EFA types, acceptance by farmers remained low. After a strong increase following the introduction of the cross-compliance mechanism to 3800 ha (approximately 0.9% of the arable land), the area of EFA decreased by one-third to 2400 ha in 2010 (Figure 17.4). The most important reason has been negative experiences with weed by some farmers, which led to a negative image of these EFA types in the farming community and the lack of adequate whole-farm advisory services (Home *et al.* 2014). The more recently introduced improved field margins, higher financial incentives and private labels have somewhat increased acceptance, but there is still a crucial lack of EFA which promote specifically the target species in arable regions.

Table 17.2 Seed mixtures (adapted to the prevailing site conditions) recommended for Ecological Focus Areas on arable land in Switzerland. Plant species are listed in alphabetical order.

	Wildflower strip		Rotational fallow		Improved field margin		Conservation headland*
	Mixture 1	Mixture 2	Mixture 3	Mixture 4	Mixture 5	Mixture 6	Mixture 7
Number of total plant species	24	37	17	21	40	35	31
Number of annual plant species	5	14	9	8	2	1	31
Number of perennial plant species	19	23	8	13	38	34	0
Characteristic plant species in the mixtures							
Agrostemma githago CH	x	x	x	x			x
Anchusa arvensis CH		x					
Camelina sativa CH		x		x			x
Campanula trachelium CH					x	x	
Centaurea cyanus CH	x	x	x	x	x	x	x
Centaurea jacea CH	x	x	x	x	x	x	
Consolida regalis CH		x					x
Daucus carota CH	x	x	x	x	x		
Echium vulgare CH	x	x	x	x	x		
Legousia speculum-veneris CH	x	x					x
Leucanthemum vulgare CH	x	x	x	x	x	x	
Melampyrum arvense CH							x
Mentha longifolia CH						x	
Papaver rhoeas CH	x	x	x	x	x		x

* Under development; CH, Swiss ecotypes.

17.3.3 Organic Farming and Integrated Production

The introduction of AES, combined with the main Swiss food retailers engagement with label products and customer demand for sustainable products, boosted organic farming and integrated production (IP). The use of pesticides in organic farming is restricted. Therefore, organic farming is one of the AES which are subsidized, while at the same time, under the organic farming label product prices are increased. At present, 6200 farms manage 12.5% of the UAA organically, and the per capita expenditure for organic farming products in Switzerland is the highest worldwide (data source: Swiss Federal Statistical Office, FSO and Research Institute of Organic Agriculture, FiBL).

'IP-SUISSE' is another label. Management restrictions for this are less severe than for organic farming and IP-SUISSE is not subsidized under the government AES. IP-SUISSE farmers aim at reducing the use of pesticides to a minimum by following rules. Nine thousand farmers produce under the IP label (managing 25% of the Swiss UAA). IP-SUISSE farmers have to comply with biodiversity standards, measured with a sophisticated point ranking system for individual measures such as the conservation of specific semi-natural habitat types on their farms. In this respect, there is an overlap with the minimum standards for cross-compliance (Birrer *et al.* 2014). Zellweger-Fischer *et al.* (2016) show an increase in the median of biodiversity values on IP-SUISSE farms of about 30% between 2010 and 2014 – a very encouraging success in biodiversity value induced by the labels and a comprehensive farmer advice service.

17.4 Ecosystem Service Provision as a New Paradigm

Ecological research soon found that EFA have the potential to promote organisms (mainly insects but also birds or fungi) that provide services to agriculture (Eggenschwiler *et al.* 2004; Pfiffner and Luka 2000; Schaffner *et al.* 1998; Schneider *et al.* 2012) (see also Chapters 7 and 8). Many studies have demonstrated the positive effects of specific plants, semi-natural habitats (EFA) and general landscape complexity on service providers such as pollinators and natural enemies of crop pests (Coudrain *et al.* 2013, 2014, 2016; Schüepp *et al.* 2013, 2014) (see also Chapters 7 and 8). Complex landscapes generally sustain larger and more diverse pollinator and natural enemy populations (Andow 1991; Bianchi *et al.* 2006; Kennedy *et al.* 2013; Le Féon *et al.* 2010 Rusch *et al.* 2010), and specific pollinating or pest parasitizing/predating taxa were often more abundant and diverse in or near semi-natural habitats (Le Féon *et al.* 2013). As an example, hoverflies are often enhanced in abundance and diversity in or near wildflower strips or hedgerows (Haenke *et al.* 2009, 2014). The main reason for these findings is that many pollinators and pest natural enemies depend on plant-provided resources during some of their life stages, and herbaceous vegetation provides shelter, overwintering sites and food sources, such as floral and extrafloral nectar, pollen or alternative hosts or prey, that may enhance their local abundance and fitness (Bianchi *et al.* 2006; Rusch *et al.* 2010).

17.4.1 Farmers Lobbying for an Ecosystem Service-Providing Ecological Focus Area

Although the promotion of ecosystem services had been listed amongst the agri-environmental objectives (FOEN and FOAG 2008), the design and implementation of EFA still targeted the original goal of biodiversity conservation. Alarming news about the declining health situation of honeybee populations then motivated a group of farmers to develop their own seed mixture of nectar- and pollen-rich flowers ('bee pasture') to improve the nutritional status of honeybees, in particular in arable areas during summer when the flowering period of crops such as canola and sunflower is over. In contrast to the available EFA, the new seed mixture was designed as a short-term element (100 days) to be sown in spring along with summer crops and to be removed after crop harvest. The farmers intensively lobbied for the integration of the honeybee flower seed mixture in the EFA scheme and, in 2015, a new EFA type called 'flower strip for pollinators and other beneficials' was introduced. This introduction of a 'functional EFA type' had been preceded by a lively and controversial debate between the farmer representatives and proponents of nature and biodiversity conservation. The latter argued that the EFA scheme should aim at the conservation of wild farmland species whilst the honeybee is a domesticated insect. They further criticized that the seed mixture consisted of a small number of – mostly – cultivated plants whilst seed mixtures recommended for arable EFA had originally consisted to a large extent of wild plants (Table 17.3). As a result of this controversial discussion, the honeybee flower seed mixture was diversified to also attract oligolectic bumble bees and solitary bees, and the introduction to EFA is currently accompanied by research which tests various types of seed mixtures with respect to their agronomic properties and their actual benefit to pollinating insects. Investigations also address the potential role of this short-term EFA type as an ecological trap and the options for longer-term (overwintering) seed mixtures (Ramseier *et al.* 2016a). In a participatory approach, farmers who implement the new functional EFA types are asked to return a questionnaire on the agronomic performance of, and their satisfaction with, the seed mixture.

Seed mixtures for the targeted promotion of specific ecosystem services need to:

- promote one or several ecosystem services (the benefit for pollinators and/or predators should be measurable and if possible obvious to the farmer)
- perform well agronomically (at least some of the sown plants have to develop quickly and in sufficient numbers to suppress weeds; ideally, the seed mixtures should be applicable to a wide range of soil and climatic conditions).

Often, increasing botanical diversity in itself is expected to enhance biological pest control (Tscharntke *et al.* 2005). However, the seed mixtures presently available do not focus on biological control (Jacot *et al.* 2008). In a comprehensive review, Andow (1991) showed that an increase in diversity does not necessarily result in a reduced level of herbivorous arthropods. This not only challenges the concept that high vegetational diversity per se benefits biological control, it also underlines that conservation biological control needs to be optimized by adding specifically selected food plants to non-crop vegetation.

17.4.2 Tailored Flower Strips for Pest Control

The effectiveness of sown flower strips in enhancing pest control depends on the design, type, attractiveness, quantity, quality and accessibility of resources as well as the timing and duration at which they are available (Wäckers and van Rijn 2012). Fundamental differences exist between annual flower strips that may provide high densities of high-quality floral resources during some weeks or months to specific guilds and perennial flower strips offering less abundant but potentially more diverse floral resources along

Table 17.3 Seed mixtures tested for functional EFA types on arable land. Plant species are listed in alphabetical order.

Promotion of	Pollinators	Predators in cabbage production	Predators in arable land*
Target pest species		cabbage moth common cabbage butterfly	aphids cereal leaf beetle
Plant species			
Agrostemma githago CH wp	x		
Anethum graveolens cp	x		x
Anthemis arvensis CH wp			x
Anthemis tinctoria CH wp	x		
Anthriscus cerefolium cp			x
Calendula arvensis cp			x
Camelina sativa CH wp			x
Centaurea cyanus CH wp	x	x	x
Centaurea jacea CH wp	x		x
Cichorium intybus CH wp	x		
Crepis capillaris CH wp			x
Coriandrum sativum cp			x
Fagopyrum esculentum cp	x	x	x
Hypochaeris radicata CH wp	x		
Lotus corniculatus CH wp	x		
Papaver rhoeas CH wp	x	x	x
Phacelia tanacetifolia cp	x		
Reseda lutea CH wp	x		x
Silene noctiflora CH wp			x
Sinapis arvensis CH wp			x
Stachys annua CH wp	x		x
Trifolium alexandrinum cp	x		
Trifolium hybridum cp	x		
Trifolium incarnatum cp	x		
Trifolium pratense CH wp	x		
Trifolium resupinatum cp	x		
Valerianella dentata CH wp			x
Vicia sativa cp	x	x	

* Currently still under testing, not yet officially available on the market. CH, Swiss ecotypes; cp, cultivated plant species; wp, wild plant species.

with structural resources and undisturbed habitats that support the long-term persistence of animal and plant populations (Pfiffner and Wyss 2004). Perennial wildflower strips have now also been shown to enhance pest control in nearby winter wheat crops (Tschumi *et al.* 2016a).

The potentially distinct requirements of service-providing guilds should be considered to promote functional diversity of communities that maximize the complementarity and stability of services (Crowder and Jabbour 2014). Semi-field and laboratory experiments on arthropod flower choice, consumption, accessibility and the effects of individual resources on fitness deliver valuable information about the potential suitability of plant species to be included in seed mixtures for promoting service providers (Wäckers and van Rijn 2012 and citations therein). Yet, as different natural enemies respond differently to floral resources, the selection of the 'right' flowering plants to optimize the species composition of flower strips requires that the performance and fitness consequences of plant species are directly assessed on multiple service-providing taxa. Additionally, service providers may not react in the same way to floral resources in the field as under semi-field or laboratory conditions (Wäckers and van Rijn 2012). Thus, the actual delivery of services must be assessed in the field.

The intrinsically rather inflexible nature of perennial flower strips, the loss of production area and the limitation of management along with increasing concerns regarding long-term weed establishment pointed out by farmers led to research turning towards annual flower strips. The focus is on designing flexible management tools for supporting natural enemies of crop pests to increase the provisioning of the associated services locally. Seed mixtures (see Table 17.3), adapted for enhancing control of cabbage pests in organic farming, showed promising results in enhancing natural enemies, and companion plants (*Centaurea cynanus*) significantly increased parasitism of cabbage pests (Balmer *et al.* 2013; Pfiffner *et al.* 2009). Mixtures adapted for arable crops were highly effective in reducing cereal leaf beetle density in nearby winter wheat and aphid abundance in potato fields (Tschumi *et al.* 2015, 2016b).

High pest reduction levels and reduced probabilities that economic pest thresholds are reached near flower strips suggest that flower strips may contribute to pesticide reduction in intensive crop cultivation (Tschumi *et al.*, 2015, 2016b, c). Further, flower strips can be valuable for low-input or organic management because they can provide one of the few effective alternatives to insecticides. However, the decision of a farmer to establish a flower strip and refrain from insecticide spraying or even to adopt an organic management scheme will largely depend on weighing the potential benefits of enhanced crop yield or spared costs for insecticide treatments against costs associated with the establishment of flower strips.

By offering a more flexible yet comparably effective management tool to farmers, annual flower strips may become a viable addition to perennial wildflower strips for agricultural practice. However, pest control benefits by perennial flower strips entail several advantages. First, agronomic and monetary investment is reduced if flower strips do not need to be renewed every year. Second, conservation benefits often increase with wildflower strip age, as the more complex structure of older wildflower areas provides key resources for wildlife that are otherwise rare in areas of intensive cropping, with positive effects, for example, on arthropod overwintering (Schmidt and Tscharntke 2005) and breeding opportunities for birds (Zollinger *et al.* 2013). Annual and perennial wildflower strips may support markedly different assemblages of arthropods and plants, with an expected much higher biodiversity support of perennial compared with annual flower strips (Frank *et al.* 2009).

Diverse resources are expected to support a higher diversity of pest antagonist communities, which may, for example through complementarity mechanisms, result in a higher and more stable provisioning of pest control services (Hegland and Boeke 2006; Jha and Kremen 2013; Naeem and Li 1997; Tilman 1996). However, more systems knowledge is needed to better understand the requirements of individual species, their interactions with the environment and amongst each other (e.g. mutual facilitation, intraguild predation, temporal availability of resources along the season, etc.).

17.5 Conclusion

AES and cross-compliance are well-established policy instruments in Switzerland, the EU and beyond. Some environmental benefits have been achieved (Flury 2005), although there is a crucial lack of systematic monitoring and evaluation (Herzog and Franklin 2016). Still, many farmers perceive cross-compliance regulations as restrictions of their professional freedom to produce agricultural commodities, which is their main motivation (Home *et al.* 2014). This is why they are open to EFA, which provide ecosystem services that support production.

This fact opens an interesting field of research with great potential for practical application to support environmentally friendly farming practices. Recent findings show a synergistic effect of effective natural pest control and optimal pollination (Sutter and Albrecht 2016) (see also Chapter 7). Understanding the contribution of existing semi-natural habitats on farmland to the promotion of natural enemies for pest control can help to strengthen this function and possibly reduce the application of pesticides (conservation biological control). Complementing conservation biological control strategies with managed habitats such as flower strips is a strategy which deserves further investigation. This type of research is ideally conducted in close collaboration with farmers (participatory research and development) (Waters-Bayer *et al.* 2015). Farmers can help to evaluate the agronomic properties of flower strips and other EFA types, including timely establishment of sown plants, suppression of noxious weeds and availability of floral resources. Researchers can help to generalize the findings (e.g. by collating information from a multitude of farmers) and to actually evaluate the effectiveness of flower strips for promoting natural enemies, suppressing pests, improving yield and promoting AEO species.

While these investigations are conducted at the local scale and target the interaction between flower strips and neighbouring crops, the implications of the planting of flower strips at the landscape scale also need to be examined. In fact, we do not know whether flower strips actually increase the population of natural enemies in the agricultural landscape or whether they just attract natural enemies from the surroundings. In the latter case, the natural enemies would benefit only the crop field next to the flower strip and would be missing in more distant fields.

The actual implementation of ecosystem service management in crop production is still rare (Lundgren 2009; Whittingham 2011). However, the Swiss example presented here shows how the aim of provisioning ecosystem services, which was often only implicitly driving AES, is becoming more explicit in political discussions. Still, it is now debated if a clearer distinction between AES protecting biodiversity ('biodiversity conservation schemes') and AES focusing on the delivery of ecosystem services ('ecosystem service schemes') is needed (Ekroos *et al.* 2014; Kleijn *et al.* 2011; Scheper *et al.* 2013). For the latter, the justification of public payments is questioned by conservationists. They argue that direct payments should be limited to measures which promote farmland biodiversity in general and target species in particular, while no financial incentives should be available for measures which support agricultural production. It would be in the interest of farmers themselves to apply those techniques, and support with taxpayer money would not be justified ('public money for public goods'). However, a reduction in the application of pesticides is an important policy goal of many governments (e.g. a planned reduction by 50% in France by 2025; http://agriculture.gouv.fr/le-gouvernement-presente-la-nouvelle-version-du-plan-ecophyto). If flower strips can contribute to reaching such goals, governments will be willing to support their implementation financially. First cost–benefit evaluations (Tschumi *et al.* 2015) indicate that, in many cases, flower strips are profitable only if they are subsidized. Additionally, we need more research on the pest control effects or pollination promotion of EFA types other than flower strips before any of them may be declared as inefficient ecosystem service providers. Also, we expect synergies between the boosting of ecosystem services and biodiversity conservation (Rey Benayas and Bullock 2012; Straub *et al.* 2008; Whittingham 2011). AES including elements specifically tailored to species of conservation concern and service providers and/or multifunctional elements which aim at maximizing both aims simultaneously may create 'win–win' situations for both biodiversity and production. Finally, if EFA are actually shown to provide ecosystem services, their acceptance by farmers will increase.

References

Andow, D.A. (1991) Vegetational diversity and arthropod population response. *Annual Review of Entomology* **36**: 561–586.

ARE (2016) Flächennutzung. Federal Office for Spatial Development, Bern, Switzerland. Available at: https://www.are.admin.ch/are/de/home/raumentwicklung-und-raumplanung/grundlagen-und-daten/fakten-und-zahlen/flaechennutzung.html (accessed 13 March 2017).

Aviron, S., Herzog, F., Klaus, K., *et al.* (2007) Effects of Swiss agri-environmental measures on arthropod biodiversity in arable landscapes. *Aspects of Applied Biology* **81**: 101–109.

Aviron, S., Nitsch, H., Jeanneret, P., *et al.* (2009) Ecological cross compliance promotes farmland biodiversity in Switzerland. *Frontiers in Ecology and the Environment* **7**: 247–252.

Balmer, O., Pfiffner, L., Schied, J., *et al.* (2013) Noncrop flowering plants restore top-down herbivore control in agricultural fields. *Ecology and Evolution* **3**: 2634–2646.

Bianchi, F.J.J.A., Booij, C.J.H. and Tscharntke, T. (2006) Sustainable pest regulation in agricultural landscapes: a review on landscape composition, biodiversity and natural pest control. *Proceedings of the Royal Society B: Biological Sciences* **273**: 1715–1727.

Birrer, S., Zellweger-Fischer, J., Stoeckli, S., *et al.* (2014) Biodiversity at the farm scale: a novel credit point system. *Agriculture, Ecosystems and Environment* **197**: 195–203.

Bundesrat (2016) *Verordnung über die Direktzahlungen an die Landwirtschaft.* SR 910.13, Swiss Federal Government, Berne, Switzerland.

Burton, R.J. and Schwarz, G. (2013) Result-oriented agri-environmental schemes in Europe and their potential for promoting behavioural change. *Land Use Policy* **30**: 628–641.
Coudrain, V., Herzog, F. and Entling, M. (2013) Effects of habitat fragmentation on abundance, larval food and parasitism of a spider-hunting wasp. *PLoS ONE* **8**(3): e59286.
Coudrain, V., Schüepp, C., Herzog, F., Albrecht, M. and Entling, M.H. (2014) Habitat amount modulates the effect of patch isolation on host-parasitoid interactions. *Frontiers in Environmental Science* **2**: 27.
Coudrain, V., Rittiner, S., Herzog, F., Tinner, W. and Entling, M. (2016) Landscape distribution of food and nesting sites affect larval diet and nest size, but not abundance of *Osmia bicornis*. *Insect Science* **23**: 746–753.
Crowder, D.W. and Jabbour, R. (2014) Relationships between biodiversity and biological control in agroecosystems: current status and future challenges. *Biological Control* **75**: 8–17.
Dobbs, T.L. and Pretty, J.N. (2004) Agri-environmental stewardship schemes and 'multifunctionality'. *Applied Economic Perspectives and Policy* **26**: 220–237.
Eggenschwiler, L., Sibylle, S., Werner, W. and Jacot, K. (2004) Floristische und faunistische Aspekte vergraster Brachestreifen. Der Anbausystemversuch Burgrain. *Schriftenreihe der FAL* **52**: 82–85.
Ekroos, J., Olsson, O., Rundlöf, M., Watzold, F. and Smith, H.G. (2014) Optimizing agri-environment schemes for biodiversity, ecosystem services or both? *Biological Conservation* **172**: 65–71.
Flury C. (2005) Agrarökologie und Tierwohl 1994–2005. *Agrarforschung* **12**: 526–531.
FOAG (2015) Agrarbericht. Bundesamt für Landwirtschaft, Berne, Switzerland. Available at: www.agrarbericht.ch/ (accessed 13 March 2017).
FOEN and FOAG (2008) *Umweltziele Landwirtschaft*. Bundesamt für Umwelt, BAFU, and Bundesamt für Landwirtschaft, BLW, Berne, Switzerland.
Frank, T., Aeschbacher, S., Barone, M., Kunzle, I., Lethmayer, C. and Mosimann, C. (2009) Beneficial arthropods respond differentially to wildflower areas of different age. *Annales Zoologici Fennici* **46**: 465–480.
Günter, M. (2000) *Anlage und Pflege von Mehrjährigen Buntbrachen unter den Rahmenbedingungen des Schweizerischen Ackerbaugebietes*. Dissertation, Universität Berne, Berne, Switzerland.
Günter, M., Schläpfer, F., Walter, T. and Herzog, F. (2002) Direct payments for biodiversity provided by Swiss farmers: an economic interpretation of direct democratic decision. Available at: https://www.cbd.int/financial/pes/switzerland-pes-oecd.pdf (accessed 13 March 2017).
Haenke, S., Scheid, B., Schaefer, M., Tscharntke, T. and Thies, C. (2009) Increasing syrphid fly diversity and density in sown flower strips within simple vs. complex landscapes. *Journal of Applied Ecology* **46**: 1106–1114.
Haenke, S., Kovács-Hostyánszki, A., Fründ, J., *et al.* (2014) Landscape configuration of crops and hedgerows drives local syrphid fly abundance. *Journal of Applied Ecology* **51**: 505–513.
Hegland, S.J. and Boeke, L. (2006) Relationships between the density and diversity of floral resources and flower visitor activity in a temperate grassland community. *Ecological Entomology* **31**: 532–538.
Herzog, F. and Franklin, J. (2016) Best practices for farmland biodiversity monitoring in North America and Europe. *Ambio* **45**: 857–871.

Herzog, F. and Walter, T. (2005) *Evaluation der Ökomassnahmen: Bereich Biodiversität/ Évaluation des Mesures Ecologiques : Domaine Biodiversité*. Agroscope FAL Reckenholz, Schriftenreihe der FAL 56/Les cahiers de la FAL 56, Zürich, Switzerland.
Herzog, F., Prasuhn, V., Spiess, E. and Richner, W. (2008) Environmental cross-compliance mitigates nitrogen and phosphorus pollution from Swiss agriculture. *Environmental Science and Policy* **11**: 655–668.
Hoag, D.L. and Holloway, H.A. (1991) Farm production decisions under cross and conservation compliance. *American Journal of Agricultural Economics* **71**: 184–193.
Home, R., Balmer, O., Jahrl, I., Stolze, M. and Pfiffner, L. (2014) Motivations for implementation of ecological compensation areas on Swiss farms. *Journal of Rural Studies* **34**: 26–36.
Jacot, K., Eggenschwiler, E., Junge, X., Luka, H. and Bosshard, A. (2007) Improved field margins for a higher biodiversity in agricultural landscapes. *Aspects of Applied Biology* **81**: 277–281.
Jacot, K., Eggenschwiler, L., Richner, N. and Schaffner, D. (2008) Botanical and social aspects of conservation headlands in Switzerland, *Bulletin IOBC/WPRS* **34**: 41–44.
Jha, S. and Kremen, C. (2013) Resource diversity and landscape-level homogeneity drive native bee foraging. *Proceedings of the National Academy of Sciences* **110**: 555–558.
Kampmann, D., Lüscher, A., Konold, W. and Herzog, F. (2012) Agri-environment scheme protects diversity of mountain grassland species. *Land Use Policy* **29**: 569–576.
Kennedy, C.M., Lonsdorf, E., Neel, M.C., *et al.* (2013) A global quantitative synthesis of local and landscape effects on wild bee pollinators in agroecosystems. *Ecology Letters* **16**: 584–599.
Kleijn, D., Rundlöf, M., Scheper, J., Smith, H.G. and Tscharntke, T. (2011) Does conservation on farmland contribute to halting the biodiversity decline? *Trends in Ecology and Evolution* **26**: 474–481.
Knop, E., Kleijn, D., Herzog, F. and Schmid, B. (2006) Effectiveness of the Swiss agri-environment scheme in promoting biodiversity. *Journal of Applied Ecology* **43**: 120–127.
Lachat, T., Pauli, D., Gonseth, Y., *et al.* (2010) *Wandel der Biodiversität in der Schweiz seit 1900*. Haupt Verlag, Berne, Switzerland.
Le Féon, V., Schermann-Legionnet, A., Delettre, Y., *et al.* (2010) Intensification of agriculture, landscape composition and wild bee communities: a large scale study in four European countries. *Agriculture, Ecosystems and Environment* **137**: 143–150.
Le Féon, V., Burel, F., Chifflet, R., *et al.* (2013) Solitary bee abundance and species richness in dynamic agricultural landscapes. *Agriculture, Ecosystems and Environment* **166**: 94–101.
Lundgren, J.G. (2009) *Relationships of Natural Enemies and Non-Prey Foods*. Springer, Dordrecht, The Netherlands.
Moser T. (2005) *Einflussfaktoren von Policy Veränderungen in der Schweizer Landwirtschaftspolitik. Eine Empirische ANALYSE am Beispiel der Öko-Qualitätsverordnung*. Lizenziatsarbeit der Philosophischen Fakultät der Universität Zürich (Schweiz), Institut für Politikwissenschaft, Abteilung Innenpolitik/ Vergleichende Politik.
Naeem, S. and Li, S. (1997) Biodiversity enhances ecosystem reliability. *Nature* **390**: 507–509.
OECD (2003) *Mesures Agro-Environnementales: Tour d'Horizon des Evolutions*. Groupe de travail mixte sur l'agriculture et l'environnement, OECD, Paris.

Pfiffner, L. and Luka, H. (2000) Erhöhte Laufkäferartenvielfalt durch biologischen Landbau, welche Arten werden dabei gefördert (Col. Carabidae)? *Mitteilungen der Deutschen Allgemeinen Angewandten Entomologie* **12**: 343–346.

Pfiffner, L. and Wyss, E. (2004) Use of sown wildflower strips to enhance natural enemies of agricultural pests. In: Gurr, G.M., Wratten, S.D. and Altieri, M. (eds) *Ecological Engineering for Pest Management: advances in habitat manipulation for arthropods.* CSIRO Publishing, Collingwood, Australia, pp. 167–188.

Pfiffner, L., Luka, H., Schlatter, C., Juen, A. and Traugott, M. (2009) Impact of wildflower strips on biological control of cabbage lepidopterans. *Agriculture, Ecosystems and Environment* **129**: 310–314.

Poiger, T., Buser, H.R. and Müller, M.D. (2005) *Evaluation der Ökomassnahmen und Tierhaltungsprogramme, Synthesebericht Bereich Pflanzenschutzmittel.* Agroscope FAW, Wädenswil, Switzerland.

Popp, H. (2013) 20 Jahre Direktzahlungen, der Weg zur Agrarreform 1992. *Yearbook of Socioeconomics in Agriculture* **2013**: 15–31.

Primdahl, J., Vesterager, J.P., Finn, J.A., Vlahos, G., Kristensen, L. and Vejre, H. (2010) Current use of impact models for agri-environment schemes and potential for improvements of policy design and assessment. *Journal of Environmental Management* **91**: 1245–1254.

Ramseier, H., Füglistaller, D., Läderach, C., Ramseier, C., Rauch, M. and Widmer Etter, F. (2016a) Blühstreifen für Bestäuber und andere Nützlinge. *Agrarforschung* 7: 276–283.

Ramseier, H., Lebrun, M. and Steinger, T. (2016b) Anwendung der Bekämpfungsschwellen und Warndienste in der Schweiz. *Agrarforschung Schweiz* 7: 98–103.

Rey Benayas, J. and Bullock, J. (2012) Restoration of biodiversity and ecosystem services on agricultural land. *Ecosystems* **15**: 883–899.

Rusch, A., Valantin-Morison, M., Sarthou, J.P. and Roger-Estrade, J. (2010) Biological control of insect pests in agroecosystems: effects of crop management, farming systems, and seminatural habitats at the landscape scale: a review. *Advances in Agronomy* **109**: 219–259.

SBV (2012) Wie ernährt sich die Schweiz? Schweizerischer Bauernverband, Brugg, Switzerland. Available at: www.sbv-usp.ch/fileadmin/sbvuspch/05_Publikationen/Situationsberichte/SB2012_de.pdf (accessed 13 March 2017).

Schaffner, D., Schwab, A., Zwimpfer, T. and Kappeler, P. (1998) Saatmischungen beeinflussen die Vielfalt in Buntbrachen. *Agrarforschung* **5**: 169–172.

Scheper, J., Holzschuh, A., Kuussaari, M., *et al.* (2013) Environmental factors driving the effectiveness of European agri-environmental measures in mitigating pollinator loss – a meta-analysis. *Ecology Letters* **16**: 912–920.

Schmidt, M.H. and Tscharntke, T. (2005) The role of perennial habitats for Central European farmland spiders. *Agriculture, Ecosystems and Environment* **105**: 235–242.

Schneider, S., Widmer, F., Jacot, K., Kölliker, R. and Enkerli, J. (2012) Spatial distribution of *Metarhizium* clade 1 in agricultural landscapes with arable land and different semi-natural habitats. *Applied Soil Ecology* **52**: 20–28.

Schüepp, C., Herzog, F. and Entling, M. (2013) Disentangling multiple drivers of pollination in a landscape-scale experiment. *Proceedings of the Royal Society B: Biological Sciences* **281**: 20132667.

Schüepp, C., Uzman, D., Herzog, F. and Entling, M. (2014) Habitat isolation affects plant-herbivore-enemy interactions on cherry trees. *Biological Control* **71**: 56–64.

Straub, C.S., Finke, D.L. and Snyder, W.E. (2008) Are the conservation of natural enemy biodiversity and biological control compatible goals? *Biological Control* **45**: 225–237.

Sutter, L. and Albrecht, M. (2016) Synergistic interactions of ecosystem services: florivorous pest control boosts crop yield increase through insect pollination. *Proceedings of the Royal Society B: Biological Sciences* **283**: 20152529.

Tilman, D. (1996) Biodiversity: Population versus ecosystem stability. *Ecology* **77**: 350–363.

Tscharntke, T., Klein, A.M., Kruess, A., Steffan-Dewenter, I. and Thies, C. (2005) Landscape perspectives on agricultural intensification and biodiversity – ecosystem service management. *Ecology Letters* **8**: 857–874.

Tschumi, M., Albrecht, M., Entling, M.H. and Jacot, K. (2015) High effectiveness of tailored flower strips in reducing pests and crop plant damage. *Proceedings of the Royal Society B: Biological Sciences* **282**: 20151369.

Tschumi, M., Albrecht, M., Bärtschi, C., Collatz, J., Entling, M.H. and Jacot, K. (2016a) Perennial, species-rich wildflower strips enhance pest control and crop yield. *Agriculture, Ecosystems and Environment* **220**: 97–103.

Tschumi, M., Albrecht, M., Collatz, J., *et al.* (2016b) Tailored flower strips promote natural enemy biodiversity and pest control in potato crops. *Journal of Applied Ecology* **53**: 1169–1176.

Tschumi, M., Albrecht, M., Dubsky, V., Herzog, F. and Jacot, K. (2016c) Nützlingsblühstreifen für den Ackerbau reduzieren Schädlinge in Kulturen. *Agrarforschung Schweiz* **7**: 260–267.

Wäckers, F.L. and van Rijn, P.C.J. (2012) Pick and mix: Selecting flowering plants to meet the requirements of target biological control insects. In: Gurr, G.M., Wratten, S.D., Snyder, W.E. and Read, D.M.Y. (eds) *Biodiversity and Insect Pests: key issues for sustainable management*. John Wiley and Sons, Ltd, Chichester, UK, pp. 139–165.

Walter, T., Eggenberg, S., Gonseth, Y., *et al.* (2013) *Operationalisierung der Umweltziele Landwirtschaft – Bereich Ziel- und Leitarten, Lebensräume (OPAL)*. ART-Schriftenreihe 18, Tänikon, Agroscope Reckenholz-Tänikon, Zurich, Switzerland.

Waters-Bayer, A., Kristjanson, P., Wettasinha, C., *et al.* (2015) Exploring the impact of farmer-led research supported by civil society organisations. *Agriculture and Food Security* **4**: 1–7.

Whittingham, M.J. (2011) The future of agri-environment schemes: biodiversity gains and ecosystem service delivery? *Journal of Applied Ecology* **48**: 509–513.

Zellweger-Fischer, J., Althaus, P., Birrer, S., Jenny, M., Pfiffner, L. and Stöckli, S. (2016) Biodiversität auf landwirtschaftsbetrieben mit einem punktesystem erheben. *Agrarforschung Schweiz* **7**: 40–47.

Zollinger, J.L., Birrer, S., Zbinden, N. and Korner-Nievergelt, F. (2013) The optimal age of sown field margins for breeding farmland birds. *Ibis* **155**: 779–791.

Part VII

Concluding Remarks, Take-Home Messages and a Call for Action

18

Environmental Pest Management: The Need for Long-term Governmental Commitment

Moshe Coll and Eric Wajnberg

18.1 The Prevalence of a Pest-centric, Bottom-up Approach to Pest Control

For thousands of years, farmers have protected their crops by combating one pest at a time, using one control method (Lewis *et al.* 1997), with very little consideration of the surrounding environment. Over the years, new control methods, tactics and technologies have been adopted. These include employment of resistant crop genotypes; pest-retarding cultivation practices such as tillage, crop rotation, timing of planting and harvest and sanitation; chemical pesticides, including new chemistries, formulations and delivery tools; biological control agents; sterile insect techniques (SIT); transgenic crops; and now transgenic pests through gene-drive mechanisms. Such pest-centric approaches have remained the dominant dogma throughout the evolution of mainstream plant protection.

Some 55 years ago, a promising attempt was made to adopt a system-wide view of pest management. In its early form, Integrated Pest Management (IPM) was intended to provide a more holistic approach to pest management (van den Bosch and Stern 1962) than that offered by the supervised control commonplace at the time (Figure 18.1a). During the following decades, some pest management programmes were developed in the spirit of IPM. However, these also tended to target a specific pest or pest group in a particular crop. IPM thus remained focused on pest populations even when area-wide approaches were adopted. Interactions between pest control measures and human and ecological environments have not been incorporated in pest management programmes. Perhaps as a result, we have failed to reduce yield losses to pests and to produce more food in sustainable and environmentally compatible ways. It has been estimated that global crop losses to arthropods, diseases and weeds increased from 34.9% in 1965 (Cramer 1967) to 42.1% in 1988–1990 (Oerke *et al.* 1994) despite continuous intensification of pest control efforts.

In light of this, it is imperative that we renew our efforts to develop and implement pest management schemes that are effective, economically viable, sustainable and safe to humans and the environment. Towards this end, chapters in the present volume review the state of our understanding of pest population management and discuss

Environmental Pest Management: Challenges for Agronomists, Ecologists, Economists and Policymakers, First Edition. Edited by Moshe Coll and Eric Wajnberg.

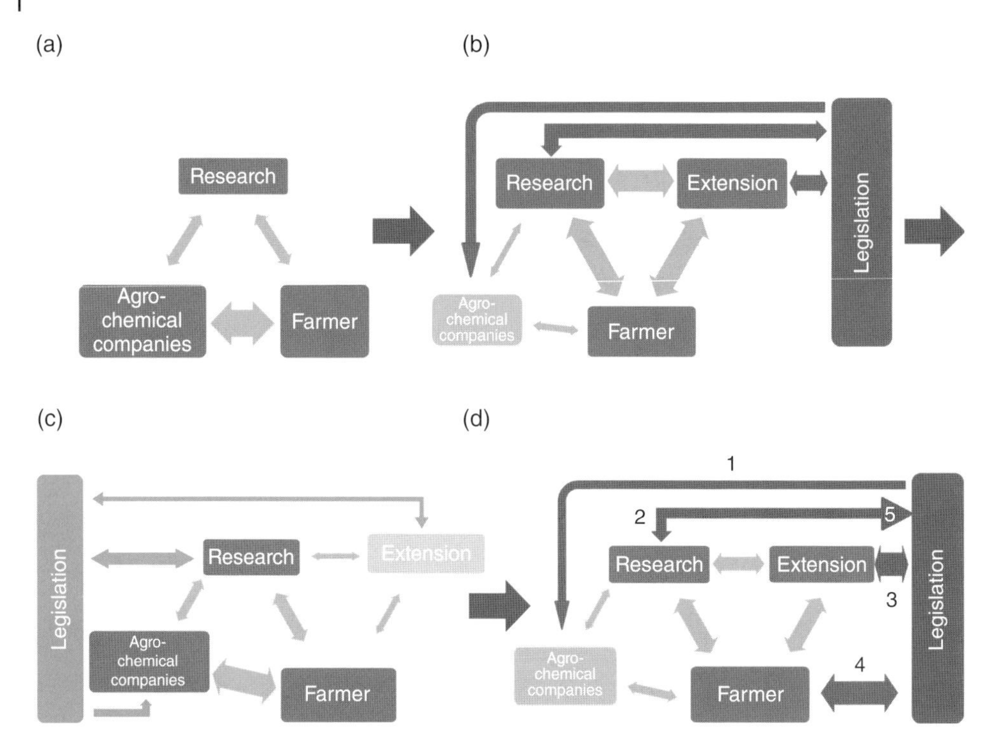

Figure 18.1 Primary players in pest control schemes over time. (a)1940s to early 1960s. (b) Mid-1960s to late 1980s. (c) Early 1990s to mid-2010s. (d) Proposed environmental pest management scheme: (1) pesticide regulation, (2) funding of invited research, (3) support for extension and farmers' participatory programmes, (4) policies to influence farmers' practices, (5) research outputs used to fine-tune governmental policies. Arrows indicate flow direction of inputs. Shade intensity of player's box and arrow width indicate relative importance of player's input.

current thinking and policy concerning the interactions among pest control actions, human health and the environment.

18.2 The Main Messages Presented in this Volume

Chapter 1: Environmental Pest Management: A Call to Shift from a Pest-Centric to System-Centric Approach

Pest control efforts have traditionally focused on specific pests in specific fields. This pest-centric approach was also commonly practised in Integrated Pest Management (IPM) programmes. We have largely failed to develop effective, safe and sustainable plant protection systems. To address this goal, a new pest management paradigm must be adopted: a system-centric approach should replace the historical bottom-up, pest-centric one. Furthermore, IPM programmes are likely to fail eventually because of the high variability and unpredictability of many interacting natural and anthropogenic factors. Therefore, goal-based environmental pest management schemes should be advanced.

Chapter 2: Approaches in Plant Protection: Science, Technology, Environment and Society

Since first proposed, the term IPM has been assigned highly diverse interpretations and meanings. Various interest groups have used the term to promote their own agendas to the point at which reduction in pesticide use and integration of multiple control tactics are no longer prioritized. The authors call for conceptual changes in IPM policy as part of a transformation of agricultural practice to systems that sustain the ecosystem services needed for viable and socially fair food production.

Chapter 3: The Economics of Alternative Pest Management Strategies: Basic Assessment

Pest control measures are prone to social conflicts since farmers act to maximize profit and are unlikely to consider the off-farm environmental consequences of their decisions. Yet many pest control practices affect neighbouring managed and unmanaged lands, and have an impact on the health of consumers and residents in nearby communities. Farmers may also fail to fully appreciate on-farm consequences of different pest control tactics.

Chapter 4: Effects of Chemical Control on the Environment

Pesticides affect the environment directly, through primary toxicity, indirectly through secondary poisoning, and in sublethal ways. In addition, constant use of pesticides leads to widespread resistance in populations of insects, weeds and micro-organisms. Resistance may be overcome by introducing new products to replace those that have become ineffective, but this solution is harmful to an environment that is already polluted with many types of toxic chemicals.

Chapter 5: Environmental Impacts of Arthropod Biological Control: An Ecological Perspective

Classic biological control has in the past been considered a safe and highly effective approach to pest management. However, in recent decades, there has been growing concern about the negative environmental impact of introduced organisms. In the release area, alien biological control agents may attack non-target organisms, thus jeopardizing biodiversity and altering the structure and function of native ecosystems. Therefore, the authors advocate that no introduction of foreign biological control agents be permitted without careful risk assessment weighing agricultural benefits and environmental risks of the proposed biological control programme and alternative pest control methods. The alternatives should also include a no-action option.

Chapter 6: Effects of Transgenic Crops on the Environment

Genetically modified crops may affect organisms in managed and natural ecosystems in a highly complex manner, both directly and indirectly. The authors of this chapter stress the need for prospective risk assessments, including quantitative uncertainty analyses.

As a whole, field impacts of genetically engineered crops on non-target organism and ecosystems have been rare. The longer temporal and larger spatial scales at which such effects may operate present further challenges for the development of comprehensive and reliable risk assessment tools.

Chapter 7: Ecosystem Services Provided by Unmanaged Habitats in Agricultural Landscapes

In many agro-ecosystems, the presence of natural and semi-natural vegetation has been shown to enhance the density and species diversity of predatory and parasitic arthropod communities. Despite this finding, most studies fall short of quantifying ecosystem services such as biological pest suppression and decrease in yield loss which are provided by native vegetation. Any suggestion concerning the manipulation of vegetation near crop fields should take into account the effects on non-target pest groups, natural enemies, pollinators, decomposers in the soil and other organisms. Unmanaged areas, for example, may serve as reservoirs for pesticide-susceptible pest populations that could then contribute to slowing the rate of resistance development.

Chapter 8: The Role of Ecosystem Disservices in Pest Management

Management of agro-ecosystems for sustainable pest management relies upon understanding the nature of interactions among multiple co-occurring ecosystem services, such as food production, biodiversity conservation, pest regulation and pollination. Each of such services has the potential for positive, negative or neutral effects on the others, but these interactions can be highly complex and their relative effects are thus difficult to quantify. This may explain the dearth of data in the literature on valuation of ecosystem disservices. In addition, patterns of ecosystem services and disservices vary greatly over local and regional scales within a landscape. This makes them even more difficult to evaluate.

Chapter 9: Effect of Climate Change on Insect Pest Management

Global warming is expected to have implications for some aspects of almost every pest control measure, from pesticide residue and toxicity patterns, through the longevity of pheromone dispensers and pheromone plume patterns, to the activity of natural enemies and their ability to locate hosts. Climate change may also affect pest and enemy development, phenology (i.e. synchronization), behaviour, reproduction, survival, etc. In this respect, we would add that global warming may also influence pest–enemy interactions by altering their geographic distribution (Schuldiner-Harpaz and Coll 2013). Finally, other global climatic changes, such as elevated levels of atmospheric CO_2, greatly influence complex crop–pest–enemy interactions (Coll and Hughes 2008).

Chapter 10: Effects of Biological Invasions on Pest Management

The global impact of invasive species in ecosystems includes changes in the structure and function of pest and natural enemy populations in agro-ecosystems. Continuous monitoring, interception efforts and trade regulatory policies are needed to protect crop plants from pests that may be transported through commodity trading and become established in non-endemic areas. These efforts require global policies and international co-operation that promote greater biosecurity in trade and travel. Similarly, multidisciplinary collaboration among researchers would help to more effectively integrate and transfer information pertinent to invasive species.

Chapter 11: Pesticides and Human Health

Pesticides undergo rigorous premarket toxicity testing with regard to carcinogenicity and other health hazards. However, these tests do not capture the full range of chronic diseases and many of the methodologies used have serious shortcomings. In addition, and because premarket testing is relatively limited in its scope, results are often obsolete by the time additional trials are conducted; many of the compounds have been taken off the shelf by then, because of declining efficacy due to resistance development, development of cheaper compounds, and other reasons. Thus, no comprehensive data are available for most widely used pesticides. The authors call for standardization of pesticide safety testing.

Chapter 12: Human Health Concerns Related to the Consumption of Foods from Genetically Modified Crops

The assessment of the health risk associated with GM foods has technical limitations that make it difficult to demonstrate that they are safe for consumption. These technical limitations can be addressed by a wide range of testing protocols which must be standardized in order to combine global efforts to ensure a safe food supply.

Chapter 13: Effectiveness of Pesticide Policies: Experiences from Danish Pesticide Regulation 1986–2015

The authors state that the sparsity of relevant comparative data on pesticide consumption across nations constitutes a critical limitation for the development of effective global pesticide policy. They call for legislators to solicit this type of data collection.

Chapter 14: Impacts of Exotic Biological Control Agents on Non-target Species and Biodiversity: Evidence, Policy and Implications

This chapter discusses major issues such as the administrative constraints on funding or on simply requiring the applicant for a natural enemy release permit to perform a postrelease validation and report the findings. The sole objective for regulators entrusted with granting release permits is to reduce risk. Postrelease monitoring of new biological control agents does not serve to mitigate risk. As a result, postrelease assessments are rare, and predictions made at the permit-granting stage remain untested. This greatly limits our ability both to assess risk and to support an effective decision-making process in the future.

Another shortcoming of most procedures for granting release permits is their focus on potential risk assessment while neither taking into account expected benefits nor weighing up the risks and benefits posed by alternative measures, including a 'do-nothing' approach. As a result, potential risks involved in the release of a biological control agent are not considered against all potential benefits or risks involved in the employment of alternative pest control measures, such as pesticide application.

Chapter 15: Pesticides in Food Safety versus Food Security

The need to provide the growing human population with sufficient, safe food of adequate nutritional quality may result in a trade-off between food safety and food security. Pesticides may help to increase food production while at the same time jeopardizing human health. The nature of such trade-offs varies among regions. Tolerance of the risk of chronic pesticide effects, for example, may be higher in regions characterized by high levels of food insecurity and/or shorter life expectancy. Therefore, the trade-off between food safety and food security can be managed only at governmental levels, where data about expected demographic changes, future food production and imports may enable the construction of models to assess the risks associated with pest control.

Chapter 16: External Costs of Food Production: Environmental and Human Health Costs of Pest Management

In many cases, some of the costs involved in implementing pest control measures are covered by neither the farmer nor by the producer of the products used. For example, the negative effects of pesticide use on human health and the environment are a burden to society, but entail no cost to pesticide users, vendors or manufacturers, all of whom benefit from the use of chemicals. The net societal good could be enhanced by governmental regulations aimed at increasing the benefits of pesticide use together with reducing their overall burden to society.

Chapter 17: The Role of Pest Management in Driving Agri-environment Schemes in Switzerland

In Switzerland, the role of pest management in agro-environmental schemes is limited to the reduction in pesticide use and in other agricultural inputs. Thus far, these schemes have failed to promote additional environmentally desirable pest control measures such as enhancement of biological control services through flower stripping, cover cropping or cultural practices. Systems currently in use also fail to externalize health and environmental costs of pesticide use.

18.3 The Role of Governments in Pest Management

The role of governmental legislation and regulatory agencies is in evidence in most of the topics reviewed in this volume, and greater involvement is often called for. This is particularly important for:

- co-ordinating health and environmental safety testing for pesticides and GM foods
- standardizing and possibly legislating a definition of IPM
- regulating postrelease assessment of biological control agents
- externalizing pesticide costs

- incorporating ecosystem services and disservices in pest management systems
- mitigating adverse effects of climate change and biological invasions
- regulating pesticide registration
- weighing food safety *versus* food security concerns
- incorporating all of these issues into agro-environmental schemes that direct more attention towards pest management concerns.

We argue that governmental commitment is critical for the sustainable employment of environmental pest management. In its early days, the IPM approach acted to displace pesticide use with other, safer pest control measures (Figure 18.1b). This was implemented through intensive public support of research, extension and participatory action research (PAR, also known as farmer participatory research, FPR) (Matteson 2000). In time, public support declined. In the USA, public funding for extension grew at the rate of 6.7% annually during the years 1915–1949, and then at 2.39% per year from 1950 to 1980 (Pardey *et al.* 2013). Public funding for extension then declined by 0.25% annually between 1980 and 2006 (Pardey *et al.* 2013). As a result, the US federal government provided 62% of the funds supporting extension in 1919, but only 21% of this funding in 2006 (Pardey *et al.* 2013). Similar trends were seen in other countries, such as the UK and New Zealand, where extension services and research were privatized and funding for farmer training was discontinued. Likewise, the most important obstacles listed by pest control practitioners and farmers to the adoption of IPM in developing countries involve lack of supportive governmental policies and farmer training (Parsa *et al.* 2014).

The vacuum created by falling public support was soon filled by the agro-chemical companies, promoting their new pest control compounds (Figure 18.1c). This is evident in a recent survey which indicates that 81% of responding extension officers in the USA are in a partnership with industry (Krell *et al.* 2016). Moreover, a significant amount of extension research is now funded by the private sector, with more than 14% of the officers acknowledging the potential risk for conflict of interest (Krell *et al.* 2016). The actual number is probably much higher.

As multinational agro-industrial conglomerates began to dominate the market, producing conventional and transgenic, herbicide-tolerant seeds, and manufacturing compatible herbicides and other pesticides, the private sector once again became a major and sometimes the sole force in pest management practice. These companies promote sales by advocating their own brand of 'IPM' ('the other IPM' *sensu* Ehler 2006), an approach that encourages the use of 'soft' pesticides as the main and often only means of pest control. Such 'soft' materials require low delivery doses of active ingredients, and have a short half-life and thus low residual effects. While the latter traits are highly desirable, the current approach is far from an integrative, sustainable and environmentally compatible strategy for pest management.

The dominance of the private sector in current pest control thought and practice is clearly evident in Krell *et al.* (2016), in which Dow AgroSciences affiliates propose that the public extension service create a partnership with the private sector to provide information to farmers (Krell *et al.* 2016). This preposterous proposal echoes the weakening of the extension service: in a 1994–1995 US Department of Agriculture survey, 69% of responding farmers reported that they obtained information from agricultural retailers and private scouting services, and only 15% from other sources, such as extension officers (Padgitt *et al.* 2000). The situation has not improved since then: 69% and

58% of Iowa farmers rely on agricultural chemical dealers for information on insect and weed management, respectively (Arbuckle *et al.* 2012).

The major reversal of the pest control approach from the original scheme of IPM back to calendar spraying is well documented, for example in the rice crop in South-East Asia (Bottrell and Schoenly 2012; Heong and Hardy 2009). While a few, mostly 'supervised control' IPM programmes are still implemented, many others have been discontinued. In California, USA, for example, almond growers have actually ceased monitoring their orchards for pests and simply spray routinely with inexpensive pesticides. Only a very few 'true IPM' programmes are now employed globally, mostly in organic farming systems, which occupy an extremely small proportion of the total arable land in the world. Therefore, in the vast majority of cropland around the world, pests are currently controlled chemically with little consideration for human and environmental health. The pest control industry once again dominates farmers' decisions by offering them new and temporarily highly effective pest control methods as they become available. These include (1) the employment of pest sex pheromones for monitoring, mass trapping and mating disruption, (2) the development of highly potent and inexpensive pesticides, and (3) the introduction of insect-resistant and herbicide-tolerant transgenic crops. In contrast, a recent study shows that IPM programmes in Asia and Africa have brought about a 30.7% reduction in pesticide use while increasing yields by 40.9% across 85 projects in 24 countries. In 30% of the cases, IPM eliminated pesticide use entirely (Pretty and Bharucha 2015). Moreover, and against the claims of the agro-chemical industry, the authors found that at least 50% of pesticide use was unnecessary.

While a 20-year-old call by Lewis *et al.* (1997) for a shift from a therapeutic to a total system approach in pest management is a step in the right direction, we argue that such a shift would be possible only through strong and permanent commitment by governments and their regulatory agencies (Figure 18.1d). At the United Nations Sustainable Development Summit in 2015, leaders of 193 countries adopted the 2030 Agenda for Sustainable Development. It includes a set of 17 Sustainable Development Goals (SDGs) to end poverty, fight inequality and injustice, and tackle climate change by 2030. Of these 17 identified goals, goal #2, 'End hunger, achieve food security and improved nutrition and promote sustainable agriculture' is directly relevant to the way in which we practise pest control. Two other goals are also pertinent to pest control: goal #12, 'Ensure sustainable consumption and production patterns' and goal #15, 'Protect, restore and promote sustainable use of terrestrial ecosystems, sustainably manage forests, combat desertification, and halt and reverse land degradation and halt biodiversity loss'. This goal-setting initiative was preceded by a National Research Council report (NRC 1996) that called for 'a paradigm shift in pest-management theory [...] that examines processes, flows, and relationships among organisms' and others that emphasized that, in its present form, crop protection treats only the symptoms of pest outbreaks instead of their causes (Zorner 2000).

18.4 Characteristics of Top-down, Environmental Pest Management

To date, some countries have adopted regulatory tools in order to achieve various agricultural and environmental goals, but these goals and approaches vary greatly among countries. Some schemes, for example, promote biodiversity conservation while others focus on agricultural productivity (see discussions in Rey Benayas and Bullock 2012, Straub *et al.* 2008, Tschumi *et al.* 2015 and Whittingham 2011). Nevertheless, pest management plays only a negligible role, if any, in these overall schemes. In Europe, for instance, conservation biological control is promoted implicitly with the objective of enhancing species diversity.

Yet synergistic promotion of ecosystem services, effective and sustainable agricultural productivity and biodiversity conservation can advance safe and environmentally compatible pest management practices. For example, increased environmental and health risk awareness in recent decades has led to a parallel increase in regulation of pesticide use and employment of genetically modified crops. Governmental involvement would also facilitate co-ordination and communication between landowners within a landscape and a thorough understanding of local and regional patterns of multi-scale ecosystem services and disservices, the provision of which is likely to be a key factor for effective and sustainable agricultural management (Bommarco *et al.* 2013; Mitchell *et al.* 2014).

However, this legislation is often handled and enforced by different governmental agencies, typically with little co-ordination among them. Intergovernmental and international co-operation is also needed in light of demographic, technological, trade, marketing and climatic considerations. This co-operation would replace the current situation in which growers, extension personnel and crop protection researchers are responsive mainly to changes in pesticide availability, due to regulatory banning and availability of new chemistries, and to the development of new technologies. Governmental involvement would also lend itself to the solicitation of invited research to fill gaps in our understanding. These new data could then be incorporated into policy decisions.

As outlined in Chapter 1, grassroots research, extension and farmer training efforts must be backed by legislative, regulatory and enforcement actions taken by governments. Governmental inputs acting to promote sustainable agricultural practices and nature conservation should have four main objectives that are currently missing in most legislation:

- the establishment of goal-based agro-environmental schemes
- externalizing the true costs of pesticide use
- strengthening of the public extension service
- soliciting goal-specific research.

Properties and methods used for the implementation of these objectives would certainly vary greatly among countries. Governmental and social structures, economic forces, traditions and other factors will shape needs, impose constraints and determine feasibility of means, and thus influence goals and approaches. In some cases, the required infrastructure already exists and needs only to be adjusted to the new objectives. The State of California, for example, charges a 'Mill Assessment' fee on pesticide sales (California Environmental Protection Agency 2016). This mechanism could be adopted to discourage pesticide use and cover health and environmental costs related to pesticide application.

For practical, marketing or ideological reasons, growers should be allowed to meet regulatory requirements in different ways: through organic farming, permaculture, IPM or by adopting just a few practices which promote desirable outcomes. Finally, centralized schemes and policies could be amended and fine-tuned as more information becomes available and with changes in agricultural production and market conditions.

Acknowledgements

We thank Ruth-Ann Yonah for her help in manuscript preparation.

References

Arbuckle, J.G., Lasley, P., Ferrell, J. and Miller, R. (2012) *Iowa Farm and Rural Life Poll.* Iowa State University Extension PM 3036, Ames, IA, USA. Available at: https://store.extension.iastate.edu/product/pm3075 (accessed 13 March 2017).

Bommarco, R., Kleijn, D. and Potts, S.G. (2013) Ecological intensification: harnessing ecosystem services for food security. *Trends in Ecology and Evolution* **28**: 230–238.

Bottrell, D.G. and Schoenly, K.G. (2012) Resurrecting the ghost of green revolutions past: the brown planthopper as a recurring threat to high-yielding rice production in tropical Asia. *Journal of Asia-Pacific Entomology* **15**: 122–140.

California Environmental Protection Agency (2016) Product Compliance Branch – Mill Assessment: Frequently Asked Questions. Available at: www.cdpr.ca.gov/docs/mill/qanda.pdf (accessed 13 March 2017).

Coll, M. and Hughes, L. (2008) Effects of elevated CO_2 on an insect omnivore: a test for nutritional effects mediated by host plants and prey. *Agriculture, Ecosystems and Environment* **123**: 271–279.

Cramer, H.H. (1967) Plant protection and world crop Protection. *Pflanzenschutznachrichten* **20**: 1–524.

Ehler, L.E. (2006) Integrated pest management (IPM): definition, historical development and implementation, and the other IPM. *Pest Management Science* **62**: 787–789.

Heong, K.L. and Hardy, B. (2009) *Planthoppers: new threats to the sustainability of intensive rice production systems in Asia.* International Rice Research Institute, Los Baños, Philippines.

Krell, R.K., Fisher, M.L. and Steffey, K.L. (2016) A proposal for public and private partnership in extension. *Journal of Integrated Pest Management* **7**: 4.

Lewis, W.J., van Lenteren, J.C., Phatak, S.C. and Tumlinson III, J.H. (1997) A total system approach to sustainable pest management. *Proceedings of the National Academy of Sciences* **94**: 12243–12248.

Matteson, P.C. (2000) Insect pest management in tropical Asian irrigated rice. *Annual Review of Entomology* **45**: 549–574.

Mitchell, M.G.E., Bennett, E.M. and Gonzalez, A. (2014) Forest fragments modulate the provision of multiple ecosystem services. *Journal of Applied Ecology* **51**: 909–918.

NRC (1996) *Ecologically Based Pest Management: new solutions for a new century.* National Research Council, National Academy of Sciences, Washington, DC, USA.

Oerke, E.C., Dehne, H.W., Shoenbeck, F. and Weber, A. (1994) *Crop Production and Crop Protection: Estimated Losses in Major Food and Cash Crops.* Elsevier, Amsterdam, The Netherlands.

Padgitt, M., Newton, D., Penn, R. and Sandretto, C. (2000) Production Practices for Major Crops in U.S. Agriculture, 1990–97. Statistical Bulletin No. 969. Available at: www.ers.usda.gov/publications/pub-details/?pubid=47139 (accessed 13 March 2017).

Pardey, P.G., Alston, J.M. and Chan-Kang, C. (2013) Public Food and Agricultural Research in the United States: The Rise and Decline of Public Investments, and Policies for Renewal.

AGree, Washington, DC, USA. Available at: www.foodandagpolicy.org/content/public-food-and-agricultural-research-united-statesthe-rise-and-decline-public-investments-a (accessed 13 March 2017).

Parsa, S., Morse, S., Bonifacio, A., *et al.* (2014) Obstacles to integrated pest management adoption in developing countries. *Proceedings of the National Academy of Sciences* **111**: 3889–3894.

Pretty, J. and Bharucha, Z.P. (2015) Integrated pest management for sustainable intensification of agriculture in Asia and Africa. *Insects* **6**: 152–182.

Rey Benayas, J. and Bullock, J. (2012) Restoration of biodiversity and ecosystem services on agricultural land. *Ecosystems* **15**: 883–899.

Schuldiner-Harpaz, T. and Coll, M. (2013) Effects of global warming on predatory bugs supported by data across geographic and seasonal climatic gradients. *PLoS ONE* **8(6)**: e66622.

Straub, C.S., Finke, D.L. and Snyder, W.E. (2008) Are the conservation of natural enemy biodiversity and biological control compatible goals? *Biological Control* **45**: 225–237.

Tschumi, M., Albrecht, M., Entling, M.H. and Jacot, K. (2015) High effectiveness of tailored flower strips in reducing pests and crop plant damage. *Proceedings of the Royal Society B: Biological Sciences* **282**: 20151369.

Van den Bosch, R. and Stern, V.M. (1962) The integration of chemical and biological control of arthropod pests. *Annual Review of Entomology* **7**: 367–386.

Whittingham, M.J. (2011) The future of agri-environment schemes: biodiversity gains and ecosystem service delivery? *Journal of Applied Ecology* **48**: 509–513.

Zorner, P.S. (2000) Shifting agricultural and ecological context for IPM. In: Kennedy, G.S. and Sutton, T.B. (eds) *Emerging Technologies for Integrated Pest Management: concepts, research, and implementation.* American Phytopathological Society Press, St Paul, MN, USA, pp. 32–41.

Index

Note: Page numbers in *italics* denote figures, where they appear outside page ranges

Environmental Pest Management: Challenges for Agronomists, Ecologists, Economists and Policymakers, First Edition. Edited by Moshe Coll and Eric Wajnberg.

d

q

r

s